Seamanship Techniques

Shipboard and Marine Operations

Other Works Published by D.J. House.

Marine Survival (3rd edition) 2011, Witherby

Navigation for Masters (4th edition) 2007, Witherby

An Introduction to Helicopter Operations at Sea: A Guide for Industry (2nd edition) 1998, Witherby

Cargo Work (7th edition, revised) 1998, Butterworth/Heinemann

Anchor Practice: A Guide for Industry, 2001, Witherby

Marine Ferry Transports: An Operator's Guide, 2002, Witherby

Dry Docking and Shipboard Maintenance, 2003, Witherby

Heavy Lift and Rigging, 2005, Brown Son & Ferguson

The Seamanship Examiner, 2005, Elsevier

Ship Handling, 2007, Elsevier

The Ice Navigation Manual, 2010, Witherby

Elements of Modern Ship Construction, 2010, Brown Son & Ferguson

Also:

Marine Technology Reference Book (Safety Chapter) edited by Nina Morgan, Butterworths

Seamanship Techniques

Shipboard and Marine Operations

D.J. House

Routledge
Taylor & Francis Group

LONDON AND NEW YORK

Fourth edition published 2014
by Routledge
2 Park Square, Milton Park, Abingdon, Oxon OX14 4RN

and by Routledge
711 Third Avenue, New York, NY 10017

Routledge is an imprint of the Taylor & Francis Group, an informa business

First edition published by Butterworth Heinemann 1987
Second edition published by Butterworth Heinemann 2001
Third edition published by Butterworth Heinemann 2004

British Library Cataloguing in Publication Data
A catalogue record for this book is available from the British Library

Library of Congress Cataloging in Publication Data
House, D. J.
Seamanship techniques / D.J. House. — Fourth edition.
    pages cm.
Includes index.
1. Seamanship. I. Title.
VK541.H85 2013
623.88—dc23   2013005300

ISBN: 978-0-415-82952-6 (hbk)
ISBN: 978-0-415-81005-0 (pbk)
ISBN: 978-0-203-79670-2 (ebk)

Typeset in Sabon
by Cenveo Publisher Services

Printed and bound by CPI Group (UK) Ltd, Croydon, CR0 4YY

Contents

Preface to the First Edition

This single-volume edition of general seamanship provides a comprehensive cover to the needs of marine students and serving seafarers. It is ideal for Merchant Navy Officers from Cadet rank to Master Mariner and incorporates all recent amendments to collision regulations.

In changing times the design and build of ships has altered and the needs of the professional mariner must be adapted to meet these modern times. However, old vessels do not disappear overnight and the old practices of basic seamanship are still required in all quarters of the globe. The practical seaman must adapt alongside a developing hi-tech industry and be able to improvise when the need arises.

This work takes account of many types of vessel engaged on many commercial trades and is expected to continue to be the accepted reference on general seamanship. It incorporates all the subjects required by the professional mariner, including: anchor work, rigging, cargo work, survival and boatwork, communications, search and rescue practice, watchkeeping, meteorology, marine instruments, tanker work and pollution, together with marine emergencies and ship-handling.

The marine industry is demanding in nature. It absorbs not only the ships which create its very existence, but also the personalities of the professional men and women cast within its perimeter. It has been my great fortune to have made the acquaintance of a number of these professionals, without whose teaching and understanding this work could never have evolved. My personal thanks are sincerely given, especially to the following:

J.W. Riley, Lt Cdr. (SCC, RNR Retd);
Mr A.R. Ollerton, Senior Lecturer, Nautical Studies, Master Mariner, DMS, AMBIM;
Mr J. Finch, Senior Lecturer, Nautical Studies, Master Mariner; and
my wife Lucia, just for being there.

Preface to the Fourth Edition

The combined volume of *Seamanship Techniques* has evolved into this fourth edition alongside a rapidly changing maritime industry. Electronic navigation charts (ENCs) are now sitting comfortably alongside 'free-fall lifeboats', while automatic identification systems (AIS) technology has become a standard bridge interpretation. However, the role of the seaman, with all the modernization taking place, continues to be led by the first principle of seamanship – 'the safety of life at sea'.

Time has changed procedures, GPS has overpowered the use of the sextant, while 'fall-preventative devices' are now recommended when launching lifeboats. The many electronic systems of radar, ARPA, ECDIS and GMDSS alongside the romance of sail are actively sharing our seas. Such a diverse arena with the majesty of passenger vessels working with bunker barges and tugs is, therefore, bound to influence the young minds of maritime personnel.

With this in mind, this latest revision needs to incorporate the latest thoughts regarding the International Safety Management (ISM) culture and the essential elements of collision avoidance. Some recent incidents like the *Costa Concordia* and the loss of the *Riverdance* remind us that it is not a perfect world. The hazards of our industry are well known and will continue to test our seafarers with extreme weather conditions. We should not be looking to enhance nature's dangers by adding human error into our environment.

The text in this book does not have all the answers. It needs the support of practical experience when handling a ship in heavy weather. The knowledge of the meteorologists and Safety Officer need to be present on the navigation bridge. The execution of the Passage Plan must have the qualities of the Officer of the Watch and the lookout. Unexpected emergencies, like an on-board fire or a need to drydock, must be catered for with positive attitudes.

It takes time to train our people and the sea is unforgiving of those with an unprofessional attitude.

Acknowledgements

I would like to express my appreciation and thanks to the following for their assistance in supplying diagrams, photographs and information relevant to this work:

Additional artwork by A. Benniston
AFA – Minerva Ltd, Marine and Offshore Division
AGA Spiro Ltd
Anker Advies Bureau b.v.
Anschutz & Co., GMBH
Ateliers et Chantiers de Bretagne – ACB
Beaufort Air-Sea Equipment Ltd
British Ropes Ltd
Bruce Anchor Ltd
Bruntons (Musselburgh) Ltd
Butterworth Systems (UK) Ltd
C.M. Hammar Handels A.B.
Creative Ropework (published by G. Bell & Sons Ltd) by Stuart E. Grainger
Dubia Dry Docks
Dunlop Beaufort Canada
Dunlop Ltd
E & FN Spon Ltd for references from *Cargo Access Equipment*
E.H. Industries Ltd
Elkem a/s Stalog Tau
Elliott Turbomachinery Ltd/White Gill Bow Thrusters
F.R. Fassmer & Co.
F.R. Hughes & Co., Ltd
General Council of British Shipping
General Council of British Shipping/MNTB
Heien – Larssen A/S
Henry Brown & Son Ltd
HM Coastguard – Maritime Rescue Sub Centre Formby, Liverpool
HMSO – British Crown Copyright Reserved
Holland Roer – Propeller, the Netherlands
Hydrographic Department of the Navy
I.C. Brindle & Co.
Imtech Marine and Industry, the Netherlands
J.M.Voith GmbH
James Robertson & Sons, Fleetwood
John Cairns Ltd (for extracts from the *International Manual of Maritime Safety* and *The S.O.S. Manual*)
Kelvin Hughes Ltd, Naval & Marine Division of Smiths Industries Aerospace
Lisnave Estaleiros Navais, S.A.
Litton Marine Systems
Lloyds Beal Ltd

Macgregor & Co. (Naval Architects) Ltd
Maritime and Coastguard Agency
Meteorological Office, U.K.
Mitsubishi Heavy Industries Ltd, Shimonoseki Shipyard and Machinery Works
MPJ Waterjets, Sweden
Negretti & Zambra (Aviation) Ltd
NEI Clarke Chapman Ltd, Clarke Chapman Marine
P & O European (Irish Sea) Ferries
Pains Wessex Schermuly Ltd
RFD Inflatables Ltd
Schilling Rudders
Shutterstock Photograph Library
Siebe Gorman & Company Ltd
Sperry (Marine Systems) Ltd
Stanford Maritime Ltd for references from *The Apprentice and His Ship* by Charles H. Cotter
Stanford Maritime Ltd, for references from *Tugs* by Captain Armitage and from *Basic Shiphandling for Masters & Mates,* by P.F. Willerton
The British Broadcasting Corporation
The Motor Ship (published by IPC Industrial Press Ltd)
The Nautical College, Fleetwood – Lancashire Education Committee
The Solid Swivel Company Ltd
The Welin Davit & Engineering Company Ltd
Thomas Mercer Chronometers Ltd
Thomas Walker & Son Ltd
United States Coast Guard
Wagner Engineering Associates Ltd
Watercraft Ltd – Survival Craft Division
Westland Helicopters Ltd
Whessoe Systems and Controls Ltd
Whittaker Corporation – Survival Systems Division
Witherby Seamanship

Additional Photography
Capt. D.A. McNamee (AFNI)
Capt. J.G. Swindlehurst, Master Mariner (MN)
Capt. K. Millar, Master Mariner (MN)
Mr. A.P.G. House (research assistant)
Mr. G. Edwards, Ch/Eng. (retd)
Mr. I. Baird, Ch/Off (MN)
Mr. J. Legge, 2nd Officer (MN)
Mr. J. Leyland, Lecturer Nautical Studies
Mr. J. Roberts, 2nd Officer (MN)
Mr. M. Croft, 1st Officer (MN)
Mr. M. Gooderman, Master Mariner (MN), BSc
Mr. P.P. Singh, Ch/Off (MN)
Mr. Z. Anderson, Ch/Off (MN)

Additional Assistance
Mr. E. Hackett, Senior Lecturer, Nautical Studies
Mr. C.D. House (IT Consultant)

About the Author

David House has now written and published 17 marine titles, many of which are in multiple editions. After commencing his seagoing career in 1962, he was initially engaged on general cargo vessels. He later experienced worldwide trade with passenger, container, ro-ro, reefer ships and bulk cargoes. He left the sea in 1978 with a Master Mariner qualification and commenced teaching at the Fleetwood Nautical College, from where he retired in 2012, after 33 years of teaching in nautical education.

The experience he gained in both a seagoing capacity and as a lecturer of marine studies led to maritime titles, for all ranks from Cadet to Master Mariner. His books are well read and respected around the world, covering such topics as ice navigation, cargo operations, communications and all areas of general seamanship.

He continues to work in the maritime field with the International Institute of Nautical Surveyors, and in a private marine consultancy role. His works are regularly updated and his books on marine survival, drydocking, ship construction, helicopter operations and anchor practice are appreciated on many bookshelves in virtually all the maritime nations.

Abbreviations

Search-and-rescue specific abbreviations can be found on p. 583 at the end of Chapter 16.

ABS	American Bureau of Shipping
AC (i)	Admiralty Class (Cast)
AC (ii)	alternating current
ACV	air cushion vessel
AHV	anchor handling vessel
AIS	automatic identification systems
ALBA	compressed air deck line
ALRS	Admiralty list of radio signals
AMD	advanced multi-hull design
AMIRIS	advanced maritime infrared imaging system
AMVER	automated mutual vessel reporting system
AP	aft perpendicular
ARCS	Admiralty raster chart service
ARPA	automatic radar plotting aid
ATT	Admiralty tide tables
AUSREP	Australian ship reporting system
aux	auxiliary
B	position of the centre of buoyancy
B/A	breathing apparatus
B/L	bill of lading
BP (i)	between perpendiculars
BP (ii)	British Petroleum
BS	breaking strain
BST	British summer time
BT	ballast tank
BV	Bureau Veritas
CABA	compressed air breathing apparatus
cc	corrosion control (LR – notation)
CCTV	closed circuit television
CD (i)	chart datum
CD (ii)	compact disc
CDP	controlled depletion polymers
CES	Coast Earth Station
CG	Coast Guard
CIE	International Commission on Illumination
CL	centre line
cm	centimetres
CMG	course made good
CML	Centre of Maritime Leadership (USA)
CMS	constantly manned station

CNIS	Channel Navigation Information Service
CO	Chief Officer
CO_2	carbon dioxide
COG	course over ground
C of B	centre of buoyancy
C of G	centre of gravity
COI	Certificate of Inspection (as issued by USCG)
ColRegs	The Regulations for the Prevention of Collisions at Sea
COW	crude oil washing
C/P	charter party
CPA	closest point of approach
CPP	controllable pitch propeller
CPR	cardiac pulmonary resuscitation
CQR	Chatham quick release
CRS	Coast Radio Station
CSH	continuous survey hull
CSM	continuous survey machinery
CSP	commencement search pattern
CSS (code)	IMO Code of Safe Practice for Cargo Stowage and Securing
CSWP	Code of Safe Working Practice
CW	continuous wave
Cwt	hundredweight
Da	draught aft
DAT	double acting tanker
dB	decibels
DB	double bottom
DBC	Dunlop Beaufort Canada
DC	direct current
DD	drydock
Df	draught forward
DGN	dangerous goods note
DGPS	differential global positioning system
Disp	displacement
Dm	midships draught
DNV	Det Norske Veritas
DNV-W1	one-man operation (DNV notation)
DOC (Alt. DoC)	document of compliance
DP	dynamic position
DPA	designated person ashore
DR	dead reckoning
DSC (i)	digital selective calling
DSC (ii)	dynamically supported craft
DSV	diving support vessel
DW (i)	dock water
DW (ii)	deadweight
DWA	dock water allowance
dwt	deadweight tonnage
E	east

EBM (EBI)	electronic bearing marker
EC	European Community
ECDIS	electronic chart display and information system
ECR	engine control room
EEBDs	emergency escape breathing device
EFSWR	extra flexible steel wire rope
ENC	electronic navigation chart
EPIRB	emergency position indicating radio beacon
ETA	estimated time of arrival
ETD	estimated time of departure
ETV	emergency towing vessel
EU	European Union
FFA	fire-fighting appliances
FLIR	forward looking infra red
FMECA	failure mode effective critical analysis
FO	fuel oil
foap	forward of aft perpendicular
FPD	fall prevention device
FPk	fore peak tank
FPSOs	floating production storage offloading system
FPV	fisheries protection vessel
FRC	fast rescue craft
FRD (Fwd)	forward
FSE	free surface effect
FSMs	free surface moments
FSU	floating storage unit
FSW	friction stir welding
FSWR	flexible steel wire rope
FU	follow-up
FW	fresh water
FWA	fresh water allowance
FWE	finished with engines
G	ship's centre of gravity
gals	gallons
GG 1	distance measured from the ship's original C of G, to a new position of the ship's C of G
GHz	gigahertz
GL	Germanischer Lloyd
GM	metacentric height
GMDSS	global maritime distress and safety system
GMT (z)	Greenwich Mean Time
GPS	global positioning system
GRB	garbage record book
GRP	glass reinforced plastic
grt (GT)	gross registered tonnage
GZ	ship's righting lever
HDOP	horizontal dilution of precision
HEX	hexagonal

HF	high frequency
HFO	heavy fuel oil
H/L	heavy lift
HLO	helicopter landing officer
HMAS	Her Majesty's Australian Ship
HMS	Her Majesty's Ship
HMSO	Her Majesty's Stationery Office
HP (i)	horse power
HP (ii)	high pressure
HPFWW	high pressure fresh water wash
HRN	house recovery net
HRU	hydrostatic release unit
HSC	high-speed craft
HSE	health and safety executive
HSSC	Harmonised System of Survey and Certification
I	intensity
IACS	International Association of Classification Societies
IALA	International Association of Lighthouse Authorities
IAMSAR	International Aeronautical and Marine Search & Rescue manual
IBC	International Bulk Chemical Code
ICAA	International Civil Aviation Authority
ICS	International Chamber of Shipping
IE	index error
IFR	instrument flying rating
IGS	inert gas system
IHO	International Hydrographic Office
IIP	International Ice Patrol
ILO	International Labour Organization
IMDG	International Maritime Dangerous Goods (code)
IMO	International Maritime Organization
INF	irradiated nuclear fuel
INS	integrated navigation system
IOPPC	international oil pollution prevention certificate
IPMS	integrated platform management system
IPS	integrated power system (controllable 'podded' propulsion)
IRF	incident report form
ISM	International Safety Management (code)
ISO	International Organization of Standardization
ISPS	International Ship and Port Security (code)
ITP	intercept terminal point
ITU (i)	International Transport Union
ITU (ii)	International Telecommunications Union
IWS	in water survey
K	representative of the position of the ship's keel
KG	distance measured from the keel to the ship's C of G
Kg	kilogram
kHz	kilohertz
kJ	kilojoule

KM	distance measured from the keel to the metacentre 'M'
kN	kilo newtons
kts	knots
kW	kilowatt
Lat	latitude
LBP	length between perpendiculars
lbs	pounds
LCB	longitudinal centre of buoyancy
LCD	liquid crystal display
LCG	longitudinal centre of gravity
LCV	landing craft vessel
LFL	lower flammable limit
LMC	Lloyd's Machinery Certificate
LNG	liquid natural gas
LOA	length overall
LOF	Lloyd's Open Form (salvage)
Lo-Lo	load on, load off
Long	longitude
LP	low pressure
LPG	liquid petroleum gas
LR	Lloyd's Register
LSA	life-saving appliances
LUT	land user terminal
M	metacentre
m	metres
MA	mechanical advantage
MAIB	Marine Accident Investigation Branch
MARPOL	Marine Pollution (convention)
mbs	millibars
MCA	Maritime and Coastguard Agency
MCTC	moment to change trim 1 centimetre
MEC	marine evacuation chute
Medivac	medical evacuation
MEPC	Marine Environment Protection Committee
MES	marine evacuation system
MEWP	mobile elevator work platform (cherry picker)
MF	medium frequency (300 kHz to 3 MHz)
MFAG	Medical First Aid Guide (for use with accidents involving dangerous goods)
MGN	marine guidance notice
MHR	mean hull roughness
MHz	megahertz
MIN	marine information notice
MMSI	maritime mobile service identity number
MN	Mercantile Marine (Merchant Navy)
MNTB	Merchant Navy Training Board
MoB	man over board
MODU	mobile offshore drilling unit
MPCU	marine pollution control unit

m rads	metre radians
MRCC	marine rescue coordination centre
m/s	metres per second
MSC	Maritime Safety Committee (of IMO)
MSI	marine safety information
MSL	mean sea level
MSN	merchant shipping notice
MV	motor vessel
MW	megawatt
N	north
NE	northeast
NFU	non-follow-up
nm	nautical miles
NOE	notice of eligibility
NP	national publication
NUC	not under command
NVE	night vision equipment
NVQ	national vocational qualification
NW	northwest
O/A	overall
OBO	oil, bulk, ore (carrier)
OiC	officer in charge
OIM	offshore installation manager
OLB	official log book
OMB	one-man bridge
OMBO	one-man bridge operation
OOW	Officer of the Watch
OPIC	oil pollution insurance certificate
ORB	oil record book
O/S	offshore
OSC (i)	on scene commander
OSC (ii)	on scene coordinator
OSV	offshore standby vessel
P	port
P/A	public address system
P & I (club)	Protection & Indemnity
PEC	pilot exemption certificate
PHA	preliminary hazard analysis
P/L	position line
ppm	parts per million
PRS	Polish Register of Shipping
PSC	port state control
PSC & RB	personal survival craft & rescue boat
psi	pounds per square inch
pts	pints
RAF	Royal Air Force
RBD	return of births and deaths
RCC	rescue coordination centre

RCDS	raster chart display system
RD	relative density
RINA	Registro Italiano Navale (Classification Society – Italy)
RMC	refrigerated machinery certificate
RMS	Royal Mail ship
RN	Royal Navy
RNR	Royal Naval Reserve
ro-pax	roll on–roll off passenger vessel
ro-ro	roll on–roll off
RoT	rate of turn
ROV	remotely operated vehicle
rpm	revolutions per minute
RS	reflected sun
RT	radiotelephone
Rx	receiver
S	south
S (Stbd)	starboard
SAR	search and rescue
SARSAT	search and rescue satellite
SART	search and rescue transponder
SATCOM	satellite communications
SBE	stand by engines
SBM	single buoy mooring
SCBA	self-contained breathing apparatus
SE	southeast
SES (i)	ship earth station
SES (ii)	surface effect ship
SF	stowage factor
SFP	structural fire protection
sg	specific gravity
shp	shaft horse power
SI	statutory instrument
SMC (i)	safety management certificate
SMC (ii)	SAR mission coordinator
SMG	speed made good
SMS	safety management system
SOG	speed over ground
SOLAS	Safety of Life at Sea (convention)
SOPEP	ship's oil pollution emergency plan
SPC	self polishing copolymer (anti-fouling paint)
SPM	single point mooring
SQU (sq)	square
SS	steam ship
SSA	Ship Building and Ship Repair Association
SSP	Siemens–Schottel Propulsion
stbd	starboard
STCW	standards of training, certification and watchkeeping
SW	salt water

SWATH	small waterplane area twin hull
SWL	safe working load
SWR	steel wire rope
SU	search unit
TBT	tributyltin
TCPA	time of closest point of approach
TEMPSC	totally enclosed motor propelled survival craft
TEU	twenty-foot equivalent unit
Tk	tank
TLV	threshold limit value
TMC	transmitting magnetic compass
TMCP	thermo-mechanically controlled process
TPA	thermal protective aid
TPC	tons per centimetre
TRC	type rating certificate
TRS	tropical revolving storm
TS	true sun
TSS	traffic separation scheme
TWI	The Welding Institute
Tx	transmitter
UAE	United Arab Emirates
UFL	upper flammable limit
UHF	ultra high frequency
UHP	ultra high pressure
UK	United Kingdom
UKC	under keel clearance
UKOOA	United Kingdom Offshore Operators Association
UKOPP	United Kingdom Oil Pollution Prevention (cert)
ULCC	ultra large crude carrier
UMS	unmanned machinery space
UN ECE	United Nations Economic Commission for Europe
USA	United States of America
USCG	United States Coast Guard
UV	ultraviolet
VCG	vertical centre of gravity
VDR	voyage data recorder
VDU	visual display unit
VFI	vertical force instrument
VHF	very high frequency
VLCC	very large crude carrier
VLGC	very large gas carrier
VTMS	vessel traffic management system
VTS	vessel traffic services
W (i)	west
W (ii)	representative of ship's displacement
W (iii)	watts
WAT	wing assisted trimaran
WBT	water ballast tank

WIG	wing in ground (effect)
W/L	waterline
WNA	winter North Atlantic
WPC	wave piercing catamaran
WPS	wires per strand
W/T (i)	wireless telegraphy
W/T (ii)	walkie-talkie radio
z	Greenwich Mean Time (GMT)

1

The Ship

Introduction

The art and science of seamanship has developed from the experience of maritime nations over many centuries. Sea travel has passed through the days of propulsion by oars, the discovery days of sail, through the advances of steam and on to the age of oil, and finally to the atomic period of advanced technology. The art of mastering the means of transportation on water, having seen the excitement of discovering new worlds and the conquering of new boundaries, has settled for the advance of trade in all directions of the compass.

The ship, once stored and provisioned, becomes the ideal in self-sufficiency, capable of the transport of cargo, livestock, troops, passengers, gas, fluids, minerals, etc. The fact that the vessel provides a source of power which can cope with varying degrees of emergency and still be able to sustain itself says a lot for the developed marine industry.

The ship is equipped with such ancillary equipment as required to be able to load and offload, in a safe condition, all cargoes and passengers as the vessel is designed to accommodate. Bearing the function of the ship in mind, it is not difficult for seafarers to realize how their characters have been influenced by the independent nature of their employment. Seamanship and the ships themselves have created the spirit of adventure that turned such men as Magellan, Drake and Nelson into more than legends.

Even in today's ships, be they of the mercantile marine or 'men of war', the same spirit prevails, and it is hoped that this book will direct the men and women who sail them safely into good seaman-like practices.

Terms and Definitions

Abeam

A bearing projected at right angles from the fore and aft line, outwards from the widest part of the ship (Figure 1.1).

Ahead

'Right ahead' is the line that the fore and aft line, if projected, would extend in front of the vessel (Figure 1.1). Opposite to the term 'astern' when used in

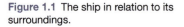

Figure 1.1 The ship in relation to its surroundings.

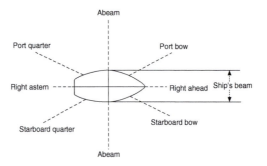

relation to relative bearings. It may also be used as an engine-room order to cause the engines to turn in order to move the ship ahead.

Amidships

The middle of the vessel in both transverse and longitudinal directions. More commonly used as a helm order, where it is usually shortened to 'midships' (see Chapter 10).

Athwartships

Defined as 'in a direction' from one side of the ship to the other, at a right angle to the fore and aft line.

Breadth

The maximum beam of the vessel measured from the outside edge of the shell plating on either side of the vessel is the *extreme* breadth (Figure 1.2).

The beam of the vessel measured amidships, between the inside edge of the shell plating on either side of the vessel, is the moulded breadth (Figure 1.2).

Camber (or round of beam)

The curvature of the deck in the athwartships direction. The measurement is made by comparing the height of the deck at the centre of the vessel to the height of the deck at the side of the vessel (Figure 1.2).

Depth

The *extreme* depth of the vessel is measured from the bottom side of the keel to the top of the deck beams, the measurement being taken at the side of the vessel.

The *moulded* depth is measured from the top side of the keel to the top of the deck beams, at the side of the vessel.

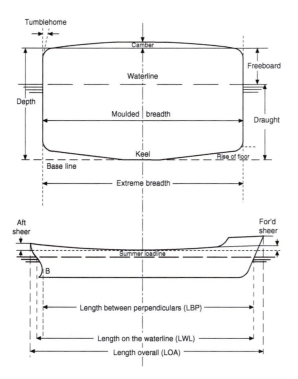

Figure 1.2 Ship's principal dimensions.

Flare

The outward curvature of the shell plating in the foremost part of the vessel, providing more width to the forecastle head and at the same time helping to prevent water coming aboard.

Fore and Aft Line

An imaginary line passing from the stem to the stern through the centre of the vessel (Figure 1.3).

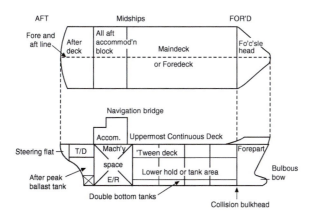

Figure 1.3 The ship in section and plan.

Figure 1.4 Keel rake.

Freeboard

This is the vertical distance, measured at the ship's side, from the waterline to the top of the freeboard deck edge. The freeboard measurement is taken at the midships point. The deck edge is marked by a painted line 25 mm × 300 mm, above the plimsoll line.

Keel Rake

The inclination of the line of the keel to the horizontal (Figure 1.4).

Length between Perpendiculars (LBP)

The distance between the forward and aft perpendiculars.

Length Overall (LOA)

The maximum length of the vessel measured from the extreme forward point of the vessel to the extreme after point (Figure 1.2).

Perpendiculars

A perpendicular drawn to the waterline from a point on the summer loadline where it intersects the stempost is called the forward perpendicular (FP).

A perpendicular drawn to the waterline at a point where the after side of the rudder post meets the summer waterline is called the aft perpendicular (AP). If a rudder post is not fitted, then it is drawn from the centre of the rudder stock.

Rise of Floor

This is the rise of the bottom shell plating above the baseline (taken from the top edge of the keel) (Figure 1.2).

Sheer

This is the curvature of the deck in the fore and aft direction, measured as the height of the deck at various points above the height of the deck at the midships point (Figure 1.2).

Ship's Beam

The widest part of the ship in the transverse athwartships direction (Figure 1.1).

Stem Rake

The inclination of the stem line to the vertical.

Tonnage

All ships constructed on or after 8 July 1982 are measured in accordance with the IMO 1969 International Conference on Tonnage Measurement. Existing ships built prior to this date were allowed to retain their existing tonnage, if the owner so desired, for a period of 12 years. Since 18 July 1994 all ships have needed to comply with the 1969 convention.

Gross tonnage (GT) is defined as that measurement of the internal capacity of the ship.
    The GT value is determined by the formula:

$$GT = K_1 V$$

when $K_1 = 0.2 + 0.02 \log^{10} V$ and V = total volume of all enclosed spaces measured in cubic metres.

Net tonnage (NT) is that measurement which is intended to indicate the working/earning capacity of the vessel. Port and harbour dues are based on the gross and net tonnage figures.
    NT for passenger ships carrying more than 13 passengers is determined by the formula:

$$NT = K_2 V_C \left[\frac{4d}{3D} \right]^2 + K_3 \left[N_1 + \frac{N_2}{10} \right]$$

NT for other vessels is:

$$NT = K_2 V_C \left[\frac{4d}{3D} \right]^2$$

where V_C = total volume of cargo spaces in cubic metres; d = moulded draught at midships in metres (summer loadline draught; or deepest subdivision loadline in passenger vessels); D = moulded depth in metres amidships; $K_2 = 0.2 + 0.02 \log^{10} V_C$;

$$K_3 = 1.25 \frac{(GT + 10,000)}{10,000}$$

N_1 = Number of passengers in cabins with not more than eight berths; N_2 = number of other passengers.

    NT is not to be taken as less than 0.30 GT. The factor $[4d/3D]^2$ is not taken to be greater than unity. The expression $K_2 V_C [4d/3D]^2$ is not to be taken as less than 0.25 GT.

Tumblehome

The inward curvature of the ship's side shell plating above the summer loadline (Figure 1.2).

Terms and Definitions Concerning Stability

Bulkhead Deck

Defined as the uppermost deck to which the watertight bulkheads are taken.

Centre of Flotation

That point in the ship's length about which the vessel will trim by the head/by the stern. In layman's terms, the tipping centre of the ship, which is very rarely the exact midships point.

Coefficient of Fineness (of the water-plane area)

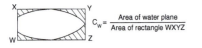

$$C_w = \frac{\text{Area of water plane}}{\text{Area of rectangle WXYZ}}$$

Figure 1.5 Coefficient of fineness.

The ratio of the water-plane area to the area of the rectangle having the same extreme length and breadth (Figure 1.5). Block coefficient of fineness of displacement is similarly applied, using the values of volume instead of area.

Displacement

The displacement of a vessel is the weight of water it displaces, i.e. the weight of the vessel and all it contains. It is the immersed volume of the ship in cubic metres × density of the water, expressed in tonnes per cubic metre. It is normal practice to regard the ship's displacement as being that displacement when at her load draught (load displacement).

Equilibrium

A body is said to be in *stable* equilibrium (Figure 1.7) if, when slightly disturbed and inclined from its initial position, it tends to return thereto. A body is said to be in a state of *neutral* equilibrium if, when slightly disturbed from its initial position, it exhibits no tendency to return thereto or to move to another new position. A body is said to be in *unstable* equilibrium if, when slightly disturbed from its initial position, it tends to move further from it.

Figure 1.7 shows a *vessel in stable equilibrium.* As the vessel heels to $\theta°$ by an external force (e.g. waves, wind), G remains in the same position and B moves to B1. A righting couple is formed, WGZ, where W is the weight effect of the ship acting through G

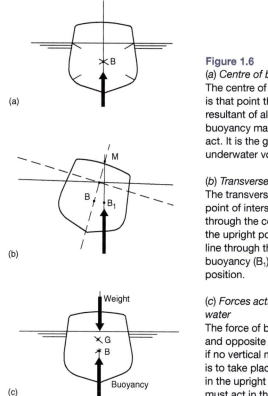

Figure 1.6
(a) Centre of buoyancy
The centre of buoyancy (C of B) is that point through which the resultant of all the forces due to buoyancy may be considered to act. It is the geometric centre of the underwater volume of the ship.

(b) Transverse metacentre
The transverse metacentre (M) is that point of intersection of a vertical line through the centre of buoyancy, in the upright position, with a vertical line through the new centre of buoyancy (B₁) in a slightly inclined position.

(c) Forces acting on a vessel in still water
The force of buoyancy must be equal and opposite to the forces of gravity if no vertical movement of the body is to take place. For the body to float in the upright position, both forces must act in the same vertical plane.

(due to gravity), GZ being known as the righting lever. In the triangle MGZ, $GZ = GM \sin \theta°$. Therefore $W \times GZ = W.GM \sin \theta°$, bringing the vessel back to the upright position. This is the situation when G is below M, i.e. *when GM is positive*.

Floodable Length

The maximum length of a compartment which can be flooded so as to bring a damaged vessel to float at a waterline which is tangential to the margin line.

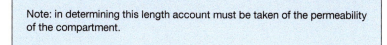

Note: in determining this length account must be taken of the permeability of the compartment.

Lightship Displacement

Lightship is defined as the extreme displacement of a ship, when fully equipped and ready for sea but without cargo, crew, passengers, fuel, ballast water, fresh water and consumable stores. The boilers are filled with water to their working level.

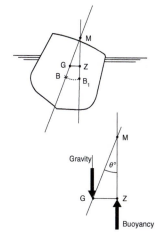

Figure 1.7 Vessel in stable equilibrium.

NB. Displacement of a vessel can be expressed as a volume, in cubic metres or as a weight determined by the volume × density of the water displaced. In sea water the density constant is taken as 1,025 kg/m³.

Load Deadweight

Deadweight is defined by the difference in tonnes between the displacement of a ship in water of a specific gravity of 1.025 at the load waterline corresponding to the assigned summer freeboard and the lightweight of the ship. It consists of the total weight of cargo, stores, bunkers, etc., when the vessel is at her summer loadline.

Margin Line

Defined by a line at least 76 mm below the upper surface of the bulkhead deck, as measured at the side of the vessel.

Permeability

In relation to a compartment space this means the percentage of that space which lies below the margin line, which can be occupied by water.

Note: various formulae within the Ship Construction Regulations are used to determine the permeability of a particular compartment.

Example values are:

- spaces occupied by cargo or stores: 60 per cent
- spaces employed for machinery: 85 per cent
- passenger and crew spaces: 95 per cent

Permissible Length

The permissible length of a compartment, having its centre at any point in the ship's length, is determined by the product of the floodable length at that point and the factor of subdivision of the vessel:

permissible length = floodable length × factor of subdivision

Reserve Buoyancy

The buoyancy of the immersed portion of the vessel is that which is necessary to keep the vessel afloat. The buoyancy of all other enclosed watertight spaces above the waterline is therefore residual buoyancy, more commonly referred to as 'reserve buoyancy'. It must be assumed that in the case of the conventionally designed ship, if water equal to the displacement and reserved buoyancy enters the vessel, it will sink. Sufficient reserve buoyancy is necessary in all seagoing vessels in order for the ship to rise quickly, owing to the lift effect, when navigating, especially in heavy sea conditions.

Subdivision Factor

The factor of subdivision varies inversely with the ship's length, the number of passengers and the proportion of the underwater space used for passengers/crew and machinery space. In effect, it is the factor of safety allowed in determining the maximum space of transverse watertight bulkheads, i.e. the permissible length.

Varieties of Ship

Ships come in all forms, and Figures 1.8–1.21 illustrate this variety. See also Plates 1–10.

Plate and Construction Terms

'A' Frame

The supporting framework for the stern tube of a twin-screw vessel; used as an alternative to a spectacle frame.

Boss Plate

A shell plate parallel to the stern tube at the level of the propeller boss.

Bulkhead

A vertical partition between compartments. May be in the fore and aft line or athwartships.

Coffin Plate

The aftermost plate of the keel, dish (coffin)-shaped to fit the stern frame.

Collision Bulkhead

A heavy-duty bulkhead in the forepart of the vessel to withstand damage after impact from collision.

Floor

A vertical athwartships member in way of the double bottom. A floor will run from the centre girder out to the margin plate on either side of the vessel. Floors may be in steel plate, solid or framed bracket form.

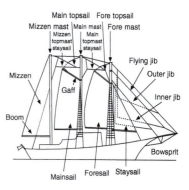

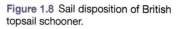

Figure 1.8 Sail disposition of British topsail schooner.

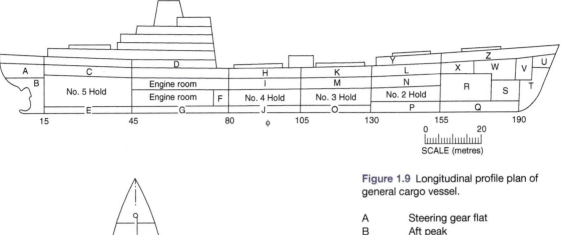

Figure 1.9 Longitudinal profile plan of general cargo vessel.

A	Steering gear flat
B	Aft peak
C	Upper 'tween deck
D	Poop 'tween deck
E	No. 5 double bottom tanks
F	Oil fuel tanks
G	Engine room double bottom tanks
H	Upper 'tween deck
I	Lower 'tween deck
J	No. 4 double bottom tanks
K, L	Upper 'tween deck
M, N	Lower 'tween deck
O	No. 3 double bottom tanks
P	No. 2 double bottom tanks
Q	No. 1 double bottom tanks
R	No. 2 cargo tank
S	Ballast tank
T	Forepeak
U	Store
V	Chain locker
W	No. 1 cargo tank
X	No. 1 upper 'tween deck
Y	No. 2 forecastle 'tween deck
Z	No. 1 forecastle 'tween deck

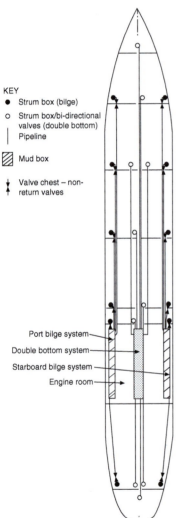

KEY

● Strum box (bilge)

○ Strum box/bi-directional valves (double bottom)

| Pipeline

▨ Mud box

↕ Valve chest – non-return valves

Port bilge system

Double bottom system

Starboard bilge system

Engine room

Figure 1.10 Typical hold bilge pumping system.

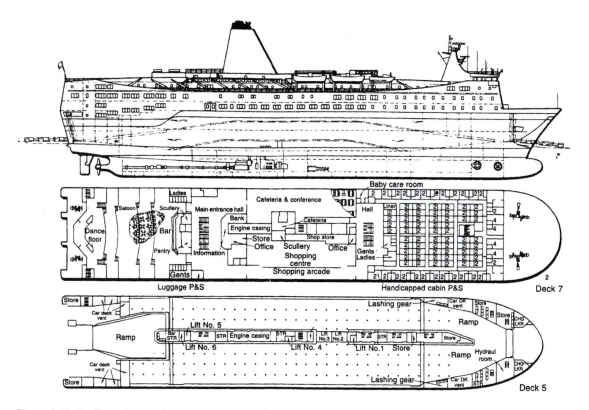

Figure 1.11 Profile and general arrangement plans of passenger/car ferry *Kronprinsessan Victoria* (15,000 gross tons).

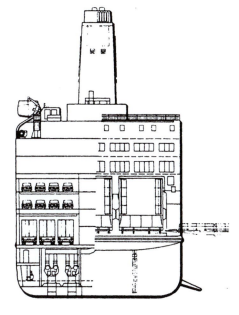

Figure 1.12 *Kronprinsessan Victoria* in section.

Length overall	150 m
Moulded breadth	26 m
Draught	6 m
Deadweight (at design draught)	3,100 tonnes
Passenger complement	2,100
Trailer capacity	70 × 18 m
Car capacity	700 × 4.5 m

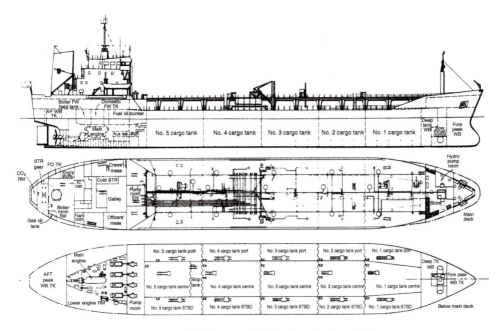

Figure 1.13 Profile and deck plans of 8,030-tonne DW Products' tanker, *Cableman.*

Frame

An internal support member for the shell plating (Figure 1.21); vessels may be framed transversely or longitudinally.

Garboard Strake

The first strake out from the keel.

Figure 1.14 Product carrier (tanker). Six tanks have heating coils, and there are four main pump rooms, with a capacity of 200 tons of water per hour. For the stainless steel tanks, centrifugal pumps can supply 30 tons per hour.

Tank coatings
P Polyurethane
S Stainless steel
X Epoxy coated
Z Zinc silicate

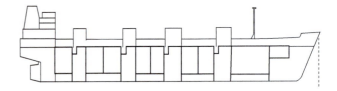

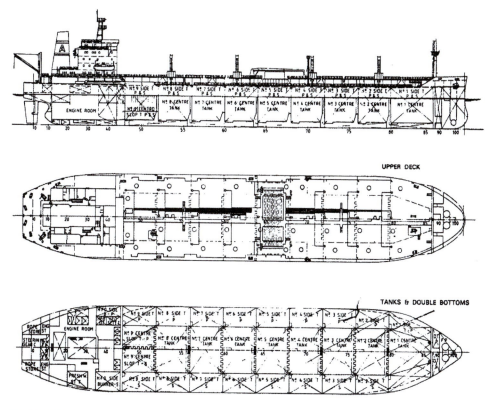

Figure 1.15 Chemical carrier.

Gusset Plate

A triangular plate often used for joining angle bar to a plate.

Intercostal

A side girder in the fore and aft line sited either side of the keel. It has an integral connection with the tank top and the ship's bottom plating and is rigidly connected by the floors.

Joggled Plating (obsolete)

A type of shell plating with an 'in/out' design at its edges. Effectively removes the need for liners, which would be required for the fitting of raised and sunken strakes. Both systems have become dated with advances in welded structures.

Keel

A centreline plate passing from the stem to the stern frame; referred to as a flat plate keel; generally of increased scantlings.

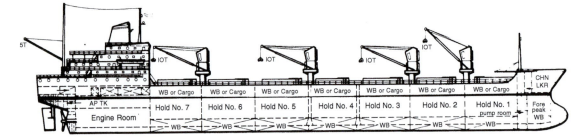

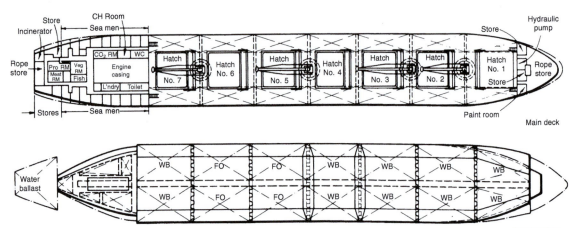

Figure 1.16 *Lok Priti* (bulk carrier).

Length, oa . . . . .172.02 m
Length, bp . . . . .162.40 m
Breadth, moulded . . . . .22.80 m
Depth, moulded . . . . .14.70 m
Draught, moulded . . . . .10.88 m
Gross register . . . . 15,638 grt
Net register . . . . 10,502 nrt
Deadweight, total at 10.67 m draught . . . . . 26,000 dwt
Deadweight, total at 10.88 m draught . . . . . 27,000 dwt

Capacities
Holds, including wing tanks and hatches . . . 35,091 m³
Ballast, including No. 4 . . . . 11,330 m³
Heavy fuel . . . . 1,530 m³
Diesel fuel . . . . 94 m³
Lube oil . . . . 135 m³

Fresh water . . . . 208 m³
Main engine . . . . GRSE-MAN-B&W K6Z78.155E
Output, mcr . . . . (7833 kW) 10,500 bhp at 122 rev/min
Trial speed, 87.5% mcr
(9,188 bhp) at loaded draught . . . . .15.00 knots
Endurance . . . . 12,000 nautical miles

COMPLEMENT
14 officers . . . . .2 cadets
8 petty officers . . . . .1 pilot
36 crewmen . . . . 1 owner
Total = 62

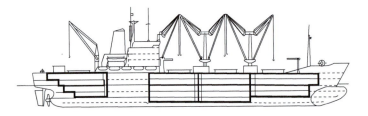

Figure 1.17 Refrigerated cargo lines (reefer): modern design.

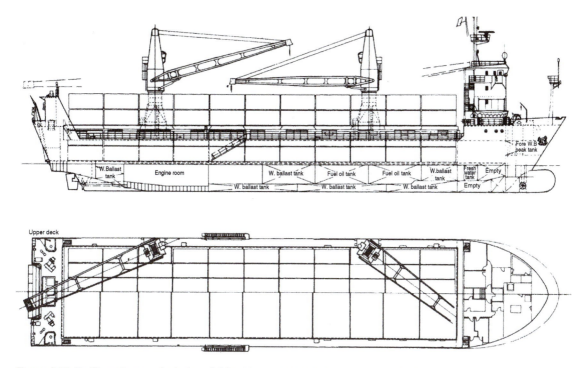

Figure 1.18 Profile and upper deck plan of CCN-Maua's 3450 DWT container ro-ro/lo-lo ship design.

Length oa . . . . 94.04 m
Length bp . . . . 80.00 m
Breadth moulded. . . . . 18.00 m
Summer draught . . . . 4.55 m
Complement 22
Vehicle stern ramp.
Two fixed deck cranes, 25 tonnes.

Duct Keel

The duct keel is a plated box/tunnelled keel allowing passage right forward. It provides additional buoyancy, together with a through passageway for cables and pipelines running in the fore and aft direction.

Lightening Holes

These are holes cut into floors or intercostals to reduce the weight content of the ship's build and to provide access to tank areas.

Longitudinal

A fore and aft strength member connecting the athwartships floors. Some vessels are longitudinally strengthened by having the frames run in a fore and aft direction as opposed to transverse framing. Additional longitudinals are to be found in areas where pounding can be anticipated when the vessel is at sea.

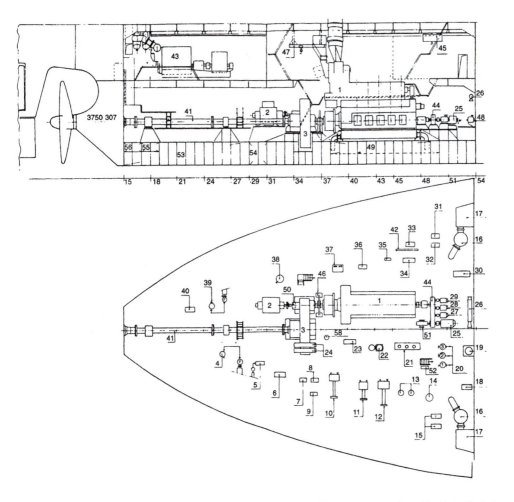

Figure 1.19 Engine-room layout and disposition of equipment on 13,230 dwt containership *Nathalie Delmas*.

Margin Plate

A fore and aft plate sited at the turn of the bilge (Figure 1.21). The upper edge is normally flanged to allow connection to the tank top plating, while the opposite end is secured to the inside of the shell plate by an angle-bar connection. It provides an end seal to the double bottom tanks, having all the floors joining at right angles up to the collision bulkhead.

Oxter Plate

A shell plate of double curvature found under the transom floor, being extended from the fore side of the sternpost in the direction of the bow.

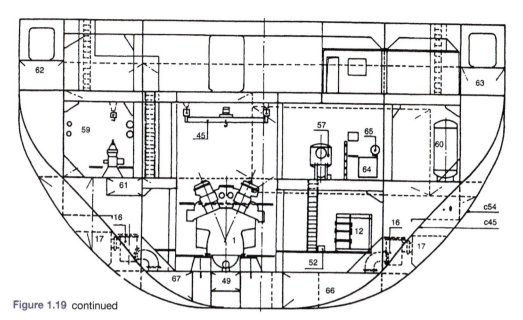

Figure 1.19 continued

Key

1 Main engine: SEMT-Pielstick 10PC4V-570
2 Main engine driven alternator: Alsthom 950 kW
3 Reduction gear/thrust bearing: ACB 386/77 rev/min
4 Sea water pump for No. 1 diesel alternator
5 Sea water pump (refrigerated provisions)
6 Main engine local control
7 Oil transfer pump
8 Main engine freshwater heating pump
9 Swimming pool pump
10 Main engine oil cooler
11 Freshwater/high temperature/main engine cooler
12 Freshwater/low temperature/main engine cooler
13 Bilge and ballast pumps
14 Ballast pump
15 Boiler feed pumps
16 Sea water filter
17 Starboard water intake
18 Evaporator ejector pump
19 Tunnel access
20 Sea water general service pump
21 Main engine oil filters
22 Main engine oil emergency pump
23 Main engine pre-lube pump
24 Reduction gear oil coolers
25 Step up gear for electric motor
26 Heel pump
27 Main engine sea water pump
28 Main engine/high temperature/freshwater pump
29 Main engine/low temperature/freshwater pump
30 Main fire pump
31 Sludge transfer pump
32 Fuel oil automatic pump
33 Diesel oil/fuel oil transfer pump
34 Fuel oil transfer pump

35 Accommodation fresh water pump
36 Oil separator pump
37 Lube oil tank and pumps for rocker arms
38 Oily water separator
39 Sea water pump for No. 2 diesel alternator
40 Bilge automatic pump
41 Shaft line
42 Level indicator panel
43 Diesel alternators: 2 × SEMT-Pielstick 6P A6L-280 engines driving Unilec 1350 kW alternators
44 Step up gear for main engine pumps: Citroen-Messian
45 Travelling crane
46 Brake: Twiflex
47 Rails for dismantling rotors of turbocharger
48 'Cocooned' spare pumps
49 Main engine oil return to ballast tank
50 Reduction gear oil pump
51 Main engine oil pump
52 Fuel oil leakage protection pump
53 Bilge water ballast tanks
54 Diesel alternators' polluted oil ballast tank
55 Sterntube oil drainage
56 Aft well
57 Evaporator
58 Reduction gear oil emergency pump
59 Fuel oil/diesel oil treatment room: Alfa-Laval separators
60 Control air tank
61 Slop tanks
62 Port side fuel oil reserve bunker for diesel alternators
63 Starboard diesel oil reserve bunker
64 Feed pressure tank
65 Drain cooler
66 Main engine oil reserve tank
67 Fuel oil/diesel oil overflow tank

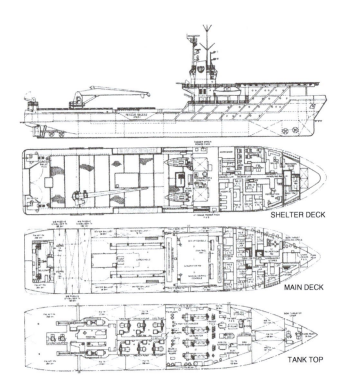

Figure 1.20 Safety/supply craft for the offshore industry.

Length . . . . .94.3 m
Moulded breadth. . . . . .19.5 m
Loaded draught. . . . . .4.65 m
Corresponding deadweight. . . . .2,997 tonnes
Gross registered tonnage 2,826
Net registered tonnage 945
Fire-fighting capability.
Pollution control facility.
Oil recovery equipment.
Operates submersibles.
Helicopter landing deck.
Hospital facility.
Command ship capability in emergency.

Panting Beams

Athwartships members in the forepart introduced to reduce the in/out tendency of the shell plating, caused by varying water pressure on the bow when the vessel is pitching.

Panting Stringers

Internal horizontal plates secured to the shell plating and braced athwartships by the panting beams.

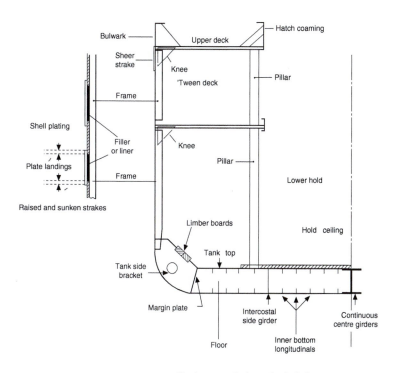

Bulwark

Upper deck

Hatch coaming

Sheer strake

Knee

'Tween deck

Pillar

Frame

Shell plating

Filler or liner

Knee

Plate landings

Pillar

Frame

Lower hold

Raised and sunken strakes

Limber boards

Hold ceiling

Tank top

Tank side bracket

Margin plate

Intercostal side girder

Continuous centre girders

Floor

Inner bottom longitudinals

The above example shows a longitudinally framed double bottom structure

Figure 1.21 Midships section through general cargo vessel (conventional structure).

Plate Landings

Refers to the shell plate. When shell plate is set in the raised and sunken method, the region where adjoining plates overlap is known as the plate landing (Figure 1.21). Generally superseded by welded flush structures.

Scantlings

Originally applied to the size of lintels in the building of wooden ships, but now used to indicate the thickness of plates, angles and flanges. Measurements of steel sections.

Sheer Strake

The continuous row of shell plates on a level with the uppermost continuous deck (Figure 1.21).

Stealer Plate

A plate found at the extremities of the vessel in the shell or deck plating. Its purpose is to reduce the width of the plating by

Plate 2 The *Queen Mary 2* lies portside to at Southampton, engaging in lifeboat launch drill

Plate 1 The motor/sailing vessel *Wind Surf* seen in the hydro lift drydock at Lisnave, Portugal

Plate 3 The cable vessel *Pacific Guardian* lies starboard side to the berth. The type of ship is distinctive, with cable fitments at the bow and the stern positions.

Plate 4 The Greek ro-ro vehicle ferry *MykonoΣ* operating in the Greek islands and Mediterranean ports.

Plate 5 A large modern oil tanker operating in Mediterranean waters, seen in the loaded condition (source: Shutterstock).

Plate 6 A modern high-speed passenger ferry.

Plate 7 A high-speed ro-ro ferry operating off the Greek coastline (source: Shutterstock).

Plate 8 A general cargo vessel working cargo alongside. Fitted with cranes and open steel hatch covers.

Plate 9 A suction dredger engaged in depositing silt (source: Shutterstock).

Plate 10 A double hull oil tanker under construction.

merging, say, three strakes into two. The single plate producing this effect is known as a stealer plate.

Bulk Carrier Construction

NB. Bulk carrier construction is expected to become all double hull structure by 2010 for new builds similar to tankers.

The framing on bulk carriers is designed as a longitudinal system in topside and double bottom tanks and as a transverse system at the cargo hold, side shell position (Figure 1.22).

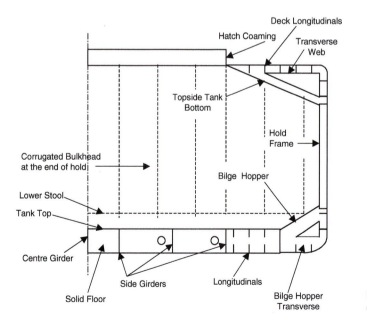

Figure labels:
Deck Longitudinals
Hatch Coaming
Transverse Web
Topside Tank Bottom
Hold Frame
Corrugated Bulkhead at the end of hold
Bilge Hopper
Lower Stool
Tank Top
Centre Girder
Side Girders
Solid Floor
Longitudinals
Bilge Hopper Transverse

Figure 1.22 Bulk carrier construction.

Main Structural Members: Compensating Stress Factors Affecting the Vessel

Beam Knees

These resist racking, heavy weights and localized stresses.

Beams

These resist racking, water pressure, longitudinal torsional stresses and local stresses due to weights.

Bulkheads

These resist racking stresses, water pressure, drydocking stresses, heavy weights, hogging and sagging, torsion stresses and shear forces.

Decks

These resist hogging and sagging, shearing, bending, heavy weights and water pressure.

Floors

Resist water pressure, drydocking stresses, heavy weights, local stresses, racking, vibration and pounding.

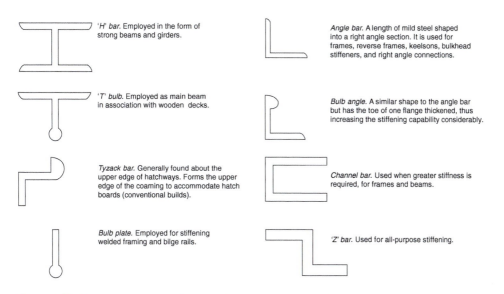

'H' bar. Employed in the form of strong beams and girders.

Angle bar. A length of mild steel shaped into a right angle section. It is used for frames, reverse frames, keelsons, bulkhead stiffeners, and right angle connections.

'T' bulb. Employed as main beam in association with wooden decks.

Bulb angle. A similar shape to the angle bar but has the toe of one flange thickened, thus increasing the stiffening capability considerably.

Tyzack bar. Generally found about the upper edge of hatchways. Forms the upper edge of the coaming to accommodate hatch boards (conventional builds).

Channel bar. Used when greater stiffness is required, for frames and beams.

Bulb plate. Employed for stiffening welded framing and bilge rails.

'Z' bar. Used for all-purpose stiffening.

Figure 1.23 General steelwork sections used in basic ship construction.

Frames

These resist water pressure, panting, drydocking and racking stresses. May be compared to the ribs of the body, which stiffen the body of the vessel. May be longitudinally or transversely constructed.

Longitudinal Girders

These resist hogging and sagging, water pressure, drydocking and pounding stresses and localized shearing stresses. Examples: keel, keelsons, fore and aft members, intercostals.

Pillars

These resist stresses caused by heavy weights, racking, drydocking and water pressure. Extensively found in general cargo vessels in lower hold structures.

Shell Plating

Steel plates of various size, which, when joined together, form the sides of the ship's hull. Plates are generally of an increased thickness (increased scantlings) in and about the keel area. The 'garboard strake' and the 'sheer strake' are also increased in thickness compared to other shell plates. Shell plating compensates for all stresses affecting the vessel. Where localized stresses are experienced, as with 'shell doors', then increased scantlings can be expected to provide the continuity of strength required. Shell plates can be identified by inspection of the 'shell expansion plan'.

Stresses in Ship Structures

It is the shipowner's responsibility to ensure that his vessel is built to a standard high enough to withstand all the stresses she may be expected to encounter. By their very nature ships are called upon to carry heavy loads, and considerable thought and experience is required to load heavy weights without causing structural damage to the vessel.

Heavy weights tend to cause a downward deflection of the deck area supporting the load (see Figure 1.24). This subsequently produces stresses, with consequent inward and outward deflections of supporting bulkheads, depending on the position of initial loading. These stresses are generally of a localized nature, in the neighbourhood of built-in structures such as windlasses, accommodation blocks, etc. and increased scantlings are the norm to prevent excessive distortion. The shipping of heavy seas may add to the load and aggravate the situation, causing unacceptable, excessive distortions.

Another form of stress comes from the water surrounding the ship, which exerts considerable pressure over the bottom and side areas of the shell plating (Figure 1.25). The pressure will increase with depth of immersion, i.e. the pressure on the bottom shell plates will exceed that on the side shell plates.

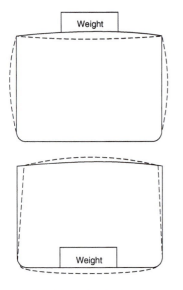

Figure 1.24 Pressure distortion due to heavy weights.

Water pressure does not maintain a constant value, and will vary when the vessel is in a seaway, especially when a heavy swell is present. Fluctuations in water pressure tend to cause an 'in and out' movement of the shell plating, with more noticeable effects at the extreme ends of the vessel. The effect of water pressure is usually more prominent at the fore end of the vessel than the after end. The general effect is accentuated by the pitching motion of the vessel and is termed 'panting'.

We have already defined panting beams, which are substantial metal beams running from port to starboard in the forepart of the vessel. They are positioned forward of the collision bulkhead to resist the in-and-out motion of the shell plating on either side of the fore and aft line. Situated at various deck levels, panting beams form a combination with panting stringers on either side in the forepart of the vessel.

A third form of stress is shearing stress in a material, which tends to move one part of the material relative to another. Consider the vessel in Figure 1.27, assumed to have loaded cargo in its centre hold, ballast water in the forepart and accommodation and machinery space all aft. Shearing forces will be experienced at points A, B, C and D. If we assumed that the various sections were free to move, then, owing to the weight distribution throughout the vessel's length, sections X, Y and Z (Figure 1.28) would have a tendency to move downwards (by the law of gravity), while sections P and Q would have a tendency to move upwards, due to the forces of buoyancy.

The two forces of gravity and buoyancy acting in opposition in this way explain the shearing stress experienced at points A, B, C and D. Shearing forces are undesirable within a ship in any shape or form, and prudent loading, together with careful ballast distribution, can reduce them.

Values of stresses incurred during the loading period may be mathematically calculated and then plotted to show the areas of stress by graph. It is worth noting that the mathematical calculations are lengthy and always leave the possibility for error. 'Stress finders' and computerized loadicators have reduced the risk of errors. There are several on the market, generally custom-made for individual vessels, and they provide the operator with such items of information as:

- bending moment
- shear stress at critical points
- mean draught
- trim of the vessel
- GM final, after loading
- deadweight.

The information supplied depends on the make and type of software employed.

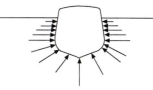

Figure 1.25 Pressure of water on ship's hull.

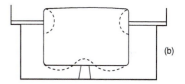

Figure 1.26 *(a) Racking* is distortion in ship's structure due to rolling action in a seaway; *(b) Drydocking* stresses are caused by pressure from drydock shores and keel blocks in local areas of the shell plating and associated strength members.

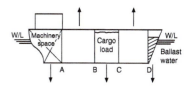

Figure 1.27 Shearing stresses.

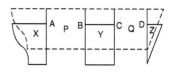

Figure 1.28 Gravity and buoyancy forces over ship's length.

Figure 1.29 Hogging.

Hogging

The length of the vessel may be considered to act like a long girder pivoted on a wave about its centre. In this position the fore and after ends of the vessel will bend downwards, causing compression forces in the keel area and tension forces at the upper deck level (Figure 1.29).

The condition is brought about by increased buoyancy forces being created at and around the midships point of the vessel. Increased gravitational force due to the metal structure of the vessel acting vertically downward occurs at the extremities of the ship. When both forces exist at the same time, e.g. as the vessel is pivoted by a wave midships, a 'hogging' condition is present.

This can be accentuated in a vessel of an all-aft design, where the additional weight of the machinery space would produce high loading in the aft part of the vessel. The condition may also be unnecessarily increased by 'bad' cargo loading in the fore and after parts of the vessel, leaving the midships area comparatively lightly loaded.

Sagging

'Sagging' is the opposite of hogging. When a vessel is supported at bow and stern by wave crests she will tend to sag in the middle. High buoyancy forces occur at the extremities of the ship. High gravitational forces, from the weight of the ship's structure, act vertically down about the midships point, in opposition to the buoyancy forces. In comparison with the condition of hogging, the vessel has a tendency to bend in the opposite direction (Figure 1.30).

Incorrect loading of the vessel or design characteristics may accentuate the condition of sagging. Watchkeeping officers should be aware of the frequency of the waves and the likelihood of this condition developing and, if necessary, take action to relieve any sagging or hogging conditions by altering the ship's course.

Due consideration at the time of loading with regard to weight distribution may alleviate either hogging or sagging. With shipbuilding producing larger and longer ships, either condition is most undesirable, as the prospect of breaking the ship's back in a heavy seaway or swell becomes a frightening reality. Prudent ballast arrangements, together with increased scantlings at the time of building, coupled with efficient ship and cargo loading, will help minimize any structural damage at a later stage due to hogging or sagging.

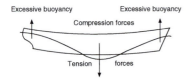

Figure 1.30 Sagging.

Torsional Stresses

With the larger build of vessel, like the ULCC/VLCC and long ore carriers, a twisting down the ship's length can be noticeable in a seaway. This is a torsional stress and may cause cracks at the ship's sides.

Loadlines

Loadlines are marks punched into and painted on the sides of British merchant vessels (Figure 1.31). These Plimsoll marks take their name from the politician Samuel Plimsoll (1824–1898), who persevered for many years before seeing a bill through Parliament in 1876, which resulted in the Merchant Shipping Act. The Act gave the Department of Trade and Industry, as we now know it, the right of inspection to ensure that a vessel should not be over-loaded beyond her Plimsoll mark or line.

Samuel Plimsoll championed the improvement of conditions for the seafarer and became the president of the Sailors' and Firemen's Union in his later years.

Assigning a Vessel's Loadline

The assigning of the vessel's loadline and the issue of the certificate is the responsibility of the Marine Authority of the country. (In the UK this is the Maritime and Coastguard Agency.) The loadline survey is conducted in accord with the International Conference on Load Lines, 1969, the Merchant Shipping (Survey and Certification) (Amendments) Regulations 2000 and the 1988 SOLAS and Load Line Protocols.

The calculation regarding the freeboard, and consequently the position of the loadlines, will be dependent on the type of vessel and its length, ships being divided into two types 'A' and 'B':

- type 'A': vessels designed to carry only liquid, bulk cargoes, e.g. tankers;
- type 'B': all other vessels not governed by the type 'A' definition.

The assigning of the freeboard will be governed by many factors and it is not within the scope of this text to detail the loadline rules. (Additional information is obtainable from Murray-Smith, 'The 1966 International Conference on Loadlines', *Transactions RINA*, vol. 11, 1969.)

With the exception of pleasure yachts, warships and the like, all British ships and the majority of vessels of other maritime nations over 80 net registered tons are obliged to be marked with statutory loadlines to ensure they are not overloaded. Various authorities assign loadlines on behalf of the British government, e.g. Det Norske Veritas (DNV), Lloyd's Register (LR).

A loadline certificate must be displayed in a prominent place aboard the vessel. The certificate is valid for five years, but an annual survey is held to ensure that the conditions of assignment and the loadline marks remain unchanged.

Should the loadline be submerged through the overloading of the vessel, so contravening the regulations, then the Master or owner is liable for a fine of £1000.00 plus £1000.00 for every

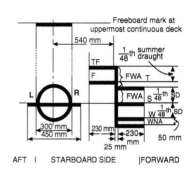

Figure 1.31 Loadlines.

LR	Lloyd's Register, Classification Society.
TF	tropical fresh
F	fresh
T	tropical
S	summer loadline draught
W	winter
WNA	winter North Atlantic

centimetre or part of 1 cm overloaded. The upper edge of the loadline marks are the recognized mark levels. The loadline itself (Figure 1.31) is punched into the shell plate and painted a distinctive colour, usually white or yellow on a dark background.

Owners of vessels may make application to the Maritime and Coastguard Agency for a vessel to be assigned an alternative tonnage. Gross and registered tonnages are assigned not only for the upper deck but also for the second deck, excluding the 'tween deck space, so treating the second deck as the upper deck level.

Once an alternative tonnage has been assigned, the tonnage mark (Figure 1.32) will be carved on each side of the vessel below the second deck and aft of the loadline disc. Should the vessel be so loaded as to submerge the alternative tonnage mark, then the normal gross and registered tonnage will apply. Should the state of loading leave the mark visible, then the modified tonnage values will remain valid.

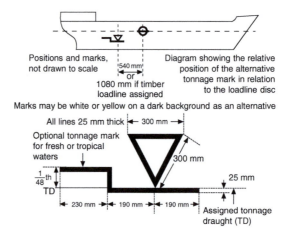

Figure 1.32 Alternative tonnage marks.

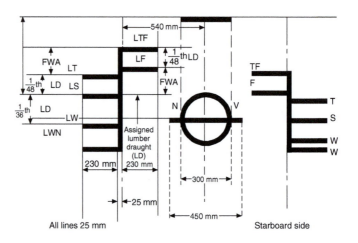

Figure 1.33 Timber loadlines.

International Safety Management

The majority of commercial shipping now complies with the International Safety Management (ISM) code. The code was adopted by SOLAS in 1994 and most vessels were participating by 2002. The purpose of the code is to ensure the safety of life at sea and avoid damage to the environment from shipping activity. It is difficult to provide a simple and definitive answer to the question what 'ISM' is, but it should be seen as possibly:

> *A safety culture for personnel both ashore and afloat, working within the perimeter of the marine industry, to provide safer ships and cleaner seas.*

Merchant Navy officers taking qualifications for Certificates of Competency are frequently asked how they know their ships are compliant with the ISM code. The answer is within the certification process and if a Safety Management Certificate and a copy of the company's Document of Compliance (DoC) are displayed (valid for five years), then the ship is known to be complying with ISM procedures.

The ship remains compliant with ISM by carrying out periodic 'audits', by internal and external auditors, to meet the requirements of the code. The 'internal auditor' is generally appointed by the company and is a person without vested interest but with marine experience, who will be unaffected by corruptive influence. The 'external auditor' is appointed by the flag state's marine authority (MCA in the UK, USCG in the United States).

The work of the auditors is intended to monitor all shipboard procedures to ensure safe operations are the norm. They will pay particular attention to planned maintenance schedules; where faults or defects occur within the ship's systems, they will issue a *non-conformity* notification, usually time related, to return the ship to a quality standard.

In the event that a serious defect is observed, which could jeopardize the overall safety of the ship or personnel on board, then the auditor may issue a *major non-conformity*, which could lead to a

---

'Computer Software'

Many vessels now employ computer loading programs to establish disposition of cargo, ballast and stores.

Such software can be beneficial in producing the ship's stability data, together with anticipated stress factors throughout the ship's length.

detention notice being placed against the ship unless the particular problem is resolved immediately.

Auditors, when carrying out their duties, will talk to any and all members of the crew to understand the practices employed on board and within the company guidelines. *A safety culture for personnel both ashore and afloat, working within the perimeter of the marine industry, to provide safer ships and cleaner seas.*

2

Anchor Work (Fundamentals)

Introduction

Anchors in one form or another are known to have been around from about 6000 BC. At this time the basic 'anchor' was most likely a basket or a type of holder for a collection of stones, probably used by the Egyptians on the River Nile and in the associated waters of the Mediterranean. Since this time many types of anchor have been developed, employing both 'hook' and 'spade' principles (e.g. Admiralty Pattern and Stockless Anchors, respectively) as well as sheer weight (e.g. clump mooring anchors).

With the increased sizes in shipping it was necessary to re-think anchor size, and subsequently examine braking systems, and the capabilities of cable lifters. Hardware has been forced to change with a modern society, no more so than with the 'offshore industry' and the developments in recovery of raw products from beneath the seas. Anchors, cables, handling techniques, associated fitments, stowage methods, etc. have all passed through major changes to provide for the needs of today's industry.

Anchors

Admiralty Pattern Anchor

Sometimes referred to as a 'fisherman's anchor', this design is still popular within the fishing industry (Figure 2.1(a)). It has been in use for many years, but because it has difficult stowage characteristics, e.g. it cannot be stowed flat with the stock in position, it has been followed by more manageable designs. Once let go, the stock, lying at a right angle to the direction of the arms/flukes, causes a fluke to dig into the sea bed. This leaves the remaining fluke exposed, and the cable may often foul it when the vessel swings. When the anchor is not in use, the forelock in the stock can be unshipped, permitting the stock to be stowed parallel to the shank.

The holding power of this anchor is generally considered to be very good indeed. The design is such that the stock is longer and heavier than the arms.

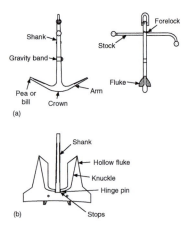

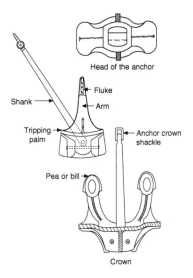

Figure 2.1 *(a)* Admiralty pattern anchor and *(b)* Admiralty cast anchor type 14 (AC14).

Figure 2.2 Hall stockless anchor.

This lends itself to the theory that the stock will be dragged flat along the sea bed, causing one of the flukes to bury itself. The angle of the stock would also be expected to turn the flukes in the direction of the sea bed as the anchor strikes the bottom. It is interesting to note that the longer the shank on these anchors, the better it holds.

The weight of the stock must be equal to 25 per cent of the weight of the anchor itself. Some stocks are designed straight if the weight of the anchor is over 12 cwt (610 kg), but a bent stock, as indicated in Figure 2.1(a) would be encountered on anchors below this weight.

The holding power of this common anchor will be, roughly speaking, 3–4 times its weight, depending on the nature of the sea bottom. It is unlikely to be seen on board merchant vessels, except possibly as a kedge anchor. The weight in any event would rarely exceed two tonnes.

The Stockless Anchor

This is by far the most popular anchor in general use today. Its principal parts are shown in Figure 2.2. The head of the anchor is secured to the shank by a hinged bolt which allows the arms to form an angle of up to 45° with the shank. Further rotation of the arms is prevented by the head meeting the shank at the built-in stops. The head of the anchor comprises the flukes, the arms and the crown, which are manufactured from cast steel, whereas the shank is made of cast steel or forged iron. The hinge bolt and the shackle are made of forged iron. The stockless anchor's greatest advantage is its close stowing properties; it is easily housed in the hawse pipe when not in use. It is easily handled for all anchor operations, and made anchor beds (used with the close stowing anchor) obsolete.

The overall size of these anchors will vary between individual ships' needs, but the head must be at least three-fifths of the total weight of the anchor. Holding power again varies depending on the nature of the bottom but, as a rule of thumb, it may be considered to be up to three times its own weight. The mariner should be aware that the rotation action of the moving arm may cause the anchor to become choked when on the sea bed so that the arms/flukes are not angled to the full amount and therefore lose the holding power effect.

Admiralty Cast Anchor

Used extensively as a bower anchor for warships, this anchor, because of good holding properties, has become very popular with the merchant service (Figure 2.1(b)). With the increase in size of ships – the large tankers of today, for example – shipowners required an anchor with greater holding power. The AC Type 14, as it was called, was developed in the United Kingdom and has the

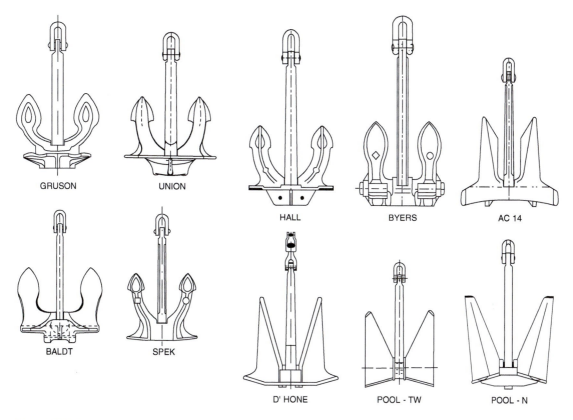

Figure 2.3 Different types of stockless anchors.

required properties. Tests show that it has more than twice the holding power of a conventional stockless anchor of the same weight. With such an obvious advantage, the Lloyd's Classification Society granted a 25 per cent reduction in regulation weight. The holding properties of this anchor are directly related to the prefabricated construction of the fluke area, the angle of which operates up to 35° to the shank. The angle of the flukes is made possible by a similar operation as with the stockless anchor, in which a hinge pin passes through the shank in the crown of the anchor (Figure 2.3).

CQR

Illustrated in Figure 2.4, the CQR is sometimes referred to as a 'plough-share anchor' or, in the United States, just as a plough anchor. It is generally used as a mooring anchor, especially for the smaller type of vessel. Holding power is again dependent on the type of ground the anchor is bedding into, but has been found to be very good. It also has extremely good resistance to drag. Like the Admiralty Pattern, it is difficult to stow. The design has been modified since its invention to incorporate a stock, and is often used as a mooring anchor (Figure 2.4). The CQR was a British invention by the scientist Sir Geoffrey Taylor, a man with little boating experience. The design showed that the application of basic principles can sometimes improve on practical experience. Small-boat owners tend to have the choice of two anchors on the market, namely the Danforth and the CQR. Both anchors have reasonable holding power, but the Danforth may have a tendency to glide, whereas the CQR will not. For its easier handling and stowing the Danforth would be more popular, but if the user decides to use an anchor for the job it is meant for, preference is generally given to the CQR.

Danforth Anchor

Generally accepted as a small-boat anchor, this anchor dominates the American boat market (Figure 2.4). A stock passes through the head of the anchor, allowing it to be stowed easily in a similar manner to the stockless anchor. Holding power is about 14.2 times its own weight. The anchor is of American design, and the idea of the stock being passed through the crown of the anchor as opposed to the top of the shank demonstrates a practical solution to the stowage problem. The stock in this position prevents the anchor being fouled on its own cable. Holding properties are good, but not as good as those of the CQR, and it has a tendency to drag or glide until the flukes bite into the sea bed. The action of this anchor is similar to that of the stockless anchor, where the tripping palms catch and cause the flukes to be angled to the shank. With the Danforth anchor, the tripping palms are generally situated closer to the centre line of the anchor. Once tripped, the spade-shaped flukes will tend to dig into the bottom.

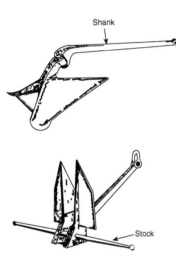

Figure 2.4 CQR anchor (*above*), Danforth anchor (*below*).

Tests on Anchors

All anchors over 168 lb (76 kg) in weight must be tested and issued with a test certificate. The weight of any anchor for the purpose of the rules and regulations governing anchors and cables shall:

- for stockless anchors include the weight of the anchor together with its shackle, if any; and
- for stocked anchors, include the weight of the anchor including its shackle, if any, but excluding the stock.

Drop Test (Cast Anchors)

Any part of an anchor over 15 cwt (762 kg) is subjected to a percussion test by being dropped both end on and side on from a height of 12 ft (3.6 m) onto an iron or steel slab. After that, the piece must be slung and hammered all over by a 7 lb (3.2 kg) sledgehammer. A clear ring must be produced to show that no flaw has developed during the percussion test.

The Bending Test (Cast Anchors)

An additional piece of metal, 20 cm long, is cast with the piece to be tested, and is cut away for the purpose of the bending test. This piece will be turned down to 2.5 cm in diameter, and bent cold by hammering through an angle of 90° over a radius of 3.75 cm. The casting will be deemed sufficiently ductile if no fracture appears in the metal.

All anchors are subject to the proof strain (Table 2.1), and subsequent proof load, but only cast steel anchors will be subjected to percussion, hammering and bending tests. Wrought iron or forged steel anchors are not subjected to these tests as they are forged from red-hot slab by hammering. All other anchors will also be annealed.

Table 2.1 Proof loads for anchors

Weight of anchor (kg)	Proof load (tonne)	Weight of anchor (kg)	Proof load (tonne)	Weight of anchor (kg)	Proof load (tonne)	Weight of anchor (kg)	Proof load (tonne)	Weight of anchor (kg)	Proof load (tonne)	Weight of anchor (kg)	Proof load (tonne)
76	3.33	700	15.20	2,300	39.60	4,700	65.10	7,200	82.60	15,000	117.70
80	3.46	750	16.10	2,400	40.90	4,800	65.80	7,400	83.80	15,500	119.50
90	3.70	800	16.90	2,500	42.20	4,900	66.60	7,600	85.00	16,000	120.90
100	3.99	850	17.80	2,600	43.50	5,000	67.40	7,800	86.10	16,500	122.20
120	4.52	900	18.60	2,700	44.70	5,100	68.20	8,000	87.00	17,000	123.50
140	5.00	950	19.50	2,800	45.90	5,200	69.00	8,200	88.10	17,500	124.70
160	5.43	1,000	20.30	2,900	47.10	5,300	69.80	8,400	89.20	18,000	125.90
180	5.85	1,050	21.20	3,000	48.30	5,400	70.50	8,600	90.30	18,500	127.00
200	6.25	1,100	22.00	3,100	49.40	5,500	71.30	8,800	91.40	19,000	128.00
225	6.71	1,150	22.80	3,200	50.50	5,600	72.00	9,000	92.40	19,500	129.00
250	7.18	1,200	23.60	3,300	51.60	5,700	72.70	9,200	93.40	20,000	130.00
275	7.64	1,250	24.40	3,400	52.70	5,800	73.50	9,400	94.40	21,000	131.00
300	8.11	1,300	25.20	3,500	53.80	5,900	74.20	9,600	95.30	22,000	132.00
325	8.58	1,350	26.00	3,600	54.80	6,000	74.90	9,800	96.20	23,000	133.00
350	9.05	1,400	26.70	3,700	55.80	6,100	75.50	10,000	97.10	24,000	134.00
375	9.52	1,450	27.50	3,800	56.80	6,200	76.20	10,500	99.30	25,000	135.00
400	9.98	1,500	28.30	3,900	57.80	6,300	76.90	11,000	101.50	26,000	136.00
425	10.50	1,600	29.80	4,000	58.80	6,400	77.50	11,500	103.60	27,000	137.00
450	10.90	1,700	31.30	4,100	59.80	6,500	78.20	12,000	105.70	28,000	138.00
475	11.40	1,800	32.70	4,200	60.70	6,600	78.80	12,500	107.80	29,000	139.00
500	11.80	1,900	34.20	4,300	61.60	6,700	79.40	13,000	109.90	30,000	140.00
550	12.70	2,000	35.60	4,400	62.50	6,800	80.10	13,500	111.90	31,000	141.00
600	13.50	2,100	36.90	4,500	63.40	6,900	80.70	14,000	113.90		
650	14.30	2,200	38.30	4,600	64.30	7,000	81.30	14,500	115.90		

Proof loads for intermediate weights shall be obtained by linear interpolation.

Marks on Anchors

Each anchor must carry on the crown and on the shank the maker's name or initials, its progressive number and its weight. The anchor will also bear the number of the certificate, together with letters indicating the certifying authority.

Anchor Certificate

After the test on the anchor is completed, an anchor certificate will be awarded. The certificate will show the:

- type of anchor
- weight (excluding stock) in kilograms
- weight of stock in kilograms (if a stocked anchor)
- length of shank in millimetres
- length of arm in millimetres
- diameter of trend in millimetres
- proof load applied in tonnes
- identification of proving house, official mark and government mark
- number of test certificate
- number of tensile test machine
- year of licence
- weight of the head of the anchor
- number and date of drop test
- name of supervisor of tests and signature.

Chain Cable Tests

Anchor cable over 12.5 mm in diameter is accepted for testing at an approved testing establishment in lengths of 27.5 m (one shackle of cable). The manufacturer will provide three additional links for the purpose of the test. These three links will be subjected to a tensile breaking stress, and if this proves to be satisfactory, then the total length of the cable will be subjected to a tensile proof test, the tests being carried out on approved testing machines. If two successive links break, the cable is rejected. Before the test on chain cable is carried out, the supervisor will satisfy himself that the quality of the material from which the cable is manufactured meets with the requirements of the anchor and chain cable regulations. After a successful test on chain cable, a certificate is awarded, stating the:

- type of cable
- grade of cable
- diameter in millimetres
- total length in metres
- total weight in kilograms

- length of link in millimetres
- breadth of link in millimetres
- tensile breaking load applied in tonnes
- tensile proof load applied in tonnes
- number and types of accessories included.

The certificate issued shall also show the:

- serial number
- name of the certifying authority
- mark of the certifying authority
- name of the testing establishment
- mark of the testing establishment, if any
- name of the supervisor of tests.

The certificate is signed on behalf of the certifying authority and the chain is marked as shown in Figure 2.5.

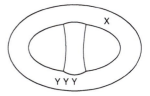

Figure 2.5 Marks on cable: X, certificate number; YYY, certifying authority.

Notes on Cable

Accessories

Anchor shackles and joining shackles are all ordered together with any additional fittings for the size of cable they are intended to work and be associated with. These accessories must be subjected to similar tensile load and proof load tests as the cable.

Material of Manufacture

Wrought iron, forged mild steel, cast steel or special-quality forged steel are used. Wrought iron is weaker than the other three materials, and is expensive to produce; consequently it is rarely seen on present-day merchant ships. Types of cable are shown in Tables 2.2 and 2.3.

Table 2.2 Types of chain cable

27.5 m = 15 fathoms = 1 shackle length

Grade	Material	Method of manufacture	Tensile range kg/mm²
1a	Wrought iron	Fire welded	31–41
1b	Mild steel	Fire welded	31–41
1c	Mild steel	Flash butt welded	31–41
1d	Mild steel	Flash butt welded	41–50
2a	Steel	Flash butt welded or drop forged	50–65
2b	Steel	Cast	50 min.
3a	Steel	Flash butt welded or drop forged	70 min.
3b	Steel	Cast	70 min.

Table 2.3 Stud link chain cable

Diam. mm	I (U1) Proof	Break	II (U2) Proof	Break	III (U3) Proof	Break	Minimum weight per shackle length (tonnes)
12.5	4.7	6.7	6.7	9.4	9.4	13.5	0.105
50	70	100	100	140	140	200	1.445
60	98.8	141	141	198	198	282	2.075
70	132	188	188	263	263	376	2.85
90	209	298	298	417	417	596	4.705
122	357	510	510	714	714	1019	8.55
152	515	736	736	1030	1030	1471	13.20

Size of Cable

The size is measured by the diameter of the bar from which the link is manufactured. Aboard a vessel, the size could be obtained from the chain cable certificate, or external callipers could be used to measure the actual cable.

Kenter Lugless Joining Shackle

The Kenter lugless joining shackle, manufactured in nickel steel, is the most popular method of joining shackle lengths of the anchor cable together. The shackle has four main parts, as shown in Figure 2.6. The two main halves interlock with the stud forming the middle of the link. All parts are held together with a tapered spile pin. This spile pin is made of mild steel and is driven into the shackle on the diagonal. A lead pellet is then forced into the inverted dovetail recess to prevent the pin from accidentally falling from the shackle.

The manufacture of the shackle in nickel steel prevents corrosion and the parts becoming frozen together. It allows the shackle to be 'broken' with relative ease when either the cable is to be end-for-ended or shackles are to be tested. When breaking the shackle, remove the spile pin by using a punch and drift (Figure 2.6). If the lead pellet has not been prised out first, be careful that it is not forced out by the percussion effect of the drift driving the spile pin, for it may emerge with considerable force. A back stop should be provided to prevent people being injured by the lead pellet being expelled from the recess.

Once the spile pin is removed, the stud can be extracted; the two halves of the shackle can then be separated by means of a top swage obtained from the manufacturer. When the shackle is reassembled, care must be taken to ream out the dovetail recess so that no residual lead is left inside. Should this not be done, then the next lead pellet inserted will not spread out and obtain a grip inside the recess.

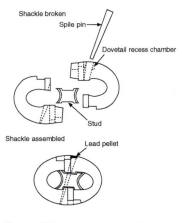

Shackle broken
Spile pin
Dovetail recess chamber
Stud
Shackle assembled
Lead pellet

Figure 2.6 Kenter lugless joining shackle.

The construction of the Kenter shackle is such that it is larger than the common links but not by so much that it will not fit into the snug of the gypsy of the windlass or cable holder. However, care should be taken that it does not lie flat on the gypsy and cause jamming.

The main advantage of this type of joining shackle is that open end links are not required, as with the 'D' lugged joining shackle. In addition, all shackle lengths are the same, which ensures smoother working in the snugs of the gypsy. The shape of the Kenter lends itself to cable working, especially around and over the bow, and the tendency for it to catch is comparatively rare. As with other accessories, these shackles are tested, but because of their type of manufacture in nickel steel, they are not heat-treated.

'D' Lugged Joining Shackle

The 'D' lugged joining shackle is used extensively for joining the cable to the anchor in more modern vessels. In the past this type of shackle was used, as the Kenter lugless joining shackle is used today, in the joining of the shackle lengths of cable together. If it is to be used for this purpose, the rounded crown part of the shackle should always face forward so that it does not foul the anchor when letting go.

It should be noted that the anchor crown shackle and the 'D' joining shackle face the opposite way to all other 'D' joining shackles in the cable. The mariner should be aware that the anchor, together with the initial joining shackle, is walked out of the hawse pipe prior to letting go (except in some cases of emergency). Consequently, the anchor crown shackle would not foul, but should other joining shackles be facing in this manner, there would be a distinct possibility of the lugs of the shackle catching on a snag in the letting go operation.

When using these types of shackle between cable length, each cable length must have an open link at the ends. This is necessary to allow the passage of the lugs through the cable.

The construction of the 'D' lugged joining shackle is illustrated in Figure 2.7, where it may be seen that the bolt, generally oval in shape, is passed through the lugs and across the jaw of the shackle. A tapered spile pin of steel, brass or wood holds the bolt in position, a lead pellet being hammered home into a dovetail recess chamber to keep the spile pin from accidently being expelled. The spile pin should be tapered to a ratio of 1:16, and wooden pins are made of ash or solid bamboo. When breaking the 'D' joining shackle, the bolt will be hammered from the unlipped end, causing the wooden spile pin to shear. Should the spile pin be made of steel, then this must be expelled by using a punch and drift in a similar manner to that described for the Kenter shackle. The steel pin is generally found in the 'D' shackle joining the anchor cable to the anchor. When assembling these shackles, it is customary to give the bolt a smear of tallow to allow easy 'breaking' at a later date.

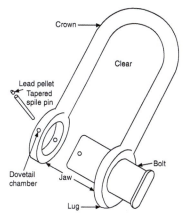

Figure 2.7 'D' lugged joining shackle.

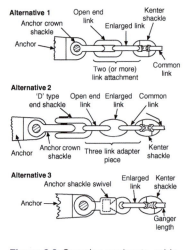

Figure 2.8 Securing anchor to cable.

Should the shackle become jammed and difficult to break, then it can be heated about the lugs. This will cause the lugs to expand, allowing the withdrawal of the bolt.

Securing and Stowage of Anchors

Alternative methods of securing anchor to cable are illustrated in Figure 2.8, and the operation of the cable in anchoring in Figure 2.9. There are many different designs of hawse pipe (Figure 2.10) in commercial use with the modern merchant vessel and the warship. The general arrangement is such that the axis of the pipe does not exceed 45° from the vertical; however, the most suitable angle is that which allows the easy lowering and restowing of the anchor. Many hawse pipe arrangements are recessed into the shell plate. This not only reduces drag effect, especially on high-speed vessels, but should contact with another vessel or quay occur, damage is considerably reduced.

Many of the modern anchors, e.g. AC14 and Bruce (see Figures 2.1 and 2.22), have incorporated an anchor bed or special stowage frame fitment about the entrance to the pipe. This usually facilitates smoother operation when letting go and better securing for the anchors when not in use.

NB. The devil's claw shown in Figure 2.9 is shown for display purpose and would not normally be secured when the anchor is deployed.

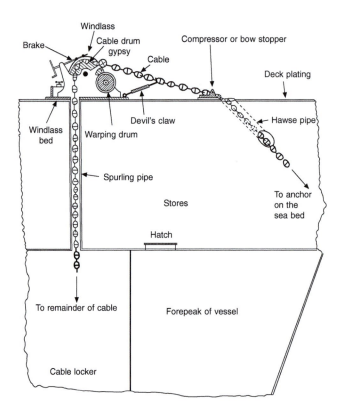

Figure 2.9 Operation of cable in anchoring.

Securing Anchor and Cable

Securing the bitter end of the anchor cable is illustrated in Figures 2.11 and 2.12, the forecastle head in Figure 2.14 and anchor securing in Figures 2.13 and 2.15. Figure 2.16 lists chain cable accessories.

Steam Windlass Operation

The following is a typical list of checks to be carried out before a steam windlass can be operated safely. You should consider what modifications to the list are needed to operate the type of windlass on your current vessel if it is different.

1   Inform the engine room of the requirement for steam to operate the windlass.
2   On the way to the windlass, ensure that the main deck steamline valve is open (this may in fact be in the engine room), and drain the deck line.
3   Check that the windlass stop valve is open (usually found under the bed of the windlass inside the forecastle head), and ensure any lashings in the chain locker are removed.
4   Open the drain cocks of the cylindrical steam chests (normally two cocks per chest).
5   Ensure that the windlass operating valve is closed (stop–start control).
6   Wait until pure steam issues from the drain cocks – not a mixture of steam and water.
7   Close the drain cocks, and steam is now at the windlass head, ready for use.
8   Ensure that the brake is firmly applied and that the windlass is out of gear.
9   Turn the windlass over by operating the start–stop valve.
10 Oil 'moving parts' as necessary to facilitate smooth running (obviously oil is applied to a stationary windlass for safety reasons).

Windlasses, winches and capstans are illustrated in Plates 11–15.

Preparing Anchor for 'Letting Go'

Once power has been obtained on deck, and the windlass has been oiled and checked, the anchors must be made ready to 'let go'. This operation must be carried out carefully and systematically to ensure that the 'letting go' operation runs smoothly. If a proper routine is established when time is not limited, the anchoring procedure is more likely to go smoothly and quickly when an emergency occurs.

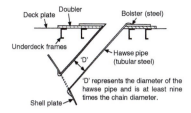

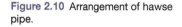

Figure 2.10 Arrangement of hawse pipe.

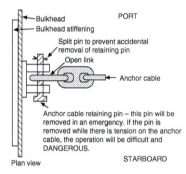

Figure 2.11 Internal securing of bitter end of anchor cable by *use of clench system inside cable locker.* In some cases the link may pass through the bulkhead, the pin being placed on the other side. It is not then necessary for a man to enter the chain locker at all in order to slip cable.

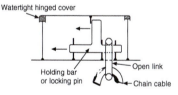

Figure 2.12 Alternative method of securing bitter end.
An external fitment is situated outside and usually above the chain locker. The hinge cover when in position prevents removal of the locking pin holding the bitter end of the cable. This method allows the cable to be slipped without any person being ordered into the locker. The locking pin is removed by a simple sliding motion once the hinged cover has been lifted. The cable is then released and the bitter end is allowed to fall back into the locker.

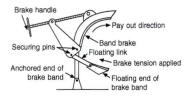

Figure 2.13 Windlass band brake system.

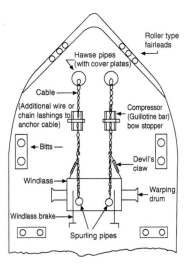

Figure 2.14 Forecastle head anchor and cable arrangement (plan view).

Figure 2.15 (a) Self-holding and automatically releasing roller bow stopper, roller manufactured and produced by Clark Chapman Ltd. (b) Self-holding and automatically releasing track bow stopper.

Once deck power is obtained, the following operations are carried out:

1   Ensure that the windlass brake is on and holding, and that the windlass is in gear.
2   Remove the hawse pipe covers.
3   Remove the devil's claw.
4   Remove any additional lashings.
5   Remove the bow stopper, guillotine or compressor.
6   Take off the brake and walk the cable back a short distance in order to break the cement pudding inside the spurling pipe. Modern ships often have spurling pipe covers instead of cement seals. If fitted, these should be removed.
7   Clear away old cement.
8   Walk back on the cable until the anchor is out, clear of the hawse pipe and above the water surface, then heave a few links back to ensure cable will run.
9   Screw the brake on hard and check that the brake is holding.
10 Take the windlass out of gear, leaving the anchor holding on the brake. Check that it is out of gear by turning power on briefly. Report to the bridge that the anchor is on the brake and ready for letting go.

Disc Brake Systems for Anchor Handling Windlass

In many cases the control of the chain is by application of the brake, but this is erratic and the response can be slow as well as generating considerable heat in the case of the conventional 'band brake' systems. In order to control the large dynamic loads anticipated during operation of the windlass, 'disc brakes' (similar to car-type disc brakes) began to evolve in the offshore industry. These were coupled with complex freshwater cooling systems which, apart from making the systems very costly and possibly not justifiable for use in commercial shipping, were quite labour intensive regarding maintenance.

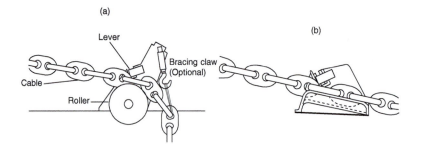

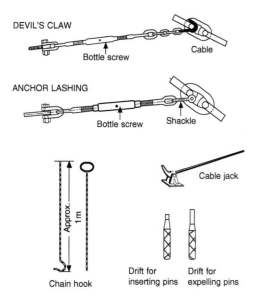

Figure 2.16 Chain cable accessories.

Disc brakes, although found to be effective, have not replaced band brakes. They still require regular, planned maintenance checks to ensure friction pads are not showing excessive wear or localized corrosion of the pinion and gear wheels is not occurring, just as the liners of 'band brakes' require similar monitoring.

It is recognized that disc brakes provide a more stable performance applied via a standard servo through hydraulic (calliper) pressure to cause the friction pads greater uniformity when engaged. They are also the probable forerunner to automatic braking systems, although these have not found great favour with mariners. Clearly the effectiveness of any braking system will be measured by the operational personnel, and their decision of when to apply the brake will influence the outcome. To this end any system should take account of when the anchor is in freefall and at what moment the velocity is observed to decrease. It is recommended that the application of the brake is not in the early stages of run-out, but more appropriately once a decreased rate of run is observed.

Braking Systems: Points of Reference

1 New liners for band brakes should be well bedded in, before use.
2 New liners may be ineffective due to heat generation, wetness or corrosion on the holding surface of the liner.
3 Walking back the anchor will restrict the run-out velocity and reduce load on the system.

Plate 11 Activating the quick release by a hammer blow to the securing of the bitter end of the anchor cables

Plate 12 A 15-tonne Stevpris anchor is recovered to the aft deck of an anchor handling vessel in the North Sea offshore operational region.

Plate 13 An electric windlass and combined mooring winch. The compressor bow stopper is seen over the anchor cable, which leads to the gypsy.

Plate 14 Centreline windlass. Cables seen are secured by devil's claws and wire/bottle screw lashings.

Plate 15 Arrangement of the starboard windlass aboard a modern ro-pax ferry with split windlass operation.

4 Noise levels in operation may be high and impose restrictions on orders being passed to brake operators.

5 Speed of the vessel over the ground should be minimized when the brake is being applied.

6 Large vessels with heavy equipment should deploy two men to operate the brake.

7 Lubrication of the windlass should take place before use.

8 Special attention to the lubrication of the pins holding the ends of the band brake should be made. They are meant to be free to permit movement of the floating end of the band.

9 The vessel should not be left 'brought up' on the brake alone. The bow stopper should also be applied.

10 If the limiting speed of the windlass is allowed to be exceeded, brake fading is expected to be experienced, which could result in the anchor not being able to be stopped.

Cable Holders

Cable holders (Figure 2.17) are often fitted to large merchant vessels as an alternative to the windlass; with recent developments these may be seen on passenger vessels. They have also been popular with warships for some considerable time because they are compact and lie low on the deck.

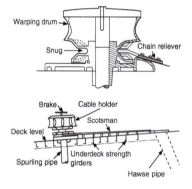

Early models employed a cable drum (gypsy) without the valuable addition of warping facilities. Modern versions include a warping drum geared to the centreline axle. This can subsequently be de-clutched when working anchor cables. A separate braking system is incorporated in each cable holder, similar to that fitted to the windlass.

Figure 2.17 Cable holders.

Anchor-securing arrangements are similar, except that the bow stopper is usually situated closer to the hawse pipe than to the cable holder. A devil's claw or slipping arrangement is sited between the bow stopper and the holder.

Where cable holders are used, the lead of cable is always close to the deck. To prevent excessive wear to deck plating from cable friction, a 'Scotsman' is a common fixture to provide the required protection.

Variations of combined capstan/cable holders are available on the commercial market, powered by steam or, more commonly, electricity. As with other similar deck machinery, additional strengthening of deck areas about operational sites is required to accommodate excessive load.

Procedure for Coming to Anchor

The preliminaries to the operation of coming to anchor include careful scrutiny of the chart of the area where the vessel is proposing to anchor, and consideration of the depth of water and the

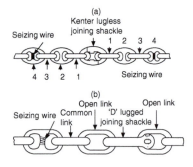

Figure 2.18 Marking anchor cable: (a) fourth shackle of cable; (b) second shackle length by means of 'D' lugged joining shackle. Open links on either side of the joining shackle are ignored for the purpose of marking cable in this case.

holding ground with the view to determining the amount of cable to use (Figure 2.18). The amount will be determined by the following:

1 depth of water
2 type of holding ground, good or bad
3 length of time the vessel intends to stay at anchor
4 sea room available for circle of swing
5 expected weather conditions
6 strength of tide, if any
7 draught and amount of hull exposed to the wind
8 type of anchor and its holding power.

These factors will vary with each case and previous experience; however, as a general rule, four times the depth of water may be taken as a working minimum. This would change, say, if the holding ground was bad, the weather deteriorating and you were expected to remain at anchor for a long period of time.

Split Windlass Arrangement

Many modern vessels with increased beam width or ferries with vehicle decks designed into the bow area have generated single (split) windlass arrangements to port and starboard (see Plate 15). Many of the split windlass designs incorporate tension drums or powered warping drums, as well as a de-clutchable gypsy (Plate 16).

Plate 16 Foredeck of a passenger vehicle ferry with split windlass arrangement incorporating a powered warping drum. An additional mooring is on the end rope drum.

Plate 17 Motorized capstan with gypsy and warping drum, seen at the aft end mooring deck (source: Shutterstock).

Bell Signals

When heaving in the cable or letting go, the bell should be struck once for every shackle's length, e.g. three shackles, three strokes of the bell.

When the anchor breaks clear and becomes 'anchor aweigh', then a rapid ringing of the bell will indicate to the bridge that the anchor is aweigh. Prudent Chief Officers tend not to ring anchor aweigh until the anchor is sighted and the flukes clear the water, in case the anchor has become fouled in any way with, say, warps or power cables.

NB. Use of bell signals is considered dated, and has generally been replaced by use of hand held walkie-talkie radios to pass communication from the forecastle to the bridge.

Marking of Anchor Cable

As the anchor is let go, the officer in charge of the anchor party will require to know the amount of cable being paid out. Each shackle length will be identified by the joining shackle, which is a larger link than the other links of the cable. The individual shackles will be distinguished by the number of studded links either side of the joining shackle. In the example given in Figure 2.18 the fourth shackle is used, and the fourth studded link from the joining shackle will be bound around the stud with seizing wire. This identification by means of seizing wire will be seen to mark the fourth shackle on both sides of the joining shackle. Seizing wire is used to enable the officer in charge to feel about the stud of the link and so locate, by his sense of touch, how far away the marked link is from the joining shackle – very useful during the hours of darkness. Seizing wire is used because it is quite robust and will stand a fair amount of wear and tear when the anchor is being let go, whereas the paint mark (see below) may tend to chip, or flake off, after a short period of time.

Plate 18 Forward mooring deck of a passenger vessel with separate port/starboard windlass.

The length of cable between the seizing wire portions is painted a bright distinctive colour, e.g. white, so that each shackle length may easily be located and acknowledged when operating anchors during the hours of darkness. Some ships often paint the joining shackle a different colour to highlight the position of the joining shackle.

If a 'D' lugged joining shackle is used to join cable lengths together (Figure 2.18(b)), then open links are found either side of the 'D' shackle. These open links must not be counted in the marking of the cable with seizing wire. Only studded links away from the joining shackle are to be counted.

Anchor cables should be checked whenever an opportunity presents itself, as in drydock where the cables can be ranged along the bottom of the dock and inspected with ease.

Clearing Away Anchors

The term 'clearing away' means preparing the anchor to let go, though different ships have different ways of operating. Most vessels are now equipped with hawse pipe covers – sliding metal covers which must be removed in order for the cable to run clear. Anchor lashings may be attached to the bow stopper or claw, or secured from deck lugs through the cable itself. These must be released and cleared away, as with the devil's claw, if fitted. The compressor or guillotine bar should be removed from the cable, together with any lashings which may have been applied inside the cable locker.

Past and Present Practice

A lashing in the cable locker served to stop the cables banging together when the ship was at sea. In bygone days the sailors used to sleep in the forecastle head area, and the banging cables tended to keep them awake. Hence they were lashed secure.

The more up-to-date thinking is that if the cable is lashed the chance of a bight of cable being buried by the remainder of the pile of cable in the locker will be reduced. This was especially so in the early days of non-self-stowing cable lockers.

Another reason, which is now by far the most popular, is that when the spurling pipes are sealed with cement, this cement plug and seal would be prevented from cracking up when the vessel was in a seaway by the secure lashing of the two cables together inside the cable locker.

Mariners should be aware that the practice of lashing cables in the locker is no longer common practice on modern vessels.

Spurling pipes must be sealed, but hinged slide design steel plates are now by far the most popular method of making them watertight. Should these steel plates not be fitted, then a pudding plug, made up of rags or cotton waste, should be forced into the aperture of the spurling pipe. Cement mix – four of sand to one of cement – should be poured over the pudding, about the anchor cable. This cement cover should be of such thickness that any movement of the anchor cable in the spurling pipe would not cause the cement to break. The purpose of the pudding is to stop the cement from dropping through to the cable locker, and also to give it something to set on.

Chain Cable/Stud Link: General Information

Volume of Anchor Chain

Anchor cable when stowed in a chain locker can be estimated at approximately 0.5 cubic metres per metric ton of chain.

Anchor Chain Renewal

Lloyd's require any length of anchor cable to be renewed if the mean diameter at its most worn part is reduced by 11 per cent below its original diameter.

Size of Cable Measurement

Cable is measured by use of external callipers. The size is found by measuring the diameter of the bar from which the link is made.

Chain Grades

This is a method of indicating the quality of steel from which the cable is manufactured. The grades have been internationally accepted and are recognized by the Classification Societies, and are listed under their regulations:

- mild steel chain
- special-quality steel chain
- extra special-quality chain (Ref. Table 3A, Lloyd's Rules).

Grade Three Cable

The lightest of the three grades of cable and ships so equipped would expect to use an increased scope when anchoring.

Wrought Iron Cable

This is the most expensive to produce and weaker than the other three qualities of forged mild steel, cast steel or special-quality forged steel. Subsequently, it is rarely seen on present-day merchant vessels. Its replacement is generally by a non-ferrous cable manufactured from aluminium bronze material.

Baldt or Dilok Cable

High-strength cable which is not easily comparable with British chain, size for size. It is widely employed on US warships and its nearest equivalent is probably cast steel chain.

Shackle Length

Anchor cable is universally manufactured and used in shackle lengths of 15 fathoms (90 feet) or 27.5 metres.

Strength

Studlink chain is 1.6–1.8 times that of the iron from which it is made. It is also 50 per cent greater than 'Open Link' cable.

Anchor Terminology

Anchor A-Cockbill

When the anchor is hanging vertically from the hawse pipe, with the flukes turned into the ship's side (Plate 19). In this position it will not stow correctly in the hawse pipe.

Anchor Aweigh

The anchor is said to be 'aweigh' at the moment it is broken out of the ground and clear of the sea bed.

Plate 19 Anchor seen a-cockbill (source: Shutterstock).

Plate 20 Single centreline anchor arrangement in the bow of a high-speed craft ferry use of combination chain cable and anchor warp.

Anchor Buoy

A buoy used to indicate the position of the ship's anchor when on the bottom (rarely used these days).

Anchor Coming Home

When the anchor is being drawn towards the ship in the operation of heaving away, by means of the windlass or cable holder/capstan, the anchor is said to be coming home. Instead of the ship being drawn towards the anchor, the reverse is happening.

Anchor Dragging

The anchor is said to be dragging when it is not held in the sea bed. It is said to bite well when it has a good hold in the ground. The vessel is 'dragging her anchor' if she moves her position while dragging the anchor over the sea bed.

Anchor Plan

This is a procedural list usually formulated by the Master and the Chief Officer prior to the ship going to an anchorage. The checklist should take account of which anchor is to be deployed and the amount of scope of cable that is expected to be used. Any associated hazards with regard to use of the anchor(s), like tidal effects and local weather conditions, would also be included. There is an example 'anchor plan' on pages 634 and 635.

Anchor Warp

The name given to a hawser or rope when it is attached to the anchor and used as a temporary cable.

Brought Up

A vessel is said to be brought up when her way has stopped and she is riding to her anchor, with the anchor holding. The terms 'come to' and 'got her cable' are sometimes used to mean the same thing. The officer in charge of an anchor party will know when the vessel is brought up, by the cable rising up from the surface towards the hawse pipe when the brake is holding it. The vessel should then move towards the anchor, causing the cable to drop back and make a catenary (Figure 2.19).

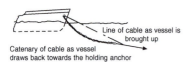

Line of cable as vessel is brought up

Catenary of cable as vessel draws back towards the holding anchor

Figure 2.19 A vessel brought up.

Cable Clench

A strong steel forged fitting in the cable locker for securing the 'bitter end' of the cable (see securing of anchors, p. 41).

Cable Jack

A device for lifting the cable clear of the deck (see anchor accessories, p. 43).

Cable's Length

A length of 600 ft, or 100 fathoms (183 m). (The term is unrelated to anchor cable and is a distance measurement.)

Cat the Anchor

The anchor is said to be catted when hung off, from what used to be called the clump cathead. More modern vessels will be fitted with a pipe lead set back from the line of the hawse pipe and used for the purpose of 'hanging off an anchor'. Found in practice when mooring to buoys by means of mooring shackles with the cable.

Chain Hook

A long iron hook used for the manhandling of cable links (see chain cable accessories, p. 43).

Cross

A cross occurs when the cables are fouled as in 'foul hawse', when the ship has swung through 180°, a cross being formed with the two cables (see Figure 2.20).

Drop an Anchor Under Foot

Letting an anchor go to the bottom, then holding on to the brake. This is sometimes done to steady the ship's head and prevent her yawing about when lying to a single anchor. Care must be taken in this operation that the second anchor is let go when the riding cable is growing (see below) right ahead, and not when it leads off the bow.

Elbow

Occurs when the cables are fouled as in 'foul hawse'. When the ship has swung through 360°, an elbow is formed in the anchor cables (see Figure 2.20).

Foul Anchor

The term used to describe the anchor when it has become caught on an underwater obstruction. The flukes of the anchor often become fouled by an old hawser or cable, obstructing its normal use.

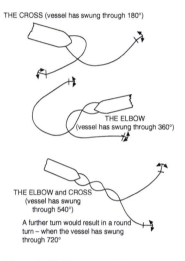

THE CROSS (vessel has swung through 180°)

THE ELBOW (vessel has swung through 360°)

THE ELBOW and CROSS (vessel has swung through 540°)

A further turn would result in a round turn – when the vessel has swung through 720°

Figure 2.20 Clearing foul hawse.

Plate 21 AC 14 anchor seen stowed in the hawse pipe and the anchor recess of the port bow of the ro-ro ferry *Seatruck Panorama*.

Foul Hawse

This term is used to describe the crossing of the anchor cables when both cables are being used at the same time, as with a running, standing or open moor, owing to the uncontrolled swinging of the vessel when anchored with both anchors (moored).

Grow

The cable is said to grow when the exposed part of the chain above the surface is seen to expand towards the anchor.

Gypsy

The vertical wheel on the windlass which the cable passes over. The cable is held in the segments of the wheel known as the 'snug'. The gypsy is held by the clutch plate (when in gear) or by the brake (when about to be let go).

Hawse Pipes

The two pipes on either bow which accommodate the bow anchors. Some vessels may be equipped with a stern anchor. The term hawse pipe is in general use for the stowage space for the anchors of a vessel.

Hove in Sight

When the anchor is hove home, it is 'sighted and clear' at the point when the anchor crown shackle breaks the surface of the water. A prudent officer would not consider that the anchor is clear until he sees that the flukes are clear. On the same basis, an officer in charge of an anchor party tends not to ring anchor aweigh until he sees the anchor is hove in sight and clear.

Joggle Shackle

This may be described as a long bent shackle, used for hauling cable round the bow (Figure 2.21). It is sometimes encountered when clearing a foul hawse or other similar operation in moving of cable.

Figure 2.21 Joggle shackle.

Kedging

Moving a vessel by means of small anchors and anchor warps is known as kedging.

Long Stay

The term applicable when the cable is leading down to the water close to the horizontal, with weight on it. A good length of the cable is exposed.

Moored

A vessel is said to be moored when she has two anchors down to the sea bed.

Ream a Shackle

To clean away any residual lead left inside the lug of a shackle after the lead pellet and spile pin have been removed, by use of a reaming tool.

Render Cable

To apply the brake lightly so that when weight comes on the cable it will run out slowly.

Round Turn

Occurs when the cables are fouled as in 'foul hawse', when the ship has swung through 720° or twice round.

Scope

This is the name given to the amount of anchor cable paid out from the hawse pipe to the anchor crown 'D' shackle.

Shackle of Cable

The length of a shackle of cable is 15 fathoms (27.5 m) (90 ft). It is defined by the length of cable between the joining shackles (previously a length of 12.5 fathoms).

Sheer

When applied to a vessel at anchor, sheer is an angular movement of the vessel about the hawse pipe point; it can be deliberately caused by applied helm to port or starboard.

Sheet Anchor

An additional anchor carried by larger vessels, a practice now discontinued (not to be confused with the spare anchor carried by the majority of vessels).

Shorten Cable

To heave in a portion of the cable, so reducing the scope.

Short Stay

The cable is said to be at short stay when the anchor is hove in close to the ship's side and not over-extended. The cable is not up and down in this position.

Snub

To snub the cable is to stop the cable running out by applying the brake. A vessel is said to snub round on her anchor when she checks the paying out of the cable by applying the brake on the windlass, so causing the cable to act as a spring, turning the bow smartly in the direction of the cable.

Spurling Pipes

Termed 'navel pipes' in the Royal Navy, the cable passes through these pipes from the windlass or cable holder to the cable locker.

Surge

To surge is to allow the cable or hawser to run out under its own weight. The term is often used when handling mooring ropes on drum ends. (You should not surge on man-made fibre ropes, because of the possibility of heat/friction causing the yarns/strands to fuse.)

Tide Rode

A vessel is said to be tide rode when she is riding at anchor head to tide.

Up and Down

The cable is said to be up and down when the angle the cable makes with the water surface is 90°, usually just before anchor aweigh.

Veer Cable

This is to pay out cable under power, by walking back the gypsy of the windlass.

Walk Back the Anchor

This means to lower the anchor under power.

Wind Rode

A vessel is said to be wind rode when she is riding at anchor head to wind.

Yaw

A vessel is said to 'yaw' when at anchor when she moves to port and starboard of the anchor position under the influence of wind and/or tide. Yawing should not be confused with sheering.

Watch at Anchor

Before the vessel is brought to an anchorage, the Master and engine-room staff should be informed of the estimated time of arrival (ETA), and time of anchoring. An anchor approach plan should be prepared, and speed reduced in plenty of time to assess the approach features and the anchorage area, including depths for echo-sounder. A responsible officer should be fully informed of details regarding amount of cable, depth of water and holding ground, and brief the anchor party accordingly. Power on deck should be obtained in ample time and the anchor walked back before the approach is started. Anchor signals or lights, which should have been tested prior to use, should be made ready.

The state of weather, with particular attention to wind, should be kept under continual observation. The state of visibility, traffic density and the proximity to navigation hazards should be assessed before entering the area for anchoring. Availability of sea room, especially to leeward, should be considered before letting go. The state of tide and current, times of high and low water, and time limits of the vessel when swinging should all be studied.

Officer of the Watch

The officer of the anchor watch should have all relevant information regarding the amount of cable paid out and the estimated position of the anchor. (An anchor buoy can indicate the approximate area of the anchor position, but is only a guide and is rarely used these days.)

The officer's duties include the checking at regular intervals of the ship's position. This may be carried out by observing compass/anchor bearings of fixed objects ashore or prominent landmarks.

These bearings laid on the chart define the vessel's position. Similar checks may be made by using radar, the bearing cursor and the variable range marker. However, the mariner should not rely solely on radar in case of instrument malfunction. Global positioning systems (GPS) would also be employed as either the primary or secondary system for ascertaining the anchored position. Transit bearings give an indication, but the objects used should be spread well apart so that any movement of the vessel would open up the transit objects quickly and allow detection of the vessel's change in position. The purpose of checking the anchor bearings is to ascertain the ship's position, to ensure that she is not dragging her anchor, so moving her position over the ground.

There are other methods of detecting whether the vessel is dragging. For example, secure a hand lead overside from the bridge wing, and let the lead sit on the bottom; if the vessel is dragging her anchor, the lead line will start to lead forward. This would indicate to the observer that the vessel was dropping away from the lead sitting on the bottom.

Transit Bearings

Extensive, practical use of transit bearings should be made by the Officer of the Watch when the vessel is at anchor, especially when 'beam transits' can be obtained. This is not to say that watch officers should rely solely on good transit marks. They should always employ whatever means are at their disposal to ascertain the ship's position, checking and double-checking at regular intervals.

General Precautions

The Officer of the Watch should also ensure that a deck watch is kept when the vessel is at anchor. Detection of the vessel dragging may be ascertained by personnel engaged on this duty 'feeling the cable'. If the vessel is dragging her anchor then vibration from the anchor bouncing over the sea bottom will travel up the length of the cable, especially if the sea bottom is of a hard, uneven nature.

The deck watch should also be aware that unauthorized personnel may try to board the vessel in certain regions of the world. Theft and piracy are rife today in some underdeveloped countries. Access is often gained by climbing the anchor cable or by grapple over the stern.

The Officer of the Watch should ensure that all anchor signals are displayed correctly, and, if oil lights are used, that these remain alight throughout the hours of darkness. Deck lights should be used whenever a vessel is at anchor, together with overside lights.

Correct fog signals should be sounded if the visibility closes in. Radar should be operational if necessary and a sharp lookout kept at all times. VHF radio should be on, and a listening watch continually kept on the local port channel or channel 16.

If in any doubt, the Master should be informed at the earliest possible moment. Should the vessel drag her anchor, the Master must be informed immediately. The engine room should be kept ready for immediate notice in order to manoeuvre the vessel out of any difficulty, should the need arise. A constant check on weather conditions should be kept, and all changes and incidents noted in the log book.

When handing over the watch to another officer, the Officer of the Watch should inform the relieving officer of all relevant details regarding the anchor and cable, weather reports, anchor bearings, ship's position, depth (from echo-sounder). The state of tide, time of expected swing and expected circle of swing should all be marked on the chart.

Another Vessel Dragging Towards your Ship

The options open to the mariner in this situation are somewhat limited. A junior officer faced with the situation should inform the Master immediately. Subsequent actions include drawing the attention of the other vessel to the fact that she is dragging her anchor, in case the incident is undetected, make ready one's own engines and send forward an anchor party.

Drawing Attention to the Situation

When vessels are in sight of one another (Rule 34(d) of the Regulations for the Prevention of Collision at Sea), or are approaching each other, and from any cause either vessel fails to understand the intention or actions of the other, or is in doubt whether sufficient action is being taken by the other to avoid collision, the vessel in doubt shall immediately indicate such doubt by giving at least five short and rapid blasts on the whistle. Such signals may be supplemented by a light signal of at least five short and rapid flashes.

The instructions give the Officer of the Watch a directive for his actions on the bridge. However, it should be borne in mind that this signal is only appropriate when the vessels are in sight of each other. When vessels' visibility is restricted (Rule 35f) a vessel at anchor may, in addition, sound three blasts in succession – one short, one prolonged, and one short blast – to give warning of her position and the possibility of collision to an approaching vessel.

Stand by Engines

Bear in mind that 'finished with engines' should not be ordered when the vessel anchors. Stand by engines must be ordered at a very early stage in order to gain any advantage, since the engine-room staff may require a period of notice to have engines ready. If this is the case, then the sooner the main machinery is made ready to manoeuvre, the better.

It could well be that the Master may decide to steam over his own anchor cable, choosing this alternative to an anchor operation. Depending on the circumstances it would be unlikely that the Master would order astern propulsion, as this would most certainly put excess strain on the cable. However, the loss of an anchor would be a secondary consideration to an imminent collision. A prudent Master would leave engines on stop to facilitate immediate manoeuvres.

Anchor Party Forward

Should the main engines not be readily available, then the moving of the vessel by means of anchors must be of prime consideration. The alternatives are (1) to pay out more cable and increase the scope, or (2) to heave in on the cable and decrease the scope. Whichever is chosen will depend on the position of the approaching vessel and the circumstances at the time.

If the cable is shortened and the ship begins to drag her anchor, it may be possible to 'sheer' her, using the helm to clear the approaching vessel. As a last resort, and if time permits, the cable could always be slipped from the locker. The end of the cable should be buoyed to aid recovery at a later date. (Time permitting such anchor operations.)

Use of VHF Radio

Attention to the fact of dragging anchor could be made by VHF radio. This is all very well if both vessels are equipped with VHF and that it is working, in a switched 'on' condition and on the same channel. But extreme care should be observed when engaged in communication by VHF that both participants have identified themselves. Mariners may find themselves in communication with a third party within range of the VHF radio who is in no way connected with the operation.

Mariners are further reminded of the wording and the content of Rule 36 of the Collision Avoidance Regulations, regarding Signals to Attract Attention, which may be made by any vessel.

Anchoring Facility for High-Speed Craft

High-speed craft must be provided with at least one anchor with an associated cable or warp length, together with means of recovery, an assumption being made that the craft would only need to anchor in the case of emergency.

Anchoring operations for the craft should be compatible with the design, so as not to cause damage to the structure during normal use.

The anchor should also have a safe means of release and adequate arrangements to be secured while the vessel is under routine operations.

The deck area employed for anchor operations (which may be an enclosed space) should accommodate adequate walkways, ventilation and illumination to permit safe working. It should also be equipped with two-way communications between the operating compartment and people operating the anchor and mooring equipment.

Chapter 6 of the HSC code makes reference to anchoring arrangements and also to towing and berthing facilities. Where provision for towing is required, an adequate bridle arrangement must be provided to satisfy the regulations, and the maximum permissible speed for towing must be inserted in the operating manual.

Mooring Anchors

Bruce (Mooring) Anchor

The Bruce anchor (Figure 2.22), probably one of the most up-to-date designs that has reached the market for a very long time, appeared in the USA about 1975, when it won an award at the Offshore Technology Conference, held in Houston, Texas. This award was to be followed by another from the Design Council in 1978 in the United Kingdom.

It is a self-orienting anchor; as it engages the sea bed, it rolls upright, irrespective of dropping attitude. It possesses absolute roll stability, the stabilizing forces being produced by the curved shape of the flukes. It is an anchor of high holding power and allows full veering capability without the risk of breaking out.

The Bruce is a one-piece design anchor, with the advantage of no moving parts. It is manufactured in heat-treated cast steel, which provides added strength. It possesses a highly efficient fluke area and subsequent fluke action, together with an optimum length of shank and fluke shank separation. The flukes are curved, there being no stock, and the anchor is reasonably compact for stowing.

Bruce anchors have been manufactured up to a size of approximately 14,500 lb (7 tons), which can replace conventional anchors of 45,000 lb (20 tons). The need for anchors of higher holding power was generated initially by the offshore oil industry, and to date the main customers for the Bruce anchor have been from this industry. However, small-boat owners are now seeming to favour this very efficient anchor.

Once the anchor has rolled shank uppermost, any further dragging will cause it to bury. The anchor is maintained roll-stable by the stabilizing forces produced by the fluke extensions. The extending fluke's insertion into the sea bed produces a higher aspect ratio than with any other anchor design. This results in very good holding

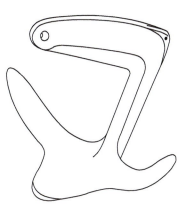

Figure 2.22 Bruce (mooring) anchor.

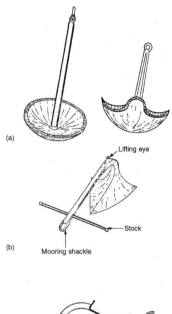

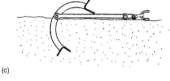

Figure 2.23 *(a)* Mushroom anchors; *(b)* improved mooring anchor; *(c)* Trotman mooring anchor.

power, together with higher tolerable cable angles than hitherto thought possible.

Three further mooring anchors are illustrated in Figure 2.23.

The 'Flipper Delta' Anchor

The Flipper Delta anchor is a high holding power anchor which has become widely employed in the 'offshore regions'. It is designed with a frame, stabilizer construction and large tripping palms. Penetration is achieved by the flukes being tripped to a standard angle of 36° away from the shank. This standard angle can be adjusted to suit specific regional bottom qualities ranging from 28° to 50° (Figure 2.24).

Examples of recommended angles are:

- soft clay: 50°
- hard bottom, like cemented sand: 28°
- general worldwide conditions: 36°.

The holding ability of the anchor is such that the additional weight of residue which accumulates inside the framed stabilized area acts to bed the anchor more solidly into position. It must be realized that this situation is highly desirable when setting anchors, but such holding power must be overcome when recovering the anchor. Added suction about the anchor body will be experienced

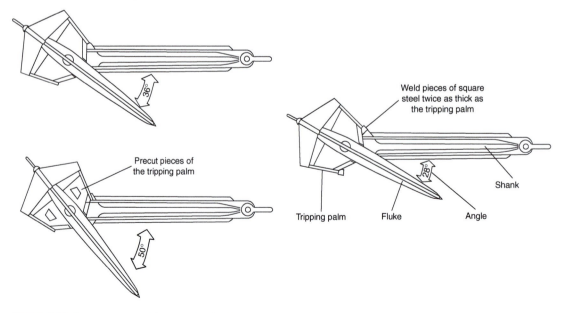

Figure 2.24 Construction and fluke angles for 'Flipper Delta' anchors.

and recovery is better achieved by avoiding a direct, vertical pull with the recovery pendant.

Breaking the 'Flipper Delta' Anchor Out

The anchor pendant is secured to a movable pad eye at the centre of the anchor crown and if this is subsequently heaved at about 15° off the vertical, astern of the anchor, the recovery could be achieved with reasonable ease (Figure 2.25).

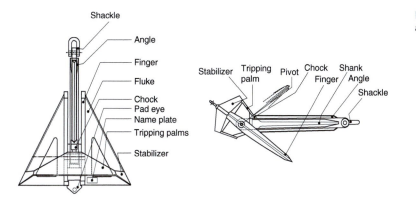

Figure 2.25 Parts of 'Flipper Delta' anchor.

3
Ropework

Introduction

Ropes of many different types and sizes have continued to play a major role in the working of ships throughout history. Clearly, the era of the sailing ships, where cordage, in virtually every form, could be found from anchor warps to ratlines, was a dominant period in history for rope manufacturers.

Times have moved on, with the container ship, the tanker, the ro–ro (roll on–roll off) vessels, where a limited amount of cordage is found on board. Of course, the standard ropes remain, namely, the mooring ropes which virtually all ships still retain. However, even these have suffered from the developments in wire ropes, tension winches and fixed mooring positioning methods.

This is not to say that ropes have disappeared – far from it – but the need for ropes has diminished. This need has been met with improved synthetic ropes, while the natural ropes are difficult to obtain and in some cases not as strong as the man-made products. No wonder that change has occurred. From the seafarer's point of view, the rigging of stages and boatswain's chairs will still arise and the mariner will still need to throw a bowline into an end, though possibly not as often as his historical counterpart.

Natural Fibre Ropes

All natural fibre rope is manufactured from manila, sisal, hemp, coir, cotton or flax fibres. The process of manufacture consists of twisting the fibres into yarns and turning the yarns in an opposite direction to establish the strands. Figure 3.1 shows a hawser laid rope being laid up to form a right-hand laid rope.

Ropes may be of a right-hand lay or left-hand lay, but the most common is right-handed. It is essential to realize that each of the

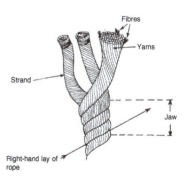

Figure 3.1 Hawser-laid rope: three strands.

components is turned (twisted) up in an opposite direction to that of its predecessor, e.g. in right-hand lay, strands are laid up right-handed (clockwise), yarns laid up left-handed, and fibres laid up right-handed.

Manila

Manila is obtained from the abaca (wild banana) plant, which grows to about 9 m (30 ft) in height, largely in the Philippine Islands, and is exported via the port of Manila, from which it acquires its name.

Manila rope is not as durable as hemp, but is most certainly more pliable and softer. It is gold-brown in colour, and never tarred. Unfortunately it swells when wet, but despite this it is considered by far the strongest natural rope made. It is very expensive and its availability will depend on the political climate. Used extensively where the safety of life is concerned, grade 1 manila is the best you can get, but because of scarcity, and, of course, the cost to shipping companies, it is a rare feature aboard merchant vessels.

An experienced seafarer likes manila, because it is the best. There are many so-called grades of manila on the market, such as sisal mixed with manila, which are not quite the same to a professional. No disrespect to these other grades of ropes, but when you need reliability with a natural fibre, this is the one to use.

Manila is an excellent fibre for the manufacture of towing hawsers, providing good spring and stretch facilities. It is not generally used for running rigging because it swells and expands, often jamming in the block.

Breaking strain and resistance to deterioration are listed in Tables 3.1 and 3.2.

Sisal

Obtained from the leaves of the plant *Agave sisalana*, a large plant of the cactus family, sisal comes largely from Russia, America, East Africa, Italy, Java and countries in Central America. The plant favours a temperate or tropical climate.

Table 3.1 Rope sizes for breaking strain of 86 tonnes

Rope type	Diameter (mm)	Circumference (in.)
Nylon	72	9
Polyester	80	10
Polypropylene	88	11
Manilla (grade 1)	112	14
Wire (6 × 24)	48	6

Substance	Resistance			
	Manilla or Sisal	Nylon	Polyester	Polypropylene
Sulphuric (battery) acid	None	Poor	Good	V. Good
Hydrochloric acid	None	Poor	Good	V. Good
Typical rust remover	Poor	Fair	Good	V. Good
Caustic soda	None	Good	Fair	V. Good
Liquid bleach	None	Good	V. Good	V. Good
Creosote, crude oil	Fair	None	Good	V. Good
Phenols, crude tar	Good	Fair	Good	Good
Diesel oil	Good	Good	Good	Good
Synthetic detergents	Poor	Good	Good	Good
Paint removers, varnish				
Solvents (chlorinated)	Poor	Fair	Good	Poor
Solvents (organic)	Good	Good	Good	Good

Table 3.2 Resistance of ropes to corrosion

Sisal rope is hairy, coarse and white. It is not as pliable as manila nor as strong. When wet, it swells up more than manila, as the water is absorbed more quickly, and it becomes slippery to handle.

It is extensively used in the shipping industry either in its own state or mixed with manila fibres, a good sisal being similar in strength to a low-grade manila. The cost of production is better suited to the shipowner, and the supply is more reliable than manila.

For handling purposes, the fibres have a brittle texture, and continued handling without gloves could cause the hands to become sore and uncomfortable. It is generally used for mooring ropes and most other general duties on board, where risk to life is not in question. Where the rope is expected to be continually immersed in water, it may be coated with a water repellent. This is a chemical coating, usually tar-based, which prevents rotting and mildew. Resistance to deterioration is listed in Table 3.2.

Coir

Sometimes referred to as grass line, coir is obtained from the fibres of coconut husks. It is mainly exported from Sri Lanka (formerly Ceylon) and ports in India.

It is a very rough, coarse and hairy rope, with only about one-quarter the strength of a hemp rope. However, it is very light and floats, possessing great elasticity. It is often used as a towing warp or mooring spring, because of its buoyant and elastic properties.

The coir fibres are short, brittle and thick. These features produce a stiff non-flexible rope, difficult to manage. Consequently they are not popular with seafarers. When used for mooring purposes, they often have a steel wire pennant attached, and the whole is known as a coir spring. These are regularly encountered where heavy swells are common, often being run from the shore as a permanent mooring picked up by the ship on berthing. This rope is never tarred, as this would weaken the fibres considerably.

If the rope becomes wet when in use, it should be dried before being stowed away. The coir fibres are found aboard many vessels in the form of coir doormats.

Hemp

Hemp is obtained from the stem of the plant *Cannabis sativa*, which yields flax for the production of canvas. (The word canvas is derived from the Latin '*cannabis*', which means hemp.) This was accepted as the best rope in the marine industry from the early developing days of sail. *Cannabis sativa* is cultivated in many parts of the world – New Zealand, Russia, China, India and the USA, for instance – but has been replaced mainly by man-made fibre ropes and manila.

The hemp fibres are a light cream in colour when supplied to the rope manufacturer. They have a silky texture and are of a very fine nature: hence the extra flexibility of the hemp rope compared to a sisal or manila.

Most hemp ropes treated in manufacture produce a tarred, brown rope which is hard and hairy to the touch. Its strength will be dependent on the place of production. Italian hemp ropes are now considered to be the best quality, having about 20 per cent greater strength than a high-grade manila. However, quality differs considerably, and hemp ropes are rarely seen at sea today except in the form of small stuff, e.g. lead line, cable-laid hemp, sea anchor hawsers, bolt rope, etc.

The advantage of hemp rope is that it is impervious to water and does not shrink or swell when wet. For this reason it was extensively used for the rigging of sailing vessels and roping sails. When used for running rigging it was preferred to manila or sisal because it did not swell up and foul the blocks. However, for vessels navigating in cold climates, hemp ropes do have a tendency to freeze up. Not all hemp ropes are supplied tarred, so the weight will vary, together with the strength.

Lay of Rope

The lay of rope is a term used to describe the nature of the twist that produces the complete rope, as we have already explained (pp. 65 and 66). The purpose of alternate twisting of fibres, yarns and strands is to prevent the rope becoming unlaid when in use.

The majority of ropes are manufactured with a right-hand lay, but left-hand laid ropes are available. The most common form of rope at sea is known as 'hawser laid rope', comprising three strands laid up right- or left-handed. Other types of lay include 'cable lay' (Figure 3.2), made of three or four hawsers laid up left-handed, and sometimes referred to as water lay, which is strictly incorrect. 'Water lay' was a rope designed to be used when wet, e.g. sounding line. Consequently it was laid up in the course of manufacture in

Shroud laid   Cable laid

Figure 3.2 Types of rope lay.

a wet condition, so allowing for shrinkage in use. Cable-laid ropes, although generally left-hand laid, may be encountered as right-hand laid (left-hand hawsers being used), but these are extremely rare.

Eight-Strand Plaited

Many mooring ropes used at sea today are 'eight-strand plaited', constructed by laying two pairs of strands left-handed, with the other two pairs right-handed. This type of lay has the advantages that it does not kink and, with eight strands, has increased flexibility. However, it is difficult to splice, and the manufacturers' instructions should be consulted.

Shroud Lay

Another type of lay found at sea is 'shroud lay' (Figure 3.2), consisting of four strands, sometimes being laid about a central heart, right-handed. As the name implies, it was used for standing rigging (the shrouds to the mast) until wire ropes came into use.

Soft-Laid

Often referred to as a long lay, soft-laid is a strong, flexible method of laying up a rope. The angle of the strand to the axis through the centre of the rope is comparatively small. It will absorb water more easily and will not be as hard-wearing as, for example, a hard-laid rope. The 'jaw' of the lay is large with a soft-laid rope.

Hard-Laid

Sometimes called short lay, when the 'jaw' of the lay is small in comparison to a soft-laid rope, hard-laid is harder wearing than the former, does not easily absorb water and tends to retain its shape better when under stress. Being hard in construction, it is not very flexible, and its breaking stress and subsequent safe working load are inferior to those of soft- or standard-laid ropes.

Standard or Plain-Laid

Standard lay may be described as a cross between hard- and soft-laid ropes. It has been found by experience to be the best in providing pliability and strength, and to be sufficiently hard-wearing and chafe-resistant to suit the industry for general-purpose working.

Sennet-Laid

Alternatively known as plaited, but not in the same way as the 'eight-strand plaited' previously mentioned, an example of sennet lay is found with the patent log line, where the yarns are interwoven, often about a central heart. This lay of rope has an effective anti-twist, non-rotational property.

Unkinkable Lay

This lay looks like standard lay, but close inspection will reveal that the yarns are twisted the same way as the strands. Left-handed in construction, it is usually ordered for a specific job, e.g. gangway falls. The advantage of this lay is that the tendency for standard lay to kink when passing through a block is eliminated.

Small Stuff

Small stuff is a collective term used at sea with respect to small cordage, usually less than 38 mm (1.5 in.) in circumference and of 12 mm diameter, approximately.

Boat Lacing

Manufactured in 14 various sizes, boat lacing is made of a high-grade dressed hemp, having a fine finish and being smooth to handle. Before the invention of man-made fibre, it was used for securing boat covers, awnings, etc. It is sold in hanks weighing from 93 g to 1.8 kg.

Marline

Marline is usually supplied in hanks by weight, tarred or untarred. It is made in two-ply, i.e. two yarns laid up left-handed, from better quality fibres than spunyarn, and produces a much neater, tighter finish to any job. It is used for seizings, serving and whipping heavy-duty ropes.

Spunyarn

Made from any cheap fibres and turned into yarns, spunyarn may have two, three or four yarns, usually laid up left-handed. The yarns are supposed to be soaked in Stockholm tar as spunyarn is used for the serving of wires, and the idea was that in hot climates the lubricant (Stockholm tar) would not run from the serving. Spunyarn is generally sold in balls of up to 3.2 kg or in coils of 6.4 kg or 25.6 kg by length or weight.

Point Line

A three-stranded manila rope, point line is made and may be ordered in three sizes, which are determined by the number of threads:

1   Circumference 1¾ in. (35 mm); diameter 11 mm; 15 threads.
2   Circumference 1½ in. (38 mm); diameter 12 mm; 18 threads.
3   Circumference 1⅝ in. (41 mm); diameter 13 mm; 21 threads.

It is used as an all-purpose lashing aboard most present vessels. Sisal very often replaces manila in so-called point line.

Many of the natural fibre ropes have become obsolete with expansion in the man-made rope sector of the industry. The loss of rigging aboard the modern vessel has also speeded up its demise.

Lead Line

Made of high-grade cable-laid hemp, it may be obtained in a size of 1¾ in. (9 mm diameter) for hand lead lines. It is supplied in 30-fathom coils for the hand lead.

Seaming Twine

Manufactured from the best flax, this three-ply twine is made up in hanks of approximately 1 lb weight and 900 fathoms length. It is used extensively for canvas work.

Roping Twine

This five-ply twine is supplied in hanks of similar length and weight to that of seaming twine. It is used for whipping the ends of ropes, worming, etc.

Signal Halyard

Often spelled halliards, this used to be three- or four-stranded dressed hemp, but this natural fibre has given way to man-made fibres such as polypropylene. It may be supplied in a variety of sizes to the customer's requirements. Plaited-laid halliards are predominant on the modern merchant vessel, being preferred because the stretch is not as great as, say, hawser lay. The word halyard was derived from the old-fashioned 'haul yard', which was previously employed on sailing vessels to trim and set the sails to the yard arms.

Synthetic Fibre Ropes

Although natural fibre ropes are still widely used throughout the marine industry, they have been superseded by synthetic fibres for

Plate 22 Tension mooring winch aboard a large tanker (source: Shutterstock).

a great many purposes. Not only do the majority of synthetic ropes have greater strength than their natural fibre counterparts, but they are more easily obtainable and at present considerably cheaper.

Breaking strain and resistance to deterioration are listed in Tables 3.1 and 3.2.

Nylon

Nylon is the strongest of all the man-made fibre ropes. It has good elasticity, stretching up to 30 per cent and returning to its original length. It is used for such functions as shock-absorbing when coupled with a mooring wire: the nylon forms a rope tail which takes the heavy shocks as a vessel ranges on her moorings. It is also used in a combination towline – one section steel wire and one section nylon rope.

Nylon ropes are light to handle, twice as strong as an equivalent-sized manila, and give the appearance of a smooth, slippery surface. They are impervious to water, have a high melting point (250°C) and in normal temperature are pliable, being suitable for most forms of rigging.

The disadvantages of nylon ropes are that they do not float, and in cold climates they tend to stiffen up and become difficult to handle. They should not be left exposed to strong sunlight or be stowed on hot deck surfaces, as their natural lifespan will be impaired. The significant point with these ropes is that they are used where great stress occurs. Should they part under such stress, there is a tendency for them to act like elastic bands, an extremely dangerous condition to be allowed to develop. The nylon rope will give no audible warning when about to part; however, when under excessive stress, the size of the rope will considerably reduce. They are difficult to render on a set of bitts, and should never be allowed to surge. Any splices in the nylon ropes will tend to draw more easily than in natural fibre when under stress. Nylon is expensive, but its life may be considered to be five times as long as its manila equivalent.

Polyester

Polyester is a heavy rope compared to nylon and not as strong, but nevertheless some of its properties make it a worthwhile rope to have aboard. It is considered to be more resistant to acids, oils and organic solvents than its nylon counterpart, while its strength remains the same whether in a dry or wet condition. It is used for mooring tails and mooring ropes.

Its disadvantages are very similar to those of nylon. It will not float. Splices must have four full tucks and may draw more easily than with a natural fibre rope when under stress. It should not be surged on drum ends. Frictional heat should be kept to a minimum when working about bitts or warping drums. The melting point is between 230° and 250°C.

Polypropylene

This is probably the most popular of the man-made fibres at sea. The ropes are cheap, light to handle, have the same strength whether wet or dry and they float. They are used extensively for mooring ropes and running rigging. The melting point is low compared to nylon (165°C). Friction-generated heat should be avoided with this man-made fibre, which is extremely susceptible to melting and fusing. Should the fibres fuse together, the rope is permanently damaged and weakened.

It is resistant to chemical attack by acids, alkalis and oils, but solvents and bleaching agents may cause deterioration. It neither absorbs nor retains water, and because of this fact has recently been used for the inner core of wire ropes, the advantage being that inner corrosion in the wire is eliminated. However, the wire would still need to be lubricated externally.

Fibrefilm, a by-product from polypropylene, is a very cheap version of the fibre. It is produced from continuous, thin, twisted polypropylene tape, and used for general lashing purposes.

Precautions When Handling Synthetic Man-Made Fibre Ropes

1   The mariner should carefully inspect a rope, both internally and externally, before it is used. Man-made fibre ropes show deterioration after excessive wear by a high degree of powdering between the strands.
2   Ropes should be kept out of direct sunlight. When not in use, they should be covered by canvas or another shield, or, if the vessel is engaged on long sea passages, stowed away.
3   When putting a splice in a synthetic fibre rope, use four full tucks, followed by two tapered tucks (strands halved and quartered). The length of the protruding tails from the completed splice should be left at least three rope diameters in length. Any tail ends of strands should be sealed by tape or similar adhesives.
4   A stopper should be of the same material as that of the rope being stoppered off, and should preferably be of the 'West Country' type. The one notable exception to this rule is that a nylon stopper should never be applied to a nylon (polyamide) rope.
5   A minimum number of turns should be used when heaving man-made fibre ropes about winch barrels or capstans. Friction-generated heat should be avoided, and to this end no more than three turns should be used on drums. Where whelped drums are being used, it may be necessary to increase the number of turns so as to allow the rope to grip; if this is the case, then these turns should be removed as soon as possible.
6   Never surge on man-made fibre rope. Should it be required to ease the weight off the rope, walk back the barrel or drum end, as when coming back to a stopper.

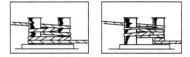

Figure 3.3 Making fast to bitts.

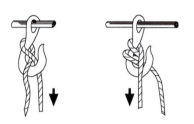

Figure 3.4 Single blackwall hitch (*left*) and double (*right*).

7 When making fast to bitts, make two round turns about the leading post, or two turns about both posts, before figure-eighting (see Figure 3.3).

Bends and Hitches

Blackwall Hitch: Single

Used as a jamming hitch, this is not in common use at sea today since it was found unreliable and had a tendency to slip. It is only effective on the larger style of hook with a wide surface area or on the very small jaw hooks (see Figure 3.4).

Blackwall Hitch: Double

This is used for the same reasons as above, but with far more confidence. Holding power is considerably better than that of a single Blackwall, and light hoists could be made with this hitch (see Figure 3.4).

Bowline

Probably the most common of all hitches in use at sea is the bowline (Figure 3.5). It is by far the best way of making a temporary eye in the end of a rope, whether it be point line or mooring rope size. It will not slip even when wet, it will not jam and it will come adrift easily when no longer required. It is commonly used to secure a heaving line to the eye of a mooring rope when running a line ashore.

Bowline on the Bight

This is one of several variants of the bowline, made with the bight of the rope, so forming two eyes (Figure 3.5). One of these eyes should be made larger than the other to accommodate the seat, while the smaller of the two eyes would take the weight under the arms of an injured person. It forms a temporary bosun's chair for lifting or lowering an injured person. It may be necessary to protect the person from rope burn or pressure by padding under the seat and armpits.

Bowline: Running

A slip knot is made by dipping the bight of rope around the standing part and securing an ordinary bowline on to its own part, so forming a running noose (Figure 3.5). It should be noted that it is a common mistake for inexperienced seafarers to assume that the tail end of rope can be passed through the eye of an ordinary bowline. This is not only inaccurate but time-consuming, especially if the length of the rope is considerable, as with a full coil.

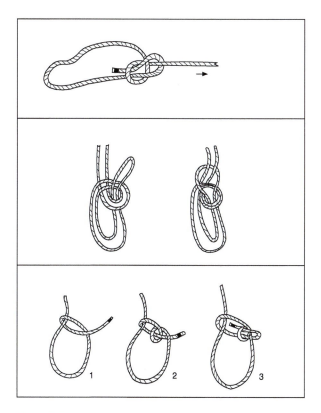

Figure 3.5 Bowline (*top*), bowline on the bight (*middle*) and running bowline (*bottom*).

Bowline: French

An alternative to the bowline on the bight, this hitch has the same function of allowing the weight of a person to be taken up by the two eyes (see Figure 3.6).

Carrick Bend: Single

Originally used for bending two hawsers around a capstan, the bend was constructed so that it formed a round knot which it was thought would not become jammed in the whelps of the capstan barrel. It is a strong, versatile bend that will not jam under strain, providing it is properly secured (Figure 3.7).

The idea of the knot is for the weight to be taken on either side; the bend should be seen to hold, and only then should the tail ends be seized to the standing parts. It is often thought that the ends should be seized immediately after securing the bend, but this is not the case; weight must first be taken and the bend will be seen to slip and close up on itself; only after this has occurred should the ends be seized.

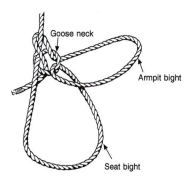

Figure 3.6 French bowline.

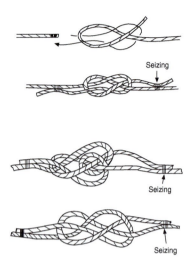

Figure 3.7 Carrick bends: single (*top*), double (*middle*) and diamond (*bottom*).

Carrick Bend: Double

This version of the carrick bend (Figure 3.7) is formed in a similar manner, except that a round turn is made about the cross of the first hawser. It is used where additional weight could be expected to bear, as in towing operations. Again the tails should be left sufficiently long so that, when the weight is taken up and the bend slips to close itself, there will be enough slack in the two tails left to seize down to the standing part. The advantage of this bend over a sheetbend is that it will easily come adrift when no longer required, whereas the sheetbend may jam and have to be cut away.

Carrick Bend: (Single) Diamond

So called because of the diamond shape formed in the middle of the bend, prior to taking weight on the two hawsers either side, it only differs from the single carrick in the fact that the tail end is not seized on the same side as in the single carrick, thus giving the appearance of being a different version of the single carrick. It is used for exactly the same purposes as above, and forms the basis for many fancy ropework knots (see Figure 3.7).

'Catspaw'

This is used to shorten a bale sling strop and is constructed by using two bights of the strop. Two eyes are formed by simply twisting each bight against itself, the same number of twist turns being applied to each bight. The two eyes so formed can then be secured to a lifting hook or joined by a securing shackle (see Figure 3.8).

The stevedore's method of 'shortening a strop' (Figure 3.9) is an alternative to the 'catspaw'. It is achieved by passing opposing bights of the strop through their own parts, effectively making an overhand knot with the bights.

Clove Hitch

A very common hitch in use at sea today, this consists of two half-hitches jamming against each other. It is a useful knot for turning about a rail and hanging things from, but unreliable, especially when the direction of weight is liable to change, which could easily cause it to slip (see Figure 3.8).

Cow Hitch

This hitch is used to form the 'bale hitch' when employing a bale sling strop. It is, however, more commonly used to hold a wire rope when constructing a chain stopper (see Figures 3.8, 3.18, 3.32).

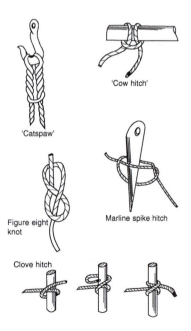

Figure 3.8 Selection of common hitches.

Figure Eight Knot

Used as a stopper knot and employed in many forms, especially at sea, this can regularly be found in the lifelines of ships' lifeboats and in the keel grablines of boats' rigging. An all-purpose knot, it prevents a rope from running through a block (see Figure 3.8).

Fisherman's Bend

This is used for securing a hawser to the ring of a buoy. The bend differs from the round turn and two half-hitches, for the first half-hitch is passed through the round turn. The second half-hitch is not always applied, but, in any event, with both the round turn and two half-hitches and the fisherman's bend, the tail end of the securing should always be seized down to the standing part (see Figure 3.11).

Marline Spike Hitch

An easily constructed hitch (Figure 3.8) much used by riggers to gain more leverage when gripping thin line or rope. It is useful when whipping or binding is required to be drawn exceptionally tight.

Midshipman's Hitch

This hitch may be used instead of a Blackwall hitch, especially when the rope being used is 'greasy'. It is a quick method of securing a rope's length to a hook (see Figure 3.12).

Reef Knot

This is basically a flat knot, ideal for securing bandages over a wound when tending injured personnel; the flat knot lies comfortably against the patient without aggravation. It is also employed in boat work for the purpose of reefing sails (see Figure 3.13).

Rolling Hitch

The rolling hitch is one of the most useful hitches employed at sea (Figure 3.10). Providing it is properly secured and the weight is against the double bight turn, the hitch should not slip. As it is a secure hitch, it is used to secure the jib halyard block to the sea anchor hawser when rigging a whip for use with the oil bag from a lifeboat.

Old sailors used to secure their hammocks by use of a rolling hitch. This prevented the hammock from sliding to and fro with the motion of the vessel when in a seaway.

Round Turn and Two Half-Hitches

This all-purpose hitch is used to secure a rope or hawser to a ring or spar. It is useful in the fact that by removing the two half-hitches,

Figure 3.9 Stevedore's method of 'shortening a strop'.

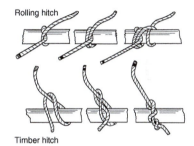

Rolling hitch

Timber hitch

Figure 3.10 Rolling hitch (*above*), timber hitch (*below*).

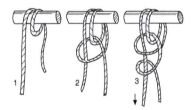

Figure 3.11 Fisherman's bend.

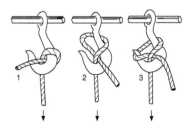

Figure 3.12 Midshipman's hitch.

Figure 3.13 Reef knot.

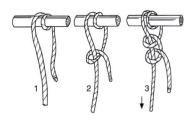

Figure 3.14 Round turn and two half-hitches.

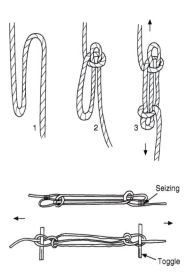

Figure 3.15 Sheepshank (*above*) and securing the sheepshank (*below*).

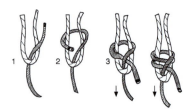

Figure 3.16 Single and double sheetbends.

the weight on the rope can still be retained and eased out by slipping the round turn. An example of this in action is seen when 'bowsing in' tackles are employed in launching ships' open lifeboats (see Figure 3.14).

Sheepshank

The sheepshank (Figure 3.15) is used generally for shortening a rope without cutting its length. It is often used in keel grablines under ships' lifeboats, and may also be employed to adjust the length of a boat's painter when the boat is tied alongside in tidal waters as the tide rises or falls.

Sheetbend: Single

This hitch is commonly used to join two ropes of unequal thickness (Figure 3.16). However, when employed for this purpose there is a tendency for it to 'jam up' after weight has been taken on the standing part. A carrick bend would be more suitable when weight, such as that consequent upon a towing operation, is expected.

Sheetbend: Double

This is used extensively when security over and above that which could be expected when employing a single sheetbend is required. It is used whenever human life needs safeguarding, such as when securing a bosun's chair to a gantline (see Figures 3.16 and 3.19).

Timber Hitch

A slip knot, in common use at sea today, the timber hitch (Figure 3.10) lends itself to gripping a smooth surface like a spar or log. It is often used in conjunction with a half-hitch. It may also be used for lifting light cases or bales, but the mariner should be aware that it is a slip knot, and once the weight comes off it, there would be a tendency for the hitch to loose itself.

Barrel Slings

See Figures 3.17 and 3.18.

Working Aloft and Overside

It should be noted that prior to rigging bosun's chairs or stages or carrying out work aloft or overside, a risk assessment should be conducted and a permit to work obtained.

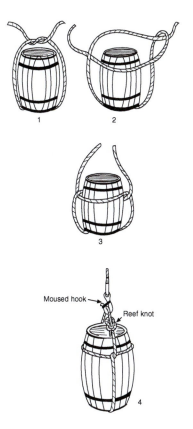

Use of a bale sling strop

Butt sling

Figure 3.18 Double barrel sling.

1 Pass the bight under the cask.
2 Pass open half-hitch over the cask with each tail.
3 Tension each tail and secure with reef knot as for single barrel hitch.
   (*Below*) Slinging a cask on its side.

Figure 3.17 Single barrel sling.

1 Pass bight under the cask and secure
  with an overhand knot above the open
  end of the cask.
2 Open up the overhand knot.
3 Take the weight on either side of the cask.
4 Secure both tails with a reef knot. Ensure
  that the reef knot is secured low to the
  top end of the cask to allow the full
  weight to be taken on the standing part.

Rigging the Bosun's Chair

See Figure 3.19. Close inspection should be made of the chair itself
and the gantline before the chair is used. The gantline should be
seen to be in good condition, and if any doubt exists, a new rope
should be broken out. The bridle to the chair should be inspected,
and particular attention paid to the internal lay and its condition.
A safety line with safety harness must always be worn when operating from a bosun's chair. This line should also be inspected before
use, then secured to a separate anchor point. When working from
a bosun's chair, the following precautions should be observed:

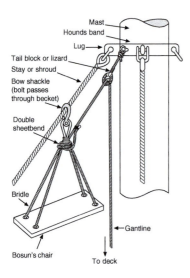

Figure 3.19 Rigging a bosun's chair for riding a stay. Bolt of bow shackle must be moused and crown bow must pass over the stay.

1   Always secure the gantline to the chair with a double sheet bend.
2   Always have the chair hoisted manually, and never heave away on the downhaul using a winch drum end.
3   Any tools, paint pots, etc. should be secured by lanyards. Any loose articles should be removed to prevent falling when aloft.
4   When riding a stay, make sure the bolt of the bow shackle passes through the becket of the bridle. This bolt should be moused.
5   Should work be required about the funnels, aerials, radar scanners and the like, the appropriate authority should be informed – engine room, radio officer or bridge, respectively.

The lowering hitch is normal when the chair is to be used for a vertical lift. The man using the chair should make his own lowering hitch, and care should be taken that both parts of the gantline are frapped together to secure the chair before making the lowering hitch.

Whether making a vertical lift or when riding a stay, ensure the tail block or lizard, whichever is being used, is weight tested to check that it is properly secured and will take the required weight.

> NB. When rigging a bosun's chair or stages, reference should be made to the 'Code of Safe Working Practice' to ensure that all safety procedures are followed.

Rigging Stages

Before rigging stages (Figure 3.20), take certain precautions:

1   Check that the stage is clean and free from grease, that the wood is not rotten and that the structure is sound in every way.
2   Check that the gantlines to be used are clean and new. If in any doubt, break out a new coil of rope. Conditions of used cordage may be checked by opening up the lay to inspect the rope on the inside.
3   The stage should be load-tested to four times the intended load (as per Code of Safe Working Practice).
4   Stages should not be rigged over a dock or hard surface, only over water. Many vessels are designed with working surfaces for painting such areas as bridge fronts. Other vessels will be equipped with scaffolding for such jobs.
5   Lizards must be in good condition and well secured.
6   Stages should not be rigged overside for working when the vessel is underway.
7   The gantlines should be of adequate length, and rigged clear of sharp edges which could cause a bad nip in the rope.
8   A correct stage hitch, together with lowering turns, must be applied.

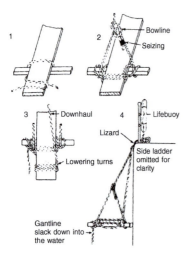

Figure 3.20 Rigging the stage.

The stage hitch should be made by the seaman going on to the stage. It is made about the end and the horns of the stage. For additional safety two alternate half-hitches should be made about the horns before tying off the bowline. This bowline is to be secured about 1.5–2 m above the stage itself to provide the stage with stability.

The lowering turns must be seen to be running on opposite sides of the stage to prevent the stage from tilting. Safety line and harness for each man should be secured to a separate point, and these must be tended by a standby man on deck. A side ladder, together with a lifebuoy, should be on site. All tools etc. should be on lanyards, and the gantlines extended down to the water.

Seizings

Flat Seizing

Make a small eye in the end of the seizing small stuff, pass the formed noose about the two ropes to be seized, then continue with about six loose turns about the two ropes. Pass the tail through the inside of the loose turns and pull the seizing taut. Pass frapping turns in the form of a clove hitch about the seizing between the two parts of rope. The seizing so formed is a single row of turns, and is used when the stresses on the two parts of the ropes are equal (see Figure 3.21).

Racking Seizing

Use spunyarn or other small stuff of suitable strength and size, with an eye in one end. Pass the seizing about the two ropes, threading the end through the ready-made eye. Use figure-eight turns between the two ropes for up to ten or twelve turns, then pass riding turns over the whole between the figure-eight turns.

Once the riding turns are completed, the seizing should be finished by passing frapping turns between the two ropes and securing with a clove hitch. This seizing is very strong and should be used when the stresses in the two ropes to be seized are of an unequal force (see Figure 3.21).

Rose Seizing

This is a means of securing an eye of a rope to a spar or other similar surface. The seizing is rove as a crossed lashing between the parts of the eye and under the spar, the whole being finished off by the end being half-hitched about the seizing under the eye.

Round Seizing

This is a stronger seizing (Figure 3.21) than the flat, and is used when the stresses on the two ropes are equal, but extra weight may

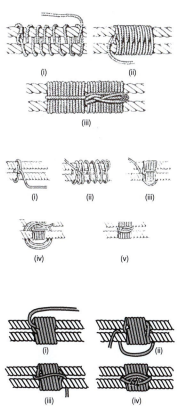

Figure 3.21 Seizings: racking (*top*), flat (*middle*) and round (*bottom*).

be brought to bear on the formed seizing. It is started in a similar manner to the other two, with an eye in the end of the small stuff. Begin as for a flat seizing and obtain a single row of turns. Work back over these turns with a complete row of riding turns. Pass frapping turns around the whole double row of turns between the two parts of the rope, finishing with a clove hitch.

Ropework and Cordage Tools

Hand Fid

A hand fid is a tapered piece of hardwood, usually lignum vitae, round in section, used to open up the lay of a rope when putting in a splice. The wood has a highly polished finish for the purpose of slipping in between the strands of the rope. The hand fid is always made of wood, not steel, so as not to cut the fibres of the rope.

Riggers' (Sweden) Fid

This is a hand fid constructed with a wooden handle attached to a U-shaped taper of stainless steel. A more modern implement than the wooden hand fid, it is suitable for ropes or wires. The U-shaped groove down the side permits the passage of the strands when splicing. The end is rounded off so as not to cut the yarns of the ropes, and the metal has a thick, smooth nature, with a blunt edge, for the same reason.

Setting Fid

This may be described as a giant version of the hand fid. It is used for splicing larger types of rope, e.g. mooring ropes, often in conjunction with a mallet to drive the taper of the fid through the strands of the rope.

Serving Board

This is a flat board, fitted with a handle for the purpose of serving the wire eye splice. The underside of the board has a similar groove to that of the serving mallet, except that the groove is 'flatter' and more open, to accommodate the broad eye of the wire rope where it has been spliced.

Serving Mallet

This wooden mallet, cut with a deep-set groove running the full length of the hammer head, is used to turn the serving (marline or spunyarn) about the wire rope. The groove accommodates the wire as the implement acts as a lever to make the serving very tight (see worming, parcelling and serving, below).

Worming, Parcelling and Serving

The purpose of the operation of worming, parcelling and serving (Figure 3.22) is threefold. First, the covering will preserve and protect the wire from deterioration (mainly due to bad weather). Second, the covering will also protect the mariner from 'jags' in the wire when handling. Third, the completed operation will produce a neat and tidy finish. Seafarers generally take pride in a clean and tidy ship, so it does help the morale of the vessel.

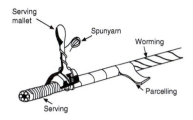

Figure 3.22 Worming, parcelling and serving.

> **Worming, parcelling and serving aid to memory**
> Worm and parcel with the lay, turn and serve the other way

Worming

In this first part of the operation a 'filler' of suitable small stuff is woven around the wire, in between the strands. This effectively prepares the way for the parcelling to produce a smooth finish, prior to serving. Marline should not be used for the worming because it is too hard and will not easily compress. When parcelled over it may cause the surface to be uneven. Small stuff suitable for the purpose of worming includes spunyarn, hemp yarns, or small rope, depending on the size of the wire being worked. The worming should be carried out in the direction of the lay of the wire.

Parcelling

This is the covering of the wire and the worming by oiled sacking, burlap or tarred canvas. The material is cut into strips up to 3 in. (75 mm) in width and turned about the wire in the direction of the lay. To ensure that the parcelling does not unravel while in the operation of serving, a lacing of sail twine may be drawn over with a marline hitch.

Cordage Splice

Back Splice

Used to stop a rope end from unravelling, a back splice performs the same function as a whipping, though it is considerably more bulky. It is formed by opening up the strands of the rope to be spliced for a convenient length and then making a crown knot.

The crown should be pulled down tight, and then the tails can be spliced into the rope against the lay, each tail being passed over the adjacent strand and under the next. The 'first tuck' is the term used to describe the 'tucking' of each tail once in this manner. A minimum of three full tucks should be made in a natural fibre rope.

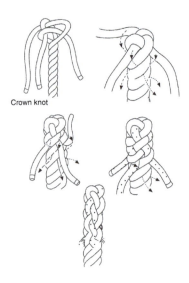

Figure 3.23 Back splice (reproduced from *Creative Ropework*).

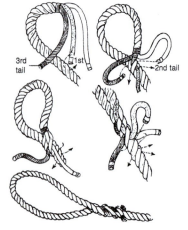

Figure 3.24 Eye splice (reproduced from *Creative Ropework*).

Eye Splice

This is by far the most widely used splice within the marine industry. The eye is made by unlaying the three strands and interweaving them into the rope against the lay. It is considered a permanent eye when completed, and if spliced without a thimble it is referred to as a soft eye splice (as opposed to a hard eye splice with a metal thimble set inside the eye). Once the first tuck is made, the normal method of passing each tail over the adjacent strand and under the next is followed.

The first tuck is made with the centre tail being spliced first at the required size of the eye, against the lay of the rope; the second tail must be spliced next, over the immediate strand in the rope and under the following one, again against the lay of the rope; and the third and final tail must be tucked on the underside of the splice against the lay, so completing the first tuck. Each tuck in the splice should be drawn tight, but care should be taken not to over-tighten the first tuck, or else a 'jaw' may result at the join of the eye to the splice.

Short Splice

This is a strong method of joining two ropes together, found in the making of 'bale sling stops' (see Cargo Handling, p. 166). The rope thickness is increased by putting in a short splice, and so it is rarely seen in running rigging, as the splice would tend to foul the block.

When making this splice, it may be necessary to whip the ends of the separate strands, and place a temporary whipping at the point where the two rope ends marry together. As more experience is gained in constructing this splice, one may probably discard the temporary whipping, unless splicing heavy-duty ropes like mooring ropes.

Long Splice

The purpose of this splice is to join two rope ends together without increasing the thickness of the rope. The splice is not as strong as a short splice and is generally used as a temporary method of joining ropes together as they pass through a block.

Examples of the use of this splice may be seen in the renewing of flag halyards, the new halyard being long spliced to the old. The old halyard is then pulled though the block, trailing the new halyard behind it. The beauty of this system is that it saves a man going aloft and rethreading the block. It is often used in decorative ropework, where the splice must be unobtrusive.

The long splice stretches over a greater length of the rope than the short splice. It is made by unlaying a strand for up to approximately 1 m (depending on the thickness of the rope), and a similar strand is unlaid from the other rope end. This single strand should then be laid in the place of its opposite number in the other tail end. This procedure is followed with all three strands of both rope ends, so the tail ends protrude from the lay of the rope at differing intervals. Each pair of tails should be finished off with an overhand knot in the way of the lay of the two ropes (Figure 3.26).

Whippings

Common Whipping

Probably the easiest of all the whippings (Figure 3.27(a)), it is not as strong as the sailmaker's whipping, and is liable to pull adrift with continual use. It is formed by frapping round the rope end and burying the end of the twine. Once sufficient turns have been taken, the pull through end of the twine is laid back down the lay of the rope. Frapping turns are then continued by using the bight of the twine. Each frapping turn made with the bight is passed about the end of the rope. When the turns have made a secure tail end finish, pull through on the downhaul of the bight and trim. There are several methods of constructing the common whipping, methods which vary with regard to the position of the whipping – whether it is made on the bight of a rope or at the tail end. Should the whipping be required in the middle of the rope, set on the bight, it would be necessary to pre-turn the whipping twine and thread the downhaul through the pre-made larger turns. Frapping could then be continued without creating 'kinks' in the twine and consequent fouling.

Any of the whippings, if constructed in a proper manner, should not be easily removed, even with regular wear and tear. This applies not only to the sailmaker's but also to the common variety.

Sailmaker's Whipping

Without doubt this is the strongest whipping in common use (Figure 3.27(b)). Should it need to be removed at a later time, it would most certainly need to be cut away.

A bight of twine is laid into the strands of the rope itself. These strands are then relaid up to form the original lay of rope, the bight of twine being left long enough to be secured by being placed about the end of the identified strand once the frapping turns have been constructed. Commence turning up the frapping turns about the tail end of the rope, having left a good length on the whipping twine. Follow the lay of the strands under the whipping and pass the bight over the same strand as shown in the figure. Draw the bight of twine tight and secure the other two ends in way of the rope lay by use of a reef knot, squeezed into the centre of the rope ends lay. The bight and long ends of the twine form a binding about the frapping turns of the whipping.

Palm and Needle Whipping

This is formed in exactly the same manner as the sailmaker's whipping, except for the fact that a sailmaker's palm and needle is

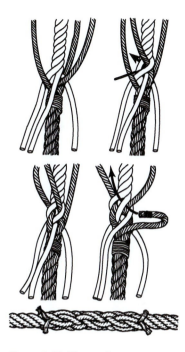

Figure 3.25 Short splice (reproduced from *Creative Ropework*).

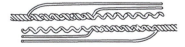

Unlay strands to different lengths

Finish by joining tails with overhand knot

Figure 3.26 Long splice (reproduced from *The Apprentice and his Ship*).

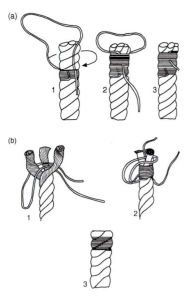

Figure 3.27   *(a)* Common whipping; *(b)* sailmaker's whipping.

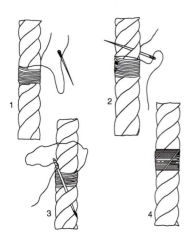

Figure 3.28 Palm and needle whipping. Used as a second whipping, set into the bight of rope to prevent lay being disturbed.

employed to 'stitch' the binding above and below the whipping (Figure 3.28). The position of the whipping is usually set well into the bight of the rope, not on the tail end, as with other more commonly used whippings. Its purpose is to add additional securing to the tail end of a rope before the end securing is placed on. It can also be used as a marker indication for set lengths of the rope.

West Country Whipping

This whipping is made in the bight of a rope and is used for marking the rope at various intervals. Although it is easy enough to construct, it is not as popular as the common whipping, which may be used for the same purpose. It would not normally be found on the tail end of a rope because the twine tends to stretch and work free with excess wear and tear; the common whipping, or better still the sailmaker's, would be stronger and more suitable for the rope end. It is made by overhand knots on alternate sides of the rope, finished off with a reef knot (see Figure 3.29).

Marrying Two Ropes Together

It is often desirable to join two ropes together, and this may be done in many ways. The first and most obvious is by use of bend or hitch – sheet-bend, carrick, reef, etc. – but this method will increase the thickness at the join. An alternative is a splice, either short, cut or two eye splices, but again the thickness of the join is prominent. For running rigging it is generally not desirable to increase the thickness, as it would run foul of the block. A long splice is another option, but this is not a strong splice.

The last option open to the seafarer is to bring the two ropes butt end on, and use a sailmaker's needle to stitch the underside of the two whippings (Figure 3.30). The stitches, made in sail twine, must be drawn very taut to keep up the pressure between the rope ends.

If the operation is being carried out on wires, then seizing wire would be used in place of sailmaker's twine. Before joining wires in this manner, ensure that the ends of the wires are securely whipped and that the whippings will not pull off. This method is extensively used for the re-reeving of new rigging, e.g. topping lifts, cargo runners, etc.

To Pass a Stopper

The purpose of the stopper is to allow the weight on a line to be transferred to bitts or cleats when belaying up. Examples of the use of stoppers may be found when the vessel is securing to a quay or wharf. They are used in conjunction with the transference of weight in the mooring rope from the windlass drum end to the bitts (bollards).

The stopper should be secured to the base of the bitts by a shackle, or around one of the bollards, so as to lead away from the direction of heaving.

Mooring ropes will use a rope stopper, either the 'common rope stopper' or the 'West Country stopper', depending on the type of lay and the material of manufacture of the mooring ropes in question. The mariner should be aware that the type of rope used for the stopper is critical and the following points should be borne in mind:

1   Use a natural fibre stopper for natural fibre ropes.
2   Use a synthetic fibre stopper for synthetic fibre ropes.
3   The stopper material should be a low-stretch material.
4   When synthetic rope stoppers are used the material should have a high melting point, e.g. polyester.
5   The stopper should be flexible.
6   Never use nylon stoppers on nylon ropes (polyamide).

The size of the rope for the stopper will vary with the type of stopper being applied, either common or West Country. In the case of the West Country stopper the size of the rope should be as near as possible to 50 per cent breaking strain of the rope it is being applied to. Table 3.3 shows sizes for the West Country stopper. The size of cordage for common stoppers should be of a sufficient equivalent.

Common Rope Stopper

This may be used on natural fibre or synthetic fibre ropes provided they are of a hawser lay. The stopper should be examined for wear and tear before use, and if there is any sign of deterioration, the stopper should be renewed. The mariner should ensure that the stopper is secured, then pass a half-hitch against the lay of the rope; the bight of the stopper between the shackle and the half-hitch should be seen to be taut. Many seafarers pass a double half-hitch (forming the first part of a rolling hitch), instead of just using the single half-hitch. The tail of the stopper is then turned up with the lay of the rope and held while the weight is transferred.

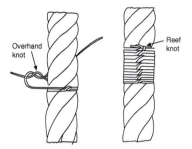

Figure 3.29 West Country whipping.

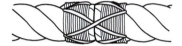

Figure 3.30 Marrying two ropes together.

Table 3.3 Rope sizes for a West Country stopper	
Diameter of mooring rope (mm)	**Diameter of stopper rope (double) (mm)**
40	20
60	32
72	36
80	40

Figure 3.31 *(a)* Common rope stopper, and *(b)* West Country (Chinese) stopper.

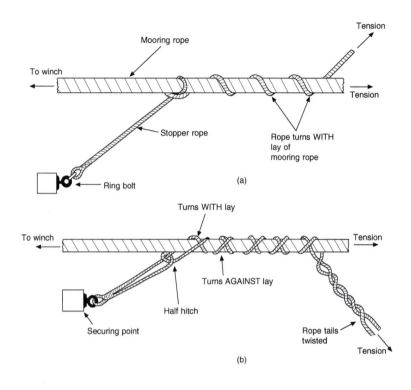

Although used extensively for mooring, the stopper is often found useful in derrick handling and towing operations (see Figure 3.31).

West Country (Chinese) Stopper

This stopper is for use on man-made fibre ropes of either a hawser or multi-plait lay. Before use, the stopper should be carefully examined for any signs of deterioration. Although most man-made fibre ropes are water-resistant, they are subject to powdering between the strands with excessive use.

The stopper is formed (Figure 3.31) of two tails of equal length secured to the base of the bitts. The tails should be half-hitched under the mooring rope to be stoppered off, and then criss-crossed on opposite sides (top and bottom) of the mooring rope. It is important to note that in the first cross of the stopper the tail nearest the rope is with the lay. When the second cross is put in on the reverse side of the mooring rope, this same tail is not the tail nearest the rope.

The criss-crossing of tails is continued about five times, then the tails are twisted together to tension up the stopper about the mooring rope.

To Pass a Chain Stopper

Chain stoppers are used for the same purpose as the common or West Country stoppers, except for the fact that they are applied to mooring wires, not ropes (see Figure 3.32).

The chain stopper consists of a length of open link chain, about 1.7 m, with a rope tail secured to the end link. The chain is shackled to the base of the bitts or to a deck ring bolt of convenient position.

The stopper is passed over the wire, forming an opened cow hitch, followed by the remainder of the chain, which is turned up against the lay. The rope tail is also turned up in the same direction, then held as the weight comes onto the stopper.

The two half-hitches of the cow hitch are kept about 25 cm (10 in.) apart. The mariner should be aware that a cow hitch is used and not a clove hitch; the latter would be liable to jam, whereas the cow hitch is easily pulled loose when no longer required. The turns of the chain are made against the lay of the wire so as not to open it up and cause distortion, and also weaken the wire.

Breaking Out Mooring Ropes

The large coil will be rotated on the swivel and turntable in the opposite direction to that in which the rope was manufactured, e.g. a right-hand laid rope should be rotated anticlockwise. The rope itself will be hauled off from the outside of the coil, flaked in long flakes down the length of the deck, then coiled down on stowage grates. A tight coil can be achieved by first starting the coil off with a cheese, then building up the coil from the outside and working inwards to the centre (see Figure 3.33).

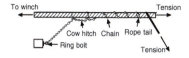

Figure 3.32 To pass a chain stopper; the example shows the stopper being passed on left-hand-laid wire.

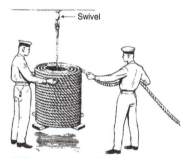

(i) Opening a new coil of small rope, from the centre of the coil.

(ii) Opening a new coil of large rope.

Figure 3.33 Breaking out a mooring rope.

4

Wirework and Rigging

Introduction

Rigging comes in many forms and has been generic to the shipping industry over thousands of years. It was easy to recognize on the tall ships, but in today's modern shipping very few vessels carry a tall mast with shrouds and stays. The style of rigging has changed to more specific needs, such as gangway requirements, cargo securings, survival craft lashings, cargo hoists and ramps, etc.

Steel wire ropes (SWRs) have become predominant, and with the industrial change away from cordage to the flexible and non-flexible wires, different skills have developed. The use of 'bull dog grips' (wire rope grips) is common, while the wire eye splices have given way to the 'ferral' and the 'Talurit clamp'.

New and improved methods continue to be sought, but in a worldwide industry the mariner needs to be capable of dealing with both the new and the old methods, wherever he may find the practice of seamanship. Early ideas still work alongside the modern concepts, and examples can be readily seen on a day-to-day basis. A crane wire built with non-rotational properties is still a crane wire, but a vastly improved wire from the original swivel and hawser laid wires. Jute hearts in flexible wires are still available, but working alongside wires with nylon hearts. Both centre cores perform the same task for flexibility and encouraging anti-corrosion within the wires.

Steel Wire Rope

An SWR is composed of three parts – wires, strands and the heart. The heart is made of natural fibre, though recently synthetic fibre has been used when resistance to crushing is required. With the many changes in the marine industry, the requirements of wire rope have altered considerably from the early production days of 1840. Then the first wire ropes, known as selvagee type ropes, were constructed of strands laid together then seized to form the rope.

Modern ropes are designed with specific tasks in mind (Table 4.1), and their construction varies accordingly. However, all wire ropes are affected by wear

Table 4.1 Steel wire rope

Type	Construction	Uses
Steel wire rope (SWR)	6 × 6 6 × 7 7 × 6	Standing rigging
Flexible steel wire rope (FSWR)	6 × 12 6 × 18 6 × 19 6 × 24	Running rigging
Extra flexible steel wire rope (EFSWR)	6 × 36 6 × 37	Running rigging, where safety of life is concerned

and bending, especially so when the ropes are operated around drum ends or sheaves. Resistance to bending fatigue and to abrasion require two different types of rope. Maximum resistance to bending fatigue is obtained from a flexible rope with small outer wires, whereas maximum resistance to abrasion needs a less flexible rope with larger outer wires.

When selecting a wire rope, choose a wire which will provide reasonable resistance to both bending fatigue and abrasion. The wire should also be protected as well as possible against corrosive action, especially in a salt-laden atmosphere. Where corrosive conditions exist, the use of a galvanized wire is recommended.

All wires should be governed by a planned maintenance system to ensure that they are coated with lubricant at suitable intervals throughout their working life. Internal lubrication will occur if the wire has a natural fibre heart, for when the wire comes under tension, the heart will expel its lubricant into the wires, so causing the desired internal lubrication.

If synthetic material is used for the heart of a wire, this also acts to reduce corrosion. Being synthetic, the heart is impervious to moisture; consequently, should the rope become wet, any moisture would be expelled from the interior of the wire as weight and pressure are taken up.

A comparison of the strengths of fibre ropes, wire ropes and stud link chain is provided in Table 4.2, which gives formulae for breaking stresses.

Construction of Steel Wire Rope

SWRs are composed of a number of thin wires whose diameter will vary between 0.26 and 5.4 mm. The thinner wires are made of hard-drawn plough steel and the thicker wires of rolled steel. The individual wires are twisted into strands about a fibre core or a steel core, or even laid up without any form of centre heart.

These strands are in turn laid up about a fibre or steel heart, or just laid up together without any centre core. The direction of

Table 4.2 Formulae for breaking stresses

Type/material	Size	Factor
	Natural fibre ropes	
Grade 1 Manila	7 mm to 144 mm	$\dfrac{2D^2}{300}$
	Synthetic fibre ropes	
Polypropylene	7 mm to 80 mm	$\dfrac{3D^2}{300}$
Polythene	4 mm to 72 mm	
Polyester (terylene)	4 mm to 96 mm	$\dfrac{4D^2}{300}$
Polyamide (nylon)	4 mm to 96 mm	$\dfrac{5D^2}{300}$
	Flexible steel wire ropes	
6 × 12	4 mm to 48 mm	$\dfrac{15D^2}{500}$
6 × 24	8 mm to 56 mm	$\dfrac{20D^2}{500}$
6 × 37	8 mm to 56 mm	$\dfrac{21D^2}{500}$
	Stud link chain	
Grade 1	12.5 mm to 120 mm	$\dfrac{20D^2}{600}$
Grade 2	12.5 mm to 120 mm	$\dfrac{30D^2}{600}$
Grade 3	12.5 mm to 120 mm	$\dfrac{43D^2}{600}$

Diameter 'D' expressed in millimetres.

Breaking stress expressed in tonnes.

laying up the wires and laying up the strands is critical. If the wires are laid in the same direction as the strands, then the hawser is said to be a 'flat strand hawser', whereas if the wires are laid up in the opposite direction to that of the strands, the wire is said to be a 'cross-laid hawser'.

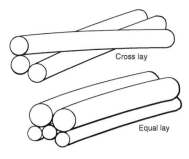

Figure 4.1 Rope lay.

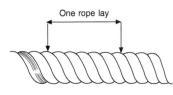

Figure 4.2 Length of lay.

Ordinary Lay

A rope of ordinary lay is one where the direction of lay of the outer layer of wires in the strands is opposite to the direction of lay of the strands in the rope. Most wire ropes are laid right-handed, but left-handed ropes may be obtained (Figure 4.1).

Variations in the length of lay (see Figure 4.2) will alter the elastic properties of the rope. For example, shortening the length of lay will increase a rope's elastic stretch properties but will reduce its breaking strain.

Equal Lay

In this type of construction the wires of a strand all have an equal length of lay. Consequently, contact between wires is of a linear nature (see Figure 4.1).

Lang's Lay

This construction is one where the outer layer of wires in a strand is the same direction as the lay of the strands of the rope. Like ordinary lay, Lang's lay is generally found as a right-handed laid rope, but may be manufactured as a left-handed one as well (Figure 4.2). It offers a greater wearing surface and can be expected to last longer than an ordinary laid rope, especially when used in work where resistance to wear is important. However, a rope of Lang's lay construction has a low resistance to unlaying, and it is usually restricted to applications where both ends of the wire are secured against rotation.

Wirex

A lay of wire laid up like a sennet laid rope, as a multi-plait. It is used extensively for crane hoist wires and lifeboat falls for its non-rotational properties and will not easily unlay once weight comes off the load.

Cross Lay

A cross lay construction is one in which the wires in successive layers of the strand are spun at approximately the same angle of lay (Figure 4.1). It follows that the wires in successive layers make point contact. Where ropes are operating over sheaves and drums, nicking of wires together with secondary bending at these points of contact occur, and failure of the wires due to early fatigue may result.

Spring Lay

The form combines galvanized wires with tarred sisal. Six ropes of tarred three-stranded sisal are inlaid with three strands of wire, each containing 19 wires per strand, and the whole is laid about a central fibre heart.

Its strength is not as great as ordinary wire rope, but the advantage of spring lay is that it is easily handled and coiled, and is about three times as strong as a grade 1 manila rope of equivalent size. Similar combination ropes are in extensive use within the fishing industry, but when encountered on merchant vessels they tend to be employed for mooring or towing springs, as they have a very good shock resistance.

Seizing Wire

This wire is usually seven-stranded, having six single wires laid about a seventh wire of the same size. It is in general use aboard most vessels where additional strength is required over and above that provided by a fibre seizing. It is specifically used for mousing shackles, marking anchor cable, etc. where the bearing surface is metal.

The general practice of good seamen is not to use seizing wire for the purpose of mousing hooks, since the shape of the bill of the hook may allow the mousing to slip off. To this end a fibre mousing is more common, either in spunyarn or other similar small stuff, depending on the size of the hook (see Figure 4.18).

Lubrication

SWRs are lubricated both internally and externally in the course of manufacture, to provide the wire with protection against corrosion. During its working life the rope will suffer pressure both externally and internally as it is flexed in performing its duty. The original lubricant may soon dry up and it will be necessary to apply supplementary lubricant at periodic intervals.

Main Core (Heart)

Within the shipping industry the majority of SWRs of the flexible nature are equipped with a hemp or jute natural fibre heart. The non-flexible wires are usually built up about a steel core. The natural fibre heart is impregnated with grease to supply internal lubrication when the rope comes under tension. An updated idea is to use a nylon heart. This acts to expunge any moisture inside the lay and so avoid corrosive activity inside the wire. External coating to these wires is the only form of lubrication.

Preforming

Preforming is a manufacturing process which gives the strands and the wires the helical shape they will assume in the finished rope. Preformed rope has certain advantages over non-preformed:

- it does not tend to unravel and is less liable to form itself into loops and kinks, making stowage considerably easier;

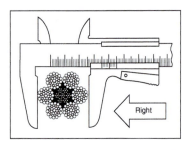

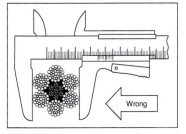

Figure 4.3 Measuring steel wire rope: by diameter of circle enclosing all strands.

- it is slightly more flexible and conforms better with the curvature of sheaves and drums;
- it provides reduced internal stresses and possesses a greater resistance to bending fatigue.

When cutting preformed wire rope, it is not essential to whip the bight either side of the intended cut, though it is good practice to do so. Whippings should be applied to all non-preformed wires when they are to be cut.

Measurement

This is carried out by the use of a rope gauge (see Figure 4.3).

Steel Wire Rope Rigging

Standing Rigging

This will be of 6 × 7 (six strands, seven wires) construction, or, with a steel core, 7 × 7 construction. For larger sizes 6 × 19 or 7 × 19 may be encountered. Examples in use would be the shrouds to port and starboard of the mast, forestay, backstay, triatic or what used to be called jumper stay, ships' wire guard rails, etc. (see Figure 4.4).

In standing rigging the wire is non-flexible, and under normal circumstances it is a permanent fixture of the vessel in that it does not or will not be moving at any time. There are exceptions to this, e.g. preventer backstays to a mast when operating a heavy lift derrick, a ship's guard rails being removed to allow access.

Running Rigging

These are flexible ropes of 6 × 12, 6 × 18, 6 × 19, 6 × 24, 6 × 36 or 6 × 37 construction. The number of wires per strand (WPS) may be as many as 91, but these ropes are generally confined to heavy industry, such as launching slipways, towage and salvage operations, as opposed to the normal working marine environment.

Running rigging examples may be seen in lifeboat falls, topping lifts for derricks and cranes, etc. As a general description, any wire, or cordage for that matter, passing over a sheave or about a drum may be classed as running rigging (see Figure 4.5).

Forestay

This is a wire stay secured to the mast table and running forward to the forecastle deck. It is usually made either of 'iron wire rope' (7 × 7) or 'steel wire rope' (7 × 7 or 7 × 19), and secured by a rigging screw in the forepart.

MOORING LINES

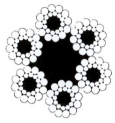

6 × 24

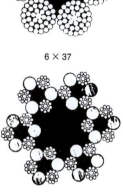

6 × 37

6 × 36

6 × 41
(Steel core)

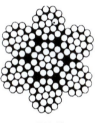

6 × 3 × 1
(Spring lashing wire)

STAYS AND SHROUDS — STANDING RIGGING

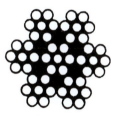

7 × 7 (Steel core)

7 × 19

CARGO LASHING WIRE

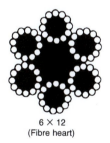

6 × 12
(Fibre heart)

Figure 4.4 Varieties of wire rope for mooring, standing rigging and cargo lashing.

It is now no longer common practice to use iron wire rope as the masts of modern vessels accommodate cargo-handling gear and the load stresses could be too great.

Topmast Stay

Topmasts are now only to be found on the tall (sailing) ships. This is a steel wire running in the fore and aft line which may be secured

CARGO-HANDLING GEAR – RUNNING RIGGING (all with fibre heart)

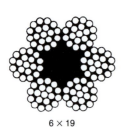

6 × 19

6 × 24

6 × 37

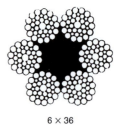

6 × 36

DECK CRANES

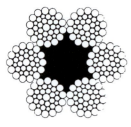

6 × 36

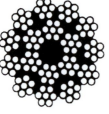

17 × 7

DERRICK TOPPING LIFTS

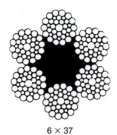

6 × 37

LIFEBOAT FALLS

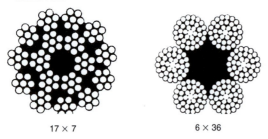

17 × 7   6 × 36

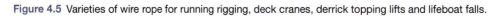

Figure 4.5 Varieties of wire rope for running rigging, deck cranes, derrick topping lifts and lifeboat falls.

either forward or aft, depending on the position of the topmast. It is secured to the hounds band of the topmast at one end, and at the other end at deck level with a rigging screw. Construction is 6 × 7 or 7 × 19.

The mariner should be aware that for maintenance purposes, or to allow greater access for cargo working, the forestay and the topmast stays are often removed; if carried, the construction and size of these stays make it essential that when the rigging screws of either stay are removed, the weight of the stay is taken up by a handy billy or equivalent purchase. The handy billy is rigged from the stay to a convenient deck bolt so as to take the weight off the rigging screw, allowing its removal and the controlled release of the stay by easing out the handy billy purchase.

Jumper Stay (obsolete)

An old-fashioned stay now very rarely seen, the jumper stay fell out of use when the signal stay and triatic stay became popular. The name 'jumper' was acquired from the early sailing ships, which carried a block secured to this stay; a single whip passed through the block and light bagged cargo was discharged by securing one end of the whip to the bags, while the other was held by a seaman who 'jumped' down to the quay from the ship's side. This action caused the cargo to be raised and discharged overside – not a very practical method by today's standards, but effective for the day.

The stay was made of steel, six strands being laid about a steel core right-handed. The jumper stay, when fitted as such, extended from the after mast to the foremast. This later wire rope standing rigging, usually of 12–16 mm diameter, gave way to the fitting of what is now termed the signal stay, which ran from the foremast to the funnel, leaving the jumper stay to become a so-called continuation from the aft side of the funnel to the after mast, until eventually it became obsolete.

Triatic Stay

This is a stay that runs from the foremast to a secure position over the monkey island/bridge. The stay was often secured to the funnel on many of the three-island type vessels, but for the stay to be effective it was necessary to reinforce the thin metal work of the funnel with a doubling plate. This reinforcement allowed the tension to be taken in the triatic stay, and permitted that stay to work in the desired direction of opposition to the forestay.

Several of the old 'Kent' class cruisers of the Royal Navy secured the triatic stay between the fore and after masts, in the place of the jumper stay. This stay was made of 1.5 in. (12 mm diameter) flexible SWR, a construction not to be expected on a merchant vessel, where the common type of triatic stay is standing rigging of a 6 × 6 or 6 × 7 nature.

Shrouds

The function of shrouds is to provide the mast with staying support to port and starboard. They are secured under the mast table (if fitted) or to the hounds band of the mainmast structure. They are now given as wide a spread as possible without interfering with derrick or crane operations, to provide not only athwartships support but a fore and aft support as well. The smaller vessel often has shrouds made of round iron bar, but the more general construction material is SWR, 6 × 7, laid about a steel core.

The number of shrouds fitted varies with the vessel's length and deck design, and with the height and expected stresses that the mast will be required to handle. Three or four shrouds are common, and more recently, where four shrouds are fitted, they are secured in pairs, providing longitudinal as well as athwartships support. When an odd number of shrouds are required, the single shroud, which is left unseized, is known as a swifter, the double pairs of shrouds being seized under the mast table.

All shrouds are secured at deck level by a rigging screw shackled into a hard eye, with a solid thimble set into each end of each shroud. The solid thimble accommodating the pin of the shackle prevents movement of the shackle within the eye of the shroud, and provides a certain amount of rigidity about the shroud. The rigging bottle screw is usually secured against movement by being locked with a locking bar or locking nuts, which prevent accidental unscrewing of the bottle screw.

Preventer Backstay

This is a general term used for describing stays that act in opposition to the general direction of weight. They are extensively used when a 'jumbo' heavy lift derrick is in use; they are then temporarily rigged to support the additional weight that the mast will bear.

The term is also used for stays having a lead opposite to the forestay, which are often led to port and starboard as with the normal shrouds. Their construction is the same as for standing rigging, and they are secured by rigging screws at deck level (see p. 128).

Rigging Fitments

Blake Slip

This is used mainly in anchor work aboard naval vessels. It is a rigid style of slip not having the versatility and hence not as popular as the senhouse slip (see Figure 4.11).

Conical Sockets

Nearly all shrouds and stays will be secured by use of a rigging bottle screw, but it is a point worth mentioning that the rigging screw (Figure 4.6) will usually be secured to the stay by one of two methods: the hard eye, with the solid thimble, is still in common use, but with better facilities ashore for engineering the use of the 'open conical socket' and the 'closed conical socket' is becoming more and more popular (Figure 4.7). The wire is sweated into the socket by heat treatment of about 300°C, at which point the socket expands and accepts the wire.

Pad Eye

The pad eye (Figure 4.8) is a common fitment aboard most modern vessels; it may be riveted, bolted or welded to the bulkhead or deck.

Quick Link

The quick link supplies a quick method of joining or repairing chains. The advantage of this link is that it may be released even when under tension, as the stress is bearing at the ends of the links, not about the screw thread.

Bottle screws (Turnbuckles)

Rigging bottle screw – employed for standing rigging, manufactured in mild steel

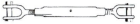

Solid thimble

Stretching screw – mild steel, jaw and bolt fittings

Stretching screw (Admiralty pattern) – galvanized, mild steel

Senhouse slip

Awning screw – galvanized steel

Figure 4.6 Bottle screws.

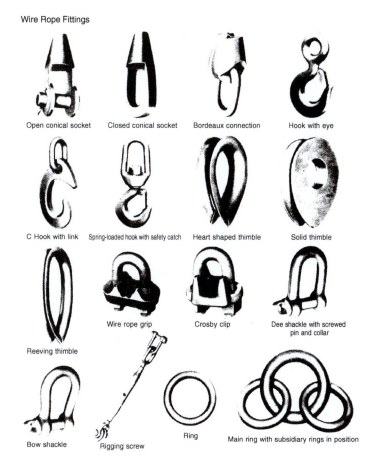

Wire Rope Fittings

Open conical socket   Closed conical socket   Bordeaux connection   Hook with eye

C Hook with link   Spring-loaded hook with safety catch   Heart shaped thimble   Solid thimble

Reeving thimble   Wire rope grip   Crosby clip   Dee shackle with screwed pin and collar

Bow shackle   Rigging screw   Ring   Main ring with subsidiary rings in position

Figure 4.7 Wire rope/rigging.

Figure 4.8 Pad eye.

Quick link – long series, wide mouth

Quick link – standard

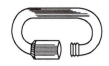

Figure 4.9 Quick link.

Figure 4.10 Screw eye bolt.

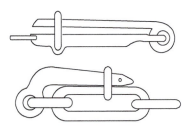

Figure 4.11 Blake slip (*above*) and senhouse slip (*below*).

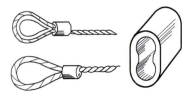

Figure 4.12 Talurit clamp.

The disadvantage is that the link cannot be used with chain made from bar which has a larger diameter than that of the link itself. However, it is manufactured in various sizes, in mild steel zinc-plated and in stainless steel (see Figure 4.9).

Screw Eye Bolt

A screw eye bolt is secured by passing the bolt through a bulkhead or deck and fixing a locking nut from the other side. It is popular as a temporary rigging fitment (Figure 4.10).

Senhouse Slip

The Senhouse slip (Figure 4.11) is a common fitment aboard most modern vessels and is encountered in a variety of sizes. Examples in use may be seen securing the web straps over life rafts or in the gripes securing lifeboats.

Talurit Clamp (ferral)

Tests have shown that use of the Talurit clamp (Figure 4.12) is probably the strongest method of putting an eye in the end of a wire. The method is most certainly stronger than an eye splice or a socket.

The ferrule, whose passages have a diameter corresponding to that of the wire rope, is fitted into position loosely about the two parts of wire. If a hard eye is required, then a thimble would be inserted at this stage. The ferrule is then compressed about the wire by a very powerful Talurit (press) machine. The disadvantage of this method is that the machines are expensive and will accommodate only up to certain size of wires. For practical purposes it is therefore a method which is generally employed as a shoreside occupation.

Thimbles

The purpose of a thimble (Figure 4.7) is to protect and reinforce the eye of the rope or wire. There are three main types.

The *open heart thimble* is probably the most widely used within the marine industry, generally in conjunction with a shackle or ring bolt. Its freedom of movement in all directions depends on the size of the shackle and the thimble.

When set into the end of a wire or rope, its construction forms what is known as a hard eye, as opposed to a soft eye, when the thimble is not in position. When inserting thimbles to make a hard eye, a tight splice will be required to retain the thimble in position. A slack splice would allow the thimble to become slack and soon drop away (see Table 4.3 for cordage).

The *round thimble* is generally found spliced into natural fibre cordage, forming a hard eye. The shape of the round thimble does not lend itself to being spliced into steel wire.

Table 4.3 Cordage table for use with open heart thimble

Rope diameter (mm)	A (mm)	C (mm)	D (mm)	G (mm)	K (mm)	Q (mm)
8	22	13	33	4	4	30
9	25	14	38	6	5	35
10	29	18	41	8	5	38
11	29	18	41	8	5	38
12	29	18	41	8	5	38
13	32	21	44	8	6	43
14	32	21	44	9	6	43
16	41	22	59	9	8	57
18	44	29	67	10	8	60
19	51	29	73	11	10	70
20	51	29	73	11	10	70
22	57	32	83	13	10	76
24	64	33	92	13	10	84
26	70	35	108	14	10	91
28	76	38	111	16	13	102
32	95	41	133	16	13	121
35	105	48	152	19	16	137
36	105	48	152	19	16	137
38	114	54	165	24	18	149
41	114	56	165	24	18	149
44	127	57	178	25	25	178
48	133	67	191	29	29	191
51	140	70	203	30	29	197
54	140	70	203	30	29	197
57	146	76	216	32	30	206
64	159	95	241	44	32	222
70	203	121	273	60	41	286

Examples in use may be encountered in the painter of a ship's lifeboat, shackled down to a ring bolt in the forepart of the bow-sheets. Another use is found in the eye of a lizard when employing stages. The heart thimble has gained in popularity because its shape is more suited to form the eye, but the round thimble has the same function of protecting the cordage.

The *solid thimble* is used only in standing rigging, such as stays and shrouds. The solid thimble does not allow any movement of the pin of the shackle connected to it, and so provides a more rigid standing structure.

Triangular Plate

Referred to by seamen as a 'union plate' or as a 'monkey face plate', it is used extensively in the rigging of two derricks in a union purchase

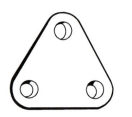

Figure 4.13 Triangular plate.

rig, taking the place of a union hook or Seattle hook. The cargo runners are shackled directly onto the plate to form the union of the whips. It is also used on a single span topping lift, providing an attachment for the chain preventer and the bull wire (see Figures 4.13 and 5.2).

Wire Splicing

There are many types of wire splices possible, but these days the short, long and cut splices are rarely seen. The eye splice, of which there are several variations, is by far the most common.

A locking splice is considered to be the most desirable for the majority of tasks encountered at sea today. The lock is formed by tucking two tails under one strand, but in opposite directions so that they cross under the tucked strands. The 'Boulevant Splice' is described in the following text.

The setting-up process of any wire splice may be compared to the foundation of a building. If it is not set correctly, the construction of the splice will very quickly fall apart.

Depending on the type of wire being spliced, whether it is of the preformed variety or not, the strands will unlay a certain amount once lengths of wire are cut from the reel or coil. Always whip the bare ends to be left on each side of the proposed cut, so preventing wastage from the coil. The strand of any wire *not* of the preformed (see p. 95) type will tend to unlay. Non-preformed wire will unlay when cut or allowed, with its bare end, to be used without a whipping.

A whipping should also be set on the wire at the point the splice is to begin. The required size of the eye should be made in the bight of the wire and both parts of wire should then be seized securely together. If a thimble is to be inserted, so making a hard eye, then this should be placed inside the bight of the eye before applying the seizing. The thimble is secured in position by light seizings about the crown and shoulders of the thimble (Figure 4.14).

The end of the wire should then be unlaid, exposing the heart. Each tail unlaid should be whipped at the end to prevent the individual wires from unlaying and distorting the strand. Each tail (unlaid strand) should be unlaid back as far as the seizing joining the two parts of wire forming the eye. The heart should not be cut away at this stage as it will be useful later in splice construction, apart from the fact that a neater splice is obtained by tucking the heart at the commencement of the splice.

Figure 4.15 shows a wire eye splice, having a lock in the first set of tucks. Number 2 and 6 tails will form the lock of the splice, and the subsequent construction will be followed by two full tucks of all six tails (unlaid strands). These tails will then be halved and two subsequent half tucks will then complete the splice.

Once the wire has been set up in the vertical position in the vice, identification of the tails may be made by pulling the heart between the two tails which are closest to the standing part of the wire. The two tails are then identified as 1 and 6, either side of

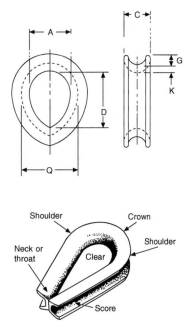

Figure 4.14 Thimble.

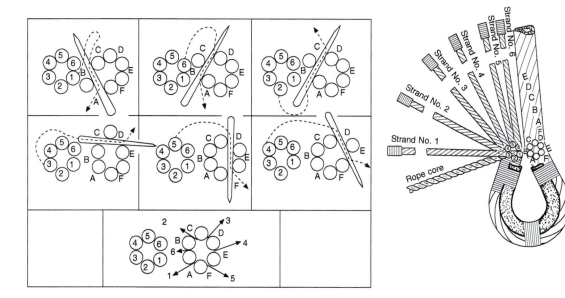

Figure 4.15 Construction of the Boulevant wire eye splice.

All wire ropes will require some preparation before splicing. Where a hard eye is to be constructed the thimble must be seized into position first. All the separate tails should have their ends whipped to prevent them unlaying during the splice operation. The standing part of the wire should normally be lashed high in the vertical to allow the seaman to work more easily around the wire. If a heavy wire is being spliced, a Spanish windlass may be required to draw both parts of the wire together around the thimble. Once both parts are closed up, then a strong seizing should join the two above the thimble position. The heart should not be cut from the wire until the splice is complete.

the heart. The remaining tails from 1 in a clockwise direction may be identified as 2, 3, 4 and 5, to end at number 6, next to the heart.

Insert the marline spike under the strand in the standing part, which is closest to the number 1 tail, and as near to the thimble as possible. Care should be taken not to over-spike the wire and cause excessive distortion. Tuck the heart under the lifted strand, then tuck the number 1 tail through the same opening but on top of the heart. Both the heart and tail 1 should be tucked in the direction of the tapered end of the spike (with the lay).

Where the heart and number 1 tail exit from the standing part identifies the opening 'A' of the wire. By moving clockwise, the openings B, C, D, E and F may also be noted. The following is the sequence of tucking the tails for the first tuck:

Tail number	Enters at	Exits at
1	B	A
6	C	B
2	B	C
3	C	D
5	D	F
4	D	E

NB. Some administrations no longer accept hand-spliced wires aboard the vessel and expect all wires to have 'Talurit clamp' eyes.

It should be borne in mind that the first tails tucked, namely 1 and 6, should be tucked in the direction of the taper of the marline spike, with the lay, whereas all subsequent tails should be tucked from the direction of the handle of the spike, against the lay. Although a difficult thing to see from the sequence of tucks on paper, in practice the mariner will clearly see that the tails 6 and 2 form the lock under strand 'B'.

After the first tuck has been completed, it should be tightened by pulling down on each tail separately. A gentle tapping of the tails to knit into the strands with the handle of the marline spike will not do any harm, but care should be taken not to overdo this, as broken wires may result.

It will be observed that after the first tuck each tail exits from the standing part from a separate opening. The splice is continued by tucking each tail over the adjacent strand and under the next – over one, under one, so to speak, as in fibre rope splicing. The second tuck is completed as each tail is tucked twice, this procedure being continued for a third and final full tuck. Each tail is tucked a total of three times.

When tucking the tails, it is important that the tails themselves do not become crossed. This can be avoided by tucking the tail on the inside, not the outside, of the next tail to be crossed.

After three complete full tucks have been inserted, and each tuck tightened up, two half tucks should be inserted to taper and complete the splice. The taper is carried out by splitting each tail in half, separating the wires of each tail into two equal sections. One section will be seen to be a tighter lay than the other, and this half of the tail is the one to be used in making the half tuck. The other will be cut off. Each tail is split in the same manner, then spliced as for the second tuck, 'over one, under one'. The operation of splitting the strands effectively reduces the cross-sectional diameter of the splice, and the two half tucks produce the desired tapered effect.

After the locking tuck, two full tucks, and the two half tucks have been inserted, and each tuck tightened up on completion, the half tails left should be chopped off as close to the standing part as possible. A cold chisel on a firm surface is effective for chopping the tails. The heart should then be 'vee'd' (cut in a V on either side of the strands next to the heart).

Stowage of Wire Hawsers

Wire hawsers are generally stowed about the cylindrical centre part of a reel specifically designed to accommodate the size of wire (Figure 4.16). Often the reels are equipped with a gearing ratio to allow easier manual operation; and, strictly speaking, if the reel is geared, it is a winch in the true sense of the word. However, it is common practice for seamen to refer to geared or ungeared drums simply as reels.

The wire should never be shackled to a reel, as this practice would prove extremely hazardous should weight come onto the

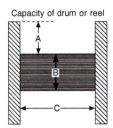

Capacity of drum or reel

Figure 4.16 Size of stowage reels. The following formula may be used to work out the rope capacity of any size drum or reel. It will produce reasonably accurate results for evenly spooled wire rope, but less accurate results for wire not so spooled.

$$\text{Formula}: \frac{A}{d} \times \frac{C}{d} \times \pi(A + B) = \text{Capacity}$$

where d = diameter of wire.

The flange A will extend beyond the outer layer of rope, and the dimension A should be taken to the outside of the rope only, not to the outside of the flange.

wire at the bare end while it was still secured to the reel. The result could well be that the reel could be torn from its deck mounting by the weight on the wire, causing excessive damage and possible serious injury. Wire hawsers should be secured by a light fibre lashing (small stuff) which would part easily under strain without ripping the reel from the deck (see Safe Handling Procedures for Wire Rope, below).

Safe Handling Procedures for Wire Rope

1 Wire ropes should be regularly treated with suitable lubricants.
2 Wires should never be used directly from the stowage reel, unless the reel is purposely designed to allow direct operation; for if the wire should foul when running, the reel and its supporting frame may be torn from the deck mountings, causing serious injury to seamen.
3 Sufficient slack should be taken from the reel to provide adequate length to cover all contingencies. Should any doubt exist as to the amount of slack that will be required, then all the wire should be removed.
4 When wires and ropes are under strain, perhaps in a towing operation, persons should stand well back in a position of safety.
5 Sharp angles on wire rope leads should be avoided.
6 When awkward leads are the only alternative when handling wires, as with cargo winch use, then snatch blocks should be suitably sited and secured sufficiently to prevent them from breaking loose.
7 When wire ropes are turned up about bitts, then the top turns should be secured against springing off, by a light lashing.
8 When wire rope is to be joined to a fibre rope, then the fibre rope should be fitted with a thimble, so preventing chafing and wear about the eye.
9 Wire ropes should never be led across fibre ropes and allowed to cause chafing.
10 Chain stoppers should be used on wire ropes.
11 Wire on drum ends should not be used as check wires.
12 Should a wire rope be used as a slip wire, then the parts of the eye should be seized together and reduced in size to allow passage through the ring of the mooring buoy.
13 When breaking out a new coil of wire, care must be taken and a turntable used whenever possible (Figure 4.17).

Mousing a Hook or Shackle Pin

The purpose of mousing a hook is to prevent the object being lifted or the hook from breaking adrift. Small stuff, for example

Coil

Turntable

Figure 4.17 Opening a new coil of steel wire rope. A two-man job, where one man will flake out the wire down the deck while the other will ensure that the turntable rotates steadily. The turntable method is popular at sea because when not in use the table lends itself to easy and convenient stowage.

If unreeling, pass a shaft through the coil of wire and pull out the wire, flaking it down the length of the deck before coiling or running it onto a stowage reel. A second man should control the rate of rotation of the coil with this method.

Should no turntable or reel arrangement be available, then no attempt should be made to unravel the wire while keeping the coil flat on the deck. This method will only result in the wire becoming excessively kinked. Roll the coil down the deck, allowing the wire to fall off in flakes.

spunyarn, is seized about the back of the hook and around the bill, effectively closing off the clear.

The objective of mousing the bolt or pin of the shackle is to prevent the shackle working itself free when in normal regular use. Seizing wire is used because of its robust character. It is passed inside the clear and through the end of the bolt in a 'figure eight', so preventing withdrawal of the pin (see Figure 4.18).

Blocks: Care and Maintenance

Before loading or discharging with ship's gear, inspect all blocks carefully, paying special attention to the following:

1 *Swivel head fittings.* Examine the nut or collar of the shank to ensure that it is securely fastened and free from any visible defect. The shank should be checked for distortion and be seen to turn freely by hand. Any clearances should not be excessive. Grease or oil the swivel fitment.

2 *Binding.* Examine the side (binding) straps for fractures or corrosion. Ensure that the block number and SWL are distinctive.

3 *Side or partition plates.* Check that there is no distortion or buckling that would allow the wire to jam between the cheeks and sheaves.

4 *Sheaves.* Examine for cracks in the metal and check that the bush is not slack in the sheave or causing excessive wear on the axle pin. Each sheave should be seen to turn freely by hand. Sheaves worn in the groove could cause excessive ropewear and should be checked whenever a new rope is to be rove.

5 *Axle pins.* Check axle pins for wear, and ensure that they do not rotate. They should be securely held in position by a holding nut. If a split pin is passed through, it should be frequently renewed.

6 *Lubrication.* Regular lubrication of all moving parts must be carried out or the life of the block and the efficiency of the rig will be impaired. Lubricants must be adequate, and any old

congealed lubricants should be removed before applying fresh grease or oils.

7   *Protection.* Blocks may be painted, provided that grease nipples are not covered or moving parts choked sufficiently to impair their function. Reference marks should be left clear.

Ordering Blocks

1   Provide the title of the block, e.g. snatch block.
2   Provide the safe working load (SWL) of the block.
3   Provide detail of internal or external binding.
4   State the diameter of the sheave.
5   State the size of wire or rope to be used with the block.
6   State what type of fitments, if any, are required.
7   State the purpose for which the block is to be employed and give a brief description of the material of manufacture, e.g. wood, metal, galvanized, etc.
8   State whether a test certificate is required.

Wood Blocks

Wood blocks are generally restricted to the tall (sailing) ships. There are many wooden blocks available on the commercial market, but their popularity has waned with the production of a stronger, more practical and cheaper metal version.

The *internal bound wood block* (Figure 4.19) comprises a wood shell built around the metal binding, which is fork-shaped and passes through the crown. This binding will have a hook or eye attached and will be left exposed over the crown, while a becket, if fitted, will protrude from the arse of the block.

The wooden shell, as with the majority of wood blocks, is made of elm, the sheaves being manufactured in brass, phosphor bronze or galvanized steel. The old-fashioned sheave used to be made of lignum vitae, but these are rarely seen on a modern vessel.

The *external iron bound block* is similar in construction to the one above, except that the binding is exposed about the shell of the wood block. The fitments may be fixed or swivel hook, or eye at the crown, an eye becket being an optional addition.

The *common block* is a very old-fashioned block and as nearly obsolete as one may encounter. It was made entirely of an elm wood shell strengthened by either a single or double strop. The strops were completely served over, with a round thimble, seized, over the crown of the block.

The *snatch block* is a single sheave block made in wood or metal. The wooden snatch block will have an internal binding, but is rarely seen in practice, having given way to metal snatch blocks capable of taking a natural fibre rope.

Other types of wood block include solid turned clump, built clump, funnel block, and the double and triple sheave blocks used for lighter lifting operations (most of which are obsolete).

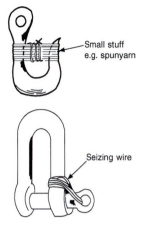

Figure 4.18 Mousing a hook or shackle pin. A spring-loaded tongue is often found on hooks which are in continual use instead of the temporary mousing, e.g. cargo hooks. (*Below*) Mousing a shackle.

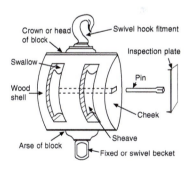

Figure 4.19 Twofold, internal bound, wood block.

Information on inspection plate: safe working loads of block (*a*) when used as an upper block and (*b*) when used as a lower block; size of rope or wire for use with the block; certificate number (block number); date of certificate of testing the block.

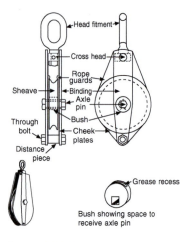

Figure 4.20 Metal block.

Metal Blocks

The *single sheave cargo block* (Figure 4.20) finds a common use at the head and heel positions of the derrick. The difference between the two is that the head block will be fitted with an oval swivel eye, and the heel block will have the duck bill fitment, to prevent toppling when in the gooseneck and fouling the runner. The head block is often referred to as a gin block, but this is not strictly correct. Calling the head block a gin probably stems from the days of coaling ships, which regularly employed a gin as a head block.

There are many types on the commercial market, one of the most common being referred to as a 'Z' block. They may also be found in topping lift purchases or the lifting purchase of a derrick.

All metal cargo blocks are now stamped on the binding with the SWL and certificate number, together with the name of the manufacturer. The block usually contains a roller bearing sheave, fixed to rotate between the cheeks about a phosphor bronze bush. The bush is held in position by the square-shaped axle pin. Sheaves are often of a self-lubrication design, having grease reservoirs cut into the bearing surface. The bush may also have a similar cut away for use as a grease cavity.

Double and treble sheave purchase blocks are of a similar construction to the single sheave, except that a partition plate separates the sheaves. They are used extensively for the heavier cargo work, being rigged in purchases for use as steam guys on direct heavy lifts. They are also in common use on 10-tonne SWL derricks, being incorporated in the lifting purchase.

The *multiple sheave cargo blocks* are used exclusively for heavy lift derricks, forming the lifting purchase and topping lifts. The number of sheaves will vary according to the design of the lifting apparatus; up to ten sheaves is not an uncommon sight when regular heavy lift work is being carried out. The following information relates to a ten-sheave block built for Costain John Brown Ltd:

- Sheaves: cast steel to take 27 mm diameter wire
- SWL of block: 102 tonnes
- Block tested to: 153 tonnes
- Weight of blocks (per pair): 6.5 tonnes (approx.).

Block Head Fittings

Designed solely for fitting into the gooseneck of a derrick, the *duck bill eye* (Figure 4.21) is attached to the heel blocks of derricks to prevent the block from toppling and fouling the cargo runner.

Probably the most popular and most practical of all head block fittings, the *oval eye* (Figure 4.22) is used for all sizes of blocks from small wood blocks up to heavy lift blocks. The length of the oval allows easy access for shackles, together with freedom of movement and secure holding.

The *round eye* may have a swivel attached or just be used as a fixed ring, generally on the smaller wood blocks rather than the larger cargo blocks (Figure 4.23).

Figure 4.21 Duckbill eye.

Figure 4.22 Oval eye.

Figure 4.23 Round eye.

A flat-sided fitment designed to accommodate supporting lugs on either side, the *stud eye* (Figure 4.24) is in common use with the heavier type of cargo blocks.

Block head fittings for heavy lift gear are shown in Figure 4.25.

Blocks and Tackles

Figure 4.24 Stud eye.

The term block and tackle, or purchase as it is sometimes called, refers to the two blocks together with the wire or rope rove between them. It is a common mistake made by young seafarers and others to refer to just a block as a tackle or pulley. It should be clearly understood that the tackle is the combination of the cordage passing over a sheave contained within the block; a second block is suspended by the standing and running parts of the cordage.

The term tackle (pronounced taykle) is illustrated (Figure 4.27) to show the hauling part as a downhaul, coming from the standing block. Where the hauling part comes from the standing block, as shown, the purchase is said to be used to disadvantage. The opposite situation is to use a purchase to advantage; Figure 4.26 shows the difference when the hauling part is led from the moving block as opposed to the standing block.

When we refer to advantage or disadvantage of a tackle, we are referring in both cases to the advantage that the purchase can give to the person carrying out the work. This may be more clearly understood if the purchase is assumed to be a simple and very basic machine which assists to move the work load.

The way the purchase is rove will affect the amount of effort that will be required to achieve the desired movement of the load.

Obviously, when any weight is lifted, either with a tackle or on a single part of rope, stresses in the cordage cause the weight to move. To this end it is possible to calculate the amount of stress in the hauling part from the formula:

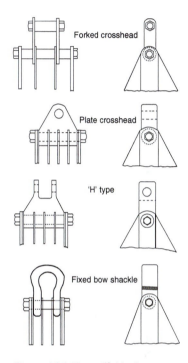

Figure 4.25 Heavy lift block fitments.

$$S = \frac{w + \dfrac{N \times W}{10}}{P}$$

where S = Stress in hauling part; N = Number of sheaves; W = Weight being lifted; P = Power gained by tackle.

Examples of the use of this formula may be found on pp. 142–145.

Tackles: advantage and disadvantage (Figure 4.26)

When a (double luff) tackle is used to advantage as illustrated, the hauling part is seen to be led from the floating or moving block. To calculate the power gained by using this tackle to advantage, a practical method can be applied by counting the moving parts of cordage, including the hauling part, which run between the blocks.

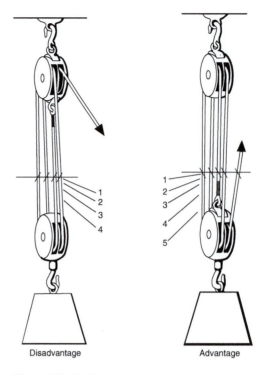

Figure 4.26 Tackles.

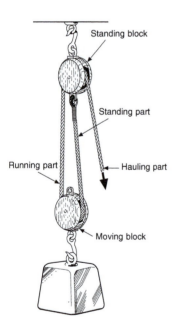

Figure 4.27 Gun tackle.

In this example the power gained is 5. The same double luff tackle is now used to disadvantage, a more common method of use at sea. Note that the hauling part is seen to lead from the standing block. When calculating the power gained in this instance, exclude the hauling part. In this example the power gained is 4.

Gun Tackle

This is a tackle comprising two single sheave blocks, originally used aboard naval vessels for hauling the guns back into position after they had recoiled on firing – hence the name. The two blocks are each fitted with a hook attachment. Because of its historic background the gun tackle was generally used for hauling a load in a horizontal direction, but there is now no reason why the gun tackle should not be used in the vertical (Figure 4.27). The power gained, or velocity ratio, by using this type of purchase will be either 2 or 3, depending on whether it is used to disadvantage or advantage.

Handy Billy

This tackle is composed of a double sheave tail block and a single sheave hook block, rove together with a small fibre rope.

The cordage should not be greater than 16 mm in circumference. The name 'handy' comes from the tackle's light weight and transportability, and because the tail attached to the double sheave block allows easy securing in convenient positions.

Jigger

This is a similar tackle to the handy billy, except that the cordage rove between the blocks is of a greater size – 20 mm in circumference – and the size of the blocks is increased accordingly. It is, therefore, a heavier-duty tackle.

Luff Tackle

This is a purchase of similar construction to the handy billy, or jigger, regarding the number of sheaves. The main differences are that the double block has no tail – instead each block has a hook or swivel becket eye – and the cordage may be up to 24 mm in diameter or similar size of wire. The power gained will be either 3 or 4, depending on whether it is used to disadvantage or advantage (see Figure 4.28).

Figure 4.28 Luff tackle.

Double Luff Tackle

Composed of two double blocks, this tackle is often referred to as a double purchase. It is a general-purpose tackle, having a velocity ratio of 4 or 5, depending on whether it is used to disadvantage or advantage (see Figures 4.26 and 4.29).

Gyn Tackle

This purchase (Figure 4.30) comprises a double and treble block, with a cordage or wire fall rove between them. The standing part is secured to the double block. The tackle produces a power gain of 6 or 5, depending on whether it is used to advantage or disadvantage.

Figure 4.29 Double luff tackle.

Threefold Purchase

A heavy-duty tackle comprising two triple sheave blocks with a rope or wire fall rove between both blocks, this purchase is used extensively in heavy lift work for both topping lift and lifting purchase (see Figure 4.31). There are two methods of reeving the threefold purchase, one with the sheaves of both blocks in the same plane and the second, more popular, method with the plane of the sheaves in each block at right angles to each other. The advantage of the latter is that when the lift is made, the lower block hangs vertically without toppling over to one side.

> Virtually all wooden blocks have been replaced by metal blocks in modern-day commercial shipping.
>
> Tackles, where used, are still employed with metal bound blocks but rigged with SWR, not cordage.

Figure 4.30 Gyn tackle.

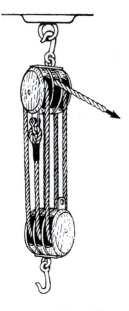

Figure 4.31 Threefold purchase.

Chain Blocks

There are several types of chain block in general use, the most common being (a) spur-geared blocks, (b) lever and ratchet and (c) wormwheel operation. They are often referred to as chain hoists, having a mechanical advantage of between 5 and 250. The lifting capability will be variable, but their use for up to 40 tonnes is not uncommon practice. Although they are usually found operating from inside the machinery spaces of vessels, for numerous duties they may be employed on deck.

With spur-geared blocks a manual drive chain turns a through spindle via geared cog wheels. A ratchet and pawl system is also incorporated so that the load may be held suspended from the load chain. This load chain is held by a sprocket arrangement which is driven by the operation of the through spindle.

The lever and ratchet types, generally used for lighter work, are smaller and permit optional positioning wherever they are required. They are usually equipped with a reversible pawl system which allows the ratchet wheel to be turned in operation in either direction.

In wormwheel operation (Figure 4.32) an endless operating chain passes over a flywheel that causes an axle fitted with a worm screw to rotate. The worm screw engages with the helical teeth of a larger gear wheel, causing the load sprocket to turn and heave on the load chain. The load chain may be led through a floating block to increase the purchase effect of the machine or, as in Figure 4.32, be led direct from the load sprocket to the lifting hook. Nearly all these types of chain block incorporate a braking system that allows the weight being lifted to be suspended.

Weston's Differential Purchase

Let us find the mechanical advantage (see Figure 4.33). Consider a load, W, being raised by the effort, P. Each of the chains A and B support ½W (½weight). Take moments about centre C. Let radii of large sheave be represented by R and small sheave by r. Then:

$$\frac{1}{2}W \times CD = (P \times CF) + \left(\frac{1}{2}W \times CE\right)$$

By transposition of the above equation:

$$P \times CF = \left(\frac{1}{2}W \times CD\right) - \left(\frac{1}{2}W \times CE\right)$$

Substitute radii R and r:

$$PR = \frac{1}{2}W(CD - CE)$$

$$PR = \frac{1}{2}W(R - r)$$

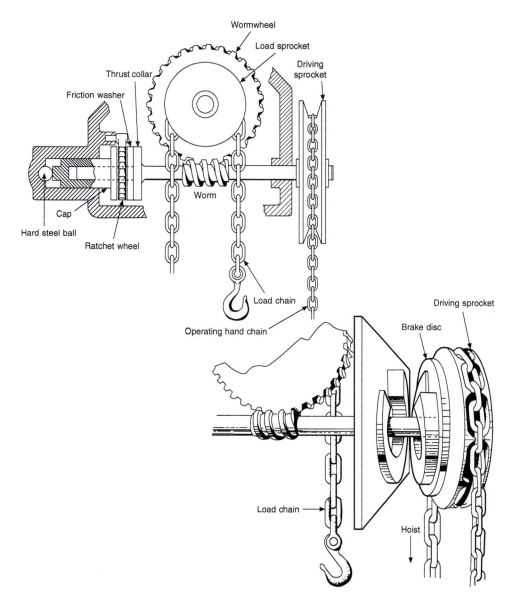

Figure 4.32 Chain block.

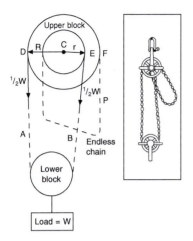

Figure 4.33 Weston's differential purchase. The upper block consists of two sheaves of different diameters, secured together. An endless chain (*right*) is rove between the upper and lower blocks. The links of the chain engage in the snug and under the rims of the sheaves, and cannot slip, so eliminating any frictional loses.

Transpose:

$$\frac{1}{2}W = \frac{PR}{(R-r)}$$

$$\frac{\frac{1}{2}W}{P} = \frac{R}{(R-r)}$$

$$\frac{W}{P} = \frac{2R}{(R-r)}$$

But:

$$\frac{W}{P} = \frac{Load}{Effort} = \text{Mechanical advantage (MA)}$$

$$\therefore MA = \frac{2R}{(R-r)}$$

Instead of radii R and *r* being used, the number of links which can be fitted round the circumference of the upper block sheaves may be substituted, as they are in proportion to the radii of the sheaves.

5
Lifting Gear

Introduction

Lifting gear has become a major casualty of modernization. Derricks have become nearly a memory, being superseded by the more versatile deck crane, container gantry cranes, and of course the ramps for unit load systems. However, some ships still retain the basic derrick rigs, often incorporated as stores derricks, or oil pipe handling gear, for tanker vessels.

Although specialized cranes, with increased safe working load (SWL) capacities dominate the general cargo market and are features of the specialized carrier, they are not in isolation of other styles of operation. Heavy lifts are still transported, and project cargoes have generated the need for the heavy lift barge and/or the specific heavy lift ships.

Cranes and derrick rigs may often work in tandem to maximize the upper load limits, and the modern ideas are designed for multiple use, e.g. double pendulum stülken derricks, capable of working two hatches as opposed to a single hold; twin cranes, each of 300 tonnes SWL rigged in tandem, to provide 600 tonnes SWL, while a single gantry will load the length of a container vessel.

With this background it might seem that stresses in wires and purchases are becoming superfluous to the ship's officer. If the officer could guarantee sailing on one type of ship, with one method of lifting, this could well be the case. However, few seafarers stay with only one ship throughout their career.

Derricks

The most widely used derricks in the marine industry are of a welded structure, consisting of either three or five welded sections of tubular steel. Wooden derricks, which generally lifted only up to three tonnes, have been superseded.

At the heel of the derrick (Figure 5.1) either a single flange or a double flange will be welded to permit attachment to the gooseneck, a through bolt passing between the gooseneck arrangement and the flange(s) of the derrick. This bolt, once secured, is guarded by a washer and split pin holding, or, in the case of heavy lift derricks, by a shallow nut and split pin. The bolt is not subjected to lateral

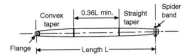

Figure 5.1 Derrick arrangement.

forces and the split-pin securing is generally an adequate method of retaining the bolt in position.

At the head of the derrick a spider band is fitted to permit the attachment of guys, topping lift and lifting purchase. This band is forged in one piece, the lugs not being allowed to be electrically welded to the band. Some heavy lift derricks have the spider band manufactured so that the lugs opposite each other are attached to a yoke piece that passes through the derrick, providing additional strength in working.

It is often the case when a derrick has a fairly considerable length, say 12 m or more, that derrick guides are fitted to prevent the cargo runner from sagging. These guides may take the form of a fixed hoop welded to the derrick, or they may be provided with a cast-iron roller. These rollers should be regularly maintained or they may cause undue chafing on the cargo runner wire.

Single Swinging Derrick

The function of the derrick is to raise, transfer and lower weights. In the shipping industry this effectively means moving goods from the quay to the vessel or vice versa.

The derrick boom is supported at the heel in a pivot arrangement known as the gooseneck, which allows elevation by means of a topping lift span. The topping lift may be of the nature of a single span or a purchase; either way the downhaul is led from the spider band of the derrick via the masthead span block (high upper support) to a convenient winch. Figure 5.2 shows a single span secured to a union plate, which also accommodates a chain preventer and a bull-rope. The bull-rope is a continuation of the downhaul for the purpose of topping or lowering the derrick.

The derrick is positioned to plumb the load by slewing the boom from port to starboard by means of a slewing guy secured on either side of the spider band. Slewing guys come in two parts, namely a cordage tackle (wire in the case of heavy lift derricks) secured to a wire guy pendant, which is shackled at the derrick head.

The derrick may be equipped with a lifting purchase or a whip (single) cargo runner. In either case, once the derrick has been plumbed at the correct height for the load, the topping lift is secured, and the downhaul of the lifting purchase is led to the winch via the derrick heel block.

Many vessels are provided with dolly winches for the sole purpose of topping and lowering derricks. Dolly winches are usually fitted with a safety bar device and leave the main cargo winch to handle the lifting purchase or runner. Other types of dolly winch are operated from the main winch, in which case combined use of topping lift and lifting purchase is not possible, the dolly winch having to be disengaged to allow separate operations to be carried out.

When the derrick is rigged in the single swinging mode, the topping lift is secured and the actual height of the derrick does not change. However, the bull-rope may be replaced by a luff tackle, with the consequence that the topping lift effectively becomes the

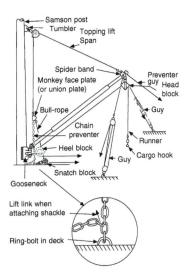

Figure 5.2 Single swinging derrick: single span topping lift with chain preventer.

downhaul of the luff tackle. If this is led to the winch direct, then the derrick is turned into a luffing derrick. With this method of rigging a second winch will be required to operate the cargo runner.

Topping a Single Span (Topping Lift) Derrick

1 Assume the derrick to be in the lowered position, secured in the crutch. Collect the chain preventer from its stored position, together with two tested shackles, a snatch block, seizing wire, marline spike and wire preventer if the derrick is to be rigged for union purchase.
2 Obtain power on deck and remove the cargo runner from the main barrel of the winch.
3 Secure the slewing guys to the spider band and stretch them to port and starboard.
4 Shackle the cargo working end of the runner to the deck, so as not to end up with the eye of the runner at the derrick head when topped.
5 Secure the bull wire to the winch barrel (assuming no dolly winch system) via the snatch block.
6 Let go the derrick head lashing or crutch clamp, and man the guys.
7 Lift the derrick clear of the crutch (float the derrick) and pass the wire preventer over the derrick head if for use with union purchase.
8 Heave on the winch, topping the derrick until the union plate (monkey face plate) is down to the snatch block.
9 Shackle the chain preventer onto the union plate, mousing the shackle.
10 Come back on the winch, lowering the derrick to the required height. Secure the chain preventer when the derrick reaches the desired working height. When shackling the chain preventer to the deck lug bolt, ensure that the shackle is clear of the next link of the preventer, so as not to foul and cause the rig to jump when under load. Mouse the shackle.
11 Remove the bull-rope from the winch and secure hand-tight about the mast cleats. This bull-rope will now provide a back-up to the chain preventer.
12 Secure guys once the derrick is slewed to the desired position.
13 Secure the cargo runner once more to the main barrel of the winch.

Topping a Derrick: Topping Lift Span Tackle

1 Assume the derrick (Figure 5.3) to be in the lowered position, secured in the crutch. Obtain lead block, chain stopper, marline spike, rope yarns and wire preventer guy if the derrick is to be used in union purchase rig.
2 Obtain power on deck and remove the cargo runner from the barrel of the winch.

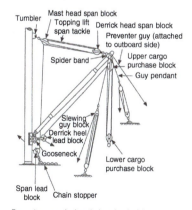

Preventer guy only rigged when the derrick is to be worked in union purchase

Figure 5.3 Single swinging derrick: topping lift span tackle.

3 Secure slewing guys to the spider band and stretch them to port and starboard.

4 Shackle the cargo working end of the runner to the deck, so as not to end up with the eye of the runner at the derrick head when topped.

5 Take the weight of the topping lift downhaul by passing a chain stopper round it. Lead the downhaul of the topping lift via a lead block onto the main barrel of the winch. Take the weight of the wire on the winch and remove the chain stopper.

6 Remove the derrick head lashing or crutch clamp, and man the guys.

7 Lift the derrick clear of the crutch and pass the wire preventer over the derrick head for use with union purchase.

8 Top the derrick up to the desired working height by heaving on the topping lift downhaul.

9 Pass the chain stopper on the topping lift downhaul once the derrick is at the required working height and the winch is stopped.

10 Ease back on the winch until the weight comes onto the chain stopper.

11 Remove the topping lift downhaul from the winch and secure it hand tight about the mast cleats. This operation should be carried out while the weight is on the chain stopper. Once completed, the stopper can be removed. When turning the wire up on to the mast cleats, make three complete turns before adding the four cross turns, the whole being secured with a light rope yarn lashing.

12 Provided a lead block is used for the downhaul of the topping lift, and not a snatch block, there is no need to remove the block from the way of the wire.

13 Secure slewing guys once the derrick is plumbed correctly, and also the cargo runner to the main barrel of the winch.

Union Purchase

This is by far the most popular rig using two derricks. It is a fast and efficient method of loading or discharging cargo. The derrick may be used in a single swinging mode when not employed in a union purchase rig, so providing versatile cargo handling over a considerable range of cargo weights.

The rigging of the union purchase rig (Figure 5.4) is arranged by plumbing the inshore derrick over the quayside, while the second derrick is plumbed over the hatch area containing the cargo. The two cargo runners are joined together at a triple-swivel hook, known as a union hook, or often referred to as a Seattle hook (Figure 5.5). The two derricks are held in position by slewing guys, which, once the derricks are plumbed correctly, are secured so that the derricks will not be allowed to move. The operation is carried out by the weight of the load being taken by one derrick and transferred via the cargo runners to the second derrick (Figure 5.5).

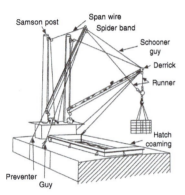

Figure 5.4 Union purchase rigged with schooner guy. For clarity, guardrails, etc. have been omitted.

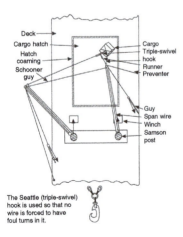

Figure 5.5 Union purchase rig (plan view).

Plate 23 Derricks rigged in union purchase either side of the amidships masthouse. All derricks are employing schooner guys and supported from Samson Posts.

It should be noted that the derricks do not move throughout the whole operation. The only moving parts are the two cargo runners led to winches.

The stresses that come into play when working this rig are considerable because of the angles made with the cargo runners. As a rough guide one-third of the SWL of the derricks may be taken as a working weight, e.g. five tonnes SWL of derricks, then 1.6 tonnes may be considered the SWL of the union purchase rig.

The union purchase rig has several variations, the main one being in the distribution and position of guys (Figures 5.4 and 5.5). An advantage with the schooner guy is that there is a saving of cordage, as only three guys are used to secure the rig, while with crossed inboard guys the total is four slewing guys.

When rigging derricks for the union purchase rig, each derrick should be topped in the normal manner (see pp. 119 and 120). The exception to this is when the schooner guy is fitted: then both derricks should be topped together, with the tension being kept on the schooner guy to prevent them splaying apart as they rise. For the operation of topping derricks with the schooner guy, more manpower is obviously required to top both derricks at once.

Preventer guys, not to be confused with slewing guys, should be passed over the derrick heads once the derricks have been floated from their crutches.

Preventer Guys

Preventer guys are to be fitted in addition to slewing guys, and their safe SWL should not be less than that indicated in

Table 5.1 Safe working load

SWL of derrick rig (tonnes)	Required SWL of each slewing guy (tonnes)
1	1
2	1.5
3	2
4	2.5
5	3
6	3.25
7 to 9.5	3.5
10 to 12.5	3.75
13 to 15	4
16 to 60	25 per cent of SWL of derrick rig
61 to 75	15
more than 75	20 per cent of SWL of derrick rig

The above table may be considered a guide only when vessels are at suitable angles of heel and trim. Under certain conditions, when additional slewing guys are attached to the lower cargo purchase block, a permitted reduction in SWL of guys is tolerated.

Table 5.1 or as found by parallelogram of forces of the rig, which-ever is the greater.

Preventers should be made of wire rope, or wire and chain con-struction, and attached to the derrick separately from the slewing guys. Deck eye plates should be so positioned as to prevent excessive guy tension building up, while keeping the working area clear for the passage of cargo slings. Preventers should be secured by use of shackles through the chain link to the eye plate on the deck, or if all wire preventers are being used, then securing is often obtained by 'ferrules' fused onto the wire at regular intervals and held by a pear link arrangement.

The preventer should be rigged with an equal tension to that of the slewing guys on the outboard side of both derricks. Should the rig become over-strained in any way, then the slewing guy will be allowed to stretch, being cordage, whereas the preventer will bear the weight and not give, being of wire or chain construction. An even tension on preventer and outboard guy is attained by securing both these guys first, and then taking the weight on the inboard guy of each derrick in turn.

Slewing Guys

Slewing guys are generally constructed in two parts: a guy pennant of steel wire rope shackled to a cordage tackle. This provides a limited amount of elasticity, allowing the guy to stretch and avoid parting under normal working conditions.

Table 5.1 is a guide to the safe working load of guys in respect of safe working loads of derrick rigs. When rigging derricks in union purchase, slewing guys and preventer guys should never be secured to the same deck eye bolt but to separate anchor points.

Safe Handling Practice for Derricks

1   All derrick rigging should be regularly maintained under a planned maintenance programme, and in any event should be visually checked for any defect before use.
2   Before a derrick is to be raised, lowered or adjusted with a topping lift span tackle, the hauling part of the topping lift should be flaked down the deck clear of the operational area. All persons should be forewarned of the operation, and to stand clear of the bights of the wire.
3   When topping lifts are secured to cleats, bitts or stag horns, three complete turns should be taken before the additional four cross turns on top. A light lashing should be placed about the whole to prevent the natural springiness of the wire causing it to jump adrift.
4   When the rig of a derrick is to be changed or altered in any way, as with doubling up, then the derrick head should be lowered to the crutch or to deck level in order to carry out alterations safely.
5   When dolly winches fitted with a pawl bar are employed, the pawl should be lifted to allow the derricks to be lowered. Any seaman designated to carry out this task should be able to give his full attention to the job and be ready to release the bar should anything untoward happen in the course of the operation. Under no circumstances should the pawl bar be wedged or lashed back.
6   Winch drivers should take instructions from a single controller, who should pass orders from a place of safety from which a clear and complete view of the operation must be available. When derricks are being raised or lowered, winch drivers should operate winches at a speed consistent with the safe handling of the guys.
7   Cargo runners should be secured to winch barrels by use of a 'U' bolt or proper clamp, and when fully extended a minimum of three turns should remain on the barrel of the winch.
8   Should it be necessary to drag heavy cargo from 'tween decks the runner should be used direct from the heel block via snatch blocks to avoid placing undue overload on the derrick boom.

Safe Handling Reminders for Union Purchase Rig

1   To avoid excessive tension in the rig the safe working angle between the married cargo runners should not normally exceed 90°, and an angle of 120° should never be exceeded.
2   The cargo sling should be kept as short as is practicable to enable the cargo to clear the hatch coaming without extending the safe working angle between the cargo runners.

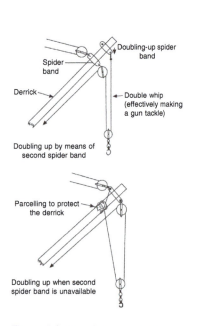

Doubling up by means of
second spider band

Doubling up when second
spider band is unavailable

Figure 5.6 Doubling up a derrick.

3   Derricks should be topped as high as practicable, and not rigged farther apart than is absolutely necessary.
4   Derricks should be marked with the SWL when rigged for union purchase. Should this not be the case, then the SWL should not be more than one-third the SWL of the derrick itself.
5   Preventer guys of adequate strength should be rigged on the outboard side of each derrick and secured to the deck in the same line and with similar tension as the slewing guy. However, they must be secured to separate pad eyes to the eyes which accommodate the slewing guys.

Doubling-Up Procedure

The cargo runner of a derrick may be doubled up when it is desired to make a lift which the rig is capable of handling safely but which exceeds the SWL of the cargo runner when rigged as a single whip (see Figure 5.6). Some derricks are equipped with a second doubling-up spider band, but this is not the case with every derrick. Obviously the doubling of the runner, making a double whip, is made very easy when the second spider band is fitted. The eye of the runner is shackled to the second band, leaving a bight between the head block and the shackled eye. A floating block is secured in the bight, effectively making the arrangement into a 'gun tackle'. Should the derrick not have the convenient second spider band, then it will be necessary to parcel the derrick with canvas and take a half-hitch with the runner around the derrick, taking the eye of the runner and securing it to the lug on the spider band that accommodates the topping lift. This effectively produces a similar bight in the wire for the floating block, as previously described.

When doubling up in this manner it will be appreciated that a snatch block used in the bight would be much simpler to rig, but it would not be as safe as an ordinary cargo block. This will necessitate the reeving of the block before completing the half-hitch about the derrick.

The half-hitch is prevented from riding down the derrick by the retaining shackle to the spider band and also by the wire biting into the parcelling that affords the derrick some protection. Once the load is off the cargo hook, the tension in the half-hitch is relieved, but owing to the weight of the wire and the floating block, it would be unlikely for the hitch to slip against the natural forces of gravity.

Yo-Yo Rig

This rig is sometimes referred to as a block in bight rig, and may be employed with two or four derricks. The purpose of the rig is to allow the loading or discharge of heavier loads than those which can be handled by the more popular union purchase rig or by a single swinging derrick.

With Four Derricks

This is probably the most popular of the two yo-yo methods (Figure 5.7). The derricks, once rigged for union purchase, do not have to be adjusted. The two cargo runners of the inboard derricks are passed through a floating block, and the two outboard derrick runners are passed through a second one. The separate pairs of runners are shackled together, as are the floating blocks, to form the union, the cargo hook being secured under the floating blocks.

The lifting operation can be started once the guys have been tightened up. The winch operators should be warned beforehand that, with the runners being shackled together, the joining shackles may run foul of the derrick head block or the floating block in the bight of the runners. An experienced winchman will identify the limits of the wire runner by marking the wire to indicate the extent the runner may be paid out or heaved in without fouling the blocks.

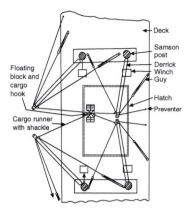

Figure 5.7 Yo-yo rigged with four derricks.

With Two Derricks

This rig uses the two inshore derricks. Each derrick in Figure 5.8 has been rigged with a gun tackle, and the moving blocks have been joined by a heavy strop supporting a floating block with a cargo hook attached. The operation of loading or discharge is carried out by slewing both derricks towards the quayside, trying to keep both the derrick heads as close together as is practicable.

The advantage of the floating block with the strop is that, should the rig suffer a winch failure, the full weight of the load will not come to bear on one derrick.

Hallen Universal Derrick

This probably represents one of the most successful advances in lifting gear over the last 20 years. The many advantages of this type of derrick make it a very popular choice with shipowners (see Figure 5.9).

The derrick is labour saving as it can be operated by one man. The lifting capacity may be up to 200 tonnes, through a working radius of 170°, being topped up to 85°. It is an extremely stable rig, being supported by either a straight mast or a Y-style mast. Stabilizing outriggers provide superior leads for the slewing operation over the greater working area. These outriggers, a recent innovation, have completely superseded the D-frame design of the early 1960s.

The complete operation of the derrick rig is handled by one man positioned at a control console. A joystick control allows topping and lowering, together with slewing to port and starboard, and a second lever operates the lifting purchase hoist. The guys of the conventional derrick design have virtually been eliminated in this design. The topping lifts have a double function of slewing the derrick as well as controlling the elevation. The topping lift wires take up to 75 per cent of the load and so provide greater safety aloft.

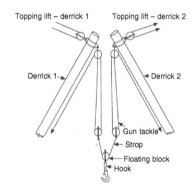

Figure 5.8 Yo-yo rigged with two derricks.

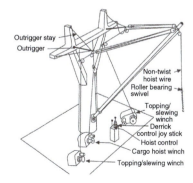

Figure 5.9 Hallen universal swinging derrick.

The whole design and operation may be compared to that of a crane, inclusive of built-in limit switches that prevent overslew and overtopping. Variations in the reeving of the topping lifts have occurred since the D-frame type, the slew tackles having been replaced by an endless fall rove to function as a conventional topping lift.

Velle Shipshape Crane

This is a derrick system (Figure 5.10) which has become increasingly popular over the last decade. The boom is fitted with a 'T' shaped yoke at its extremity for the purpose of fitting four short steel wire 'hangers'. This bridle arrangement allows very wide slewing angles because the topping lift falls act to aid recovery when the derrick is slewed outboard. The yoke also provides the securing points for the two hoist wire leading blocks. The separation between the leading blocks allows a sympathetic motion between the load on the hook and the derrick head and so reduces pendulous swinging of the load.

Luffing and slewing motions of the rig are controlled by two winches, each equipped with divided barrels. The luffing winch accommodates the fall wires, being turned up onto the barrels in the same direction, so allowing both to lengthen or shorten as

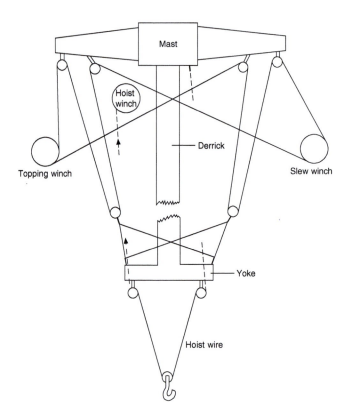

Figure 5.10 Velle shipshape crane.

desired, while the wires on the slewing winch are turned up in opposite directions. As rotation occurs one fall shortens while the other pays out, so slewing the derrick to port or starboard.

The advantages of this type of rig are that cargo-handling speed can be increased as the derrick can engage in luffing and slewing operations at the same time while under full load. It has also been shown to be a very stable rig in operation, being controlled by a single operator using a joystick lever control similar to the Hallen derrick.

Heavy Lift Procedures

Before beginning a heavy lift operation the officer in charge should make sure that the lift can be carried out in a safe and successful manner. Depending on the load to be lifted, the vessel can be expected to heel over once the lift moves off the fore and aft line. Therefore, heads of departments should be given ample warning of an expected list before the operation begins.

The ship's gangway should be lifted clear of the quayside, and all fore and aft moorings tended, to ensure no damage is incurred by the heeling angle of the vessel. The critical times are when the load is overside and the vessel is at maximum angle of heel, and once the load is landed and the vessel returns to the upright position.

The vessel's stability should be thoroughly checked before starting the operation, with particular regard to free surface in tanks. When the lift is taken up by the derrick, the rise in the ship's centre of gravity (C of G) should be such that she is not rendered unstable. (The effective C of G of the load acts from the derrick head position, above the C of G of the ship, once the load is lifted.)

All rigging must be examined by the officer in charge, and any preventer backstays to the supporting mast structure should be secured in position prior to lifting the load. Correct slings should be used on the load, together with beam spreaders if required. Steadying lines should be secured to all four corners of the load, and these should be substantial enough to control oscillations when lifting from ship to quay and vice versa.

Steam guys or power guys should be rigged and tested to ensure correct leads. The lifting purchase should be seen to be overhauling, and winches should all be in double gear.

The lugs on the load itself should be checked before securing slings to ensure that they are adequate to handle the load stress. Extreme care should be taken with crated heavy objects. Shippers are known to crate loads without reinforcing the crate itself, and the possibility of having the load fall from the bottom of the crate is a real one.

Landing the load onto a truck or flat top rail car may cause lateral drag on the vehicle as the weight comes off the derrick; the vessel may return sharply to the upright position, accentuating this effect. To alleviate the situation, the offshore guy could be eased out as the load lands and the lifting purchase with the topping lift

Conventional Heavy Lift (Jumbo) Derricks

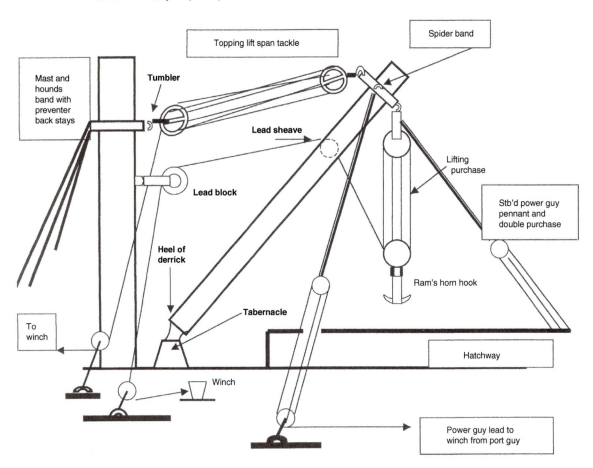

Figure 5.11 Terminology and basic working design of a conventional heavy lift, shipboard derrick found up to about 150 tons SWL.

should be veered smartly. It is essential that competent winch drivers are operating the lifting purchase and the guys, and that throughout the operation they are under the control of a single person.

The terminology of a heavy lift derrick is shown in Figure 5.11.

Rigging of a Heavy Lift (Jumbo) Derrick

This operation is generally carried out with the derrick (Figure 5.12) in the vertical position while clamped against the mast. Special lugs are secured to the mast to facilitate the raising of the topping lift blocks to the required positions. The topping lift is often left in the reeved condition, in place between the derrick and the mast, and in this case the rigging is usually protected by a canvas covering.

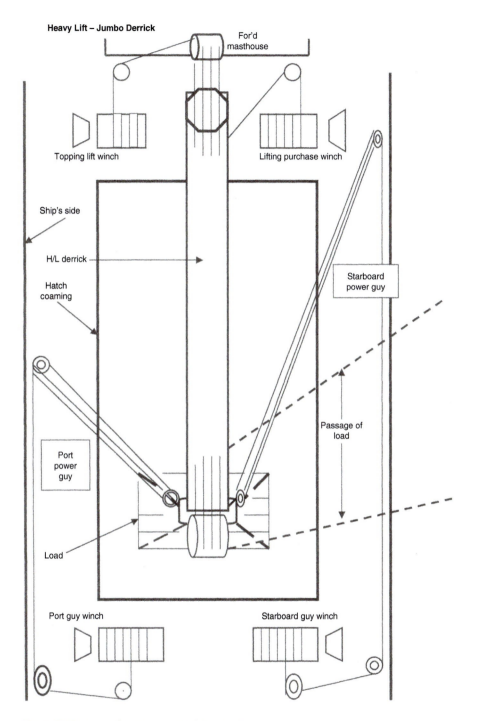

Heavy Lift – Jumbo Derrick

For'd masthouse

Topping lift winch

Lifting purchase winch

Ship's side

H/L derrick

Starboard power guy

Hatch coaming

Passage of load

Port power guy

Load

Port guy winch

Starboard guy winch

Figure 5.12 Heavy lift jumbo derrick (plan view).

Precautionary Checklist for Heavy Lift Derrick Operation

1   Carry out a 'risk assessment' prior to commencing the operation to ensure that all possible areas of hazard are taken into account and that all risks are at an acceptable, tolerable level.

2   Ensure that the stability of the vessel is adequate to compensate for the anticipated angle of heel that will be experienced when the load is at the maximum angle of outreach. All free surface elements should be reduced or eliminated if possible to ensure a positive value of 'GM' throughout the operation.

3   Any additional rigging, such as 'preventer backstays' should be secured as per the ship's rigging plan.

4   A full inspection by the officer in charge of all guys, lifting tackles, blocks, shackles and wires should be conducted prior to commencing the lift. All associated equipment should be found to be in correct order with correct SWL shackles in position and all tackles must be seen to be overhauling.

5   Men should be ordered to lift the gangway from the quayside, and then ordered to positions of standby, to tend the vessel's moorings at the fore and aft stations.

6   The ship's 'fenders' should be rigged overside to prevent ship contact with the quayside at the moment of heeling.

7   Ensure that the deck area, where the weight is to be landed (when loading) is clear of obstructions and the deck plate is laid with timber bearers (heavy dunnage) to spread the weight of the load. The ship's plans should be consulted to ensure that the limitations of the load density plan and deck load capacity are not exceeded.

8   Check that the winch drivers are experienced and competent and that all winches are placed into double gear to ensure a slow-moving operation.

9   Remove any obstructive ship's side rails if appropriate and check that the passage of the load from shore to ship is clear of obstructions.

10 Release any barges or small boats moored to the ship's sides before commencing any heavy lift operation.

11 Secure steadying lines to the load itself and to any saucer/collar connection fitment attached to the lifting hook.

12 Inspect and confirm the lifting points of the load are attached to the load itself and not just secured to any protective casing.

13 Ensure that the area is clear of all unnecessary personnel and that winch drivers are in sight of a single controller.

14 Set tight all power guys and secure the lifting strops to the hook and load, respectively.

15 When all rigging is considered ready, the weight of the load should be taken to 'float' the weight clear of the quayside (loading). This action will cause the vessel to heel over as the full weight of the load becomes effective at the head of the derrick boom.

> NB. Some lateral drag movement must be anticipated on the load and it is important that the line of plumb is not lost with the ship heeling over.

16 Once the load is suspended from the derrick and the Chief Officer can check that the rigging of the equipment is satisfactory, then the control of the hoist operation can be passed to the hatch controlling foreman.

Assuming that all checks are in order, the Chief Officer would not normally intervene with the lifting operation being controlled by the hatch foreman unless something untoward happened that warranted intervention by the ship's officer. This is strictly a case of too many cooks and could spoil a safe loading operation.

> NB. The main duties of the Chief Officer are to ensure that the vessel has adequate positive stability; this can be improved by filling double bottom water ballast tanks. Additionally, the Chief Officer should ensure that the derrick is rigged correctly and that all moving parts are operating in a smooth manner.

Stability Changes: Heavy Lifts

It is realized from the onset that once a heavy lift is taken up by a crane or derrick, the C of G of the load is deemed to act from the head of that derrick or crane jib. When calculating the ship's stability criteria, this assumption is, for all intents and purposes, like loading a weight above the ship's C of G.

Mariners who find themselves involved in ship stability calculations will appreciate that when a weight is loaded on board the vessel, a movement of the ship's 'G' will be expected. This movement (GG1), will be in a direction towards the weight being loaded. It therefore follows that once a weight is lifted and that weight effectively acts from the head of the derrick, the ship's position of 'G' will move upwards towards this point of action.

Plate 24 Heavy lift being made by floating sheer legs.

The outcome of lifting the load and causing an upward movement of 'G' is to cause G to move towards the metacentre (M). This action would be to effect a reduction in the ship's GM value (GM = metacentric height) (Figure 5.13(a)).

Once the weight of the load is taken by the ship's derrick, Chief Officers should appreciate that the ship's 'G' will rise towards 'M', possibly even rising above 'M', causing an unstable condition. It would therefore make sense to lower the position of 'G' in anticipation of the rising 'G' prior to a heavy lift being made (Figure 5.13(b)).

The Chief Officer would normally be charged with the task of ascertaining the maximum angle of heel that would affect the vessel during the period of lifting.

If the GM can be increased before the lift takes place, i.e. by filling double bottom tanks, the angle of heel can be seen to be less (Figure 5.14).

Stresses Experienced by the Derrick

When about to commence a lift operation with any derrick it must be anticipated that the lifting equipment will experience certain stresses, these are as follows.

Tensile stresses

Tensile stresses will affect the topping lift and the main purchase wire ropes. Slewing guy pennants or power guy tackles are often brought under tension, but because of their composition and indirect angles have a tendency to absorb shock loading better.

Mast, fore and aft stays, when rigged, and the shrouds to port and starboard will all experience tension. The side arms of shackles will also experience tensile forces.

Compression forces

Compression forces affect the mast and the lower supporting structure to the mast. The length of the derrick receives compressive forces acting throughout its length, especially in the area of the tabernacle/shoe and derrick heel. Steel decks can also expect to be depressed when landing heavy weights.

Shear forces

Shear forces directly affect shackle pins and axle bolts of blocks passing through sheaves. Goosenecks are also affected by shearing forces. Depending on the direction of weight, lugs in deck positions and on 'spider bands' or 'hounds bands' may suffer similar forces.

The Stülckenmast: Cargo Gear System

The heavy lift Stülcken systems are noticeable by the prominent angled support mast structure positioned to either side of the ship's

Heavy lift vessels have an extensive tank system. This allows the loading deck to be submerged to allow the load to be floated over the deck. Tanks are then emptied, causing the vessel to rise with the project cargo on deck.

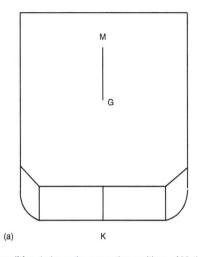

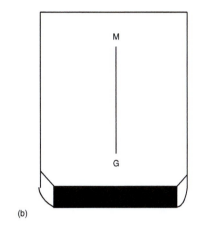

(a)   K

(b)

Condition 1 shows the respective positions of M, the Metacentre. G, the ship's Centre of Gravity, and K, the position of the keel with the vessel in an upright aspect.

Condition 2 the vessel is still in the upright, but the double bottom tanks have been filled, adding weight below 'G'. This action causes 'G' to move down and generates an increase in the ship's GM value.

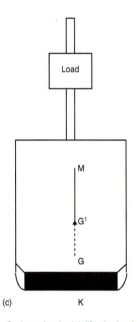

(c)   K

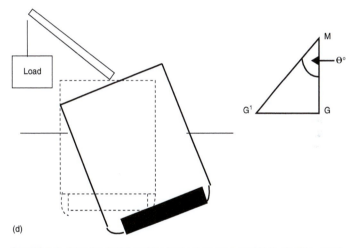

(d)

Condition 4 where the derrick and the load are swung over side causing the vessel to heel over to θ°.

Condition 3 where the derrick lifts the load on the centre line of the vessel would cause 'G' to move upwards, towards the new G1 position (vessel stays in the upright). Double bottoms full and pressed up, eliminating any possibility of free surface effects.

Figure 5.13 Balancing a ship's centre of gravity.

Figure 5.14 Small and large GM responses.

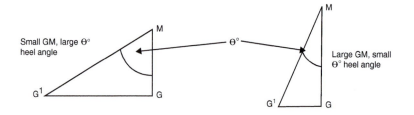

Small GM, large Θ° heel angle

Large GM, small Θ° heel angle

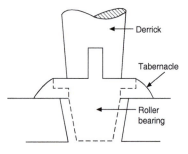

Derrick

Tabernacle

Roller bearing

Figure 5.15 Stülcken derrick. The heel is set on a tabernacle that has a roller bearing for derrick movement.

centre line. The main boom is usually socket mounted and fitted into a tabernacle on the centre line. This positioning allows the derrick to work two hatches forward and aft and does not restrict heavy loads to a single space, as with a conventional derrick (Figure 5.15).

The Stülcken posts, set athwartships, provide not only leads for the topping lifts and guy arrangement, but also support smaller five- and ten-ton derricks with their associated rigging. The posts are of such a wide diameter that they accommodate an internal staircase to provide access to the operator's cab, usually set high up on the post to allow overall vision of the operation.

The rigging and winch arrangement is such that four winches control the topping lift and guy arrangement while two additional winches control the main lifting purchase. Endless wires pay out/wind onto the winch barrels by operation of a one-man, six-notch controller.

Various designs have been developed over the years, and modifications have been added. The 'double pendulum' model, which serves two hatches, operates with a floating head which is allowed to tilt in the fore and aft line when serving respective cargo spaces. A 'ram's horn hook', with a changeable double collar fitting, is secured across the two pendulum lifting tackles. The system operates with an emergency cut-off which stops winches and applies electromagnetic locking brakes (Figure 5.16).

Plate 25 The heavy lift vessel *Super Servant 3* engages in project cargo operations by transporting the offshore barge *Al-Baraka 1*.

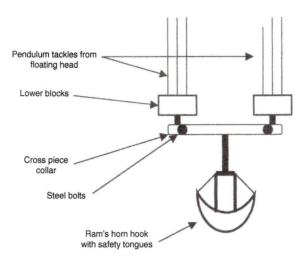

Pendulum tackles from
floating head

Lower blocks

Cross piece
collar

Steel bolts

Ram's horn hook
with safety tongues

Figure 5.16 Collar and ram's horn hook arrangement for use with double pendulum Stülcken derrick.

Stülcken derrick rigs are constructed with numerous anti-friction bearings which produce only about 2 per cent friction throughout a lifting operation. These bearings are extremely durable and do not require maintenance for up to four years, making them an attractive option to operators.

The standard wires for the rig are 40 mm and the barrels of winches are usually spiral grooved to safeguard their condition and endurance. The length of the span tackles is variable and will be dependent on the length of the boom.

Although Stülcken rigs still remain operational, their use has diminished with the improved designs of heavy lift vessels, which

Double pendulum type
capacity SWL 300 tons.

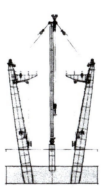

Split purchase type
capacity SWL 105 tons.

Figure 5.17 Stülcken masts for heavy-lifting operations.

Both types are manufactured by Blohm and Voss. Each system may be fitted with mast cranes or light derricks for the handling of smaller cargo.

Figure 5.18 Stülcken mast: pivot type.

Capacity SWL   250 tons
Length of derrick   30 m
Outreach   up to 14.15 m
Operation with up to 5° list and/or –2° trim

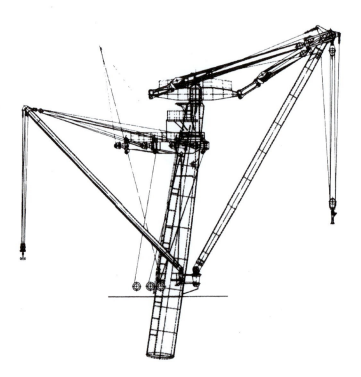

tend to have dominated the project cargo section of the industry over the last decade.

Cranes

The crane, although a standard piece of port or harbour equipment, has been incorporated aboard the modern cargo vessel with successful results. Not only is the crane a labour-saving device (only one driver per crane), but the manoeuvrability of the cargo hoist is much greater than that of a derrick (see Plate 26).

Most shipboard cranes can be fitted to swing through 360°, but for the purpose of safe handling limit switches often act as cut-outs to stop the jib of the crane fouling obstructions. Limit switches are also fitted to the luffing operation of the jib, as well as the cargo hoist wire, to prevent offsetting of the jib-boom and the cargo hook fouling the upper sheave(s) of the hoist.

All cranes are provided with individual motors to permit luffing, slewing and cargo hoist operation. They can operate against an adverse list of approximately 5° together with a trim of 2°. Twin cranes may operate independently or be synchronized to work under one driver from a master cabin.

The seafarer should be aware that there are many types of crane on the commercial market, and their designs vary with customer

Plate 26 Deck crane with safe working load of 25 tonnes, on board the general cargo vessel *Scandia Spirit.*

requirements. As a general rule cranes conform to the following design:

- *Machinery platform.* Accommodates: the DC generator; gearboxes for luff, slew and hoist operations; the slewing ring; and the jib foot pins.
- *Driver's cabin.* Integral with the crane structure. Welded steel construction with Perspex windows. Front windows to open. Internal lighting.
- *Jib.* Of welded steel construction. Supporting upper sheaves for topping lift and cargo hoist.
- *Sheaves.* Mounted in friction-resistant bearings.
- *Topping lift and hoist.* Galvanized steel wire ropes with a minimum breaking strain of 180 kgf/mm^2. Depending on the SWL of the crane, this breaking strain could be greatly increased.

Cranes and Derricks: Advantages and Disadvantages

Cranes

Advantages in use are:

- ability to plumb over the lifting point;
- single-man operation, controlling luffing, slewing and hoisting;
- straight lift means that SWL is usually adequate.

Disadvantages in use are:

- complexity of operation requires lengthy maintenance;
- SWL decreases with jib radius because the span becomes less effective as it approaches the horizontal;
- large amount of deck space required for installation.

Derricks

Advantages in use are:

- simplicity of component parts;
- ability to change rig to suit loading/discharging requirements;
- maintenance is minimal, provided that winches are good.

Disadvantages in use are:

- deck is cluttered with guyropes and preventers;
- operation usually requires two winch drivers and a hatchman;
- time delays in changing derrick rig for different cargoes.

Hallen derrick

Advantages in use are:

- simplicity of components in comparison to a crane;
- single-man operation, controlling luffing, slewing and hoisting;
- can be used against a 15° list, and can lift its full capacity down to a 15° angle above the horizontal;
- comparatively clear decks – no guyropes or preventers;
- up to 200 tonnes capacity derrick operates with speed appropriate to light loads. Only cargo hoist needs changing for different load requirements.

Plate 27 Heavy lift cranes work in tandem on project cargoes; remote control of both cranes is a general feature by the ship's Chief Officer when working in this manner.

Stülcken derrick

Advantages in use are:

- can be used at the *two* hatches forward and aft of derrick rig;
- topping lift also acts as guys, as in Hallen derrick;
- conventional lighter derricks can be fitted either side;
- single-man operation with mobile control unit.

Velle crane

Advantages in use are:

- simple components in comparison to crane – similar to Hallen derrick;
- single-man operation, controlling luffing, slewing and hoisting;
- arrangement damps pendulation/rotation of load, which allows the operator to luff and slew at the same time, with quicker handling;
- comparatively clear decks – no guyropes or preventers.

Derrick Tests and Surveys

It is a requirement of most national regulations that cargo-handling gear should be inspected once a year by the Chief Officer (annual inspection), in addition to the usual working checks by the Officer of the Deck. The cargo gear would also be thoroughly examined under survey every five years.

The surveyor at the five yearly inspections will pay particular attention to all associated fittings on the derrick, mast and deck. He will check for any excessive wear or corrosion, and may carry out hammer tests. All blocks, shackles, links, chains and wires will be examined to ensure that they are all in a satisfactory condition. Should any component have suffered damage, this should be replaced and, provided the component is individually tested, a retest on the rig is not required.

When a survey inspection takes place, the gear will be given a more detailed examination, and a drilling test may be required. It is recommended that the derrick should be retested at the third and each subsequent survey inspection.

Where the SWL of a derrick exceeds 15 tonnes, the proof load (see Table 5.2) is to be applied by hoisting movable weights by the

> Proof load is defined by the SWL + percentage tonnage to which the lifting equipment is tested – e.g. 60 tons SWL, derrick proof load = 66 tonnes.

Table 5.2 Tests on derricks

Safe working load	Proof load
Up to 20 tonnes	25 per cent in excess of SWL
Exceeding 20 tonnes but not exceeding 50 tonnes	5 tonnes in excess of SWL
Over 50 tonnes	10 per cent in excess of SWL

cargo purchase, and with the weights in the hoisted position the derricks are to be swung as far as possible in both directions. Where the SWL is 15 tonnes or less, the proof load may be applied, if desired, by means of a spring or hydraulic balance.

Calculating Stresses in Derricks by Empirical Formula

The Cargo Officer should be aware that when using the empirical formula, shown with the following examples, the additional effort applied to the hauling part to overcome friction has always been taken as one-tenth. This may not necessarily always be the case. Cargo-handling gear may achieve efficient bearings, but this cannot be guaranteed 100 per cent, and the allowance for friction should be based on the advice of the manufacturer.

> The rigging of sheer legs and gyns has become generally obsolete.
>
> However, it is still used as a rigging exercise when training cadets or junior seaman.

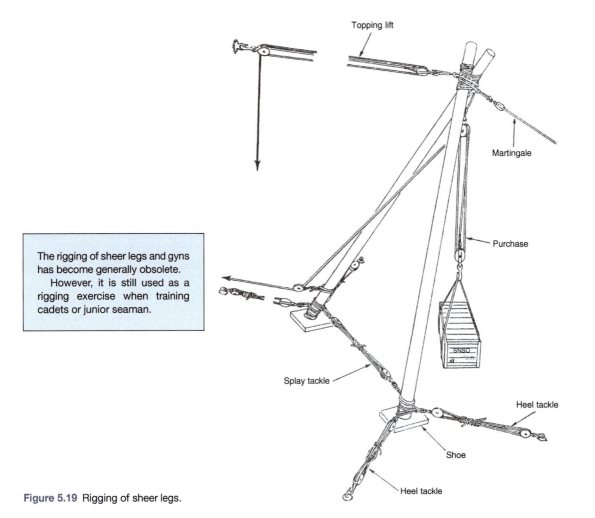

Figure 5.19 Rigging of sheer legs.

When calculating the size of wire or rope to use in a tackle, the SWL is taken as one-sixth of the breaking strain. However, industry may in practice operate SWLs of one-fifth of the breaking strain.

It should also be remembered that less friction is encountered when using sheaves of a large diameter than sheaves of a small diameter. Similarly, less friction is found when using a thinner rope than a thicker rope. Consequently, maximum advantage is gained by the use of the larger sheaves and the thinner rope.

Should a heavy lift be required, the officer should bear in mind that a multi-sheave block (over four sheaves) will have a considerable weight of its own. The rig will add additional weight to the load and could effectively neutralize any mechanical advantage gained by the use of a heavy-duty rig.

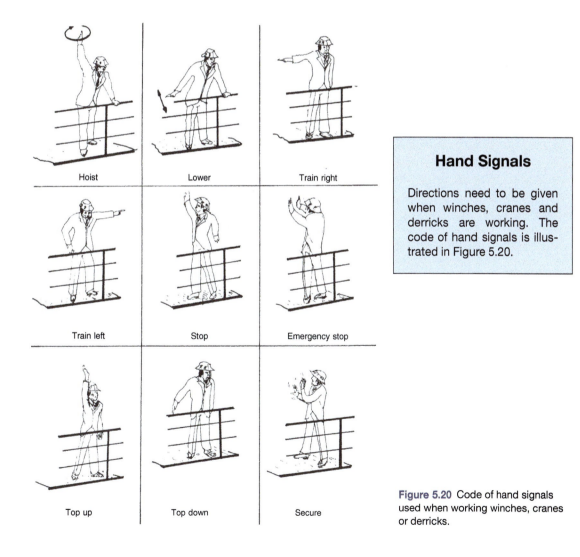

Hand Signals

Directions need to be given when winches, cranes and derricks are working. The code of hand signals is illustrated in Figure 5.20.

Figure 5.20 Code of hand signals used when working winches, cranes or derricks.

Example 1

Calculate the size of manila rope to use if the stress on the hauling part will be not greater than three tonnes.

Use the formula $\dfrac{2D^2}{300}$ as the breaking strain for manila rope.

$$\dfrac{\text{Breaking strain}}{6} = \text{Safe working load}$$

As the given stress on the hauling part = three tonnes, the SWL = three tonnes.

$$\dfrac{2D^2}{300} = 3 \times 6$$

Transpose

$$2D^2 = 18 \times 300$$
$$D^2 = \dfrac{18 \times 300}{2}$$
$$D^2 = 2700$$
$$D = \sqrt{2700}$$

Diameter of rope = 51.96 mm = 52 mm

Example 2

Calculate the size of flexible steel wire rope to use if the stress on the hauling part of the wire is 3.75 tonnes. The construction of the wire is 6 × 24 WPS. Use the formula:

$$\dfrac{20D^2}{500}$$

as the breaking strain for steel wire rope.

$$\dfrac{\text{Breaking strain}}{6} = \text{Safe working load}$$

As the given stress on the hauling part = 3.75 tonnes, SWL = 3.75 tonnes.

$$\dfrac{20D^2}{500} = 3.75 \times 6$$
$$20D^2 = 22.5 \times 500$$
$$D^2 = \dfrac{22.5 \times 500}{20}$$
$$D = \sqrt{562}$$

Diameter of wire = 23.7065 mm = 24 mm

Example 3

Calculate the stress on the hauling part of a gyn tackle rove to disadvantage (3 and 2, sheaves), and used to lift a load of one tonne (see Figure 5.21).

$$S = \frac{W + \dfrac{n \times W}{10}}{P}$$

where S = stress in hauling part; W = load being lifted; n = number of sheaves in tackle; P = power gained.

$$S = \frac{1 + \dfrac{5 \times 1}{10}}{5}$$

$$= \frac{3}{10}$$

$$= 0.3 \text{ tonnes}$$

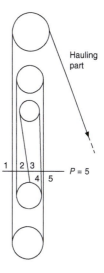

Figure 5.21 Gyn tackle rove to disadvantage.

Example 4

Calculate the stress on the hauling part of a double luff tackle rove to advantage (double purchase 2 and 2 sheaves), when used to lift a load of five tonnes (see Figure 5.22).

$$S = \frac{W + \dfrac{n \times W}{10}}{P}$$

where S = stress in hauling part; W = load being lifted; n = number of sheaves in tackle; P = power gained.

$$S = 5 + \frac{4 \times 5}{10}$$

$$= \frac{7}{5}$$

$$= 1.4 \text{ tonnes}$$

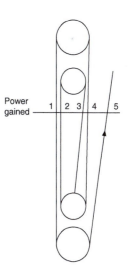

Figure 5.22 Double luff tackle rove to advantage.

Use of the above empirical formula in the examples shown takes into consideration that the allowance for friction is about one-tenth of the load to be lifted for each sheave in the purchase. The allowance for friction is added to the load that is to be lifted. Effectively the increase in the load on the hauling part necessary to overcome friction is spread almost equally between the several parts of the purchase.

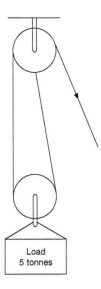

Load
5 tonnes

Figure 5.23 Gun tackle rove to disadvantage.

Example 5

A five-tonne load is to be lifted by means of a gun tackle used to disadvantage. Find the approximate stress on the hauling part of the tackle, and the minimum size of wire you would use in the tackle (see Figure 5.23).

$$S = \frac{W + \dfrac{n \times W}{10}}{P}$$

where S = stress in hauling part; W = load being lifted; n = number of sheaves in tackle; P = power gained.

$$S = \frac{\dfrac{5 + 2 \times 5}{10}}{2}$$

$$= \frac{6}{2}$$

$$= 3.0\,\text{tonnes}$$

$$\frac{BS}{6} = SWL$$

$$\frac{20D^2}{500} = 3.0 \times 6$$

$$20D^2 = 18 \times 500$$

$$D^2 = \frac{18 \times 500}{20}$$

$$D = \sqrt{450}$$

$$= 21.213 = 22\,\text{mm}$$

Example 6

A ten-tonne load is to be lifted by means of a double luff tackle used to disadvantage, with a gun tackle rigged to advantage secured to the hauling part of the luff tackle. Calculate the stress on the hauling part of the gun tackle. (Note that the power gained by the combination of two purchases is approximately equal to the product of their separate powers (4 × 3 = 12) (see Figure 5.24).

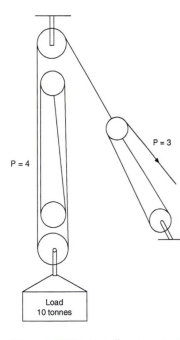

P = 3

P = 4

Load
10 tonnes

Figure 5.24 Double luff and gun tack combined rig.

$$S = \frac{W + \dfrac{n \times W}{10}}{P}$$

where S = Stress in hauling part; W = load being lifted; n = number of sheaves; P = power gained.

$$S = \frac{10 + \dfrac{6 \times 10}{10}}{12}$$

$$S = \frac{16}{12}$$

$$= 1\tfrac{1}{3} \text{ tonnes}$$

Calculation of Stresses Involved in the Use of Derricks

It is not the intention of this text to involve the reader in physics, but it is considered necessary that the Cargo Officer is aware of the methods of using the parallelogram of forces to obtain the stresses that may be encountered in practice when using derrick rigs. Assuming a single swinging derrick, the areas of stress when lifting a load will be:

- the stress in the downhaul of the lifting purchase;
- the resultant load on the head block;
- the tension in the topping lift span;
- the resultant thrust on the derrick;
- the stress on the heel block;
- the stress on the span block.

The stress in the downhaul of the lifting purchase

This may be found by use of the empirical formula as shown with previous examples.

$$S = \frac{W + \dfrac{n \times W}{10}}{P}$$

To obtain the resultant load on the head block (derrick head shackle)

The construction of a parallelogram of forces similar to that of ABCD should be made with scaled values for the stress in the hauling part of the lifting purchase (AD) being resolved against the scaled value of the weight being lifted (AB).

Once these two forces are resolved then the resultant force at point 'A' (derrick head) is represented by the scale value AC. AD represents the calculated stress in the hauling part of the lifting purchase. AB represents the weight/load being lifted. (This is always acting vertically due to the force of gravity.) By resolving the forces AD and AB into the parallelogram ABCD, AC represents the resultant stress acting on the derrick head shackle.

To obtain the tension in the topping lift span

By employing the vector produced, which represents the stress on the derrick head shackle (AC from Figure 5.25) and the length of the derrick itself, the tension in the topping lift span can be obtained by *scaled construction*.

By construction AC represents the resultant stress acting on the derrick head shackle, caused by the load AB acting vertically down. By construction of the parallelogram ACFE, FC parallel to AE represents the tension in the topping lift span (Figure 5.27).

To obtain the thrust on the derrick

The reader should bear in mind that the forces which produce thrust in the derrick are the tension in the topping lift span and the total load on the head shackle. By use of Figure 5.27, where AC represents the resultant load on the derrick head shackle, AE represents the tension in the topping lift span.

In parallelogram ACFE, AF represents the thrust on the derrick (by scale measurement).

To obtain the stress on the heel block

A vector is a quantity having magnitude and direction. By scaled construction similar to the parallelogram PQRS, the vectors

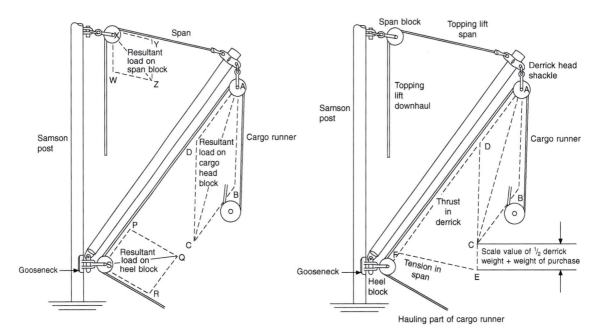

Figure 5.25 Derrick work calculations.

representing the stress in the wire should be plotted from the position of the heel block (Figure 5.25):

- SP represents the stress in the downhaul of the cargo runner, acting in the direction of the derrick.
- SR represents the stress in the cargo runner acting in the direction of the winch.
- SQ represents the resultant stress on the heel block.

To obtain the stress on the span block

By scaled construction similar to the parallelogram WXYZ, the vectors representing the stress in the downhaul of the topping lift span and the tension in the topping lift span should be plotted from the position of the span block (Figure 5.25):

- WX represents the stress in the downhaul of the topping lift.
- XY represents the tension in the topping lift span.
- XZ represents the resultant stress on the span block.

With all calculations obtained by use of scaled diagrams it should be remembered that the larger the scale the more accurate the results will be.

Notes for Solving Problems

1 Always state clearly the scale of the values being used, and display this statement with all diagrammatic calculations.
2 Assume that all wires run in the same direction as a conventional derrick unless another rig is specified. For example, the cargo runner may be assumed to run in a parallel direction to that of the derrick, so that scaled values may be laid off in the same line.
3 Where a purchase is being used, the empirical formula should be applied to obtain the stress on the downhaul of the purchase. This would become the laid-off value for the purpose of the scaled value and the construction of the parallelogram, as with, for example, a span tackle topping lift.
4 Unless otherwise stated the allowance for friction per sheave should be taken as 10 per cent of the load being lifted.
5 Always use the largest scale possible, as this will tend to eliminate minor diagrammatic errors.

The Register of Ship's Lifting Appliances and Cargo-Handling Gear

Previous legislation required that a ship's officers kept certificates and maintenance records of lifting gear employed aboard the vessel. This documentation and monitoring of lifting appliances has now

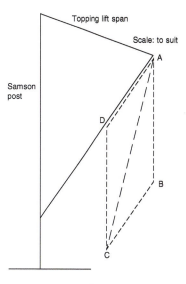

Figure 5.26 Parallelogram to assess stress on head block.

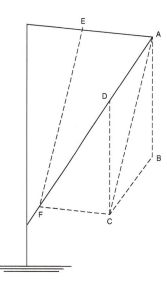

Figure 5.27 Parallelogram to assess tension in topping lift span.

been formalized within 'The Register of Ship's Lifting Appliances and Cargo-Handling Gear'. The register is kept and maintained by the ship's Chief Officer and is an item of considerable importance because of the requirements of ISM auditors.

The register contains all the certificates of derricks/cranes, blocks, shackles, wires, chains, hooks, etc. How the register is maintained by the Chief Officer can vary, but it would normally be a file system by hatch or by item, i.e. all the certificates of wires, all the certificates of shackles. Alternatively, all the certificates of equipment at No. 1 hatch, all the equipment at No. 2 hatch, and so on.

Cargo Documentation

It should be realized that when Chief Officers hand over to relief officers, one of the main areas of concern will be the cargo documentation of the vessel. Not only would the cargo plan, cargo manifest, cargo securing manual, mate's receipts, bills of lading, etc., be part of the handover, but also the Register of Lifting Appliances and Cargo-Handling Gear, along with the rigging plan and the planned maintenance schedule.

The Cargo Securing Manual

The manual is individual to a particular ship and is the product of the IMO Publication, Regulations V1/5 and V11/6 of the SOLAS convention and the Code of Safe Practice for Cargo Stowage & Securing (CSS code).

The Cargo Securing Manual (CSM) is approved by the Marine Administration for all cargoes other than solid or liquid bulk cargoes. The manual must suit the stability and trim criterion of the vessel and is not meant to infringe on the ship's loadline requirements.

It is meant to specify cargo securing methods for the type of cargoes that the vessel can be expected to transport. It is not meant to replace the interests of good seamanship and is designed to provide guidance to prevent movement of cargo parcels when the vessel is at sea and subject to transverse, longitudinal and vertical forces.

The manual contains four chapters: general; securing devices and arrangements; stowage and securing of cargo; and a supplementary section concerning different types of vessel, e.g. containers.

The contents will include a deck plan of securing devices like pad eyes and ring bolts, bulwark and bulkhead securings, together with their respective SWL, the disposition and strength test details of portable lashings, such as chains and load binders, together with illustrations and examples of their use.

The manual should also include tables and diagrams affecting the calculations and accelerations of forces acting on cargo parcels.

Definitions Respective to the Manual

- *Cargo securing devices*: fixed and portable devices used to secure and support cargo units.
- *Maximum securing load*: the allowable load capacity for a device used to secure cargo to a ship – that is, the SWL of the device.
- *Standardized cargo*: cargo for which the ship is provided with an approved securing system based on specific cargo unit types.
- *Semi-standard cargo*: cargo for which the ship is provided with a securing system which is capable of securing a limited variety of cargo units such as vehicles or trailers.
- *Non-standard cargo*: cargo which requires individual stowage and securing arrangements.

6

Cargo and Hatchwork

Introduction

The twenty-first century has produced a shipping industry which tends to specialize far more than it ever did in the past. Container vessels are involved in the carriage of most, if not all, general cargoes. Trailer units in the roll on–roll off (ro-ro) ferry trades have also replaced many of the short-haul cargoes carried by the smaller coastal vessels, while dry and liquid bulk products tend to find shipment in the designated carriers for LNG, oil, grain, coal, etc.

The outcome of such specialization has been that the conventional hatch has virtually disappeared and modern steel hatch covers dominate the dry cargo sector of the industry. Cranes have generally replaced many of the smaller shipboard derrick rigs, now 'gantry rigs', both ashore and afloat, handle the container traffic. Very few individual cargo loads are now being slung and hoisted over the ship's sides.

The design of ro-ro vessels has changed cargo access thinking, and although some vessels have a dual capability of load on, load off combined with roll on–roll off, these tend to be in the minority. The medium heavy lift operation has in many cases been taken up by the ro-ro vessel working in conjunction with ground-handling equipment, like low loaders. This handling method employs the 'bow visor' and/or the 'stern ramp', but loads are restricted to the size of the access points. The larger, heavier, exceptional 'project cargo' tends to be transported by the designated heavy lift ship.

NB. Exception is ship stores, and most vessels inclusive of the tankers carry a stores derrick or crane for convenience.

Conventional Hatch

The conventional hatch is all but obsolete, with possibly an exception in small coastal or barge traffic. The hatch was covered by 'hatch boards' or the larger 'hatch slab' set into ridged 'king beams' with a 'queen beam' set midway underside of the boards. These were then covered by three canvas tarpaulins, respectively tabled, tucked into a side coaming and wedged cleating arrangement (Figure 6.1).

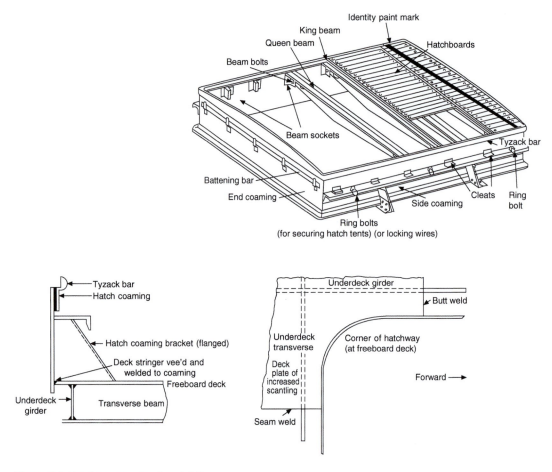

Figure 6.1 Hatchway construction detail.

Conventional Hold

Construction

The hatchway entrance is a cut-away from the upper deck stringer plates. The corners of the hatchway are cut on the round to provide continuity of strength and prevent shearing stresses causing cracks athwartships and bending forces causing cracks in the fore and aft line. The corner turns of the hatchway are often fitted with reinforcing bars to prevent loading and racking stresses (Figure 6.1).

Tank Top Ceiling

This is a wooden sheathing over the double bottom tank tops, usually in way of the hatch, providing the tank tops with some protection from wear and tear. The ceiling also assists ventilation and drainage of cargoes, and with many cargoes relieves the necessity for laying of double dunnage.

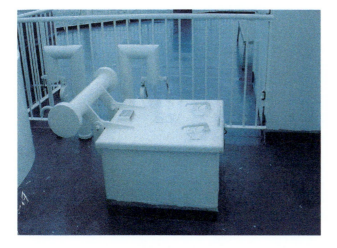

Plate 28 A booby hatch with counterweight provides access for cargo holds and emergency escapes.

This wood covering may come in one of two forms – either wide flats, laid on bearers which leave space for liquids to drain off to the bilges, or set close into a composition of cement and Stockholm tar. When bulk cargoes are being regularly carried, the second method is often employed, as the drainage spaces tend to become choked when the first method is used.

It is not uncommon to see the most modern vessel with no tank top ceiling at all, but in this case the tank top itself is normally protected by having increased scantlings.

The turn of the bilge construction is shown in Figure 6.2.

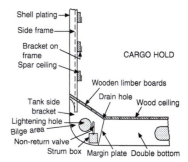

Figure 6.2 Turn of bilge construction.

Spar Ceiling

This may be in the form of horizontal or vertical wooden battens to keep cargo off the steel work of the ship's side. Contact between the shell plate and the cargo tends to lead to excessive cargo sweat damage; to prevent this occurrence a spar ceiling, sometimes referred to as cargo battens, is secured in cleats throughout a cargo hold and 'tween deck.

Limber Boards

These are wooden boards similar to hatch boards that cover the bilge bays, which are situated at the bottom sides of the lower holds. These bays run the full length of the hold and should be regularly inspected for their cleanliness. The boards are supported by the tank side brackets between the floors and the frames.

Bilge Suctions: Strum Box

The bilge suction (Figures 6.3 and 6.4) is usually found in the aftermost bay of the hold. Vessels normally trim by the stern, so that

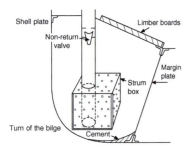

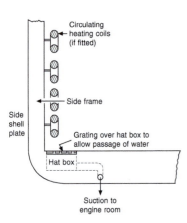

Figure 6.3 Bilge suction: strum box.

Figure 6.4 Deep tank suction: hat box.

this aft siting is best for drainage within the confines of the hatch. Scupper pipes tend to drain direct from the after part of the 'tween deck into the bay containing the strum box.

A non-return valve must be fitted clear of the strum, and in the more modern vessels this valve is situated clear of the bilge area. The purpose of the non-return valve is to prevent accidental run back from the pumps, which may cause flooding in the hold. The suction end of the pipe is kept clear of obstructions by the strum box arrangement built about the pipe opening. This strum box is so constructed as to allow the passage of water but not the passage of solids, which could interfere with suction. The sides of the strum are either slotted or hinged to a framework which will allow the box itself to be dismantled for cleaning and maintenance. The whole bay containing the strum is covered by limber boards.

General Cargo Vessel Deep Tanks

General arrangements vary, especially in the securing of the deep tank lids and the number of tanks constructed. It is normal to find deep tanks in pairs or, if situated in a large hatch, then 2×2 pairs, to port and starboard. They are extensively used for bulk cargoes such as grain or chemicals, but very often fitted with steam-heated coils for the carriage of such things as 'tallow'. They may also be used to take on extra ballast when the vessel is in a light condition.

Hat box pumping arrangements are operated from the ship's engine room and the lines are fitted with a blanking off fitment when required. Most systems allow for gravity filling and tanks are all fitted with air and sounding pipes (Figure 6.4).

Steel Hatch Cover

The more modern type of cargo vessel will be equipped with one of the many types of steel hatch covers that are present on the commercial market (Figure 6.5). The many advantages with this style of cover far outweigh the disadvantages. They are fast in closing or

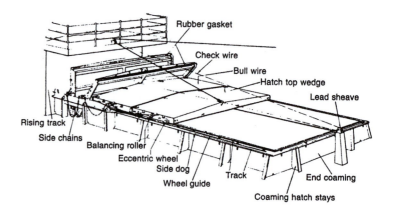

Figure 6.5 Single-pull steel hatch cover.

opening, and the latest versions are so labour saving that one man could open up all the hatches of a ship in the time it takes to strip a single conventional hatch. Their structure, being of steel, is extremely strong and generally forms a flush surface in 'tween deck hatches, providing ideal conditions for forklift truck work. Steel covers may be encountered not just at the weather deck level, but throughout a vessel, inclusive of 'tween decks. Hydraulic operated covers are simple in operation, but should hydraulic fluid leak at any time, cargo damage may result. The direct pull type must be operated with extreme care, and all safety checks should be observed prior to opening the chain-operated types.

Steel covers are illustrated in Figures 6.5 to 6.11 and Plates 29 to 34.

Opening Single-Pull Macgregor Steel Hatch Cover

1 Release the side securing lugs, ensuring they are correctly stowed in flush position with the track.
2 Clear away any hatch top wedges between hatch sections.
3 Rig the check wire to the lug of the leading hatch section and turn up the bight of the wire onto cleats or bitts.
4 Rig the bull wire so as to provide a direct pull to the winch from the leading edge of the hatch cover.

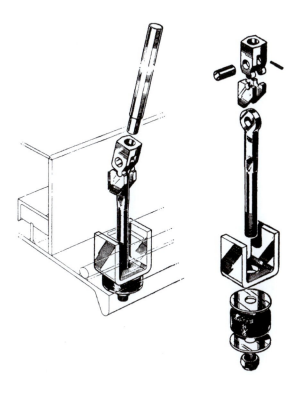

Figure 6.6 Securing steel covers: Cleating (dogging) arrangement.

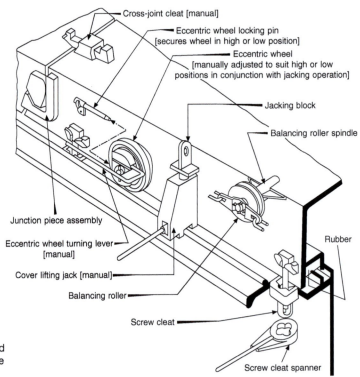

Figure 6.7 Steel hatch cover securing and operational fittings, seen with jack in place for lifting cover.

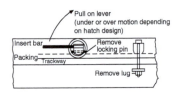

Figure 6.8 Turning down eccentric wheels.

5   Complete all work on top of the hatch covers. Check that the track ways are clear of all obstructions, such as pieces of dunnage, etc.

6   Turn down the eccentric wheels by use of bar levers, or by using the jacks under the hatch cover sections.

7   Check that the locking pins are securely replaced in the eccentric wheels once the wheels have been turned down to the track in such a manner that they will not slip out when the wheel rotates or when the hatch is in the vertical stowed position.

8   Ensure that all personnel are aware that the hatch cover is about to open, and that the stowage bay for the covers is empty and clear to allow correct stowage of the sections.

9   Have a man stand by to ease the check wire about the bitts, and, just before hauling on the bull wire, remove the locking pins at the ends of the leading hatch section.

10   Heave away easily on the bull wire once the locking pins are removed, taking the weight of the leading hatch section.

11   Ease out on the check wire as the bull wire heaves the hatch open (Figure 6.5).

12   Once all hatch sections are in the stowed vertical position, the bull wire should not be removed until the securing chains from a fixed point are in position to hold back the hatch sections in the stowage bay area.

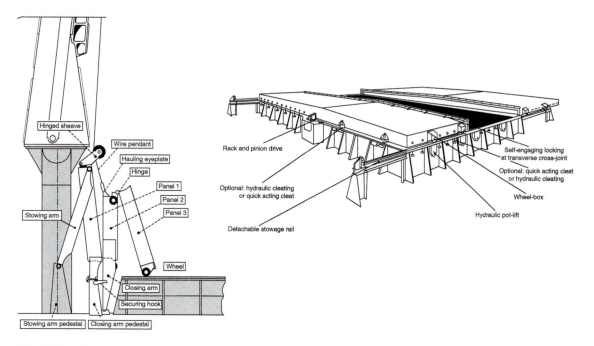

Figure 6.9 Direct pull weatherdeck hatch covers (*above*). Side rolling covers (*right*). Rack and pinion drive, with hydraulic lifting and cleating.

13   Clear away the check wire, coiling it down to one side of the hatch. Do not attempt to detach the check wire from the lug of the leading edge of the hatch.

General Cargo Terminology

Bale Space

This is internal volume measured to the inside edges of the spar ceiling, beams, tank top ceiling and bulkhead stiffeners (a spar ceiling is often referred to as cargo battens).

Broken Stowage

This is the unfilled space between packages; this tends to be greatest when large cases are stowed in the end holds, where the shape of the vessel fines off.

Deadweight Cargo

This cargo measures less than 40 cu. ft per ton (1.2 m³ per tonne), and freight is paid on the actual weight.

(a)

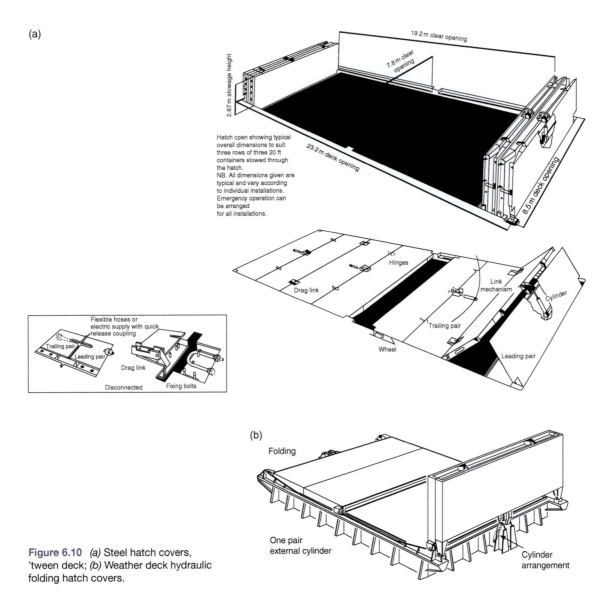

19.2 m clear opening

7.8 m clear opening

2.87 m stowage height

23.2 m deck opening

8.5 m deck opening

Hatch open showing typical
overall dimensions to suit
three rows of three 20 ft
containers stowed through
the hatch.
NB. All dimensions given are
typical and vary according
to individual installations.
Emergency operation can
be arranged
for all installations.

Hinges

Drag link

Link mechanism

Cylinder

Trailing pair

Wheel

Leading pair

Flexible hoses or
electric supply with quick
release coupling

Trailing pair

Leading pair

Drag link

Disconnected

Fixing bolts

(b)

Folding

One pair
external cylinder

Cylinder
arrangement

Figure 6.10 *(a)* Steel hatch covers,
'tween deck; *(b)* Weather deck hydraulic
folding hatch covers.

Dunnage

This is material used when stowing cargo to protect it from contact
with steelwork, other cargoes, or any possibly damaging influences.
Tank tops are usually covered with a double layer of dunnage
wood, the bottom layer running athwartships to allow drainage to
bilges, and normally being more substantial than the upper layer,
e.g. 2 × 2 in. (see Figure 6.12).

Additional dunnage is soft, light wood, dry and free from stains,
odour, nails and large splinters. New timber should be free of resin
and without the smell of new wood. Materials also used for similar
purpose are matting, bamboo or waterproof paper.

Grain Space

This is the total internal volume of the compartment, measured from the shell plating either side and from the tank top to under-deck. This measurement is used for any form of bulk cargo that could completely fill the space, an allowance being made for space occupied by beams and frames.

Measurement Cargo

This is cargo measuring 40 cu. ft per ton (1.2 m^3 per tonne) or more. The standard is used for comparatively light cargo on which freight is paid on space occupied.

Figure 6.11 Roll stowing covers: Rolltite. Originally designed by Ermans and under manufacture by Macgregor.

Plate 29 Stackage steel hatch covers aboard a modern coastal vessel. The ship is fitted with its own travelling mini-gantry crane.

Plate 30 Macgregor steel hatch covers shown in the stowed upright position, with locking devices in place.

Plate 31 Hydraulic-operated M-type steel hatch covers. Forward end open. Aft covers seen in place.

Plate 32 A coastal vessel approaches the berth with Macgregor steel hatch tops fully closed.

Plate 33 Fully open hydraulic hatch tops seen aboard the *Baltiyskiy-110*.

Plate 34 Horizontal, stackable steel hatch covers. The ship is equipped with its own mini-track-operated gantry hoist designated solely for lifting, transporting and stacking of the covers (gantry not shown).

Plate 35 Grain elevator and tractor engaged in clearing the remnants of bulk grain cargo.

Stowage Factor

The volume occupied by unit weight, this is usually expressed in cu. ft/ton or m³/tonne, no account being taken of broken stowage.

Duties of the Junior Cargo Officer

When carrying out cargo watch duties the company policy is usually adopted and the 3rd Officer or 2nd Officer could be given the responsibility of the supervision of the loading or discharge of the vessel. The duties involved would include any or all of the following:

1   Prior to commencing loading cargo, and on the orders of the Chief Officer, the hold spaces would be inspected to ensure they are ready in every way to receive cargo. They should be cleaned throughout, with the bilge pump suction tested. The hold lighting should be inspected and dunnage correctly laid if required.

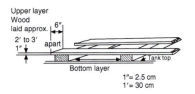

Upper layer
Wood
laid approx. 6″
2′ to 3′
1″ apart
Bottom layer
Tank top
1″= 2.5 cm
1′= 30 cm

Figure 6.12 Use of dunnage.

Plate 36 A container vessel lies alongside a container terminal loading dock. The ship's own cranes are seen turned outboard so as not to interfere with the shoreside loading cranes.

Plate 37 A five-hatch general cargo vessel seen alongside as an 'arrived ship' with all hatches open and ready to work cargo.

2   Hatch covers should be opened (weather permitting) and secured in the correct stowage position.

3   The derricks or the cranes should be rigged for correct operation.

4   'Tween deck guard rails should be seen to be in place and secure, in respect of 'tween deck vessels.

5   Regular checking of the holds should take place to ensure they are being loaded or discharged correctly. Notes should be taken of cargo parcels being loaded to be added to the cargo plan. If discharging, note should be taken of any damaged cargo to be reported to the Chief Officer.

6   All times of cargo work and stoppages are to be recorded in the deck log book.

7   Attention should be given throughout operations to prevent pilferage, smoking in cargo spaces and damage to the vessel.

8   Where 'general cargo' is being loaded, visits to the shoreside warehouse to ensure continuity of flow and special cargoes to be catered for should be anticipated.

9   Special cargoes which have specific requirements, such as tallying, lock-up stow, hazardous goods, port separation or other similar needs, should be catered for, and the Chief Officer kept informed on progress.

10 The overall deck security should be monitored throughout, inclusive of gangway and hold access points, fire precautions, ship's draught and moorings, together with tonnages and capacities loaded per space. Once operations are complete (for the day) the hatches should be secured and locked.

Cargo Duties of the Chief Officer

It would be normal practice for the Chief Officer to be provided with the cargo manifest by the ship's agents at the last port of discharge or the next loading port. Inspection of this would subsequently provide details on all cargo parcels, inclusive of the tonnages and the destination of said cargo. This would then allow a loading plan to be constructed, preferably prior to arrival. In any event, the Chief Officer would note special items on the manifest, in particular:

- hazardous cargoes;
- heavy lift cargo;
- valuable/special cargoes.

The Chief Officer is responsible for the safe handling, loading and discharge of the ship's cargo, and to achieve this safely and satisfactorily he would carry out the following activities and duties:

1 Having obtained the details of tonnages/capacities of cargo from the manifest, the Chief Officer would develop an appropriate loading plan for the ship to give a hold distribution of the cargo and to allow multiple hatch discharge.
2 The Chief Officer would carry out a ship stability assessment, taking into account the appropriate tonnages, together with bunker capacity, stores and ballast arrangements. This assessment would be expected to confirm an appropriate 'GM' for the vessel, and ensure that the vessel does not infringe loadline regulations while on passage.
3 Prior to commencing loading cargo the Mate would order the cleaning of all cargo spaces, the testing of bilges, checks on hold lighting, ventilation and general hold conditions, inclusive of spar ceiling and dunnage arrangements.

> NB. One of the Chief Officer's tasks is to order clean dunnage meeting ship/cargo requirements.

4 The Chief Officer would act in a supervisory role of a junior Cargo Officer and ensure that the cargo plan was being correctly constructed during any loading period.
5 Where special cargo is to be loaded, the Chief Officer would be expected to take a positive role to ensure adequate loading facilities are provided and correct stowage is available, e.g. heavy lift derricks rigged correctly, tally clerks available if required, etc.
6 Documentation on specific cargoes, such as hazardous parcels, livestock or valuables would be obtained and administered by the Chief Officer.
7 The Ship's Mate would monitor the draught and ensure that the loadline regulations are not infringed, causing the marks to be submerged below the summer loadline. He/she would order the density of dock water to be obtained and take into account the Dock Water Allowance.

Plate 38 Dry cargo vessel with prominent deck cranes deployed (source: Shutterstock).

8   Any damaged cargo received on board would be noted and the bill of lading would need to be endorsed. Similarly, on receipt of special cargo the Chief Officer may be called on to issue a 'Mate's Receipt' for cargo delivered on board.

9   In conjunction with junior Cargo Officers the Mate would ensure that all cargo is stowed correctly and secured in the stowed position against all expected ship movement when at sea.

10 His/her duties include the despatch of all the relevant cargo documentation, including the cargo plan, to the company agents prior to sailing departure.

Hold Preparation

1   The compartment should be swept clean, and all traces of the previous cargo removed. The amount of cleaning is dependent on the nature of the previous cargo: some cargoes, such as coal, will require the holds to be washed before the carriage of a general cargo. Washing is always carried out after the compartment has been swept. Drying time for washed compartments must be allowed for before loading the next cargo; this time will vary with the climate, but two to three days must be expected.

2   Bilge areas should be cleaned and all 'bilge suctions' seen to be working satisfactorily. All 'holes' in rose boxes should be clear to allow the passage of water and the lines' non-return valves seen to be in a working condition. Should the bilges be contaminated from odorous cargoes, it may become necessary to 'sweeten' them by a wash of chloride of lime. This acts as a disinfectant as well as providing a coating against corrosion.

3   The fire/smoke detection system should be tested and seen to function correctly.

4   The hold's drainage system and 'tween deck scuppers should be clear and free from blockage.

5   The spar ceiling (cargo battens) should be examined and seen to be in a good state of repair.
6   Steel hatch covers should be inspected for their watertight integrity about any joints. If hard rubber seals are fitted these should be inspected for deterioration.
7   Ladders and access points should be inspected for damage and security.
8   Hold fitments such as built-in lighting and guard rails should be checked and seen to be in good order.
9   Soiled dunnage should be disposed of. New dunnage, clean and dry, should be laid in a manner to suit the next cargo, if needed.
10   Hold ventilation system should be operated to check fan conditions.

Additional for Special Cargoes

1   *Grain.* Limber boards should be plugged and covered with burlap. This prevents grain blocking bilge suctions, while at the same time allowing the passage of water.
2   *Coal.* The spar ceiling should be removed and covered (most bulk cargoes require this).
3   *Salt.* Metalwork should be whitewashed.

Stowage Methods

Bagged Cargo (Paper Bags)

These should be stowed on double dunnage. Ideally the first layer should be stowed athwartships on vessels equipped with side bilge systems. Steelwork should be covered by brown paper or matting to prevent bags making contact. Torn bags should be refused on loading. Canvas rope slings should be made up in the hatchway centre to avoid dragging and bursting bags. Hooks should never be used with paper bag cargoes. When stowing, bag-on-bag stow is good for ventilation, whereas bag on half-bag is poor for ventilation but good for economical use of space.

Barrels

Stowed 'bung' uppermost on wood beds, in a fore and aft direction. 'Quoins' are used to prevent movement of the cargo when the vessel is in a seaway. Barrels should never be stowed more than eight high.

Coal (Bulk)

Check that bilge suctions are in working order and that limber boards are tight fitting. Remove all spar ceiling, stow in the 'tween deck and cover with a tarpaulin or other similar protection. Plug 'tween deck scuppers. Remove all dunnage and make arrangements for obtaining temperatures at all levels if engaged on a long voyage. Ensure that the coal levels are well trimmed and provide the compartment with surface ventilation whenever weather conditions permit.

Copra

As it is liable to spontaneous combustion, it should be kept dry and clear of steelwork surfaces, which are liable to sweat. Copra beetle will get into any other cargoes which are stowed in the same compartment.

Cotton

Bales are liable to spontaneous combustion, so the hold must be dry and clean, free of oil stains, etc. Adequate dunnage should be laid and all steelwork covered to prevent contact with cargo. Wet and damaged bales should be rejected at the loading port.

Hoses and fire appliances should be on hand and readily available during the periods of loading, fire wires being rigged fore and aft.

Edible Oils

Deep tank stow, for which the tank must be thoroughly cleaned, inspected and a certificate issued by a surveyor.

Heating coils will be required, and these should be tested during the period of preparation of the space. All inlets and outlets from the tank should be blanked off. Shippers' instructions with regard to carriage temperatures should be strictly adhered to. A cargo log of these temperatures should be kept. Extreme care should be taken on loading to leave enough 'ullage' for expansion of the oil during passage. Overheating should never be allowed to occur, as damage to the oil will result.

Flour

Susceptible to damage from moisture or by tainting from other cargoes, it should never be stowed with fruit, new timber or grain. Should a fire occur during passage, 'dust explosions' are liable from this cargo.

Fruit

Usually carried in refrigerated spaces, especially over long sea passages, it may also be carried chilled under forced ventilation. However, regular checks should be made on ventilation system and compartment temperatures. This cargo gives off CO_2 and will consequently require careful ventilation throughout the voyage.

Glass (Crates)

Crates of glass should never be stowed flat, but on their edge, on level deck space. Plate glass should be stowed athwartships and window glass in the fore and aft line, each crate being well secured by chocks to prevent movement when the vessel is at sea. Overstowing by other cargoes should be avoided.

Vehicles

These should be stowed in the fore and aft line, on level deck space. They should be well secured against pitching and rolling of the vessel by rope lashings. Fuel tanks should be nearly empty. Close inspection should be made at the point of loading, any damage being noted on acceptance.

Cargo Handling

Use of Snotters

Rope or wire snotters are in common use when general cargo is discharged. Wire snotters are probably the most widely used, but care should be taken that when using them as illustrated in Figure 6.13, the wire is not allowed to slip along the surface of the steel. This possibility can be eliminated by spreading the area of pressure by inserting a dunnage piece between wire and cargo. Snotters should be secured on alternate sides, passing eye through eye to provide stability to the load.

Use of the Bale Sling Strop

A bale sling strop is more commonly known as a sling or even just a rope strop. It is an endless piece of rope whose ends have been joined by a short splice, used extensively for the slinging of cases or bales – hence its full title (see Figure 6.14).

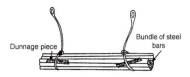

Figure 6.13 Use of snotters.

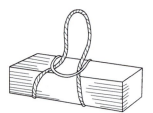

Figure 6.14 Use of bale sling strop.

Plate 39 Modern general cargo vessel *Centaurus* fitted with deck cranes.

Plate 40 Logs being loaded up to the hatch coaming by use of wire snotters.

Palletization

This is a most convenient pre-package cargo-handling technique (Figure 6.15). Separate slings of cargo are made up before the vessel berths, which speeds up turnaround time, so saving the shipowner considerable port costs. The cargo is generally stacked on wood pallets, which allows easy handling by the use of forklift trucks. The upper layer of cargo packages are often banded or the full load may be covered by protective polythene. This securing acts as a stabilizing factor when the load is being hoisted, as well as an anti-theft device while the pallet is being loaded, stowed or discharged.

The slings are usually made of steel wire rope (SWR), having four legs secured to a lifting ring. Each pair of wire slings holds a steel lifting bar, which is used to lift the ends of the pallet and its cargo.

Each load is usually squared off, and bound secure to reduce broken stowage within the hold, especially so when the vessels are of a flush deck and square corner construction. The pallets cause a certain amount of broken stowage, but this has become an accept-able factor compared to costs of lengthy handling procedures.

Cargo Nets

Fibre rope cargo nets (Figure 6.16) are in general use throughout the marine industry and are extensively used for such cargoes as mail bags, personal effects, etc. where the extra strength and wear resistance of a wire rope net is not required.

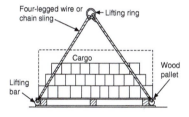

Figure 6.15 Use of pallets.

Figure 6.16 Cargo nets.

Figure 6.17 Timber dogs.

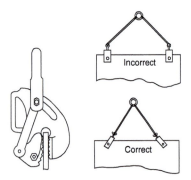

Figure 6.18 Use of plate clamps.

Wire rope cargo nets are designed for longer life, and are stouter than fibre nets. They carry a bigger load with greater safety, and tend not to distort under the most difficult conditions.

Fibre rope nets are generally of a knotted mesh, but may be woven. The mesh of a wire rope net will contain a specially designed clip at every cross, to provide reinforcement for the net as well as protecting the wire from wear.

Timber Dogs

Timber dogs are used purely for the lifting of short, heavy logs. The weight of the log causes the sharpened dogs to exercise an even greater grip when inserted into the grain end of the timber. Extreme caution should be observed with this method of lifting, to ensure that the point of the dog is well embedded before starting the lift (see Figure 6.17).

Plate Clamps

If the construction of the plate will permit this method of lifting, then it should be employed. Whether or not the construction of the plate structure lends itself to the use of shackles and slings, or to plate clamps, only one plate should be lifted at any one time.

When lifting with plate clamps (Figure 6.18), loads must not exceed the marked capacity of the clamp, and the jaws must be as narrow as possible for the plate thickness. Before lifting the plate it should be checked to ensure that it is properly gripped, and under no circumstances should packing be used between the jaws and the plate. When two clamps are to be used, they should be inclined and secured in the line of the sling, once the slack has been taken out of the slings.

Slinging Sheet Metal

In this operation plate dogs (Figure 6.19) or can hooks (Figure 6.20) can be used. They are based on a similar holding operation, where the hooks or dogs are tensioned together by a single chain sling (per pair) drawing them tight about the load. The purpose of the adjustable spreader (Figure 6.19) is to prevent the two slings closing up and disturbing the stability of the load.

Use of Chain Slings

Chain slings (Figure 6.21) are used for such heavy types of load as metal castings. Extreme care should be observed with any load, but even more so with a heavy lift, especially if chain slings are employed. There is a tendency for links in the sling to kink inside each other, and if the sling is pulled clear, the links or any kinks in the chain could cause the load to tip, with possible dangerous consequences. It should be remembered that a kink in a chain is a severe weakening factor and should be avoided at all costs.

Timber bearers to provide a clearance for the sling to be safely released should be used when landing loads of this nature.

Ventilation

Natural

This is the most common form of ventilation, when cowls (Figure 6.22) are trimmed into the wind to take in outside air, and trimmed back to wind to allow the air circulation an exit from the hold. Fans may be incorporated into this cowl ventilator system, especially for the lower hold regions where fans assist delivery and air extractors assist the exhaust system. Cowls may also be fitted with manually operated closure flaps.

Forced

More recent developments in ventilating systems have led to air being pre-dried before entering the hold. In some cases the temperature of the air as well as its humidity may be controlled before entering the compartment (Figure 6.23). This artificial or forced ventilation has become increasingly popular because, when properly used, it can almost prevent any sweat damage to cargo.

Refrigerated Cargoes

Refrigerated cargoes include meat carcasses, carton (packed) meat, fruit, cheese, butter, fish and offal. Ships are specifically designed for their carriage, with separate spaces in holds and 'tween decks, each fitted with suitable insulation and individual control of ventilation. Ordinary general cargoes may be carried in the spaces at other times, the temperature being regulated accordingly for the type of cargo being carried.

Insulation around a compartment consists of either a fibreglass or polystyrene type of packing over the steelwork of the vessel, with an aluminium alloy facing. This insulation is comparatively fragile and requires regular inspection and maintenance.

Cooling a compartment on modern vessels is achieved by circulating pre-cooled air by means of fans. The air is cooled by an ordinary refrigeration plant employing a refrigerant with the most practical qualities – namely a high thermal dynamic efficiency, low costs, low working pressure, low volume non-toxicity, non-inflammability, non-explosivity and ready availability from numerous sources.

Typical Refrigerants

- *Carbon dioxide* (CO_2). Non-poisonous, odourless, with no corrosive action on metal. It has a low boiling point but a high saturated pressure.

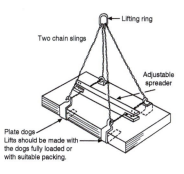

Figure 6.19 Use of plate dogs.

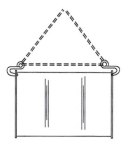

Figure 6.20 Use of can hooks.

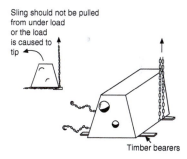

Figure 6.21 Use of chain slings.

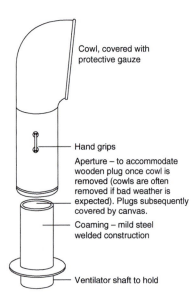

Figure 6.22 Cowl ventilators.

- *Ammonia (NH₃)*. Poisonous vapour, and therefore requires a separate compartment of its own. It will corrode certain metals, e.g. copper. Has a lower saturated pressure than CO_2.
- *Freon (CCl_2F_2)*. Non-poisonous, non-corrosive and has a low saturated pressure. By far the most popular in modern tonnage.

Properties of a Good Insulating Material

1   *Odour.* All material used should be odourless to prevent tainting of cargoes.
2   *Vermin.* The material should be of such a nature, or so treated, that it will not harbour vermin.
3   *Moisture.* The material should not readily absorb moisture.
4   *Fire.* Insulation material should be non-combustible, if possible, but at least fire-resistant.
5   *Cost.* The financial outlay must be considered in view of the quantity of material required.
6   *Weight.* Not as important as one might think for merchant vessels; however, for ports with shallow water this would become a factor for consideration.
7   *Maintenance.* Costs of installing and maintaining the insulation in good condition should be considered at the building/fitting-out stage.
8   *Settling.* Value of the material is lost if, after settling, the air pockets left will necessitate repacking.
9   *Durability.* Must be considered in comparison to the life of the vessel.
10   *Strength.* A great advantage would be if the material was of such quality as to withstand impact when loading or discharging.

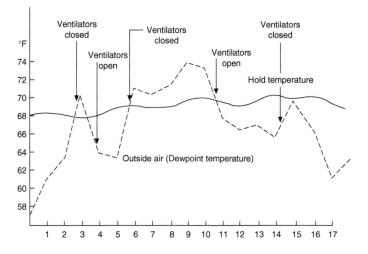

Figure 6.23 Hold temperature and outside dewpoint temperature graph.

Loading

Absolute cleanliness is required during the loading of refrigerated cargo, and the following points should be observed:

1 The compartment should be cleaned of all debris and previous cargo.
2 The deck should be scrubbed and the bulkheads and deck wiped with a light disinfectant.
3 All bilges must be cleaned and bilge suctions tested.
4 'Tween deck scuppers must be tested, together with all 'U' brine traps.
5 Bilge plugs should be inspected and sealed. The cover plug over bilge suction may be left off for the purpose of survey.
6 Fans must be checked for direction of air flow.
7 Bare steelwork must be insulated.
8 All odours must be cleared from the compartment.
9 All outside ventilation must be shut down.
10 Pre-cooling of the compartment must take place before the cargo is received, times being noted in the cargo log or deck log book.
11 Before loading, the compartment should be surveyed. The surveyor's comments, together with the opening temperature of the chamber, should be recorded in the Mate's deck log book.

Any dunnage required for the cargo should be of a similar standard of cleanliness as that of the compartment. All slings, chains, etc. should also be clean and pre-cooled in advance of cargo reception.

Carriage of Goods in Deep Tanks

Deep tanks are cargo compartments that may be used for the carriage of dry or liquid cargoes. They are usually found in dry cargo vessels at the bottom of one of the holds, forming what would normally be the lower hold portion of the hatch. Some vessels were built with deep tanks either side of the shaft tunnel (three-island type vessels), where they ran from the midships machinery space, aft.

The openings into the tank are:

- main lid
- manhole entrance
- ventilator trunkings
- sounding pipe (usually in the hat box or well)
- ullage pipes
- bilge suction line (into the hat box or well)
- ballast line
- CO_2 fire extinguishing line (not always fitted)
- steam inlet pipes for heating coils.

When the tank is to be used for dry cargo, the following actions should be carried out before loading the cargo:

1   Open CO_2, if fitted.
2   Blank off ballast line.
3   Check bilge suction and leave the bilge line open.
4   Blank off steam inlet to heating coils. Coils may sometimes be removed.
5   Open or close ventilator trunks, as required.

When the tank is to be used for liquid ballast, the following actions are necessary:

1   CO_2 lines should be blanked off.
2   Bilge line opened.
3   Steam inlet to heating coils sealed off.
4   Ventilator trunks opened.
5   Ballast bend fitted.
6   Main lid hard rubber packing should be inspected and checked for deterioration. If found to be in good condition, the locking bolts should be seen to be well screwed down to obtain even pressure on the seal. Manholes should be treated in a similar fashion.

Preparation of Deep Tanks to Receive Liquid Cargo

Tanks must be tested by a head of water equal to the maximum to which the tank may be subjected, but not less than 2.44 m above the crown of the tank. The rubber seal should be inspected for any signs of deterioration about the perimeter of the main lid. Any rubber gaskets about the inspection manholes should be seen to be in good order and to make a good air/water seal.

After the tank has been tested, it should be thoroughly cleaned and sealed. No rust spots or oil patches etc. should be visible. Hat boxes and wells should be meticulously cleaned and sealed off, and ballast and CO_2 lines blanked off. Pressure valves should be fitted into ventilators and the steam coils fitted and tested.

Once all preparations have been completed, the tank must be inspected by a surveyor before loading and a certificate of the tank's condition will be issued.

Table 6.1 Cargoes carried in deep tanks

Product	Specific gravity	Cu.ft per tonne
Coconut oil	0.925–0.931	38.8
Palm oil	0.920–0.926	38.9
Palm nut	0.952	37.5
Tallow	0.911–0.915	39.4
Whale oil	0.880–0.884	40.76

Dangerous/Hazardous Cargoes

This section applies to dry cargo/container ships or ro-ro vessels. In the event of any dangerous goods or harmful substances being carried aboard the vessel, reference should be made to 'The International Maritime Dangerous Goods (IMDG)' code. Additionally, the Chemical Data Sheets contained in the Tanker Safety Guide (Gas and Chemical) issued by the International Chamber of Shipping may be appropriate.

Such goods/substances must be classified, packaged and labelled in accordance with the Merchant Shipping Regulations. Such trailers or vehicles should be given special consideration when being loaded and inspected for leakage prior to loading on the vessel. Such vehicles/containers should also be provided with adequate stowage that will provide good ventilation in the event of leakage while in transit, e.g. upper deck stowage, exposed to atmosphere is recommended as a general rule.

Deck Officers should pay particular attention to the securing of such transports to ensure negative movement of the unit. Special attention should also be given to the securing of adjacent units to prevent escalation of cargo shifting in a seaway. Tank vehicles may not necessarily be carrying hazardous goods, but any spillage of the contents could act as a lubricant on surrounding units and generate a major cargo shift on ro-ro vessels in heavy seas.

In the event that a cargo parcel/unit is found to be 'leaking' or have exposed hazards, the nature of the cargo should be ascertained and personnel kept clear of the immediate area until the degree of hazard is confirmed. In any event, the unit should not be accepted for shipment and rejected until satisfactorily contained.

Where a hazardous substance is discovered at sea to be a threat to personnel, full information should be sought as soon as possible. Any action taken would depend on the nature of the substance and the

Plate 41 Modern vehicle ro-ro ferry operating with Norfolk shipping line.

emergency actions stipulated in carriage instructions. It may become prudent to seek additional instructions from the manufacturer of the substance and act accordingly.

> NB. With reference to Regulation 54 of SOLAS (1996 Amendment) in ships having ro-ro cargo spaces, a separation shall be provided between a closed ro-ro cargo space and the weather deck. The separation shall be such as to minimize the passage of dangerous vapours and liquids between such spaces. Alternatively, separation need not be provided if the arrangements of the closed ro-ro space are in accordance with those required for the dangerous goods carried on the adjacent weather deck.

The Shipping Procedure for the Loading and Transportation of Hazardous Goods

The shipping procedure for hazardous/dangerous goods is:

1   The shipper is responsible for obtaining the export licences for the shipment.
2   The shipper is also responsible for marking and labelling the goods to be shipped in accordance with the IMDG Code.
3   The shipper would then be in a position to contact the shipping company's agents and must provide: the number of packages, their weight, the value, the volume and any special requirements that may be required for the cargo.
4   Customs clearance would be required, and the goods may be liable to inspection.
5   The bill of lading is also liable to be endorsed, especially if packages are damaged and are rejected.
6   The goods would be listed in the ship's manifest and on the ship's cargo plan.
7   Ship's officers would check the details of the goods, including the labelling, the respective United Nations number, the condition of the package, together with any special stowage requirements prior to loading the cargo.
8   Throughout this procedure the ship's Master has the right to accept or reject the cargo prior to loading.

Once stowed on board the vessel the IMDG Code requirements would be followed throughout the period of the voyage.

Monitoring of Hazardous Cargoes

Different operators monitor the shipment of hazardous goods in various ways, but each vessel needs to be fully aware of the position of the cargo, its class and the emergency procedures that are involved if transport difficulties arise. Ferry operators tend to identify on a cargo stowage outline the position of the 'special unit', and the relevant details are recorded once it has arrived for shipment. Figure 6.24 is included as a guide.

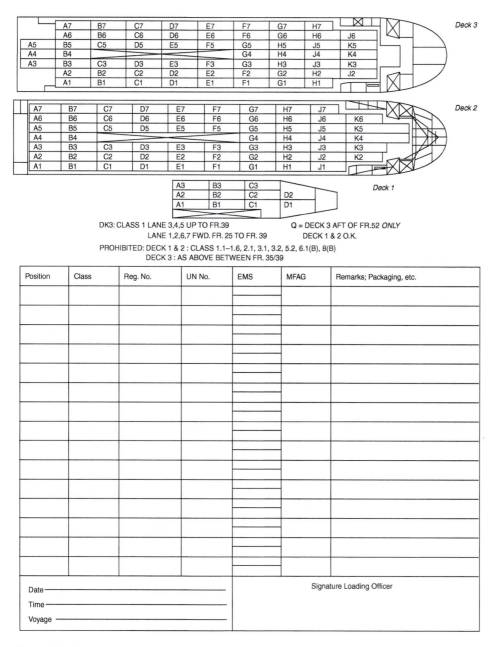

DK3: CLASS 1 LANE 3,4,5 UP TO FR.39   Q = DECK 3 AFT OF FR.52 *ONLY*
        LANE 1,2,6,7 FWD. FR. 25 TO FR. 39   DECK 1 & 2 O.K.
PROHIBITED: DECK 1 & 2 : CLASS 1.1–1.6, 2.1, 3.1, 3.2, 5.2, 6.1(B), 8(B)
        DECK 3 : AS ABOVE BETWEEN FR. 35/39

Position	Class	Reg. No.	UN No.	EMS	MFAG	Remarks; Packaging, etc.

Date ——————————————
Time ——————————————
Voyage ——————————————

Signature Loading Officer

Figure 6.24 Hazardous cargo stowage plan.

Cargo Plans

General Cargo Vessels

A ship's cargo plan shows the distribution as well as the disposition of all parcels of cargo aboard the vessel. The plan is usually formulated from the workbooks of the Deck Officers, a fair copy being produced before departure from the final port of loading. This allows copies of the plan to be made before the vessel sails. The copies are forwarded to agents at ports of discharge to allow the booking and reservation of labour, as appropriate.

The cargo plan should include relevant details of cargoes, i.e. total quantity, description of package, bales, pallets, etc., tonnage, port of discharge, identification marks and special features if and when separated. The port of discharge is normally 'highlighted' in one specific colour, reducing the likelihood of a parcel of cargo being overcarried to the next port. Cargoes which may have an optional port of discharge are often double-coloured to the requirements of both ports.

The plan should be as comprehensive as space allows. Consequently, abbreviations are a common feature, e.g. Liverpool as L'pool, 500 tonnes as 500 t, cartons as ctns, cases as c/s, and heavy lift 120 tonnes as H/L, 120 t. Additional information, such as the following, generally appears on most plans:

- name of the vessel
- name of the Master
- list of loading ports
- list of discharging ports, in order of call
- sailing draughts
- tonnage load breakdown
- hatch tonnage breakdown
- voyage number
- total volume of empty space remaining
- list of dangerous cargo, if any
- list of special cargo, if any
- statement of deadweight, fuel, stores, water, etc.
- details of cargo separations
- recommended temperatures for the carriage of various goods
- Chief Officer's signature.

The plan provides at a glance the distribution of the cargo and shows possible access to it in the event of fire or the cargo shifting. Its most common function is to limit overcarriage and the possibility of short delivery at the port of discharge. It also allows cargo operations, stevedores, rigging equipment, lifting gear and so on to be organized without costly delays to the ship.

Roll On–Roll Off System

Ro-ro methods of handling cargo have developed from the original container idea of a door-to-door service for the shipping customer. The concept is based on a quick turnaround, making the delivery not only fast and efficient, but very economical. A larger type of vessel has recently been constructed for the more lengthy voyage, with the combined rapid turnaround producing high-yield profits.

Plate 42 Bow visor arrangement on the ro-pax vessel *Jupiter.*

Plate 43 Stern door arrangement of a passenger vehicle ferry (source: Shutterstock).

Ro-ro ships are usually built with extensive fire-prevention systems, including total CO_2 flooding to all garage spaces, automatic sprinkler/water mist and/or water curtains, usually a foam installation, together with conventional water hydrants. Fire precautions are maintained to a high degree, with no smoking on vehicle decks, private cars to have limited fuel in tanks, etc. Regular drills and fire patrols are maintained and a smoke detection system must be employed.

Construction of this type of vessel usually includes such special features as longitudinally strengthened decks, clear of obstructions

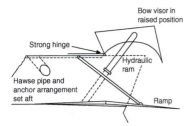

Figure 6.25 Visor type: bow door and ramp. All openings are usually above the waterline and the door is kept watertight by hard rubber packing and hydraulic pressure on closing.

such as pillars and spar ceiling; a forced ventilation fan system for the purpose of clearing exhaust fumes; internal ramps to upper decks and/or cargo lifts (electro/hydraulic) to lower levels; and a bow visor (Figure 6.25) with vehicle ramps fore and aft, allowing access into main garage areas. Built-in overhead lighting, welded struts for tyre grips and anti-roll ship stabilizers are also regular features of the ro-ro trade.

Hull Openings

Since the *Herald of Free Enterprise* incident, everybody has been aware of the dangers of water on vehicle decks, and the changes in regulations have incorporated a variety of measures to ensure safer operations (Figure 6.25). It is now a requirement that all openings, by way of the bow or stern doors, or side shell doors, are monitored by open/shut sensors and that relevant information is transmitted to a light sentinel on the ship's bridge. These openings must also be covered by CCTV and real-time visuals transmitted to the bridge.

Masters and ship's officers are subsequently in a position to observe all of the hull openings and ensure that the sentinel is showing that all access points are closed and watertight prior to taking the vessel off the berth. These regulations also cover passenger gangway access and store reception, and shell doors set into the hull sides.

Sentinels for monitoring watertight integrity tend to lead to simplicity with conventional red or green lights to show open or shut, respectively, for each hull access position. Pressure sensors being directly linked to the door's locking device, usually by either locking pins or cleating arrangements, register open or shut conditions. Although different operating methods for closures exist, the popular method is by hydraulics; positive pressure is placed on the seals before shutting down hydraulic systems.

Passenger Roll On–Roll Off Stability

Free surface effect

Vehicle decks usually extend throughout the length and breadth of this type of vessel, without restriction. Such design makes them vulnerable to the effects of free surface in the event of flooding taking place. In such an event the ship could rapidly lose all stability and capsize.

Regulations changed in 1990 to cause increased stability standards for passenger ships. More recently, ro-ro passenger vessels in northern Europe have been required to meet higher standards of stability criteria – with 50 cm of water on the deck – and some form of subdivision may well be required. IMO SLF 40 recommends that the main ro-ro deck be divided longitudinally into three compartments by two watertight bulkheads.

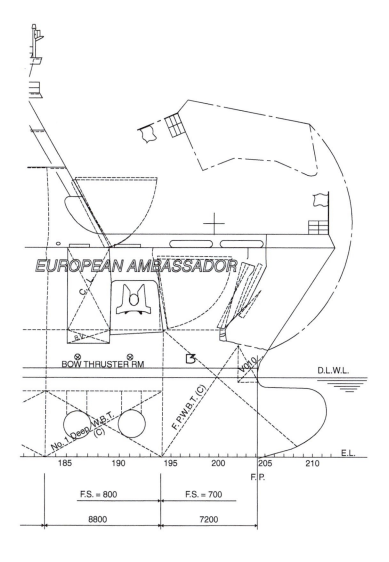

Figure 6.26 Example of a hull bow visor opening.

Precautionary measures include automatic draught gauges at the stem and the stern with remote readouts. These are expected to guard against flooding of the vehicle deck when the vessel is in port. Also, a loading computer must be accessible to the ship's officers in port to allow the rapid calculation of stability before the ship sails. Heavy ro-ro cargo units must be weighed ashore and such information passed to the ship's officers. Such units would normally be chained to the deck once loaded. Enclosed cargo spaces above the bulkhead deck are further required to have increased drainage requirements.

The overall purpose of such regulation changes is to ensure that such vessels leave port with sufficient stability to operate safely in adverse weather conditions or withstand the possible effects of collision and the loss of watertight integrity of the hull.

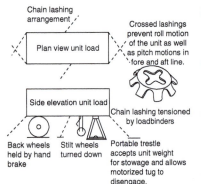

Figure 6.27 Ro-ro securing methods.

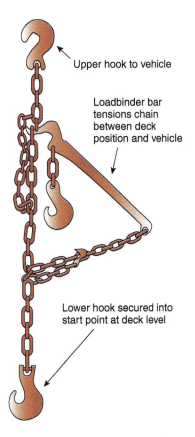

Plate 44 Example of a chain lashing to ro-ro units.

Securing Methods

All unit loads are disengaged from the loading tugmaster and secured by a minimum of six chain lashings (40 ft unit). Additional lashings would be secured to heavier or longer/wider loads. Each chain lashing is tensioned and locked by a bar lever or loadbinder. Various ports have their own systems: for example, some ports do not cross the end lashings on units, but leave them just in the fore and aft line (see Figure 6.27).

Lashings are secured to the deck in 'star insert' or 'star dome' securing points. A club foot is locked into the point while the other end of the lashing hooks onto the lugs of the unit. The star insert points are flush to the deck and are preferable to the raised dome securing points, which is illustrated.

List of Relevant Container Definitions and Terms

Administration

This refers to the government of a contracting party, under whose authority containers are approved.

Approved

This means approved by the administration.

Approval

This means the decision by the administration that a design type or a container is safe within the terms of the present convention.

Cargo

This is any goods, wares, merchandise and articles of every kind whatsoever carried in the containers.

Cell

This is space which could be occupied by a single vertical stack of containers aboard a container vessel. Each stowage/hatch space would contain multiple cells, each serviced during loading/discharging by 'cell guides'.

Cell Guide

A cell guide is a vertical guidance track which permits loading and discharge of containers in and out of the ship's holds in a stable manner.

Container

A container is an article of transport equipment:

- of a permanent character and accordingly strong enough to be suitable for repeated use;
- specially designed to facilitate the transport of goods by one or more modes of transport, without intermediate reloading;
- designed to be secured and/or readily handled, having corner fittings for these purposes;
- of a size such that the area enclosed by the four outer bottom corners is either,
  (i) at least 14 m² (150 sq ft); or
  (ii) at least 7 m² (75 sq ft) if it is fitted with top corner fittings.

The term 'container' includes neither vehicles or packaging. However, containers when carried on chassis are included.

Container spreader beam

This is the engaging and lifting device used by gantry cranes to lock on, lift and load containers.

Corner fitting

This is an arrangement of apertures and faces at the top and/or bottom of a container for the purposes of handling, stacking and/or securing.

Existing container

This is a container which is not a new container.

Flexible boxship

This is a container vessel designed with flexible-length deck cell guides, capable of handling different lengths of containers, e.g. 20 (6), 30 (9) and 40 feet (12 metres).

Gantry crane

This is a large heavy-lifting structure found at container terminals employed to load/discharge containers to and from container vessels. Some container vessels carry their own travelling gantry crane system on board.

Hatchless holds

This is a container ship design with cell guides to the full height of the stowage without separate or intermediate hatchtops interrupting the stowage.

International transport

This is transport between points of departure and destination situated in territory of two countries to at least one of which the present (CSC) convention applies. The present convention will also apply when part of a transport operation between two countries takes place in the territory to which the present convention applies.

Karrilift

This is the trade name for a mobile ground-handling container transporter. There are many variations of these container transporters found in and around terminals worldwide. Generally referred to as 'elephant trucks' or 'straddle trucks'.

Lashing frame/lashing platform

This is a mobile or partly mobile personnel carrier by which lashing personnel can work on twist-locks at the top of the container stack without having to climb on the container tops.

Maximum operating gross weight

This is the maximum allowable combined weight of the container and its cargo.

Maximum permissible payload (P)

This is the difference between the maximum operating gross weight or rating and the tare weight.

New container

This is a container the construction of which was commenced on or after the date of entry into force of the present convention.

Owner

This is the owner as provided for under the national law of the contracting party or the lessee or bailee, if an agreement between the parties provides for the exercise of the owner's responsibility for maintenance and examination of the container by such lessee or bailee.

Prototype

This is a container representative of those manufactured or to be manufactured in a design type series.

Rating (R)

See maximum operating gross weight.

Stack

This is the deck stowage of containers in 'tiers' and in 'bays'.

Safety approval plate

This is an information plate which is permanently affixed to an approved container. The plate provides general operating information inclusive of: country of approval and date of manufacture, identification number, its maximum gross weight, its allowable stacking weight and racking test load value. The plate also carries 'end wall strength', 'side wall strength' and the maintenance examination date.

Tare weight

This is the weight of the empty container, including permanently affixed ancillary equipment.

TEU

Twenty-foot equivalent unit. Used to express the cargo capacity of a container vessel.

Type of container

This is the design type approved by the administration.

Plate 45 The container vessel *P&O Nedlloyd Susana* lies portside to the gantry cranes in Lisbon, Portugal.

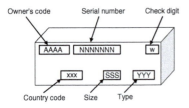

Figure 6.28 Container markings.

Type-series container

This is any container manufactured in accordance with the approved design type.

Loading Containers

The order of loading when the large container vessels are carrying currently up to 9500 TEUs must be well planned and considered as a detailed operation. Planners are usually employed ashore to provide a practical order of loading, particularly important when the vessel is scheduled to discharge at two or three or more terminal ports (Figure 6.32).

Once loading in the cell guides is complete, the pontoon steel hatch covers, common to container vessels, are replaced and secured. Containers are then stowed on deck in 'stacks' often as high as six tiers. The overall height of the deck stowage container stack may well be determined by the construction of the vessel. It must allow sufficient vision for bridge watchkeepers to be able to carry out their essential lookout duties. The stability criteria of the vessel, when carrying containers on deck, must also be compatible with the stowage tonnage below decks.

Any deck stowage requires effective securing, and this is achieved usually by a rigging gang based at the terminal. As the 'stack' is built up, each container is secured by means of specialized fittings between containers themselves and to the ship's structure.

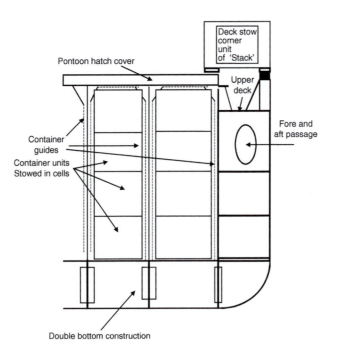

Figure 6.29 Container vessel construction.

Container Tonnage

The container is probably the most common unit load system in operation today. The introduction of standard-size containers took place in the late 1950s, and the container trade has flourished ever since. Most cargoes are shipped in container form, including heavy steel and liquids. Suitable refrigerated containers may also be used for the carriage of frozen and chilled foodstuffs, their plant power supply being connected to the vessel's main electrical source.

Containers of all sizes are generally loaded by a shoreside gantry crane travelling the length of the quay on trackways. These cranes are usually equipped with automatically controlled lifting mechanisms to facilitate the lifting and loading of units. The jib section of the gantry crane is lowered from the stowed elevated position after the vessel has berthed alongside. It would appear that in container operations one of the more sensitive areas for accidents to ships' personnel is ashore in the container stowage area. Straddle trucks, often referred to as elephant trucks, used for the transportation of containers from the park to the gantry crane, are driven by drivers in a highly elevated position. The field of view is somewhat restricted by the structure of the trucks, making the area extremely dangerous for unauthorized personnel. (Some terminals now use driverless transports.)

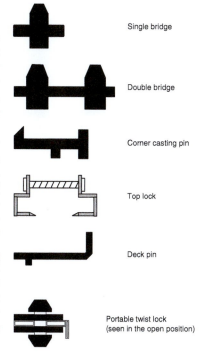

Single bridge

Double bridge

Corner casting pin

Top lock

Deck pin

Portable twist lock
(seen in the open position)

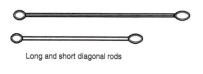

Long and short diagonal rods

Bottle Screw or Turnbuckle

Figure 6.30 Container lashing fitments.

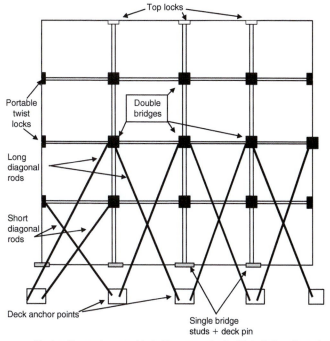

Short and long rods secured by bottle screw or turnbuckle to deck anchor points

Figure 6.31 Container deck stowage.

Plate 46 Ready to discharge containers (source: Shutterstock).

Plate 47 Container vessel alongside gantry cranes (source: Shutterstock).

Plate 48 Shoreside terminal showing container stacks (source: Shutterstock).

Plate 49 Loading containers into hatches (source: Shutterstock).

Plate 50 Cell guides exposed during the loading process of a container vessel. Containers above deck seen lashed in the stowed position.

The disadvantage of the container trade is that an empty container with no load to refill it becomes a liability, left at the wrong end of the trade route. Consequently, the majority of container-designed vessels will at some part of the voyage carry to and fro some empty units. This means a limited loss of revenue to the ship-owner, though a necessity for the continuation of the operation.

Ship's officers should be aware that containers should be loaded in an even manner, both athwartships and fore and aft, to maintain the stability of the vessel. Not all containers are in a fully loaded condition, so the centre of gravity (C of G) of the containers will vary. This will affect the final C of G of the vessel on completion. The problem is that unless each container is opened up on loading, the ship's personnel have only the shore authority's word with respect to the weight and C of G of the container. However, it should be remembered that most container terminals have means (weigh bridge) of checking container weights.

Personnel engaged in working container units should make additional reference to MGN 157.

Loadicators and Loading Plan Computers

Many ships are now equipped with loadicator systems or a loading computer with appropriate software. It is usually a conveniently sited visual display for the Master and the Loading Officers and is gainfully employed on ro-ro vessels, tankers and bulk carriers.

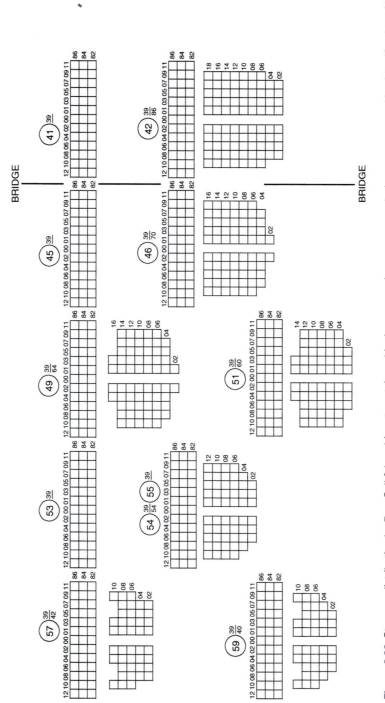

Figure 6.32 Stowage distribution by Bay, Cell & Level (most container shipping companies are now employing computer tracking identification of individual containers).

Plate 51 Twenty- and forty-foot container units stowed in the cell guides aboard a large container vessel.

Plate 52 Container lashing bars, seen secured from the pontoon to the second tier of deck containers.

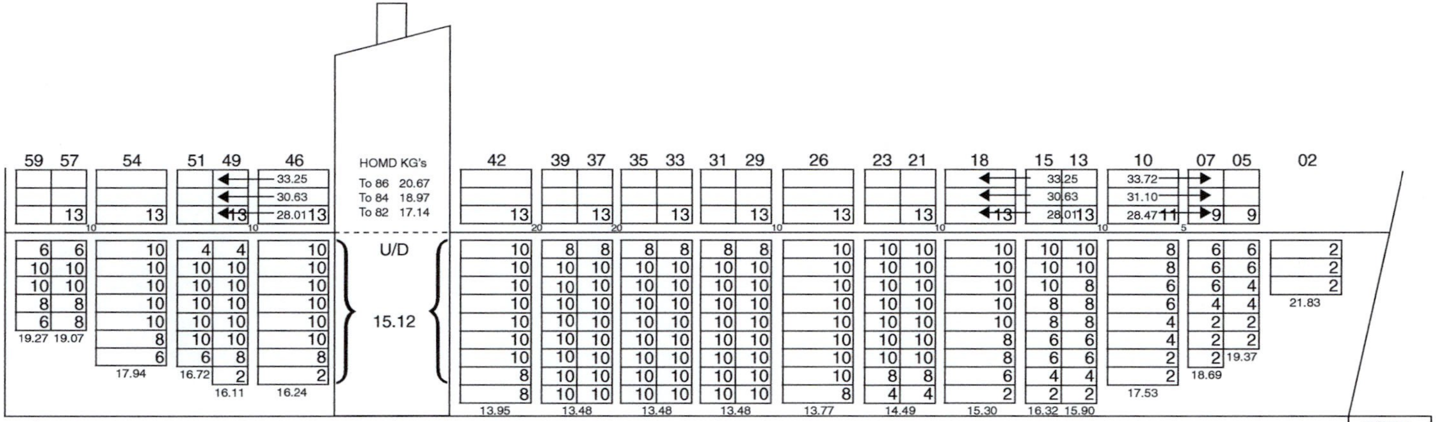

The modern type of container vessel will normally operate a container 'box' tracking system which allows continuous monitoring of any single container at any time during its transit. The plan allows a six figure number to track and identify its stowage position aboard the vessel. Distinct advantages of such a system tend to satisfy shipper enquiries as well showing that the shipping company is efficient in its business.

Other aspects of security are also clearly beneficial in a security conscious age.

An example tracking system could be typically:

The first two numbers of the six digit number = the identification of the 'bay' of stowage.

The second two numbers = the 'cell' of stowage.

The last two numbers = the level/tier of stowage.

Figure 6.33 Example container ship: cargo plan.

Plate 53 The container hoist engaged in lifting the pontoon hatch covers clear of the cellular holds.

Plate 54 Vehicles loading onto a ro-ro passenger ferry (source: Shutterstock).

The system should ideally be interlinked with the shoreside base to enable data transmissions on unit weights, tonnages or special stow arrangements. The computer will permit the location and respective weights of cargo/units to be entered quickly and provide values of limiting 'KG' and 'GM', together with deadweights at respective draughts/displacements. It will also have the capability to provide a printed record of the state of loading and show a visual warning in the event of an undesirable stability condition or overload occurring.

Distribution of the ship's tank weights, stores and consumables affecting final calculations and total displacement are also identifiable within the completed calculations; the primary aim of the loading computer is to ensure that the vessel always departs the berth with adequate stability for the voyage. If this situation can be achieved quickly, costly delays can be eliminated and safety criteria are complied with.

The data required to complete the stability calculations would need to be supplied by the shoreside base with regard to cargo weights. This, in turn, would be certified by the driver for ro-ro unit loads, obtaining a load weight certificate authorized by an approved 'weigh bridge' prior to boarding the vessel. Draught information will inevitably come from a 'draught gauge system' for the larger vessel and be digitally processed during the period of loading.

The ship's personnel could expect to become familiar with manipulation of the changing variables very quickly alongside the fixed-weight distribution throughout the ship (Figure 6.34). This would permit, in general, few major changes to the programme, especially on short sea ferry trade routes where limited amounts of bunkers, water and stores are consumed and values stay reasonably static.

Fixed weights are applicable to a variety of vehicles and, as such, where units are pre-booked for the sea passage an early estimate of the ship's cargo load and subsequent stability can often be achieved even prior to the vessel's arrival.

The loadicator programs provide output in the form of:

- shear forces and bending moments affecting the vessel at its state of loading;
- cargo, ballast and fuel tonnage distributions;
- a statement of loaded 'GM', sailing draughts and deadweight.

Bulk Carriers

Additional safety measures came into force for bulk carriers in July 1999 to enhance the structural safety and the resistance to flooding of bulk carriers. Reference to these changes can be made from MGN 144, and SOLAS Chapter 12 (adopted 1997 conference).

The regulations affect new bulk carriers over 150 m in length carrying high-density dry bulk cargoes over 1,000 kg/m^3. Existing vessels, if affected, have until they are 17 years old to comply if they are carrying cargoes with a density of 1,780 kg/m^3.

8,100 TEU design

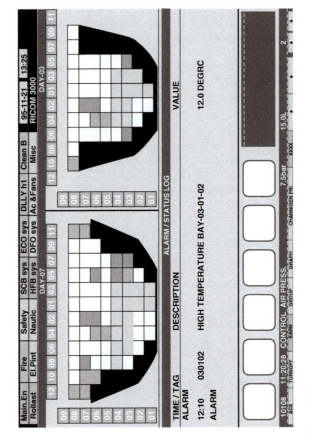

A typical container loadicator display which is capable of monitoring reefer units and their respective temperatures.

Loadicator displays similar to the container model are also available for tank level gauging for tanker vessels when loading/discharging and Roll On–Roll Off vessels for unit load operations.

Figure 6.34 Large container ship profile.

The regulation applies to:

- bulk carriers with the traditional structure of single deck, upper wing tanks and lower hopper tanks in cargo spaces;
- ore carriers with two longitudinal bulkheads and double bottom in cargo areas carrying ore cargoes in centre holds only;
- combination carriers, e.g. tankers capable of carrying dry cargo in bulk.

Bulk carriers must comply with a mandatory enhanced survey programme. No bulk carrier over ten years old may carry a high-density bulk cargo unless she has undergone a periodical survey or an equivalent survey of her holds.

Bulk carriers are subject to damage survivability checks were a collision to occur. Concerns are aimed at structural strength when high-density cargoes are carried, leaving large empty spaces in the cargo holds, so permitting a greater volume of sea water to enter in the event of a breach of the hull. To this end additional strengthening may be required for double bottoms and transverse bulkheads, especially in way of the forward hold(s).

Cargoes with a declared density in the range of 1,250 kg/m^3 –1,780 kg/m^3 must have their density verified by an accredited testing organization. Where imposition of loading/operating restrictions applies, as per regulation XII/6.3, ships must be marked with a triangle on the side shell of the vessel's hull. This means that restricted loading of cargoes with a density of 1,780 kg/m^3 and above is applicable.

SOLAS Guidance MSC 277 (85), effective 1 July 2010 requires that loading instrument and water ingress pumps, together with alarms, are fitted for vessels over 500 tons and over 100 m in length carrying dry bulk cargoes (water ingress monitors).

Tankers

Tanker cargo plans are constructed on the same principle as plans for general cargo ships. However, by the very nature of the cargo, it is only necessary to show disposition of the tank cargoes at one level. The plan proves especially useful when a number of differing grades or types of cargoes are to be loaded.

The plan should contain relevant information for the Loading/Discharge Officer and should include the following:

- grade of liquid
- weight of cargo in the tank
- the ullage of the tank
- volume and relative density at a specific temperature
- the carriage temperature
- slack tanks' identification
- empty tanks' identification

Plate 55 Fore end view of tanker vessel at sea; pipeline and ventilation arrangements illustrated.

- loaded draughts
- deadweight
- tonnage load breakdown
- Chief Officer's signature.

Colour schemes are usually employed to highlight the grade of cargo, the danger from contamination being greater than that from overstowage and overcarriage of cargo. It is not uncommon to see pipelines overprinted on the plan, enabling Cargo Officers to see clearly which lines are to be used for specific parcels of cargoes. This addition also lessens the risk of contamination.

Further reading on cargoes/handling can be obtained in House, D.J. (1998), *Cargo Work*, 7th edition, Kemp and Young.

7

Boatwork and Life-Saving Appliances

Introduction

The 1974 SOLAS convention and its two protocols of 1978 and 1988, together with amendments of May 2000 as adopted by resolution MSC 92 (72), entered into force on 1 January 2002, and are applicable to ships constructed on or after 1 July 1998.

Several major changes have taken place in the field of survival craft and life-saving appliances over the last decade. Account has been taken of different hull constructions, particularly relevant to high-speed craft and associated problems present with high freeboard vessels.

Manufacturers have devised novel ideas to satisfy increased legislation, and many old-fashioned ideas have been scrapped altogether. Equipment in motor propelled survival craft has been amended to suit the twenty-first century. Sails have been removed, automatic release systems have been established and free-fall boats are no longer considered abnormal.

With so much development riding on the back of the ever larger ship, every seafarer must be prepared to see increased changes in life-saving appliances. It should be realized that it is essential that the level of training and familiarization keeps pace with modernization.

Selected Terminology and Definitions

Active survival craft

This is a survival craft propelled by an engine.

Anti-exposure suit

This is a protective suit designed for use by rescue boat crews and marine evacuation parties.

Certificated person

This is a person who holds a Certificate of Proficiency in Survival Craft as issued by the marine authority or recognized as valid by the administration.

Code

This refers to the International Life-Saving Appliance Code, 2003 edition (Resolution MSC 48(66)), as required by Chapter III SOLAS.

Detection

This is the determination of the location of survivors or survival craft.

Embarkation ladder

This is the ladder provided at survival craft embarkation stations which permits safe access to survival craft after launching.

Fall-preventative devices

Following several accidents involving the launching of lifeboats fitted with on-load-release mechanisms the IMO (2009) has approved proposed guidelines for the fitting and use of fall-preventative devices (FPDs). These devices can be in the form of:

1   a locking pin designed to prevent the on-load release activating until it is removed, so preventing accidental release; or
2   a strop or sling to provide an alternative weight path, by being secured to the fall and to a boat lug fitment.

The effect is to provide a secondary weight path in the event the release hook fails. The fitting of FPDs is only considered as an interim risk-mitigating measure, being used with existing on-load-release systems, until these release systems can be changed to become compliant. Mechanisms that do not comply with the Life Saving Appliances (LSA) code must be replaced no later than the next scheduled drydocking of the ship (2010).

NB. The author would draw the attention of mariners to the need to maintain the lifeboat fall wires in good condition. In the event that the fall wire were to part, FPDs would not stop the boat from free falling.

Float-free launching

This is the method of launching a survival craft whereby the craft is automatically released from a sinking ship and is ready for use.

Free fall

See pages 222–227 for relevant definitions affecting free-fall lifeboats.

Immersion suit

An immersion suit is a protective suit which reduces the body heat loss of a person wearing it in cold water.

Inflatable appliance

This is an appliance which depends on non-rigid gas-filled chambers for buoyancy and which is normally kept deflated until ready for use.

Inflated appliance

This is an appliance which depends upon non-rigid, gas-filled chambers for buoyancy and which is kept inflated and ready for use at all times.

Launching appliance

This is a means of transferring a survival craft or rescue boat from its stowed position safely to the water.

Marine evacuation system

This is an appliance used for the rapid transfer of persons from the embarkation deck of a ship to a floating survival craft.

Novel life-saving appliance or arrangement

This is a life-saving appliance or arrangement which embodies new features not fully covered by the SOLAS text or by the code, but which provides an equal or higher standard of safety.

Passive survival craft

This is a survival craft which is not propelled by an engine.

Positive stability

This is the ability of a craft to return to its original position after the removal of the heeling moment.

Recovery time (for a rescue boat)

This is the time required to raise the boat to a position where a person on board can disembark to the deck of the ship. The recovery time includes preparation time on board the rescue boat for passing the painter and the time to raise the boat. It does not include the time to lower the launching device into a position to recover the boat.

Rescue boat

This is a boat designed to rescue people in distress and marshal survival craft.

Retrieval

This is the safe recovery of survivors.

Retro-reflective material

This is a material which reflects in the opposite direction a beam of light directed on it.

Survival craft

This is a craft capable of sustaining the lives of people in distress from the time of abandoning the ship.

Thermal protective aid (TPA)

This is a bag or suit made of waterproof material with low thermal conductance.

General Requirements for Lifeboats

1   All lifeboats shall be properly constructed and have ample stability in a seaway with sufficient freeboard when fully loaded with their full complement of persons and equipment. All lifeboats shall have rigid hulls and shall be capable of maintaining positive stability when in an upright position in calm water, fully loaded as described and holed in any one location below the waterline, assuming no loss of buoyancy material and no other damage.
2   Lifeboats should be of sufficient strength to:
    (a) enable them to be safely lowered into the water when loaded with their full complement and equipment;
    (b) be launched and towed when the ship is making headway at a speed of five knots in calm water.
3   Hulls and rigid covers shall be fire retardant or non-combustible.
4   Seating shall be provided on thwarts, benches or fixed chairs fitted as low as practicable in the lifeboat and constructed so as to be capable of supporting the number of persons, each weighing 100 kg.

5 Each lifeboat shall be of sufficient strength to withstand a load without residual deflection on removal of that load:

(a) In the case of boats with metal hulls, 1.25 times the total mass of the lifeboat when loaded with its full complement of persons and equipment.

(b) In the case of other boats, twice the total mass of the lifeboat when loaded, as stated.

Mariners should note that this requirement does not apply to rescue boats.

6 The strength of each lifeboat when fully loaded and fitted with skates or fenders where applicable should be capable of withstanding a lateral impact against the ship's side at an impact velocity of at least 3.5 m/s and also a drop into the water from a height of at least 3 m.

7 The vertical distance between the floor surface and the interior of the enclosure or canopy over 50 per cent of the floor area shall be:

(a) not less than 1.3 m for a lifeboat permitted to accommodate nine or fewer persons;

(b) not less than 1.7 m for a lifeboat permitted to accommodate 24 persons or more;

(c) not less than the distance as determined by linear interpolation between 1.3 and 1.7 m for a lifeboat permitted to accommodate between nine and twenty-four persons.

General Information Regarding Lifeboats

Access into lifeboats

1 Every passenger ship lifeboat shall be so arranged that it can be rapidly boarded by its full complement of persons. Rapid disembarkation shall also be possible.

2 Every cargo ship lifeboat shall be so arranged that it can be boarded by its full complement of persons in not more than three minutes from the time the instruction to board is given. Rapid disembarkation must also be possible.

3 Lifeboats shall have a boarding ladder that can be used on either side of the lifeboat to enable persons in the water to board. The lowest step of the ladder shall not be less than 0.4 m below the lifeboat's light waterline.

4 The lifeboat shall be so arranged that helpless people can be brought on board either from the sea or on stretchers.

5 All surfaces on which persons might walk shall have a non-skid finish.

6 No lifeboat shall be approved to accommodate more than 150 persons.

Lifeboat buoyancy

All lifeboats shall have inherent buoyancy or shall be fitted with inherently buoyant material which shall not be adversely affected by sea water, oil or oil products, sufficient to float the lifeboat with all its equipment on board when flooded and open to the sea. Additional inherent buoyancy material equal to 280 N of buoyant force per person shall be provided for the number of persons the lifeboat is permitted to accommodate. Buoyant material, unless in addition to that required above, shall not be installed external to the hull of the boat.

Lifeboat freeboard and stability

All lifeboats, when loaded with 50 per cent of the number of persons the lifeboat is permitted to accommodate in their normal positions to one side of the centre line will be stable and have a positive GM value, and have a freeboard measured from the water-line to the lowest opening through which the lifeboat may become flooded, of at least 1.5 per cent of the lifeboat's length or 100 mm, whichever is greater.

Each lifeboat without side openings near the gunwale shall not exceed an angle of heel of 20° and shall have a freeboard measured from the waterline to the lowest opening through which the life-boat may become flooded of at least 1.5 per cent of the lifeboat's length or 100 mm, whichever is greater.

Lifeboat markings

A lifeboat should be permanently marked with the dimensions of the boat and the number of persons it is permitted to accommo-date. The name and port of registry of the ship to which the lifeboat belongs shall be marked on each side of the lifeboat's bow in block capitals of the Roman alphabet. The mark for identifying the ship to which the boat belongs and the number of the lifeboat shall be visible from above.

Parts of the Lifeboat

The modern lifeboat is normally made of metal or fibreglass, with some wooden internal fitments. The days of all-wood construction have all but disappeared, but the names of the parts of the wooden boat have been handed down to their modern counterparts (see Figure 7.1) and wooden boats will continue to exist for many years worldwide. Lifeboats are illustrated in Plates 57–66.

Gunwale Construction

The upper part, previously referred to as the gunwale capping, was a fore and aft member which provided additional strength and pro-tection to the gunwale itself and the sheer strake. It was pierced at regular intervals to accommodate the crutches, or cut away when rowlocks were fitted.

With extensive changes in the builds and designs of modern lifeboats, many of the old wooden boat terms have been dropped from everyday use. However, totally and partially enclosed boats manufactured in glass reinforced plastic (GRP) have retained the old terminology as and where appropriate.

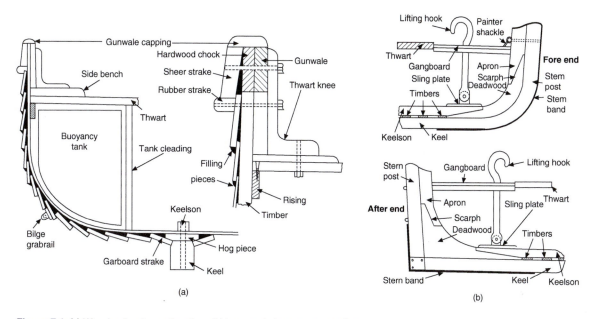

Figure 7.1 (a) Wooden boat construction, (b) bow and stern cross-section.

The gunwale was a strength member running in a fore and aft direction and backed by a hard wooden chock up to the filler piece and the sheer strake. This build-up has been replaced in modern boats by metal or fibreglass moulded into a rounded section, so obtaining a prefabricated strength section. The rubber is incorporated on the outside of this section in the early course of manufacture.

Plate 56 Boat deck and survival craft launching exercise from the *Queen Mary 2* while portside to in Southampton. The boats being launched are of the partially enclosed type and a semi-rigid rescue boat is seen at the surface. The starboard rescue boat is also seen stowed in its launching position at the boat deck level

Side benches and thwarts are still manufactured in wood in many cases, and secured by metal knees. The ribs of the old wooden boats, known as timbers, have been rendered obsolete by the increased strength of materials used in modern manufacture; however, the rising, another fore and aft member, is still fitted in most new builds. Previously used to lash the lanyards of the crutches to and support the thwarts, the rising still performs the same function, except that crutches are often on short chains and secured under the gunwale area.

Fore and After End Construction

The stem and stern posts, fore and aft, are the most upright components of the boat, and are now generally moulded section in GRP. When they were made of wood, it was common practice to have a protective stem band of metal over the prow and a wooden apron piece backing inside the boat. The apron was used to secure the planking on either side. New construction is no longer fitted with the apron as such, and the moulded sections are welded or fused directly as the stem and stern posts.

The bow and stern sheet areas still incorporate lower breast hooks, but these generally form an extension of the side benches. The gangboard is a wood or metal board joining the port and starboard sides of the breast hook, together with the stem and stern bars, to the forward and aftermost thwarts. The lifting hook passes up through the gangboard.

Fitments will vary from boat to boat, but the majority are supplied with a ring bolt or eye on the inside of the stem post to facilitate securing the permanent painter. They also have a fairlead sited on the gunwale for use with the sea anchor.

Internal Construction

The *keel* of the more modern boat is made of metal or reinforced glass fibre section. Running in the fore and aft line, it forms a natural extension between the stem and stern posts. The athwartships strength, previously provided by timbers, is now obtained by metal ribs welded or riveted from the keel up to the gunwale. These ribs still retain the term 'timbers', but as they are manufactured, if fitted, in metal, the term is technically incorrect. They are not as frequent as the old wooden timbers used to be because additional strength is obtained by the use of metal/GRP instead of wood.

Covering the keel and the timbers are the wood bottom boards of the boat, providing a safe walking area. When movement about the boat becomes necessary, then safety for the boat and the individual is improved by walking on the bottom boards. It is bad practice to step on the thwarts, and once on board, individuals should endeavour to lower the centre of gravity (C of G) of the boat by keeping their own weight as low as possible. The bottom boards generally tend to be secured in between the sides of the buoyancy tanks, which are positioned hard up against the outer hull on either side of the boat.

The number of thwarts in the boat will be determined by the boat's length. They not only provide the seating within the boat, but also act as athwartship strength members, tying the upper sides of the boat together. Similarly, the side benches, used to seat survivors in a crowded boat, also act as stringers, strengthening the upper parts of the boat in the fore and aft line. It should be noted that circumstances may dictate the use of the bottom boards as seats: for example, when beaching a lifeboat.

Additional fitments may come in the form of wood stretchers for use by crew members when rowing: these are often adjustable for the length of leg of the individual.

Types of Wood Construction

Ships' lifeboats are no longer of wood construction.

Carvel

In this form (Figure 7.2) the planks are laid end on to each other, as opposed to the overlapping of clinker construction. A flush finish is obtained by caulking between the joins of the planks where they butt together. Not the most stable or the strongest build of boat, it is nevertheless tried and tested, and withstands considerable wear.

Clinker

This is the most stable of all the builds of wooden boats (Figure 7.2). The planks overlap, providing an anti-roll property and double strength at the joins of the planks. Filler pieces are used to effect a tight fit plank-on-plank, and copper nails to secure the planks to the timbers. The build is popular for boats under 10 m in length because of the stabilizing factor of the overlapping planks.

Water resistance to the hull is extensive and the speed attained by the boat under the same conditions would not be the same as that of the carvel-built boat. The clinker is easy to repair should the outer planks become damaged; a section of the plank is simply removed and replaced, with little of the trouble that it would take to repair a carvel-built boat.

Double-Diagonal

This is the strongest of the wooden-built boats and the most expensive (Figure 7.3). It is double-planked laid diagonally from the keel, forming a double-skin effect. Two layers of strakes are laid in the direction of opposing diagonals, with a layer of calico or canvas between the plank layers. This flax skin is usually glued to the inner skin and painted, providing a waterproof seal in its own right.

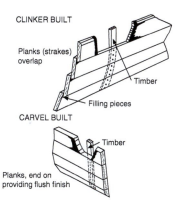

Figure 7.2 Clinker and carvel building. Planks are secured to timbers by the use of copper or similar type nails. Watertightness is achieved by the use of washers, known as 'roves', for each nail.

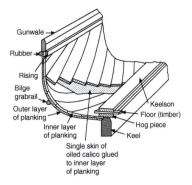

Figure 7.3 Carvel double-diagonal building.

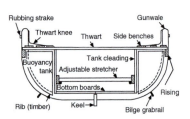

Figure 7.4 Cross-section of modern lifeboat.

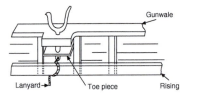

Figure 7.5 Crutch.

The outer layer of planking is butted edge to edge so as to give a carvel effect, reducing water frictional resistance.

This build of boat was used where additional strength was required, as with air/sea rescue craft. Economic reasons alone have restricted its use and the expertise in building this type of hull is becoming a thing of the past. Modern methods of obtaining strength in construction are by the use of glass fibre, with, in some cases, the incorporation of a wooden framework in the early stages of building.

Boat Fitments

A cross-section of a modern lifeboat appears in Figure 7.4.

Bilge Grabrail

Made either of metal or wood, this rail is fitted at the turn of the bilge on the outside of the boat's hull. Its length will vary, but is approximately two-thirds of the boat's length. Its purpose is to provide a handhold for persons in the water, and it would without doubt provide a useful footrest if trying to right a capsized boat.

Buoyancy Chambers

These are sealed air tanks fitted under the side benches, providing the internal buoyancy of the boat. They are held in position by a wooden supporting frame called the 'tank cleading'. The total volume of the chambers should be equal to at least one-tenth of the volume of the boat.

They are manufactured from non-corrosive metals such as copper or galvanized steel. Copper and brass are expensive, so the galvanized metals are more popular today. The chambers are often coated with varnish or boiled linseed oil to protect them from corrosion.

Crutches

The crutch (Figure 7.5) is a metal 'U'-shaped support to accommodate the loom/shaft of the oar when rowing. A metal pin supports the 'U' and the pin is fitted into a socket set into the gunwale. To prevent loss each crutch is secured by a lanyard to a fore and aft member of the boat, e.g. stringer or rising. The crutches are positioned just aft of the pulling thwarts to give crew members comfortable pulling positions.

Lifting Hooks

Normally these are single hooks at each end of the boat, facing inwards, to accommodate the blocks of the falls. Some boats in passenger vessels may be fitted with a ram's horn style of lifting hook, but the majority of vessels are equipped with the single lifting hook.

They are secured to a lugged sling plate referred to as a 'keel plate'. This plate in turn is held fast by holding down bolts passing through to the keel. The rigidity to the structure is improved by the metal stem from the hook being supported by a gangboard.

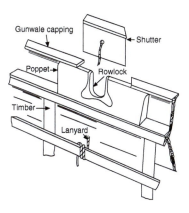

Figure 7.6 Rowlock.

Rowlock

The term rowlock is widely misused in many boating circles for 'crutch'. The rowlock (Figure 7.6) is in fact a cut-away made in the gunwale of the boat to perform the same function as the crutch, hence the common mistake. Larger builds of boat may have the rowlock cut into a wash strake, set over and above the gunwale. When not in use, as when the boat is sailing, the rowlocks are covered by a shutter/poppet arrangement, so leaving a flush gunwale capping which prevents the sheets becoming fouled if the boat is under sail.

Skates

A ship's lifeboats, except the emergency boat, on passenger vessels are equipped with skates or fenders to help them skid down the ship's side when being launched. The skates are secured on the inboard side of the lifeboat, and perform no function other than assisting the successful launch. To this end, once the vessel has sunk they may be discarded. However, circumstances may dictate that, after the boat is launched, the possibility of returning to the parent vessel may arise. Should this be the case, then the skates may be required a second time.

They are normally made of wood or metal, and hinged to meet the curved lines of the boat's hull. A wire passing under the keel secures the skate in position on the offshore side of the boat.

A rescue boat may be a designated lifeboat; as such it may also be fitted with skates.

Small Gear Locker

This is a watertight tank containing the food rations and all other small gear carried in the boat. There may be as many as three separate lockers, each clearly labelled and provided with screw sealing lids. Some types of locker are provided with metal keys.

Thole Pins

A single thole pin used in conjunction with a grommet was the old-fashioned method of securing an oar when rowing or sculling. A more popular method today would be to use two thole pins (Figure 7.7) and place the oar between them. However, even this use of thole

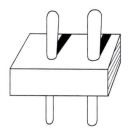

Figure 7.7 Thole pins.

pins has become obsolete and they are rarely used in conjunction with oars.

Thole pins are, however, usefully employed in a sailing boat, for crew to take a turn about them with the sheets and halyards. Modern sailing rigs often use miniature thole pins in and about the mast thwart. They are usually made of brass, not wood. Heavy wooden thole pins are to be found aboard the larger sailing craft, schooners and the like.

Water Containers

Metal or fibre glass tanks of a non-corrosive nature, water containers are usually sited under the thwarts of the boat. Each tank is fitted with a large filling hole, which is also used, with the dipper, to extract the water. The water should be frequently changed to ensure that it remains in a 'sweet condition', and to this end a drain plug is fitted to the lower part of each tank. It is common practice to wash out the tank before refilling (alternative – canned water).

Standard Lifeboat Equipment

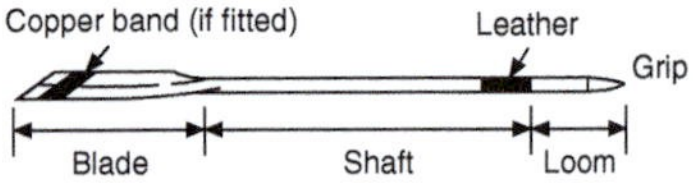

Figure 7.8 Oar.

All ship's lifeboats are now motorized. As such reliance on oars has diminished and sails have been totally removed.

1   Except for free-fall lifeboats, sufficient buoyant oars to make headway in calm seas. Thole pins, crutches or equivalent arrangement shall be provided for each oar provided. Pulling oars are normally between 3.05 and 4.26 m in length (10–14 ft), are generally made of ash or elm wood and are stowed with their blades facing forward.

A steering oar, which is no longer specified, if carried, is usually approximately (12 in.) 0.3 m longer than the pulling oars. Its blade faces aft and is usually coated with a distinctive colour. It is used extensively to provide additional leverage in order to steady the boat's head when used in conjunction with the sea anchor.

2   Two boat hooks, to be left unlashed and ready for use in fending away from the ship's side.

3   A buoyant bailer and two buckets. These are secured by lanyards to the structure of the boat. Buckets are usually of a two gallon size (9 l) and manufactured in galvanized iron or rubber, stowed either end of the boat.

4   A survival manual.

5   Two axes (hatchets) stowed one at each end of the boat. It is common practice to cover the metal head of the axe with a canvas protective cover to prevent the metal from pitting and corroding.

6   A jack-knife to be kept attached to the boat by a lanyard. The blade normally incorporates a tin opener and screwdriver, and a small hand spike is usually attached.

7   Two buoyant rescue quoits, attached to not less than 30 m of buoyant line. These are normally stowed in the small gear locker.

8   Six doses of anti-seasickness medicine and one seasickness bag for each person the boat is permitted to accommodate. The medicine is normally in tablet form.

9   A manual pump (unless the lifeboat is automatically self-bailing). Usually fixed to the structure of the boat. It is fitted with an easily removed cover to allow cleaning and the suction end contains a gauze filter to avoid blockage of the system.

10 A sea anchor of adequate size fitted with shock-resistant hawser and a tripping line which provides a firm hand grip when wet. The strength of the hawser and the tripping line shall be adequate for all sea conditions.

11 Four rocket parachute flares which comply with the regulations.

12 Six hand flares (red) which comply with the regulations.

13 Two buoyant smoke floats (orange) which comply with the regulations.

14 One waterproof electric torch suitable for Morse signalling, together with one spare set of batteries and one spare bulb in a waterproof container.

15 One whistle or equivalent sound signal. Normally of plastic construction of the non-pea design. This will allow its use in cold weather without discomfort to the user.

16 One daylight signalling mirror with instructions for its use for signalling to ships and aircraft (see life raft equipment list for use).

17 An efficient radar reflector, unless a survival craft radar transponder is stowed in the boat.

18 One copy of the life-saving signals table, prescribed by regulation V/16 on a waterproof card or in a waterproof container.

19 Two efficient painters of a length equal to not less than twice the distance from the stowage position of the lifeboat to the waterline in the lightest seagoing condition or 15 m, whichever is the greater. One painter attached to the release device, placed at the forward end of the lifeboat, must be capable of being released when under tension. The other painter shall be firmly secured at or near the bow of the lifeboat ready for use.

20 A binnacle containing an efficient compass which is luminous or provided with suitable means of illumination. In a totally enclosed boat the binnacle shall be permanently fitted at the steering position, in any other lifeboat it shall be provided with suitable mounting arrangements.

When setting up a boat's compass, the mariner should bear in mind that it must be visible to the coxswain and a fore and aft line may have to be set up between the stem and stern to provide reference for means of aligning the boat's head to the lubber line.

21 Sufficient tools to allow minor adjustment to the engine and its accessories.

22 Portable fire extinguishing equipment suitable for extinguishing oil fires.

23   A searchlight, capable of effectively illuminating a light-coloured object at night having a width of 18 m at a distance of 180 m for a total period of six hours and of working continuously for not less than a three-hour period.

24   Thermal protective aids which comply with the regulations, in sufficient number for 10 per cent of the total number of persons that the boat is permitted to carry, or two, whichever is greater.

25   A watertight receptacle containing a total of three litres of fresh water for each person the lifeboat is permitted to accommodate. One litre of this amount may be replaced by a de-salting apparatus capable of producing an equal amount of fresh water in two days.

26   A rustproof dipper with a lanyard, used for extracting fresh water from the containers. The lanyard should be long enough to reach the bottom of any water tank.

27   A rustproof graduated drinking vessel.

28   Three tin openers.

29   One set of fishing tackle.

30   A food ration totalling not less than 10,000 kJ for each person the lifeboat is permitted to accommodate. These rations shall be kept in airtight packaging and be stowed in a watertight container.

31   A first aid outfit in a waterproof case capable of being closed tightly after use.

All items of equipment of the lifeboat, with the exception of the two boat hooks, should be secured by lashings or kept in storage lockers, or secured by brackets or other similar mounting arrangement. Considerable changes in standard equipment have taken place with the 1983 amendment to the SOLAS convention.

The Sea Anchor

The sea anchor is a cone-shaped canvas bag, open at both ends, one end being much larger than the other (680 mm in diameter). It is usual to stream the sea anchor over the bow when heaving to, allowing the boat to ride head to wind and sea. However, it may be used as a drag effect over the stern when the boat has a following sea and is running before the wind.

To stream the sea anchor:

1   Bring the boat head to wind.
2   Unship the tiller and rudder.
3   Rig the steering oar over stern through the grommet.
4   Stream the sea anchor over the bow, paying out hawser and tripping line from the same side of the boat, if practical.

Lifeboat (Engine) Propulsion

In accordance with the 1983 amendment to the SOLAS convention 1974, every lifeboat shall be powered by a compression ignition engine.

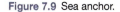

Figure 7.9 Sea anchor.

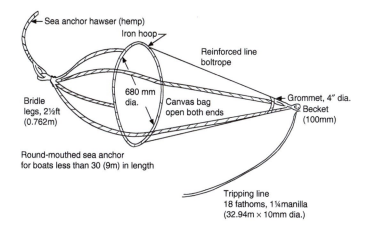

Sea anchor hawser (hemp)

Iron hoop

Reinforced line
boltrope

680 mm
dia.

Canvas bag
open both ends

Bridle
legs, 2½ft
(0.762m)

Grommet, 4″ dia.
Becket
(100mm)

Round-mouthed sea anchor
for boats less than 30 (9m) in length

Tripping line
18 fathoms, 1¼manilla
(32.94m × 10mm dia.)

No engine shall be used for any lifeboat if its fuel has a flashpoint of 43°C or less (closed cup test).

1   The engine shall be provided with either a manual starting system or with a power starting system with two independent rechargeable energy sources. Any necessary starting aids shall also be provided. The engine starting systems and starting aids shall start the engine at an ambient temperature of –15°C within two minutes of commencing the start procedure unless, in the opinion of the authority having regard to the particular voyages in which the ship carrying the lifeboat is constantly engaged, a different temperature is appropriate. The starting systems shall not be impeded by the engine casing, thwarts or other obstructions.

2   The engine shall be capable of operating for not less than five minutes after starting from cold with the lifeboat out of the water.

3   The engine shall be capable of operating when the lifeboat is flooded up to the centre line of the crank shaft.

4   The propeller shafting shall be so arranged that the propeller can be disengaged from the engine. Provision shall be made for ahead and astern propulsion of the lifeboat.

5   The exhaust pipe should be so arranged as to prevent water from entering the engine in normal operation.

6   All lifeboats shall be designed with due regard to the safety of persons in the water and to the possibility of damage to the propulsion system by floating debris.

7   The speed of a lifeboat when proceeding ahead in calm water, when loaded with its full complement of persons and equipment and with engine-powered auxiliary equipment in operation, shall be at least six knots and at least two knots when towing a 25-person life raft loaded with its full complement of persons and equipment, or its equivalent. Sufficient fuel, suitable for use throughout the temperature range expected in the area in which the ship operates, shall be provided to run the fully loaded boat at six knots for a period of not less than 24 hours.

8   The lifeboat engine transmission and engine accessories shall be enclosed in a fire-retardant casing or other suitable arrangements providing similar protection. Such arrangements shall also protect persons from coming into accidental contact with hot or moving parts and protect the engine from exposure to weather and sea. Adequate means shall be provided to reduce noise from the engine. Starter batteries shall be provided with casings which form a watertight enclosure around the bottom and sides of the batteries. The battery casing shall have a tight-fitting top which provides for necessary gas venting.

9   The lifeboat engine and accessories shall be designed to limit electromagnetic emissions so that the engine operation does not interfere with the operation of radio life-saving appliances used in the lifeboat.

10   Means shall be provided for recharging all engine-starting, radio and searchlight batteries. Radio batteries shall not be used to provide power for engine starting. Means shall be provided for recharging lifeboat batteries from the ship's power supply at a supply voltage not exceeding 55 V, which can be disconnected at the lifeboat embarkation station.

11   Water-resistant instructions for starting and operating the engine shall be provided and mounted in a conspicuous place near the engine controls.

Propulsion Requirements for Totally Enclosed Boats

The engine and transmission shall be controlled from the helmsman's position. The engine and its installation shall be capable of

Plate 57 Totally enclosed motor propelled survival craft, stowed in davit arrangement. Launch control wire clearly seen above the coxswain station (source: Shutterstock).

running in any position during capsize and continue to run after the lifeboat returns to the upright or shall automatically stop on capsizing and be easily restarted after the lifeboat returns to the upright. The design of the fuel and lubricating systems shall prevent the loss of fuel and the loss of more than 250 ml of lubricating oil from the engine during capsize.

In the case of air-cooled engines, they shall have a duct system to take in cooling air from, and exhaust it to, the outside of the boat. Manually operated dampers shall be provided to enable cooling air to be taken from, and exhausted to, the interior of the lifeboat.

Totally Enclosed Survival Craft

These craft (Figure 7.10) are invariably made of GRP. They have proved themselves in practice to be a worthwhile advance in the marine field of survival craft, especially with regard to heat-resistance and the exclusion of toxic fumes. These qualities are now statutory for parent vessels engaged in the tanker trades.

Totally enclosed survival craft are self-righting, even when fully laden. They contain their own internal air supply, together with a pump for providing an external water mantle to reduce fire hazards.

Tests have shown that a comfortable temperature is maintained inside the craft when outside temperatures have exceeded 1150°C. This desirable quality has been achieved by high standards of fire-resistant resin in the glass fibre construction, in conjunction with the sprayed water mantle from the built-in nozzles about the exterior hull.

These craft are popular not just in oil tankers, but also in the offshore oil industry on rigs and platforms. When launching takes place, all hatches should be closed. The release gear may be operated from inside the craft, reducing the risk of accident from floating blocks, which may occur with the conventional davit and open boat systems.

The craft are manufactured with varying carrying capacity and engine size. An example of the performance and capacity of the

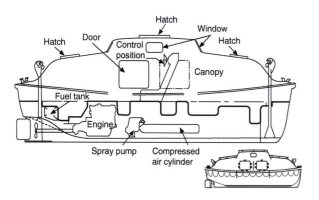

Figure 7.10 Totally enclosed lifeboat.

Plate 58 Partially enclosed rescue/
lifeboat.

totally enclosed survival craft, as built by Aberglen–Welin, is as
follows:

Capacity	Length (m)	Speed (knots)
21	6.2	6
35	7.3	6
50	8.0	6
66	8.75	6

Speeds can be increased by installing a larger engine, but a loss
in carrying capacity would result.

Plate 59 A 60-man partially
enclosed Harding lifeboat seen in the
stowed position with the self-launch
control system seen fixed above the
coxswain's position.

Plate 60 Totally enclosed lifeboat seen in the battened down situation at sea.

The adopted amendments to SOLAS 1974 have made additional requirements for passenger vessels and cargo ships, which became effective from 1 July 1986 for new tonnage.

Passenger Ships

Those engaged on international voyages which are not short international voyages shall carry lifeboats complying with the requirements of: regulations for partially enclosed boats, self-righting partially enclosed boats and totally enclosed boats on each side of such aggregate capacity as will accommodate not less than 50 per cent of the total number of persons aboard.

Plate 61 Totally enclosed boat in davit arrangement.

Substitution of lifeboats by life rafts will be permitted provided that there shall never be less than a sufficient number of lifeboats on each side of the vessel to accommodate 37.5 per cent of the total number of persons on board. Life rafts will be served by launching apparatus.

In addition, there should be life rafts of such aggregate capacity as will accommodate 25 per cent of the total number of persons on board. These life rafts should be served by at least one launching device on each side, which may be those as stated above.

Cargo Ships (Except Oil Tankers, Chemical Tankers and Gas Carriers)

One or more lifeboats should be carried complying with the requirements of regulations for totally enclosed boats, of such aggregate capacity on each side of the ship as will accommodate the total number of persons on board.

Cargo ships, if operating under favourable climatic conditions and in suitable areas, may carry boats which comply with regulations for self-righting partially enclosed boats, provided the limits of the trade area are specified in the cargo ship safety equipment certificate, and life raft or life rafts are capable of being launched either side and are of such aggregate capacity as will accommodate the total number of persons on board. If the life rafts cannot be readily transferred on each side, the total capacity available each side shall be sufficient for all persons on board.

Alternative for Cargo Ships

One or more totally enclosed boats capable of being 'free-fall launched, over the stern'. The aggregate capacity of such craft must accommodate all persons on board. In addition there should be life rafts either side of the ship to accommodate all persons on board, and at least one side of the ship shall be served by a launching device.

Cargo Ships Less than 85 Metres in Length

Cargo ships other than oil tankers, gas carriers and chemical tankers may comply with the regulations by carrying one or more life rafts on either side of such aggregate capacity for all persons on board. Unless these life rafts can be readily transferred for launching either side, additional life rafts shall be provided so that the total capacity available on each side will accommodate 150 per cent of the total number of persons on board.

Oil Tankers, Chemical Tankers and Gas Carriers

Those carrying cargoes having a flash point not exceeding 60°C (closed cup test) shall carry fire-protected, totally enclosed boats.

In the case of chemical and gas carriers carrying cargoes which give off toxic vapours, the lifeboats carried must be equipped with a self-contained air-support system which complies with the regulations.

Requirements for Totally Enclosed Lifeboats

Every totally enclosed lifeboat shall be provided with a rigid enclosure. The enclosure shall be so arranged that:

1　It protects the occupants against heat and cold.
2　Access into the lifeboat is provided by hatches which can be closed to make the boat watertight.
3　Hatches are positioned so as to allow the launching and recovery operations to be performed without any occupant having to leave the enclosure.
4　Access hatches are capable of being opened and closed from both inside and outside and are equipped with means of holding them securely in the open positions.
5　It must be possible to row the lifeboat.
6　It is capable, when the lifeboat is in the capsized position with the hatches closed and without significant leakage, of supporting the entire mass of the lifeboat, including all equipment, machinery and its full complement of persons.
7　It includes windows or translucent panels on both sides which admit sufficient daylight to the inside of the lifeboat with the hatches closed to make artificial light unnecessary.
8　Its exterior is of a highly visible colour and its interior of a colour which does not cause discomfort to the occupants.
9　Handrails provide a secure handhold for persons moving about the exterior of the lifeboat and aid embarkation and disembarkation.
10　Persons have access to their seats from an entrance without having to climb over thwarts or other obstructions.
11　The occupants are protected from the effects of dangerous sub-atmospheric pressures which might be created by the lifeboat's engine.

Capsize and Re-righting

The boats shall all be fitted out with safety seatbelts designed to hold a mass of 100 kg when the boat is in the capsized position. To this end it is essential that the occupants, once embarked, are securely strapped into the seated areas to ensure the self-righting property of the boat becomes a viable proposition. Also, all hatches and access doors are battened down and are seen to be in a watertight condition.

The design of the boats should be such that in the capsized situation the boat will attain a position which provides an above-water escape. Exhausts and engine ducts will be so designed as to prevent water entering the engine during a capsized period.

Launching when Parent Vessel Is Making Way

Cargo ships of 20,000 tons gross tonnage and upwards should have lifeboats capable of being launched, where necessary utilizing painters, with the ship making headway at speeds up to five knots in calm water.

Release Mechanism

Every lifeboat to be launched by a fall or falls shall be fitted with a release mechanism which complies with the following:

1   The mechanism shall be so arranged that all hooks release simultaneously.
2   The mechanism shall have two release capabilities, namely:
    (i)   a normal release capability which will release the craft when waterborne or when there is no load on the hook;
    (ii) an on-load release capability which will allow the release of the craft when load is on the hooks. This will be so arranged as to release the boat under any condition from no load with the boat in the water, to when a load of 1.1 times the total mass of the lifeboat (fully loaded) is acting on the hooks. The release mechanism should be adequately protected against accidental or premature use.
3   The release control should be clearly marked in a contrasting colour.
4   The mechanism shall be designed with a safety factor of 6 based on the ultimate strength of materials used, assuming the mass of the boat is equally distributed between falls.

Painter Release

Every lifeboat shall be fitted with a release device to enable the forward painter to be released when under tension.

Lifeboats with Self-Contained Air-Support Systems

Lifeboats with self-contained air-support systems shall be so arranged that when the boat is proceeding with all entrances and openings closed, the air inside the lifeboat remains pure and the engine runs normally for a period of not less than ten minutes. During this period the atmospheric pressure inside the boat shall never fall below the outside atmospheric pressure, nor shall it exceed it by more than 20 millibars. The system shall be provided with visual indicators to indicate the pressure of the air supply at all times.

Fire-Protected Lifeboats

A fire-protected lifeboat, when waterborne, shall be capable of protecting the number of persons it is permitted to accommodate when

Seamanship Techniques

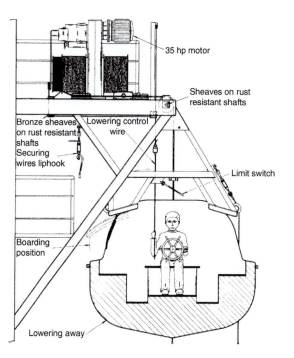

Figure 7.11 Watercraft-Schat launching system.

subjected to a continuous oil fire that envelops the boat for a period of not less than eight minutes.

Water Spray Systems

A lifeboat with a water spray system shall comply with the following:

1 Water for the system shall be drawn from the sea by a self-priming motor pump. It shall turn 'on' and turn 'off' the flow of water over the exterior of the lifeboat.
2 The sea water intake shall be so arranged as to prevent the intake of flammable liquids from the sea surface.
3 The system shall be arranged for flushing with fresh water and allowing complete drainage.

Partially Enclosed Boats (as Defined by Chapter 4, Section 4.5 of the Code for International Life-Saving Appliances 2003)

1 Partially enclosed boats must comply with the general requirements for lifeboats.
2 Every partially enclosed boat shall be provided with effective means of bailing or be automatically self-bailing.

3 They shall be provided with permanently attached, rigid covers extending over not less than 20 per cent of the boat's length from the stem, and 20 per cent of the length from the most after part of the boat. The lifeboat will be fitted with a permanent attached foldable canopy which together with the rigid covers completely encloses the occupants of the boat in a weather-proof shelter and protects from exposure. The arrangement of the canopy will meet the following:

- The canopy must be provided with adequate rigid sections or battens to permit the erection of the canopy.
- It must be easy to erect by not more than two persons.
- It must be insulated to protect the occupants against heat and cold, having not less than two layers of material separated by an air gap or other efficient means of insulation. Means must be provided to prevent the accumulation of water in the air gap.
- Its exterior should be of a highly visible colour and the interior colour should not cause discomfort to the occupants.
- It has entrances at both ends and on each side provided with efficient adjustable closing arrangements which can be easily and quickly opened and closed from inside or outside so as to permit ventilation but exclude the sea water, wind and cold. Means shall also be provided for holding the entrances securely in the open and closed positions.
- With the entrances closed it admits sufficient air for the occupants at all times.
- It has means for collecting rainwater.
- The occupants can escape in the event of the lifeboat capsizing.

4 The interior of the lifeboat should be of a highly visible colour.
5 The radio installation required by the regulations shall be installed in a cabin large enough to accommodate both the equipment and the operator. No separate cabin is required if the construction of the lifeboat provides a sheltered space to the satisfaction of the certifying authority.

Marine students should note that the above is for partially enclosed lifeboats, and not self-righting, partially enclosed boats, which are covered by Regulation 43.

Lifeboat Additional Fittings

In accordance with the amendments of the SOLAS 1974 convention:

1 Every lifeboat shall be provided with at least one drain valve fitted near the lowest point in the hull, which shall be automatically open to drain water from the hull when the lifeboat is not waterborne and shall automatically close to prevent entry of water when the lifeboat is waterborne. Each drain

valve shall be provided with a cap or plug to close the valve, which shall be attached to the lifeboat by a lanyard, chain or other suitable means. Drain valves shall be readily accessible from inside the lifeboat and their position shall be clearly indicated.

2 All lifeboats shall be provided with a rudder and tiller. When a wheel or other remote steering mechanism is also provided the tiller shall be capable of controlling the rudder in case of failure of the steering mechanism. The rudder shall be permanently attached to the lifeboat. The tiller shall be permanently installed on or linked to the rudder stock; however, if the lifeboat has a remote steering mechanism, the tiller may be removable and securely stowed near the rudder stock. The rudder and the tiller shall be so arranged as not to be damaged by operation of the release mechanism or the propeller.

3 Except in the vicinity of the rudder and propeller, a buoyant lifeline shall be becketed around the outside of the lifeboat (see ropework in lifeboats, pages 226–228).

4 Lifeboats which are not self-righting when capsized shall have suitable hand holds on the underside of the hull to enable persons to cling to the lifeboat. The handholds shall be fastened to the lifeboat in such a way that when subjected to impact sufficient to cause them to break away from the lifeboat, they break away without damage to the lifeboat.

5 All lifeboats shall be fitted with sufficient watertight lockers or compartments to provide for the storage of the small items of equipment, water and provisions required by the regulations. Means shall also be provided for the storage of collected rainwater.

6 Every lifeboat shall comply with the GMDSS requirements and have use of VHF radiotelephone apparatus. Lifeboat/rescue boats of passenger ships would have a fixed radio installation. Other craft would employ portable two-way 'walkie-talkies'.

7 All lifeboats intended for launching down the side of a ship shall have skates and fenders as necessary to facilitate launching and prevent damage to the lifeboat.

8 A manually controlled lamp visible on a dark night with a clear atmosphere at a distance of at least two miles for a period of not less than 12 hours shall be fitted to the top of the cover or enclosure. If the light is a flashing light, it shall initially flash at a rate of not less than 50 flashes per minute over the first two-hour period of operation of the required 12-hour operation period.

9 A lamp or source of light shall be fitted inside the lifeboat to provide illumination for not less than 12 hours to enable reading of the survival and equipment instructions; however, oil lamps shall not be permitted for this purpose.

10 Unless expressly provided otherwise, every lifeboat shall be provided with effective means of bailing or be automatically self-bailing.

Plate 62 Open boat seen at surface level immediately after launch. The falls and manropes are seen hanging vertically from the davits and span. The painter is stretched forward while crew exercise with the house recovery net bringing in a casualty from the water.

11 Adequate viewing forward, aft and to both sides of the lifeboat must be provided from the control position to allow safe launching and manoeuvring.
12 Each seating position in the boat should be clearly indicated.

Free-Fall Lifeboats

Free-fall launching is defined as that method of launching a survival craft whereby the craft with its complement of persons and equipment on board is released and allowed to fall into the sea without any restraining apparatus.

Definitions for Use with Free-Fall Lifeboats

Effective clearing of the ship

This is the ability of a free-fall lifeboat to move away from the ship after free-fall launching without using its engine.

Float-free launching

This is the method of launching a survival craft whereby the craft is automatically released from a sinking ship and is ready for use.

Free-fall acceleration

This is the rate of change of velocity experienced by the occupants during launching of a free-fall lifeboat.

Plate 63 Free-fall lifeboat, mounted at the stern of the cargo vessel *Scandia Spirit*. Such totally enclosed boats require means of recovery and are generally supplied with a small derrick or davit arrangement of adequate SWL to achieve this. Weight in these boats when operated fully loaded, depending on size and manufacturer, is generally between 2.5 and 3.5 tons.

Free-fall certification height

This is the greatest launching height for which the lifeboat is to be approved, measured from the still water surface to the lowest point on the lifeboat when the lifeboat is in the launch configuration.

Launching ramp angle

This is the angle between the horizontal and the launch rail of the lifeboat in its launching position with the ship on an even keel.

Launching ramp length

This is the distance between the stern of the lifeboat and the lower end of the launching ramp.

Required free-fall height

This is the greatest distance measured from the still water surface to the lowest point on the lifeboat when the lifeboat is in the launch configuration and the ship is in its lightest seagoing condition.

Water entry angle

This is the angle between the horizontal and the launch rail of the lifeboat when it first enters the water.

Life-Saving Appliances

Free-Fall Lifeboats: Relevant Detail

Capacity of free-fall lifeboats

The capacity of the boat will be determined by the number of persons that can be provided with a seat that will not interfere with the means of propulsion or the operation of any of the lifeboat's equipment.

Where a free-fall lifeboat is carried for launching over the stern of the ship, it must be of such aggregate capacity as will accommodate the total number of persons on board.

Performance

The free-fall launched boat must be capable of being launched with its full complement, against a list of 20° and a trim of 10° on the parent vessel.

The craft must also be capable of being launched with the occupants positioned forward to cause the C of G to be in the most forward position. In a similar manner the boat must be capable of being launched with all persons aft, to cause the C of G to be in the most aft position. Additionally, it must be capable of being launched with an operating crew only.

The free-fall height shall never exceed the free-fall certification height.

Construction and protection against harmful accelerations

The free-fall boat must be constructed in a manner to withstand a free-fall launch with its full complement and equipment from a height of at least 1.3 times the free-fall certification height. The construction should be such as to protect the occupants from harmful accelerations resulting from being launched in the free-fall manner.

Release system

The release of the free-fall boat will be by two independent activation systems operated from inside the boat, the operational aspects of which will be clearly marked in contrasting colours. The release must be capable of operation, with the boat in any condition of loading, up to 200 per cent load.

The system must be protected against accidental release and must also be designed to permit testing without release of the boat.

Operational detail

Free-fall lifeboats must comply with the regulations for totally enclosed lifeboats, inclusive of a self-contained air supply and fire protection water spray system where required.

Launching

The free-fall lifeboat must be capable of being launched not only in the free-fall mode, but also by a secondary method by falls in conditions of unfavourable trim of up to 2° and list of up to 5°. If the power for operation of this secondary launch system is not gravity, then the appliance must be connected to both the ship's mains and the emergency power supply.

The boat must also be capable of, and designed to float off from, its stowed position automatically.

The boat must be capable of being launched with the full complement, within a period of ten minutes from the time the abandon ship signal is given.

Testing and Drills with Free-Fall Lifeboats

New boats

Every new free-fall lifeboat should be loaded to 1.1 times its related load and launched by free fall with the ship on an even keel and in its lightest seagoing condition. It is also a requirement that the free-fall boat can be recovered and restowed in its correct launching position and properly secured.

Lifejackets for use with free fall

Lifejackets used with free-fall lifeboats and the manner in which they are carried or worn shall not interfere with the entry into the lifeboat, the occupants' safety or the operation of the boat.

Abandon ship drills: cargo ships

These must be conducted every month, and within 24 hours of the ship leaving port if more than 25 per cent of the crew have not participated in a drill on board that particular ship the previous month.

Where a new crew is engaged or when a ship enters service, such drills must take place before the vessel sails, although the administration may accept other arrangements for those classes of vessel which would find this impractical.

NB. Passenger vessels of Class I, II, II(A) and III must carry out weekly drills for fire and abandonment.

The drill must include the lowering of the boat into the water where free-fall launching is impracticable, provided the lifeboat is free-fall launched with its operating crew on board and manoeuvred in the water at least once every six months. However, where this is not practical the administration may extend this period to 12 months provided arrangements are made for simulated launching which will take place at intervals of not more than six months.

Regulatory Requirements for Free-Fall Lifeboats

Where an approved free-fall lifeboat complies with the code, the vessel must additionally carry one or more compliant inflatable or rigid life rafts on each side of the ship and of such aggregate capacity as will accommodate the total number of persons on board. The life rafts on at least one of the sides of the vessel must be served by a launching device. Additionally, it must also carry an approved rescue boat.

Ships less than 85 m in length with a free-fall lifeboat (except oil/chemical tankers and gas carriers) must carry inflatable or rigid life rafts to accommodate the total number of persons on board on either side. Where the life rafts are stowed in a position for easy side-to-side transfer, then the capacity provided may equal 150 per cent of persons on board.

Boat Rigging

Ropework in Lifeboats

Painters

Standard equipment must include two painters, both stowed in the forward part of the boat. One of these shall be permanently secured

Plate 65 Free-fall lifeboat stowed at the aft end launch position (source: Shutterstock).

to the boat and coiled down on top of the bottom boards or in the bow sheets. The second painter should be secured to the release device at or near the bow, ready for immediate use.

Both painters should be of a length equal to not less than twice the distance from the stowage position of the lifeboat to the waterline when the vessel is at her lightest seagoing condition or 15 m, whichever is the greater. The size of painters is normally 20–24 mm manila or equivalent synthetic cordage.

Man-made fibres may be used for life-saving appliances provided it has been approved by the appropriate authority. Observation of an approved man-made rope will show a coloured thread/yarn passing through the lay of the rope. The idea is based on the 'Rogues' Yarn' method of identifying the various dockyards from which ropes originally came, and so prevent theft between ships.

NB. Following revision of regulations lifeboats must now be fitted with a motor. Mast and sails are therefore no longer carried as standard equipment.

Mast and sail detail has been retained within the text to provide general seamanship information.

Buoyant (becketed) lifeline

Each lifeboat will be provided with a buoyant lifeline becketed around the outside of the boat, except in the vicinity of the rudder and the propeller. These are often manufactured in a synthetic material having a wood hand grip rove in the bight. If natural cordage is used it is normally of 16 mm size, beckets being spaced

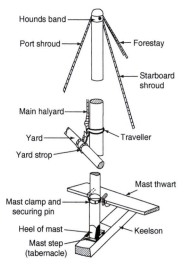

Figure 7.12 Mast and rigging.

approximately 60 cm apart. Its purpose is to provide hand holds for survivors in the water. With this idea in mind, the height of beckets should be just clear of the water surface when the boat is fully loaded.

Keel grab lines (if fitted)

These are fitted to assist the righting of a capsized, conventional boat. They are secured inside the boat, on either side, passing from gunwale to gunwale under the keel. They will normally be secured having 'figure-eight' knots on either side of the boat to provide hand holds, together with a sheepshank directly under the keel. This sheep shank can be released when the boat is in a capsized condition and the increased bight of the line can be used in con-junction with an oar to form a Spanish windlass and so lever the boat over into a correct upright position. Keel grab lines are usually of 20 mm manila or suitably approved cordage.

Lifelines

Not fewer than two lifelines are required for partially enclosed boats. These must be secured to the span between the davit heads and should be of sufficient length to reach the water with the ship in its lightest condition under unfavourable conditions of trim, with the ship listed not more than 20° either way. They should be of an approved cordage 20 mm in size, and seized to the span not less than 30.5 cm away from the davit heads so as not to foul the fall wires.

Lifeboat falls

Falls shall be constructed in corrosion-resistant extra flexible steel wire rope (EFSWR) having rotation-resistant properties.

An example in use is 'Kilindo', 18 × 7 or Wirex 17 × 7. They are multi-strand wires that involves laying up round strands in the opposite direction to the previous layer of strands. Although termed a non-rotating rope, this is not strictly accurate because the separate layers of strands do twist, but each layer of strands turns in an opposing direction, giving a balance effect when hoisting/lowering.

Lifeboat falls shall be long enough for the survival craft to reach the water with the ship in its lightest seagoing condition, under unfavourable conditions of trim and with the ship listed not less than 20° either way.

Maintenance

Falls used in launching shall be renewed when necessary due to deterioration, or at intervals of not more than five years, whichever is earlier.

Rate of descent

The speed at which the survival craft or rescue boat is lowered into the water shall not be less than that obtained from the formula:

$$S = 0.4 + (0.02 \times H)$$

where S = speed of lowering in metres per second; and H = height in metres from davit head to the waterline at the lightest seagoing condition.

The maximum lowering speed is established by the authority, taking note of the design of the craft, the protection of its occupants from excessive forces and the strength of the launching appliance (taking into account inertia forces during an emergency stop). Means must be included in the system to ensure that the speed is not exceeded.

Bowsing in Tackles

Small rope tackles, usually double luff, these are secured between the foot of the davit aboard the parent vessel and the loose linkage under the floating block. Their purpose is to relieve the weight from the tricing pendants and allow the conventional boat to be eased out away from the ship's side during the lowering operation to the waterline (see Figure 7.13).

The tackles are rove to disadvantage, with the downhaul leading into the boat. When they are secured, it is normal to use a round turn with two half-hitches on the bight. This will enable the two men manning the tackles at each end to slack away together on the round turn, and also check the motion of the boat should it be going off in an uneven manner.

Each block is fitted with a hook/swivel fitment to allow securing in an easy manner with minimum loss of time.

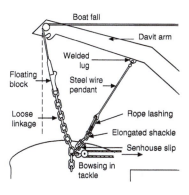

Figure 7.13 Use of tricing pendant.

Tricing Pendant

This is a short length of SWR, having a senhouse slip and a rope lashing at one end, with a shackle secured to the underside of the davit arm at the other end (Figure 7.14). The purpose of the tricing pendants is to 'trice the boat into the ship's side', to allow persons to board the boat safely. This precaution is particularly important if the parent vessel has an adverse list, which would cause the boat to be slung in the vertical away from the ship's side.

The pendants are secured between the linkage directly under the floating block to the underside of the davit. The senhouse slip is held secure by a wooden pin, which will not rust or jam and can easily be broken to release. The reason the rope lashing is incorporated into the make-up of the pendant is that it can be cut in an emergency. Tricing pendants should be released once the bowsing in tackles are secured. Survivors should then board while the weight

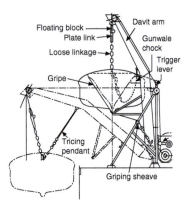

Figure 7.14 Pivot gravity davit.

of the boat is bowsed in. The tackles can then be paid out and released before lowering.

Gripes

These are constructed of SWR that has a rope lashing and a bottle screw with senhouse slip incorporated in its length. The purpose of the gripes is to hold the boat firmly against its stowage chocks and keep it in a secured stowed position in the davits.

There are several methods of griping the boats against the davit chocks, one system being shown in Figure 7.14. It will be seen that the gripe wire is secured to the trigger lever, passing over the gunwales of the boat and then being secured via lead sheaves to the inside foot of the davit.

The gripes are cleared by compressing the two parts of the senhouse slip and clearing away the securing link. Once the slip and bottle screw are released, the gripe can be passed back over the boat and the end cleared. However, some gripes are interconnected to the trigger system, and the men in the boat clearing the gripes over the gunwales should in fact check and report that the triggers have fallen and the davits are clear to lower away. The coxswain in charge of the launching operation should also check that the securing end of the gripe does not fall under the lowering davit arm, causing buckling or fouling the lowering operation.

In resecuring the gripes the bottle screw arrangement will need to be opened up in order to pass the senhouse slips. Once this is done, the gripes can be re-tensioned by use of the screw. A rope lashing is incorporated for the same purpose as with the tricing pendant, so that it may be cut in an emergency.

Launching Procedure

Gravity Davits

These davits (Figure 7.14) operate on the principle of the boat's own weight doing the work to bring about the launch. The construction of the davits includes a safety device, usually a trigger arrangement attached to the gripes; when launching, care should be taken to check that these triggers are cleared before proceeding.

Gravity davits must be fitted with SWR falls and operated by a controlled winch. The rate of descent of the boat is separately controlled by a centrifugal brake. A main ratchet-type brake is also incorporated in the more modern designs, and it can hold the boat at any stage of lowering. This may be operated, in some cases, from inside the craft itself, so that the launching cycle can be carried out remotely, thus saving time once waterborne for taking on personnel.

The majority of gravity davits are fitted with tricing pendants, and the boat must be equipped with means of bowsing in against the ship's side to permit the removal of the pendants before embarking personnel.

Plate 66 No. 6 lifeboat being launched down the portside of a passenger cruise liner (source: *Marine Survival*, published by Witherby Seamanship).

The davits will successfully launch the boat against a 20° adverse list in the following way:

1 Two men should be ordered into the boat, to ship the plug and check that the painter is rigged in a correct manner. (Painter is passed inside the fall and outside everything else, and secured well forward.) Once all work inside the boat is complete, these two men should be seen to sit down in the boat and hold on to the lifelines.

2 The coxswain should check that the harbour pins are out.

3 The gripes should be slipped and any triggers checked to see that they are clear, the gripes being passed down to deck level, clear of the boat.

4 A winchman must be ordered to stand by to lower the boat down to the embarkation deck.

5 Check that the overside is clear, then lower away by lifting the brake handle. The boat should descend from the davits until the tricing pendants take the boat's weight and draw the boat into the ship's side.

6 The bowsing in tackles should be rigged in such a manner that the downhaul is secured in the boat with a round turn and two half-hitches, on the bight about the linkage on the end of the falls.

7 Have the two men in the boat slip the tricing pendants once both ends of the boat are securely bowsed in.

8 The remainder of passengers and boat's crew should now be embarked and seated as low as possible in the boat.

9 Ease out on the bowsing in tackles and allow the boat to come away from the ship's side, then let go the tackles from inside the boat and throw them clear, back towards the parent vessel.

> Where on/off load release hooks are employed, fall pre-ventative devices should be seen to be in place before all boat launching exercises are commenced

10   Order the winchman to lower the boat with a run. Ship the tiller.

11   Unless release gear is fitted to the boat, it is more practical to lower the boat into a trough of a wave. As the crest of the wave brings the boat higher, this will allow the falls to become slack, which will in turn allow easy slipping from the lifting hooks. Once the falls are clear, the boat falls away from the ship's side as the wave drops away. Should quick release gear be fitted to the boat being launched, then it would be more practical to slip the release mechanism as the boat takes the crest of a wave. As the wave drops away into a trough, so it takes the boat away from the ship's side with it.

> NB. Boats on port side are numbered 2, 4, 6, 8, 10, 12 from forward to aft.

The time of the boat becoming waterborne is the most critical, and many serious accidents have occurred in the past. The floating heavy blocks of the falls are a major cause of the accidents, as they are in the direct vicinity of the boat in the water and they oscillate wildly at head height. A prudent coxswain will endeavour to clear the area as soon as possible. An alternative means of reducing this danger is to secure light lines to the floating blocks of the falls and manning them by additional personnel on deck. Once the boat has slipped the falls, these blocks can be pulled up clear, back aboard the parent vessel out of harm's way. Personnel in the boat could also wear prefabricated crash helmets as an additional safeguard.

Launching Stations

Launching stations shall be in such positions as to ensure the safe launch of survival craft, having particular regard to clearance from the propeller and steeply overhanging portions of the hull. As far as possible survival craft, except free-fall craft, should be arranged to allow launching down the straight side of the ship. If positioned forward, they shall be located abaft the collision bulkhead and in a sheltered position.

The stowage arrangement shall be such that it will not interfere with the launching operation of other survival craft or rescue boat at any other station. Craft should be stowed as close to the accommodation as possible, and their muster and embarkation areas should be adequately illuminated, supplied by an emergency source of electrical power.

Each launching station, or every two adjacent launching stations, should have an embarkation ladder which complies with the regulations. This ladder should be constructed in a single length and reach from the deck to the waterline in the lightest seagoing condition under unfavourable conditions of trim and with an adverse list of 15° either side. These ladders may be replaced by approved devices which provide access to survival craft when waterborne; however, at least one embarkation ladder would still be a requirement, on either side.

Survival craft should be maintained in a continuous state of readiness, so that two crew members could prepare for embarkation and launching in less than five minutes (fully equipped). They should be attached to their respective launching devices and positioned so that in the embarkation situation they are not less than 2 m above the waterline when the ship is in the fully loaded condition, and listed up to 20° either way. Lifeboats for lowering down the ship's side should be positioned as far forward of the propeller as is practical.

On cargo ships of 80 m in length and upwards but less than 120 m in length each lifeboat shall be stowed so that the after end of the lifeboat is not less than the length of the lifeboat forward of the propeller.

On cargo ships 120 m or over and passenger ships of 80 m and over each lifeboat shall be stowed so that the after end of the boat is not less than 1.5 times the length of the lifeboat forward of the propeller.

Launching Appliances

Information regarding launching and embarkation appliances is covered by Regulations 16 and 17, in Section I, Chapter III of the International Convention for the Safety of Life at Sea.

Salient points have been extracted below.

1   Every launching appliance, together with its lowering and recovery gear, should be so arranged that the fully equipped survival craft or rescue boat can be safely lowered against a trim of 10° and a list of 20° either way (*a*) when boarded by its full complement, from the stowed position; and (*b*) without persons on board.

2   A launching appliance shall not depend on any means other than gravity or stored power which is independent of the ship's power supply to launch the survival craft in the fully loaded condition.

3   Launching must be possible by one person from a position on the ship's deck, and that person should be capable of keeping the survival craft or rescue boat visible throughout the launching process.

4   Winch brakes of launching appliances should be of sufficient strength to withstand the static test and the dynamic test.

5   Structural members, blocks, falls, links, pad eyes and all fastenings shall be designed with not less than a minimum factor of safety on the basis of the maximum working load assigned and the ultimate strength of the material used for construction. A minimum factor of safety of 4.5 shall be applied to all davit and winch structural members, and a minimum factor of safety of 6 shall be applied to all falls, suspension chains, links and blocks.

6   The lifeboat launching appliance should be capable of recovery of the lifeboat with its crew. Operating speed should not be less than 0.3 m/s.

> It is now a requirement that lifeboats are moved from the stowage chocks on the davits weekly.

7   Every launching appliance shall be fitted with brakes capable of stopping the descent of the survival craft or rescue boat, holding it securely with its full complement of persons and equipment. Brake pads shall, where necessary, be protected from oil and water.

8   An efficient hand gear shall be provided for the recovery of each survival craft or rescue boat.

9   Where davit arms are recovered by power, safety devices shall be fitted which will automatically cut off the power before the davit arm reaches the stops in order to prevent over-stress on the falls, unless the winch is designed to prevent such overstressing.

10 Falls shall be of rotation-resistant SWR. They should wind off the drums at the same rate when lowering and wind on to the drums evenly at the same rate when hoisting (multiple drum winch).

Survival Craft, Launching and Recovery Arrangements

Launching appliances complying with the regulations shall be provided for all survival craft except the following:

1   Survival craft that are boarded from a position on deck that is less than 4.5 m above the waterline in the lightest seagoing condition and that either:
   (a) have a mass of not more than 185 kg; or
   (b) are stowed for launching directly from the stowed position under unfavourable conditions of trim of up to 10° and with the ship listed not less than 20° either way.

2   Survival craft having a mass of not more than 185 kg and which are carried in excess of the survival craft for 200 per cent of the total number of persons on board the ship.

Each appliance provided must be capable of the launching and recovery of the craft. Throughout any launch or recovery operation the operator of the appliance is able to observe the survival craft.

During the preparation and launching operation, the survival craft, the launching appliance and the water area to which the craft is being launched shall be adequately illuminated by lighting supplied from the emergency source of electrical power, required by the regulations. Preparation and handling of survival craft at one launch station shall not interfere with the handling of any other survival craft or rescue boat.

The release mechanism used for similar survival craft shall only be of one type carried aboard the ship.

Passenger ships

Each survival craft provided for abandonment in passenger ships by the total number of persons on board shall be capable of being launched with their full complement of persons and equipment within a period of time to the satisfaction of the authority, from the time the abandon ship signal is given.

Cargo ships

With the exception of survival craft mentioned in 1(*a*), all survival craft required to provide for abandonment by the total number of persons on board shall be capable of being launched with their full complement of persons and equipment in the shortest possible time from the time the abandon ship signal is given.

Lifeboats of the partially enclosed type, if carried, shall be provided with a davit head span, fitted with not fewer than two lifelines of sufficient length to reach the water with the ship in its lightest seagoing condition, under unfavourable conditions of trim and with the ship listed not less than 20° either way.

Lifeboat launching appliances for oil tankers and gas carriers, with a final angle of heel greater than 20°

These shall be capable of operating at the final angle of heel on the lower side of the ship.

> NB. Rescue boats must be capable of being launched and recovered within a five-minute period.

Launching Stations

Embarkation ladders

Hand holds shall be required to ensure a safe passage from the deck onto the head of embarkation ladders and vice versa.

Construction of the ladder

The steps shall:

- be made of hardwood, free of knots or other irregularities, smoothly machined and free from sharp edges and splinters, or of a suitable material of equivalent properties;
- be provided with an efficient non-slip surface either by longitudinal grooving or by the application of an approved non-slip coating;
- be not less than 480 mm long, 115 mm wide and 25 mm in depth, excluding the non-slip surface or coating;
- be equally spaced not less than 300 mm or more than 380 mm apart and secured in such a manner that they will remain horizontal.

The side ropes shall consist of two uncovered manila ropes not less than 65 mm in circumference on each side. Each rope should be continuous, with no joints below the top step. Other materials may be used provided the dimensions, breaking strain, weathering, stretching and gripping properties are at least equivalent to those of manila rope. All rope ends shall be secured against unravelling.

Taking the Boat Away from the Ship's Side

This is always a dangerous operation, for conditions at sea level may not always be apparent to a person standing up high above the water, as on the bridge of an oceangoing vessel. It is always preferable for the parent vessel to provide a lee, if possible, for the launching of the boat, so giving a limited amount of shelter from the wind. In fact, the parent vessel may be either stopped in the water and making no way, or underway at reduced speed. The launching of a boat with the parent vessel at any speed over 3–4 knots must be considered extremely hazardous and should not be attempted under normal circumstances.

Parent Vessel Stopped

Boat under oars

This is the more acceptable condition for launching a boat, but care should be taken with the effects of swell and wind when the falls have been released. Figure 7.15 indicates the use of 'bearing off ' with the looms of the outboard oars. Although this is a practical method, use of 'springing off ' by pulling the painter down the inboard side of the boat can prove just as effective. Springing off may be the only alternative should the boat be in the water under the curved lines of the stern of the parent vessel. There could be the distinct possibility of breaking the looms of the oars by trying to bear away with them in or around this stern area.

The bowman should endeavour to combine his action of letting go the painter and springing off by pulling it down the inboard side and bearing off by use of the boathook. The bowman's efforts, together with the combined weight of the outboard oarsmen, should turn the bow of the boat far enough away from the ship's side to enable the inboard oarsmen to down their oars and give way.

After the outboard oarsmen have borne away on the looms of their oars, a prudent coxswain will order them to hold water. With additional use of the rudder he will try to bring the fore and aft line of the boat at a broad angle to the ship's side, so gaining sea room with any forward motion of the boat.

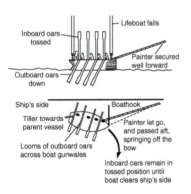

Figure 7.15 Taking boat away from ship's side of parent vessel stopped, under oars.

Parent Vessel Making Way

Boat under power

Once the boat falls have been released and the boat is held on the painter, push the tiller towards the ship's side. This action effectively gives the boat a sheer. Keep the painter taut until the boat reaches a point of maximum sheer, then briefly alter the position of the tiller so that the bow cants inwards towards the parent vessel. The results of this action will be for the painter to become temporarily slack, which will permit its easy slipping. Push the tiller towards the ship's side and gain sea room (see Figure 7.16) and move the gear of the engine from neutral to engage forward motion.

Boat Recovery in Heavy Weather

If a boat is lowered at sea for a specific job, under normal circumstances that boat must be recovered before the voyage can proceed. This operation may become extremely hazardous with a heavy swell running or with a rough sea. To this end a recommended method of recovery is given below.

Preparation

Secure a wire pendant to an accessible point on the davit arms (Figure 7.17, section 1). Extreme care must be taken to ensure that the strop and the wire pendant, together with any shackles used, are of sufficient strength to accept the weight of the fully laden boat. The boat falls should be retrieved at deck level and nylon rope strops shackled to the linkage from the floating blocks, since rope strops are easier to handle than chain in the confines of the oscillating conditions of the boat in the water. The wire pendant and the boat falls, together with the nylon strops, should be set up above the waterline as in Figure 7.17, section 1. Ensure that the strop is also of adequate strength to support the full weight of the laden boat.

Hoisting

If the operation is taking place on a Class 1 passenger vessel, then the sequence of actions are made easier by the use of the ram's horn lifting hook, a standard fitment in the emergency boats of passenger vessels. However, where a single lifting hook is to be used, as with Class 7 vessels, then the method of recovery can be achieved in the following way: fit both nylon strops over the lifting hooks fore and aft in the boat and hoist the boat clear of the water until the floating blocks are 'block on block' with the davit head (Figure 7.17, section 2).

It is at this stage that the wire pendant is secured to each of the lifting hooks, on top of the nylon strops. If a ram's horn hook was being used, then the opposing half of the hook would accommodate the pendant. The idea at this stage is to transfer the weight of the boat from the falls to the wire pendants, so that the boat falls may be correctly secured.

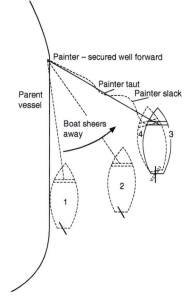

Figure 7.16 Taking boat away from ship's side when parent vessel is making way.

1 Boat launched and secured by painter alone, tiller towards the parent vessel.
2 Boat sheers away from ship's side by continued effect of the tiller being towards the ship.
3 A position of maximum sheer is reached, tiller is eased to amidships, maintaining a forward motion of the boat and keeping the painter taut.
4 Tiller is briefly pushed away from the parent vessel, allowing the painter to go slack as the bow of the boat turns in towards the parent vessel. At this point, slip the painter.

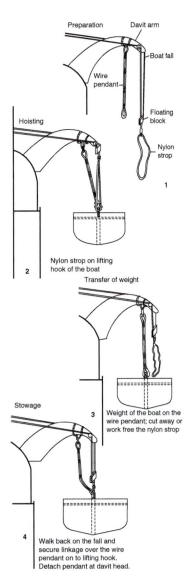

Figure 7.17 Boat recovery in heavy weather.

Transfer of Weight

This can only be achieved as shown in Figure 7.17, section 3 if the wire pendant is long enough to reach from the davit head to the lifting hook in the boat when the floating blocks are hard up at the davit head. By walking back on the boat fall, the weight comes onto the pendant and the strop becomes slack. This is the time either to cut away the strop at the hook or unshackle the other end from the linkage of the floating block.

Stowage

Continue to walk back on the falls to enable the open links to be slipped over the wire pendants and the lifting hooks, hoist away and take the weight on the falls. Detach the pendants from the davits and restow the craft.

Beaching a Lifeboat

This is always a dangerous operation and should be carried out during the hours of daylight only. If approaching the shoreline at night it is recommended to wait until daybreak. All preparations should be made well outside the line of surf and a plan of approach should be well thought out in advance.

Beaching under Power

Depending on the sea conditions and the design of the boat, beaching under power can be made either head on or stern on to the beach. Mariners are naturally adverse to obstructing the propeller, especially so if the boat is to be used again at a later time. A method of running before the surf and beaching bow first is a viable alternative to the 'stern first approach', provided the boat's speed can be employed to equal the rate of the following sea. This approach will, by its very nature, be a fast exercise, even with engine power at slow speed, the character of the surf dictating the rate of approach. Also, if there are obstructions on the approach there is far less time to take avoiding action, even if sighted.

The method should in any event never be attempted with boats with a transom stern. Coxswains if approaching bow first will require considerable experience and use decisive judgement once inside the line of surf.

Beaching under power, stern first, is slower and must generally be considered safer. Use of the sea anchor over the bow with combined use of the boat's engines should provide the required force to keep the bow end on to the direction of surf. Pulling oars should be kept ready in the event of engine failure and for use prior to striking the beach itself. An obvious danger of the propeller turning on the final approach must be considered. This is especially so if it is the intention to put men over the side, once the boat enters shallows.

The propeller should always be stopped before people are despatched to drag the boat up to the beach.

Ideal Conditions

The ideal conditions for beaching would be a gentle sloping beach, sandy and free from rocky obstructions, with little or no surf and calm weather conditions. As all these factors are unlikely to occur, the coxswain should minimize the risk of injury by keeping all non-essential personnel low down in the boat, seated on the bottom boards and away from the forward and after sections.

Once the boat comes stern on to the beach all persons should disembark over the stern, never over the bow. A bowman should keep the tension on the sea anchor hawser to prevent the boat from broaching too. As soon as practical after people are ashore the sea anchor should be tripped and the boat together with its equipment should be salvaged.

Boat Handling and Safe Procedures

Responsibilities of the Coxswain

1 To check that all crew members and passengers are wearing lifejackets, and that these are secured in a correct manner.
2 To ensure that all crew members and other personnel are correctly attired, preferably in soft-soled shoes, warm clothing and oilskins or immersion suits.
3 To inspect the boat before embarking personnel and ensure that all equipment and the boat's condition are in good order.
4 To maintain authority and make all orders in a clear manner to ensure the safe handling of the boat.

Plate 67 Open boat seen showed in gravity davits (source: Shutterstock).

5   To check overside that the launch area is clear and free of obstructions.

6   To carry out an orderly safe launch, take the boat away from the ship's side and carry out any operations in a correct manner.

7   Throughout all boat operations the coxswain's responsibility is for the safety of his own crew; any decisions taken should bear this in mind, at all times.

Methods of Attracting Attention

1   Use of the orange smoke canister. Thrown overboard, down-wind, this is most effective for attracting the attention of a rescue aircraft.

2   Use of the red hand flare. Hold at arm's length, overside, to reduce the risk of fire in the survival craft. Extreme care should be taken in its use with the rubber fabric of the buoyancy chambers of a life raft. It is most effective for attracting rescue aircraft.

3   Use of the rocket parachute flare. Hand held, this activates at about 300 m and will burn for 40 seconds, producing 30,000 candela. Most effective use is to attract surface rescue vessels.

4   Emergency use of: EPIRB, SART or portable VHF radio-telephone.

5   Raising and lowering the arms is an international distress signal, but only effective at close range and best used in conjunction with another signal.

6   Transmission of SOS by any available means, e.g. by use of the torch.

7   Burning rags, showing flames from a bucket or other improvised holder. If burning a small quantity of oil, then black smoke becomes the focal point, easily seen by a rescue aircraft by day.

8   Heliograph, to direct the rays of the sun. This is effective for aircraft or surface rescue operations, but the range is limited and it is effective only on sunny days.

9   A square flag, having above or below it a ball or anything resembling a ball. These two distinctive shapes, seen at a distance, are an international signal of distress.

10 A gun or other explosive signal, or the continuous sounding of the whistle.

Dangers in an Open Boat

These come mainly from exposure and capsizing, with subsequent drowning or injury to the occupants. Experienced handling of the boat with correct use of its equipment can limit the possibility of disaster. Prudent use of the 'sea anchor' will go a long way to increase the chances of survival.

Sail Theory

Tacking

This operation is carried out when a boat under sail wishes to change her course from the port tack to the starboard tack, or vice versa (Figure 7.19). It is sometimes referred to as going about, and is not always a practical method of altering course, especially if the wind is either too strong or too light. The operation entails passing the bow through the wind, so bringing the wind direction to the opposite side of the boat, thus changing tack.

In order to complete the operation successfully it will be necessary for the boat to have enough way on her to carry the bow through the wind. To this end it may be required to 'up helm' and allow the boat to 'pay off' from the wind and increase her speed before attempting to 'go about'.

Once the boat comes head to wind, the experienced sailor will adjust the weight distribution in the boat by transferring the passenger(s) to the new weather side. It may also be prudent to back the jib to assist the bow through the wind.

Wearing

This operation is carried out when it is considered dangerous to tack or conditions make it impractical to do so (Figure 7.20). The result of wearing is to alter the course of the boat by passing the stern through the wind.

The main feature of the operation is that when the wind is on the quarter, the mainsail is lowered to avoid 'gybing'. As the stern passes through the wind and the wind direction effectively acts on the opposing quarter, the mainsail is reset.

Gybing

Should this method of changing the course of the boat be employed, then extreme care must be taken to control the operation. The main dangers of an uncontrolled gybe are that a person may be knocked overboard by the lower boom swinging across the boat or that the boom may react so dramatically when caught by the wind that the boat is dismasted or capsized, especially if great care is not taken in handling the main sheet.

If the gybe is carried out in a controlled manner, the effect of the boom crossing from one side to the other may be cushioned by reducing the slack in the sheets as the stern passes through the wind. In addition, the speed of the manoeuvre could be reduced, providing an easier resultant motion.

Running Before the Wind

This is a term used to describe the boat when she is sailing with the wind from dead astern (Figure 7.21). She is said to be running

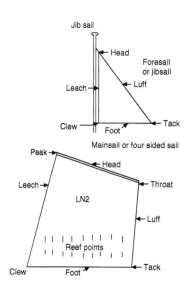

Figure 7.18 Sails.

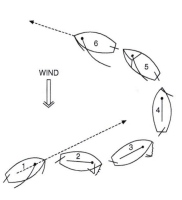

Figure 7.19 Tacking.

1 Boat on port tack.
2 Up helm, to increase the way on the boat.
3 Down helm, let fly jib sheet, take in on main sheet, hauling mainsail aft.
4 Boat head to wind; bow passes through wind, aided by backing the jib sail.
5 Bow passes through wind; ease out on mainsheets to fill mainsail.
6 Trim sheets of jib and main sails, set course on starboard tack.

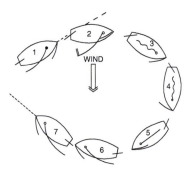

Figure 7.20 Wearing.

1 Boat on port tack.
2 Up helm, ease out main sheets.
3 Wind on port quarter, mainsail lowered.
4 Boat continues to make headway on jib sail; stern passes through wind.
5 Wind on starboard quarter, reset mainsail, trim jib sail.
6 Ease up helm and trim jib and main sails.
7 Set course on starboard tack.

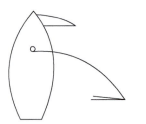

Figure 7.21 Running before the wind. The speed of the boat is reduced to that attainable with the wind on the quarter. An accidental gybe is possible, should a wind change occur. The sea anchor streamed over the stern, providing the drogue effect, will limit the risk of broaching to.

before the wind with her sails out on the same side, the sheets being at right angles to the fore and aft line.

Goose-winging

This is a similar condition to running before the wind, except that the sails are out on opposite sides, producing a greater sail area exposed to the wind (Figure 7.22). The sheets are again set at right-angles to the fore and aft line, a style that appears popular with yachtsmen.

Reefing

This is a procedure for reducing the sail area should the wind increase in strength to such a point as to make sailing under full canvas a dangerous proposition. To sail under strong wind conditions with full canvas is to invite capsizing, with all its serious consequences.

The mariner should bear in mind that if the wind increases to such strength that it becomes extremely hazardous to continue to sail, then the alternative would be to heave to and ride to the sea anchor. Reefing would be carried out in order to keep sailing in a safe manner, e.g. when making a landfall.

The reefing procedure is as follows:

1 Down helm, bring the boat head to wind and let fly the sheets.
2 Stream the sea anchor, steadying the boat's head.
3 Lower the mainsail, and detach the yard strop from the traveller.
4 Lay the yard on the side benches, and clear the foot of the sail.
5 Secure the luff earring to the tack cringle.
6 Secure the leach earring to the clew cringle.
7 Gather up the foot of the sail and tie the reef points from forward to aft.
8 Rehook the yard strop onto the traveller and reset the sail.
9 Trip the sea anchor and retrieve it, then resume the course.

In boats fitted with booms the reef points should be passed around the foot of the sail, never around the boom.

> Sail theory has been retained within this text for the benefit of yachtsmen and seafarers engaged on the 'tall ships'.

Sail Terminology

Beating

This is a general term used to describe the passage of a boat which is working her way to windward by a series of alternate tacks.

Broad reach

A boat is said to be sailing on a broad reach if the wind is just abaft its beam.

> It should be noted that with the mast and sails removed from lifeboats, professional seafarers may have limited interest in sailing. However, they will still encounter sailing vessels and it is in the interests of general seamanship to retain this section.

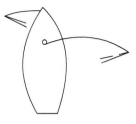

Figure 7.22 Goosewinging.

Close hauled

A boat is close hauled when she is sailing as close to the wind as she can possibly get. Some modern sailing vessels may point as close as three points to the wind, but it is convention that a boat can effectively sail no closer to the wind than six points of the compass.

Close reach

A boat is said to be on a close reach if the wind is just forward of her beam.

Luff

This is an expression to describe the boat's head moving closer to the direction of the wind, e.g. 'to luff up' to wind.

Pay off

This is an expression used to describe the boat's head moving further from the direction of the wind.

Stays

This condition, often referred to as 'in irons', occurs when a boat is changing from one tack to another, or when the wind drops and allows a vessel to come head to wind, where she will neither pay off to port or starboard tack. If the head fails to pay off onto the opposite tack during an attempt to tack, and falls back to its original position, then the boat is said to have 'missed stays'.

8

Survival Craft and Practice

Introduction

Novel life-saving appliances are now well and truly embedded in the industry. No longer do ships just carry the statutory lifeboat and lifebuoy. The MES and the MEC are now well established. Automatic, inflated lifejackets, reversible life rafts, effective rescue boats and the associated equipment of today's commercial shipping are now predominant in all sectors of maritime activity.

One of the early developments for the offshore industry was the Survival Systems International (previously Whittaker Capsule). When marine authorities saw the round boat, they could not have realized they were probably looking at the tip of the iceberg.

Development in the maritime sector has surged ahead at a pace which for a time seemed to outstrip the speed of training our personnel. Fortunately, more of our seafarers are gaining the benefits from better equipment, new fabrics in protective suits, etc. Tighter safety legislation and the introduction of ISM are all helping to make the mariner's task safer.

It has always been a practical industry, but it continues to move forward with the education of our men and women and the application of common sense. No more so than when dealing with life-saving appliances and making our crews prepared for that eventual, hopefully rare incident.

Survival Systems International

This is an alternative to the lifeboat or life raft for use by oil rigs and production platforms. It provides protection from fire, toxic fumes, explosion, swamping, capsize and exposure, in much the same way as the totally enclosed lifeboat.

It is manufactured in a fire-retardant fibreglass reinforced plastic, having an external international orange-coloured surface. It is now built in three sizes to accommodate 21, 34 and 50 persons, and each capsule is fitted with all standard equipment for basic survival needs. The structure incorporates a ventilation system, a fire sprinkler system for the exterior, seat belts and a 40 HP Westbeck, water-cooled diesel engine.

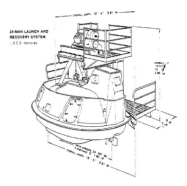

Figure 8.1 Survival Systems International (PL34) launch and recovery system.

Launching is achieved by a winch and single cable system from platform level. An independent release from the fall is possible by operation of an offload release gear.

The launching and recovery of the capsule is operated by a 25 HP electric motor. Speed of lowering is approximately 135 ft/min., and hoist speed is about 60 ft/min.

The wire cable is 2.22 cm diameter, galvanized steel wire rope (SWR). Its construction is 3 × 46 torque balanced, non-rotating, and has a breaking strength of 84,000 lb. A similar design has been adopted for a larger model to hold 50/54 people, the wire cable size being increased to 2.54 cm in diameter to support the extra weight.

Plate 68 Survival Systems International operational at sea in active training capacity.

The Inflatable Life Raft

Construction

There are several manufacturers of life rafts who supply inflatables to the merchant and military services throughout the world, e.g. Beaufort, RFD, Dunlop and Viking. The size of rafts will vary with customer requirements, but their capacity will be not less than six persons. The largest rafts currently in use are of a 150-man size, employed in marine evacuation systems.

Every life raft should be capable of withstanding exposure for 30 days of sea conditions. This is not to say that provisions and water would last for this period of time. Standard rafts should be robust in construction to be launched from a height of 18 m, and when inflated be able to withstand repeated jumps onto its surface from heights of up to 4.5 m.

The main buoyancy chamber should be divided into two compartments each being inflated via a non-return valve. Each compartment should be capable of supporting the full complement of the raft in the event of damage to either buoyancy chamber.

> NB. Jumping onto life raft canopies is not recommended and unless the raft is close, boarding should be achieved via the water so as not to damage the fabric.

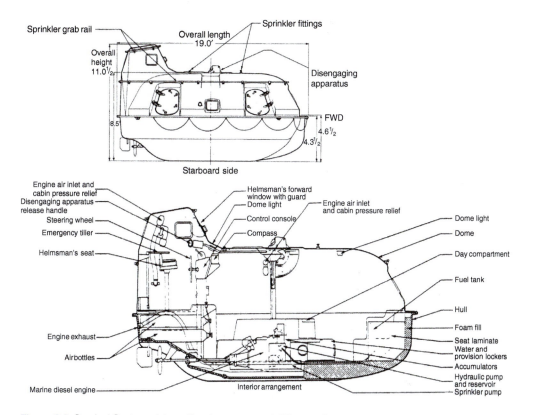

Figure 8.2 Survival Systems International arrangement: 34-man size.

The canopy should protect the occupants from exposure and should automatically set in place when the raft is launched. It should protect from heat and cold by two layers, separated by an air gap, and means to prevent water accumulating inside the air gap. The exterior should be of a highly visible colour, while the interior colour should not cause discomfort to the occupants.

It should be provided with rain catchment area(s) and with at least one viewing port. Entrances should be clearly indicated and fitted with efficient adjustable closing arrangements. A ventilation system should be provided which allows the passage of sufficient air but excludes the passage of sea water and cold.

General particulars regarding the overall construction of the raft should include sufficient headroom for sitting occupants under all parts of the canopy. All materials used in the manufacture should be: rot-proof, corrosive-resistant, unaffected by sunlight and not unduly affected by sea water, oil or fungal attack. Retro-reflective material should be prominently displayed to assist in detection of the raft.

Life Raft Fittings

The fittings should permit the raft to be towed at a speed of three knots in calm waters when loaded with its full complement and

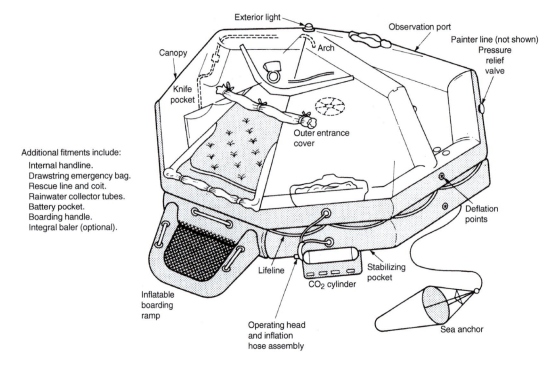

Exterior light

Observation port

Painter line (not shown)
Pressure
relief
valve

Canopy

Arch

Knife
pocket

Outer entrance
cover

Additional fitments include:
Internal handline.
Drawstring emergency bag.
Rescue line and coit.
Rainwater collector tubes.
Battery pocket.
Boarding handle.
Integral baler (optional).

Deflation
points

Inflatable
boarding
ramp

Lifeline

Stabilizing
pocket

CO_2 cylinder

Operating head
and inflation
hose assembly

Sea anchor

Figure 8.3 The RBM inflatable life raft.

having one of its 'sea anchors' streamed (towing bridle/patch). Provision will also be made to fit becketed lifelines both inside and outside every life raft.

The total weight of each raft, unless it is to be launched by an approved launching device, will not exceed 185 kg, inclusive of the case and all fitments.

Inflation of the life raft should be with a non-toxic gas and should take place within a period of one minute when at an ambient temperature of 18–20°C or within a three-minute period at a temperature of –30°C. Normal practice to cause inflation is by

Plate 69 Racked inflatable life rafts stowed aboard a passenger cruise liner (source: Shutterstock).

'tugging' on the painter line. The painter's length should be not less than twice the distance from the stowed position to the waterline, when the vessel is at its lightest seagoing condition. (SOLAS requires 15 m painter length.)

The floor of the raft may inflate automatically, but provision must be made for deflation and inflation of the floor by the occupants in order to provide insulation for the occupants.

Illumination

Illumination shall be provided by a manually controlled lamp fitted to the top of the life raft canopy, visible on a dark night with a clear atmosphere at a distance of at least two miles for a period of not less than 12 hours. If the light is a flashing light it shall flash at a rate of not less than 50 and not more than 70 flashes per minute for the operation of the 12-hour period. The lamp shall be provided by power from either a sea water activated cell or a dry chemical cell and shall automatically light when the raft inflates. The cell shall be of a type that does not deteriorate due to damp or humidity inside the stowed life raft.

A manually controlled lamp shall be fitted inside the life raft, capable of continuous operation for a period of at least 12 hours. This lamp will light automatically when the life raft inflates. Its intensity should be sufficient to allow the reading of the survival instructions.

Access and Boarding of Inflatable Life Raft

Every life raft which accommodates more than eight persons must have at least two entrances, which are diametrically opposite. One of these entrances must be fitted with a semi-rigid boarding ramp to allow persons to board from the sea. Past experience has shown that survivors wearing standard bulky lifejackets experienced extreme difficulty in boarding the raft from the water, especially after using valuable energy in swimming towards the craft.

Entrances which are not provided with the ramp should be fitted with a boarding ladder, the lowest step being situated not less than 0.4 m below the raft's lightest waterline. Means should also be provided inside the raft to allow people to pull themselves into the raft from the ladder.

Capacity of Life Rafts

The number of people which a life raft shall be permitted to accommodate shall be equal to the lesser of:

1   the greatest whole number obtained by dividing by 0.096 the volume measured in cubic metres of the main buoyancy tubes

(which for this purpose shall include neither the arches nor the thwarts if fitted) when inflated;

2   the greatest whole number obtained by dividing by 0.372 the inner horizontal cross-sectional area of the life raft measured in square metres (which for this purpose may include the thwart or thwarts if fitted) measured to the innermost edge of the buoyancy tubes; or

3   the number of persons having an average mass of 75 kg, all wearing lifejackets, that can be seated with sufficient comfort and headroom without interfering with the operation of any of the life raft's equipment. (Maximum capacity must not exceed 150 persons.)

Markings on the Life Raft

Each life raft should carry the following markings:

- the maker's name or trade mark;
- a serial number;
- the date of manufacture;
- the name of the approving authority;
- the name and place of the servicing station where it was last serviced;
- the number of persons it is permitted to accommodate over each entrance in characters not less than 100 mm in height of a contrasting colour to that of the raft.

Life Raft Equipment

1   One buoyant rescue quoit, attached to not less than 30 m of buoyant line. Used to assist the recovery of additional survivors.

2   One safety knife of the non-folding type, having a buoyant handle and lanyard attached. It should be stowed on the exterior of the canopy near to that point to which the painter is secured.

    In addition, a life raft which is permitted to accommodate 13 persons or more shall be provided with a second knife which need not be of the non-folding type.

3   One buoyant bailer for every life raft which is permitted to accommodate not more than 12 persons. Two buoyant bailers for life rafts which accommodate 13 persons or more.

4   Two sponges, one being theoretically for drying out the floor of the raft, while the other may be used to collect condensation from the inner canopy.

5   Two sea anchors, often called drogues. Each fitted with a shock-resistant hawser and tripping line. The strength of hawsers and tripping lines should be adequate for all sea conditions. Sea anchors shall be fitted with swivels at each end of the line and will be of a type that is unlikely to foul inside out between its shroud lines. One of the sea anchors should be

permanently secured in such a manner that when the life raft inflates it will be caused to lie oriented to the wind in the most stable of manners.

The purpose of the drogues is to restrict the drift rate of the life raft and reduce the risk of capsize. It is especially required if the life raft is engaged in helicopter operations.

6   Two buoyant paddles used to assist the manoeuvring of the raft away from the ship's side. It should be noted that it is extremely difficult to give directional force to a circular raft by use of the paddles alone and seafarers may find it more helpful to weight the sea anchor and throw it in the direction in which the raft is required to travel, so drawing the raft through the water towards the drogue. The paddles may also make useful splints for administration of first aid to possible broken limbs of injured parties.

7   Three tin openers. These may be incorporated with the safety knives.

8   One whistle or equivalent sound signal.

9   Two buoyant smoke signals.

10   Four rocket parachute flares.

11   Six hand held flares.

12   One waterproof electric torch, suitable for Morse signalling. It should be supplied with one spare set of batteries and one spare bulb in a waterproof container.

13   One efficient radar reflector.

14   One heliograph (daylight signalling mirror). This is a silvered sheet of metal which fits into the operator's hand. It is used to reflect the sun's rays in the direction of a potential rescue aircraft or surface vessel. In marine use it is a means of attracting attention and it would be extremely unlikely that the instrument could be used to transmit Morse code successfully from a small boat or raft which would probably be moving in an erratic manner in swell and/or sea.

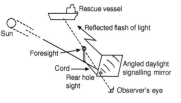

Figure 8.4 Use of heliograph.

15   One set of fishing tackle, usually comprising a fishing line and six hooks. Any fish caught should be treated with reserve as it may be of a poisonous variety. Fish, in any case, are not recommended to be eaten unless a plentiful supply of water is available. Should the raft become badly torn fishing gear can be used for repair.

16   One copy of the lifesaving signals, referred to in Regulation V/16, on a waterproof card or in a waterproof container.

17   Thermal protective aids which comply with Regulation 34, in sufficient numbers for 10 per cent of the people that the life raft is permitted to accommodate, or two, whichever is the greater.

18   A food ration totalling not less than 10,000 kJ for each person the life raft is permitted to accommodate. These rations should be contained in airtight packages and stowed in watertight containers.

19   Fresh water in watertight receptacles amounting to 1.5 litres per person that the raft is permitted to accommodate, of which

Plate 70 Rack stowage of inflatable life rafts aboard a passenger vehicle ferry.

0.5 litres per person may be replaced by a de-salting apparatus capable of producing an equal amount of water in two days.

20   One rustproof graduated drinking vessel (usually plastic).
21   One set of instructions on survival.
22   One set of instructions on immediate actions.
23   Anti-seasickness medicine sufficient for at least 48 hours (seasick tablets) and one seasickness bag for each person the raft is permitted to accommodate.
24   One first aid outfit in a waterproof case capable of being closed tightly after use.

A basic first aid kit should contain:

- collapse revivers (six capsules of fragrant ammonia) contained in a tin;
- 25 compound codeine tablets;
- two standard dressings 20 × 15 cm;
- two standard dressings 15 × 10 cm;
- five triangular bandages of 140 cm base;
- white absorbent gauze;
- four compressed roller bandages;
- calico bandage, unbleached;
- one packet of compressed cotton wool (4 oz);
- six safety pins;
- one tube of soft petroleum jelly;

- one pair of scissors, stainless steel;
- 60 energy tablets;
- one set of instructions;
- one capsule of silica gel to keep the contents dry and free of moisture.

Inflatable life rafts must carry in addition a repair outfit for repairing punctures to the buoyancy chambers and a topping up pump or bellows.

Markings on the Container of the Life Raft

Each life raft container shall be marked with:

- the maker's name or trade mark;
- serial number;
- the name of the approving authority and the number of persons it is permitted to carry;
- SOLAS;
- the type of emergency pack enclosed in the raft;
- the date when last serviced;
- length of painter;
- the maximum permitted height of stowage above the waterline (dependent on the drop test height and length of painter);
- the launching instructions.

Launching the Inflatable Life Raft

The web straps securing the raft in its stowage cradle should be released by slipping the manually operated senhouse slip. The raft container should then be manhandled to the launching position by the ship's side. If side rails are in position, these should be removed to facilitate the launch.

The painter from the raft should be checked to see that it is well secured to a strong point of the vessel. Where a hydrostatic release

Plate 71 Rack of six inflatable Beaufort life rafts seen aboard an Australian passenger ferry.

unit is fitted, the painter and 'D' ring should be inspected to ensure they are well fast.

Pull out a limited amount of the painter from the container, check that the water surface is clear of other survivors, or debris, then throw the raft away from the ship's side. Inflation will be caused by a sharp 'tug' on the fully extended painter, causing the CO_2 gas bottle to be fired.

Boarding the Raft

Seafarers should endeavour to board the raft while they are dry; should they try to jump directly into the raft, they may cause serious damage to the raft, other occupants and themselves.

Precautions

Before boarding the raft, remove all heavy shoes, sharp objects, etc. from the person to prevent accidental puncture of the fabric. Once inside the raft, check for leaks and ensure that excess carbon dioxide gas is vented clear of the inner canopy.

Mariners should always remember to try to board in the dry condition.

Rigid Life Rafts

All rigid life rafts shall comply with the general requirements specified by the regulations regarding life rafts.

Construction

The buoyancy of the rigid life raft shall be provided by approved inherently buoyant material placed as near as possible to the

Plate 72 Inflatable life raft secured with a hydrostatic release unit.

periphery of the life raft. The buoyant material shall be fire retardant or be protected with fire-retardant covering.

The floor of the raft shall prevent the ingress of water and shall effectively support the occupants out of the water and insulate them from cold. The stability should be such that unless it is capable of operating safely whichever way up it is floating, it must be either self-righting or can be readily righted in a seaway and in calm water by one person. Once loaded with its full complement of persons and equipment it must be possible to tow the raft in calm water at a speed of up to three knots.

Access into the Rigid Life Raft

At least one entrance shall be fitted with a rigid boarding ramp to enable persons to board the raft from the sea. In the case of a davit-launched life raft having more than one entrance, the boarding ramp shall be fitted at the entrance opposite to the bowsing and embarkation facilities. Entrances not provided with a boarding ramp shall have a boarding ladder, the lowest step of which shall be situated not less than 0.4 m below the life raft's lightest water-line. There should also be means inside the raft to assist persons to pull themselves inside from the ladder.

Capacity: Rigid Life Rafts

The number of persons that the life raft is permitted to accommodate shall be equal to the lesser of:

- the greatest whole number obtained by dividing by 0.096 the volume measured in cubic metres of the buoyancy material multiplied by a factor of 1 minus the specific gravity of the material;
- the greatest whole number obtained by dividing by 0.372 the horizontal cross-sectional area of the floor of the life raft measured in square metres; or
- the number of persons having an average mass of 75 kg, all wearing lifejackets, who can be seated with sufficient comfort and headroom without interfering with the operation of any of the life raft's equipment.

Fittings

The life raft shall be fitted with a painter system, the breaking strength of which, inclusive of its means of attachment and excepting any weak link required, shall be not less than 10.0 kN for rafts permitted to accommodate nine persons or more, and not less than 7.5 kN for other rafts.

They should also have a manually controlled lamp, visible at a distance of at least two miles and capable of 12 hours of operation, fitted to the top of the raft canopy. If this is a flashing light it should operate at not less than 50 and not more than 70 flashes per minute for the 12-hour period. The lamp should be powered by a sea water activated cell or a dry chemical cell which shall automatically light when the canopy is set in place. This cell should not deteriorate due to dampness or humidity while in stow.

A manually operated lamp should also be positioned inside the life raft, capable of at least 12 hours of operation and activated automatically.

Markings on Rigid Life Raft

The life raft shall be marked with:

- the name and port of registry of the ship to which it belongs;
- the maker's name or trade mark;
- a serial number;
- the name of the approving authority;
- the number of persons it is permitted to accommodate over each entrance in characters not less than 100 mm in height of a colour contrasting with that of the life raft;
- SOLAS;
- the type of emergency pack enclosed;
- length of painter;
- the maximum permitted height of stowage above the waterline (drop test height);
- launching instructions.

The Davit-Launched Life Raft

Every davit-launched life raft shall comply with the general requirements regarding life rafts, and in addition davit-launched rafts shall be used with an approved launching appliance and shall:

- when the life raft is loaded with its full complement of persons and equipment, be capable of withstanding a lateral impact against the ship's side at an impact velocity of not less than 3.5 m/s and also the drop into the water from a height of not less than 3 m, without damage that will affect its function;
- be provided with the means for bringing the life raft alongside the embarkation deck and holding it securely during embarkation.

Every passenger ship davit-launched life raft shall be so arranged that it can be boarded by its full complement of persons. Every cargo ship davit-launched life raft shall be so arranged that it can be boarded by its full complement of persons in not more than three minutes from the time the instruction to board is given.

Davit-Launched Inflatable Life Rafts

When suspended from the lifting hook or bridle, these shall withstand a load of:

- four times the mass of its full complement of persons and equipment at an ambient temperature and a stabilized raft temperature of 20 ± 3°C, with all relief valves inoperative; and
- 1.1 times the mass of its full complement of persons and equipment at an ambient temperature and a stabilized raft temperature of –30°C, with all relief valves operative.

Rigid containers for life rafts to be launched by a launching appliance shall be so secured that the container or parts of it are prevented from falling into the sea by container retaining lines during and after inflation and launching of the contained raft.

Davit-Launched Rigid Life Rafts

A rigid life raft for use with an approved launching appliance shall, when suspended from its lifting hook or bridle, withstand a load of four times the mass of its full complement of persons and equipment.

Davit-launched survival craft muster and embarkation stations shall be so arranged as to enable stretcher cases to be placed in the survival craft. They shall also be stowed within easy reach of the lifting hooks unless some means of transfer is provided which is not rendered inoperable within the limits of trim and list specified by the regulations.

Where necessary, means shall be provided for bringing the davit-launched life raft against the ship's side and holding it alongside so that persons can be safely embarked.

Plate 73 Davit-launched life raft station aboard a passenger ferry.

Plate 74 Davit-launched life
raft released to surface (source:
Shutterstock).

Every life raft launching appliance shall comply with the general requirements (paragraphs 1 and 2) for launching appliances, except with regard to use of gravity for turning out the appliance, embarkation in the stowed position and recovery of the loaded raft. The arrangement should not allow premature release during lowering and shall release the raft when waterborne.

Comment on Davit-Launched Life Raft

The davit-launched life raft system is designed to be a speedy method of evacuation for would-be survivors. The obvious advantage over the inflatable raft is that persons can board in the dry condition without running the risk of having to enter the water, bearing in mind that the body loses heat approximately 26 times faster in water than when in a dry condition. Also, casualties can be lifted directly into the raft at the embarkation deck level without the risks of getting close to the surface.

Its operation usually takes the form of several rafts stowed in racks. These can all be launched by recovery of the fall/release hook by means of the tricing line without the necessity for turning the davit back inboard for each raft. The fall returns to embarkation deck level after each launch by the winch.

Great care should be taken with every system and the manufacturer's instructions closely followed. The real danger of confusing similar types of systems and causing inflation at the wrong moment could cause bowsing lines to part and damage the raft and render the raft incapable for use in the intended way.

Procedure for Launching Davit-Launched Life Raft

The mariner's attention is drawn to the following guidelines for launching davit-launched life rafts. The text should only be accepted in

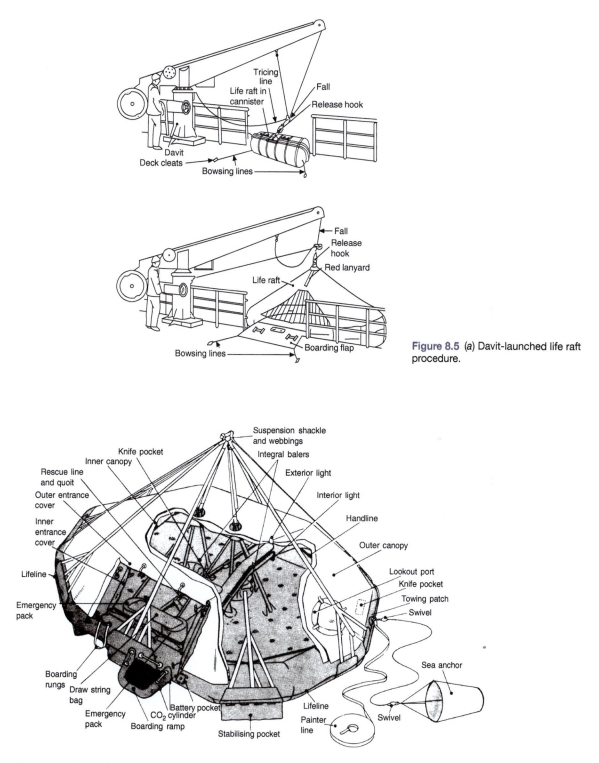

Figure 8.5 (*a*) Davit-launched life raft procedure.

Figure 8.5 (*b*) Beaufort davit-launched inflatable life raft.

a general form as the methods of launch will vary depending on the size and manufactured type of raft/davit. It is also pointed out that different manufacturers have a wide range of fitments which are not necessarily carried on all rafts, and may be described by different names.

1   Manhandle the life raft in its canister under the wire fall of the davit.
2   Pull off the sealing patch of the canister and pull out the securing shackle of the life raft.
3   Secure the fall hook to the exposed shackle.
4   Pull out and tie off the container retaining lines to the side rails.
5   Pull out and secure the bowsing in lines to the deck cleats provided.
6   Pull out the short painter and tie off at the embarkation deck.
7   Hoist the life raft canister clear above the deck.
8   Turn out the davit by the handle provided to its designed limit (usually about 70° off the ship's fore and aft line).
9   Inflate the life raft by giving a sharp tug on the painter. (The two halves of the canister should fall away to either side of the ship being retained by the 'container retaining lines' secured at the rails.)
10 Tension the bowsing in lines to ensure that the life raft is flush against the ship's sides.

> NB. Some life rafts have bowsing lines secured to a boarding flap.

11 The person in charge of the life raft should then carry out internal checks on the condition of the buoyancy chambers for any defects and ensure that the inside is well ventilated and not containing CO_2.
12 A further check must be made overside to ensure that the water surface beneath the life raft is cleared of debris and it is safe to launch.
13 Load the life raft in a stable manner, checking that all personnel have no sharp objects on their person.
14 When fully loaded, detach the bowsing in lines and the painter and cast them into the access point of the raft.
15 Commence lowering the life raft on the fall towards the surface.
16 Approximately 2–3 metres from the surface, the person in charge of the life raft should pull down on the lanyard from the release hook.

> NB. This effectively unlocks the hook arrangement, but the sheer weight inside the life raft does not allow the shackle to release from the hook at this time.

17 Continue to lower to the surface. As the buoyancy affects the underside of the life raft the hook arrangement (spring-loaded) retracts from the shackle to effect the release of the life raft.

Plate 75 Davit-launched life raft station aboard a ro-pax ferry.

Once clear of the immediate area life rafts should be joined together with about a 10 m length between them to avoid snatching in a seaway.

> NB. It would be considered prudent for the person in charge of the life raft to have two well-built individuals on the paddles to manoeuvre the craft away from the ship's side as soon as the life raft reaches the surface and releases.

Mills Atlas RFD Release Hook

In this release mechanism (Figure 8.6), designed for use with the davit-launched life raft, each hook is subjected to a static load test of just over 5.75 tonnes (5,842 kg), while the safe working load (SWL) of the unit is just over 2.25 tonnes (2,286 kg). The operating lanyard is tested separately by a static load test of 500 lb (227 kg).

The hook unit it secured to the shackle protruding from the valise and locked in position by means of an internal safety catch. When the raft starts to descend, the lanyard can be operated from within the raft. This lanyard only unlocks the safety catch, and the hook will not release the raft until the raft itself becomes waterborne.

The principle of operation is that the weight of the raft maintains the hook in the closed position. However, when the load suspended from the hook, namely the raft, falls below 30 lb (13.5 kg), the hook is allowed to open, so releasing the raft. The apparent loss in weight of the raft is effective as soon as it becomes waterborne.

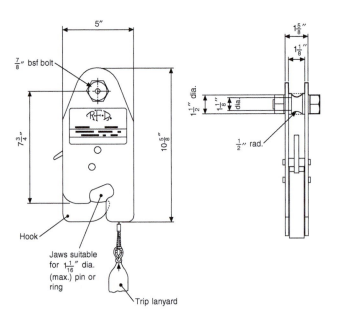

Figure 8.6 Release hook for the securing of davit-launched life raft.

Hydrostatic Release Unit

The hydrostatic release unit (HRU) (Figure 8.7) is an extension of the securing lashings over the inflatable life raft. The lashing itself must incorporate a senhouse slip to allow manual release.

Should the vessel suffer a disaster and subsequently sink, then the release unit will automatically activate under the surface of the water, at not more than 4 m depth, when the pressure is sufficient to release the draw bolt. Once the draw bolt is released, the web strap lashing no longer retains the raft canister in its cradle support. The life raft is free to float to the surface, extracting the length of painter as it rises clear of the sinking vessel.

Plate 76 Hamar disposable hydrostatic release unit.

The painter is secured to a 'D' ring, which was previously retained by the draw bolt, the arrangement providing a weak link securing which will part after the painter becomes fully extended and the pressure increases, causing the SWL/breaking strain to be exceeded.

Increased tension on the painter/weak link components will effectively cause inflation of the raft, provided that the vessel continues to sink further. The raft would not become inflated if the vessel sank and settled on the sea bed at a depth less than the length of the painter; and inflation would take place on the surface only when the painter is fully extended and the canister started to 'snatch' over the painter's length.

By releasing the senhouse slip arrangement the raft may be launched manually and inflated in the normal manner, or released for despatch ashore for annual servicing.

Float-Free Arrangements for Life Rafts

1 The breaking strength of the painter system, with the exception of the weak link arrangement, shall be not less than 10.0 kN for rafts carrying nine persons or more and not less than 7.5 kN for other life rafts.
2 A weak link system, if used, shall break under a strain of 2.2 ± 0.4 kN.
3 Any weak link should not be broken by the force required to pull the painter from the raft. If applicable, the weak link should be of sufficient strength to permit inflation of the raft.
4 HRUs, if used, should not release the raft when seas wash over the unit and they should be fitted with drains to prevent water accumulation inside the hydrostatic chamber.

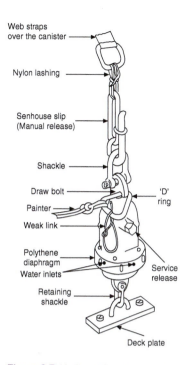

Figure 8.7 Hydrostatic release unit (permanent fixture type).

Plate 77 Inflatable life raft stowed in cradle and fitted with diposable hydrostatic release unit.

Plate 78 An RFD throwover inflatable life raft seen stowed in its cradle and fitted with the Hamar disposable hydrostatic release unit.

5 Release units should be constructed of compatible material so as to prevent malfunction. Galvanizing or other forms of metallic coating on parts of the HRU shall not be accepted.

6 The life raft should be permanently marked on the exterior with its type and serial number.

7 Either a document or identification plate stating the date of manufacture, type and serial number will be provided.

8 Any part connected to the painter system shall have a strength of not less than that required for the painter.

NB. Different types of life rafts are now found within the marine environment. These include the 'Heli-raft' used extensively by aircraft and open reversible rafts by high-speed craft. Other examples are in use with MES & MEC operations.

The mariner should note that where a survival craft requires a launching appliance and is also designed to float free, the float-free release of the survival craft from its stowed position should be automatic.

Servicing of Hydrostatic Release Units (Non-disposable)

Hydrostatic release units shall be serviced at intervals not exceeding 12 months. However, in cases where it appears proper and reasonable, the authority may extend this period to 17 months. Servicing must be carried out at an approved servicing station by properly trained personnel.

NB: Disposable hydrostatic release units are now available. These have an active life cycle of two years and are then replaced.

Additional Fitments to Raft

Water Catchment Areas

These are situated on the outer canopy to collect rainwater and supplement the basic water ration supplied with the raft. The catchment areas will differ in shape from manufacturer to manufacturer,

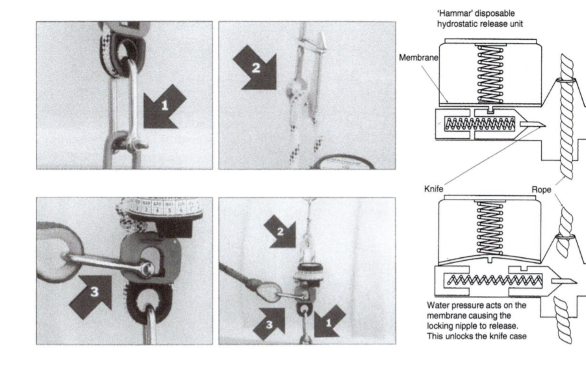

Figure 8.8 Hydrostatic release unit: how it works.

but the purpose of each is the same – to collect rainwater and deliver it to a collection tube inside the life raft. Survivors should be careful to wash off any salt content from the catchment area before drinking the newly acquired rainwater.

Double Doorway Entrance

The covering over the entrance is manufactured as two separate sheets of weatherproof fabric, the outer cover being rolled down and secured to the buoyancy chambers by Velcro strips or tape positioned at intervals across the raft entrance. The inner cover is opened upwards from the buoyancy chamber to be secured inside the upper canopy of the raft by a toggle. The inner cover will have two water pockets, one at each side, equipped with a draw string. The purpose of the water pockets is to allow any water being bailed from the raft to be expelled over the side without opening up the entrances and losing the internal warmth of the raft.

Abandoning Ship

The order to abandon the vessel will be passed by word of mouth from the Master, or the most senior officer, but this order will only be given as the very last resort. It should be borne in mind that the mother ship is the very best life-support craft available to you, and it should only be left when all hope of remaining safely afloat has been lost.

After the order has been passed to abandon the vessel, apprehension with regard to the future will be the main source of concern to all. Survival craft should be prepared for launch if time allows, and the vessel should be manoeuvred to a suitable position to permit safe launching. Headway on the vessel should be reduced and, if possible, taken off completely to allow safe launching of survival craft.

Once survival craft are in the water the following actions and procedures should take place:

Initial vital actions to protect against exposure	1 Cut painter 2 Stream drogue 3 Close entrance 4 Maintain craft	so far as possible simultaneously in group situations
Cold climates		
Cut painter	Assist other survivors to board craft. Use safety knife provided. Manoeuvre clear from ship's side or obstructions.	
Stream drogue	Reduces the rate of drift – allows survivors to reach craft – helps keep craft at location of casualty – ideally holds entrances at angle to weather – helps to prevent capsizing.	
Close entrance	Keeps out water (sea or rain) and wind. Allows natural body heat of survivors to warm interior air. When warm and atmosphere heavy and uncomfortable, adjust ventilation. A very small opening should be sufficient.	
Maintain craft	Insulation – inflate the floor of the raft. Seaworthiness – remove excessive water; check for damage; repair or plug if necessary; check position of inflation (topping up) valves.	
Tropical climates	It is unlikely that it will be necessary to close up or insulate. It will be necessary to keep cool and avoid exposure to the sun. However, keep water out.	
Secondary vital actions	The above are essential actions to combat exposure. The following actions should also be taken as soon as possible (they are not necessarily in priority order).	
Seasick remedy	Life rafts in particular are known to make even the best sailors seasick. Seasickness is incapacitating and may destroy the will to survive, and to carry out survival procedure you need to be physically fit and mentally capable. It is therefore *imperative* to take seasickness pills as early as possible. Seasickness is not only a physical handicap, but valuable body fluid may be lost. The pills themselves will make your mouth feel dry, but resist the urge to drink.	
Injured survivors	Maintain a clear airway; control bleeding. Treat injured survivors using the first aid kit. Instructions for use are included.	
Bail out	Remove any water with bailer and dry out with sponges.	
Warming up	If people are chilled or shivering, get everybody to huddle together but do not upset trim – the closer they get, the warmer they get. Cover with spare clothing. Sit on lifejackets as extra insulation if necessary. Loosen any constriction on feet. Keep wriggling toes and ankles to reduce chance of getting cold injuries. Change lookouts if necessary to prevent frostbite on exposed skin.	

Congregation of rafts	Join up and secure with other craft – mutual aid. In cold weather, get maximum numbers together for warmth. Two or more crafts are easier to find than one.
Search for survivors, lookout	Listen for whistles: post lookouts to look for survivors, signalling lights and lights of other rafts, ships or aircraft. Lookout to collect useful debris, etc. Assist survivors by using the quoit and line, thereby avoiding swimming. Raft may be manoeuvred using drogue or paddle.
Handbook	Read the survival craft handbook for further guidance on actions to be taken etc.
Morale and will to survive	Cold, anxiety, hunger, thirst, effects of seasickness all work against the will to survive. Keep spirits up. Maintain confidence in rescue. Firm but understanding discipline. Keep a lookout for signs of abnormal behaviour and avoid doing things which annoy others. Physically restrain delirious people; bear in mind the effect on other survivors.
Subsequent actions	The initial and secondary actions combat the immediate threats to survivors. This section concerns subsequent actions and survival craft routine.
Leader	Appoint or elect a leader.
Sharp objects	Collect sharp objects or potential weapons.
Roll call	Have a roll call to muster survivors.
Routine	Establish a routine and allocate duties to survivors – ration keeper – lookouts – repair party – bailers, etc. It is important to keep the minds of survivors occupied while avoiding unnecessary exertion.
Watches	One-hour watches in pairs – one outside and one inside. *Duties outside* – lookout for ships, survivors, aircraft, etc.; gather useful wreckage. *Duties inside* – maintain raft (bailing, drying, ventilation, buoyancy tubes, etc.); supervise raft management while others rest; attend to injured persons; look after equipment, valuables, etc.
Raft management	
Protection	*Cold climates* Keep warm and dry. Adjust ventilation to minimum required. Huddle together for warmth if necessary. Carry out simple exercises to avoid cold injury: open and clench fists, stretch limbs; wriggle toes, ankles, fingers, wrists. This will maintain circulation but not waste energy. Avoid exposure – rotate lookout to avoid exposure. *Hot climates* Arrange ventilation. Avoid exposure (sunburn). Reduce need for water by: avoiding unnecessary exertion (no swimming); maintain a through breeze; check position of drogue; keep outside of raft wet; wet clothing by day.
Location	Keep lookouts. Have location aids readily available. Keep rafts congregated. Drogues or sea anchor will reduce rate of drift from casualty area.
Water/food	Issue rations *after* first 24 hours at set times during the day – sunrise, midday, sunset. Method of issuing rations. Collect rainwater whenever possible. Eat fish only if plentiful supply of water available. Precautions in hot weather – see 'Protection'. *Do not issue* rations during first 24 hours except to injured people (if conscious). *Rainwater* should be collected from the outset.
Rations	The rations provided in the craft are based upon the above knowledge and instructions for their use *must be followed*. The length of time that the rations will last will depend upon the number of occupants. However, the minimum with a full complement is four days.

Issue of rations	*Do not issue water during the first 24 hours.* The body is already full of water – if more is put in it will be wasted in the form of urine. After 24 hours the body will be drier and will absorb any water that is drunk, just as a sponge will hold water but a wet sponge will not hold any more. Only if a person is injured is it permissible to give him a drink in the first 24 hours to replace fluid lost due to bleeding or burns and *only if he or she is conscious.* After 24 hours, issue a full ration three times daily at sunrise, midday and sunset. Don't be tempted to give more rations than necessary. Make sure the carbohydrate food is taken. In prolonged survival, wait until the fourth day before reducing the daily ration if absolutely necessary and then only by half. It is essential to supplement rations with rainwater from the outset.
Supplement rations	Supplement the basic supply of water whenever you can. Rainwater should be retained. Condensation is a possible supply.
Sea water/urine	*Do not drink sea water or urine.* Madness or death follows rapidly upon the drinking of sea water or urine.
Fish flesh	*Do not eat fish flesh* unless you have an abundant supply of water. Protein foods such as fish tend to consume vital body fluids.

Acknowledgement: the author would like to thank the Merchant Navy Training Board for their permission to reproduce part of the *Basic Sea Survival Course Instructor's Manual.*

Marine Evacuation System

Designed initially to provide an effective means of evacuating passengers and crew from high freeboard vessels of the ferry/passenger class into life rafts, the marine evacuation system (MES) consists of a double-track inflatable slide with an integral boarding platform, mounted in a deck stowage container. The life rafts drop into the water adjacent to the platform (see Figure 8.9).

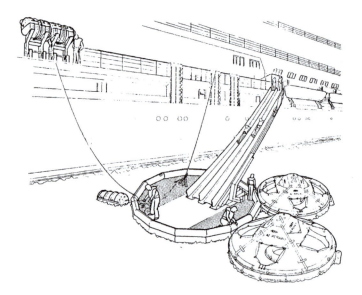

Figure 8.9 Marine evacuation system.

Each life raft will carry a designated number of persons and is contained in a weathertight container. These rafts are inflated from the boarding platform at the end of the slide, then linked to the platform by bowsing in lines to facilitate safe boarding.

The operation starts with the removal of a cover in order to release an outboard door, which is unlocked from inside the ship, and the slide and platform are pushed away from the ship, causing them both to inflate (see Figure 8.10). The inflating agent is a supply of nitrogen stored in cylinders in the deck-mounted container.

The boarding platform varies in diameter and is normally manned by crew members who will secure the rafts to the sides of the platform before embarkation. The number of life rafts that the system can use may be adapted to the passenger-carrying capacity of the parent vessel.

In extreme emergency an inflated slide and platform can be released from the vessel to serve as an additional flotation aid

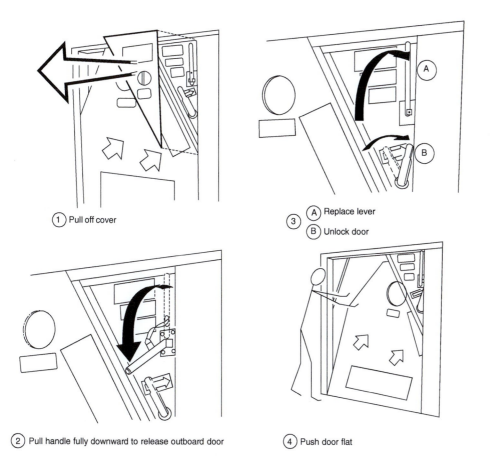

① Pull off cover

③ Ⓐ Replace lever
   Ⓑ Unlock door

② Pull handle fully downward to release outboard door

④ Push door flat

Figure 8.10 Operation of marine escape system.

for survivors. Each vessel is equipped with two systems, one on each side. New tonnage under construction is presently being fitted out with four systems, two on each side. Regulations stipulate that total evacuation must take place within 30 minutes of the alarm being raised, and this system should more than meet these requirements.

Evacuation-Slide Launching and Embarkation

Every evacuation-slide system should conform to the general requirements for launching appliances, and in addition:

- the evacuation slide shall be capable of being deployed by one person at the embarkation station;
- the evacuation slide shall be capable of being used in high winds and in a seaway.

The system is based on the rapid escape systems applicable to aircraft. The operation, which has been approved by the administration (Figure 8.10), becomes operational by activating a slide out from the ship's side. This slide is variable in length and runs onto a landing platform, the platform being an integral part of the slide. Evacuation from the parent vessel takes place by persons sliding down the double trackway to the landing platform, then embarking into life rafts.

The slide unit makes an angle of approximately 35° to the horizontal and is stowed in a box unit 8 × 8 × 3 ft, weighing 8,000 lbs. The compact unit is stowed at deck level and the number of life rafts will be variable depending on the passenger-carrying capacity of the vessel using the system.

Plate 79 Viking MES deployed showing the slide to the surface. A large 45-man life raft is secured alongside the boarding platform.

MES Variations

The larger passenger ferry has been equipped with the MES for many years and the conventional vessel suited the standard double-track slide. However, with customized hull designs, operating under the HSC code, a positive market for the shorter single-track MES became viable, especially with the smaller catamarans and the lower freeboard, fast mono-hull craft.

Several manufacturers realized the potential and designed mini-systems to accommodate deck heights of between 1.5 m and 4 m above the waterline. These are also equipped with reversible life rafts in order to comply with IMO Resolution MSC 81(70), in force since 2000. The life rafts with virtually all mass evacuation systems have moved towards large-size capacity survival craft – 100-, 128- and 150-person sizes would not be unusual today. Such high numbers of accommodated persons and the weight factor involved clearly does not intend to include moving the survival craft far from the on-scene situation that generates MES use. This would also seem compatible with range limitations imposed within the vessel's 'permit to operate'. Survival craft are, in the main, operated within close proximity of land because that is where the mother ship's designated route must lie.

Inspection and Maintenance of MES

Under the regulations, life-saving appliances must be inspected on a monthly basis and a report of the inspection should be recorded in the ship's log book. All inflatable equipment like lifejackets, life rafts, rescue boats and MES must be serviced at 12-monthly intervals by an approved service agency. This period may be extended to 17 months by the administration in the event that the 12-month service proves impractical.

Plate 80 Double chutes deployed with life rafts at the waterside. Example configuration of MEC system as manufactured and produced by Dunlop/Beaufort, Canada (source: *Marine Survival*, published by Witherby Seamanship).

NB. On installation, at least 50 per cent of MESs are subject to a harbour trial deployment. One of these systems must be deployed with at least two of the associated life rafts to establish correct launching, bowsing in and inflation procedures have been correctly installed.

Additionally, or in conjunction with the service arrangements, an MES should be deployed from the ship on a rotational basis at intervals as agreed with the administration, provided that each system is deployed at least once every six years.

Drills on ships fitted with MES must include exercising of the procedures up to the actual point of deployment. Personnel who are designated to participate as a member of the deployment party of an MES must, where possible, be party to a full deployment to the water, at intervals not exceeding two years.

Marine Evacuation Chute System

Several companies have now manufactured the marine evacuation chute (MEC) system, including DBC of Canada and RFD in the UK. Although variations of operation differ slightly between manufacturers, the principle of safe and fast evacuation is the same.

How It Works

A single or double near-vertical fabric chute is deployed from an upper embarkation deck by the action of a single crew member. The chute will then act as a 'feeder' for evacuees to descend to an inflated boarding platform or, in the case of the RFD model, directly into large-capacity inflated life rafts.

Plate 81 Single chute shown deployed with the surface landing platform. Life rafts are stowed aft of the MEC position (source: *Marine Survival*, published by Witherby Seamanship).

Systems generally require the entrance and exits of the chutes to be manned to ensure a rapid throughput of personnel towards the survival craft. Once fully loaded these would be manoeuvred away from the distressed vessel by the rescue boat(s) (Figure 8.11).

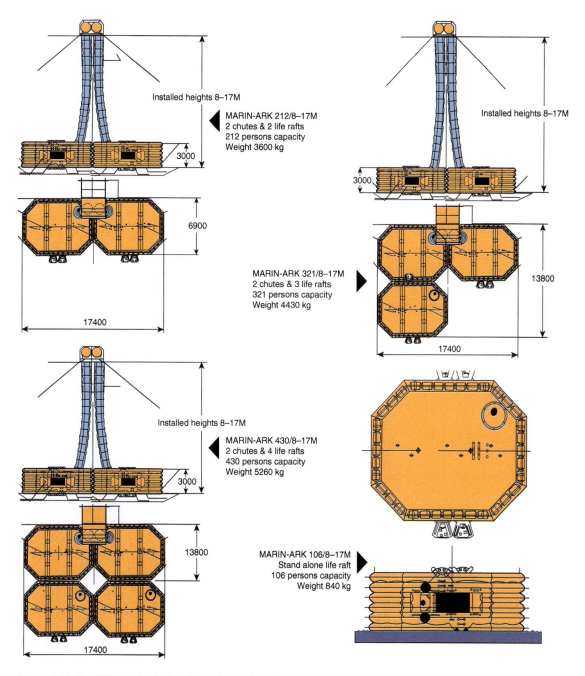

Installed heights 8–17M

MARIN-ARK 212/8–17M
2 chutes & 2 life rafts
212 persons capacity
Weight 3600 kg

3000

6900

17400

Installed heights 8–17M

3000

MARIN-ARK 321/8–17M
2 chutes & 3 life rafts
321 persons capacity
Weight 4430 kg

13800

17400

Installed heights 8–17M

MARIN-ARK 430/8–17M
2 chutes & 4 life rafts
430 persons capacity
Weight 5260 kg

3000

13800

17400

MARIN-ARK 106/8–17M
Stand alone life raft
106 persons capacity
Weight 840 kg

Figure 8.11 The RFD 'Marin-Ark' configuration options: double chute with two, three or four fully reversible life rafts providing speedy, dry-shod evacuation.

Once passengers enter the MEC they descend in a zigzag pattern and then transfer to a life raft at the surface, via the landing platform.

The life rafts, which are bowsed into the platform, once loaded, are released and pulled away by the rescue boat.

The actual controlled passive descent takes about 2–3 seconds from the point of entry through an approximate 13 metre fall, to attain the rafts at the surface.

Righting a Capsized Life Raft

In the event of the life raft inflating in a capsized condition, the mariner's priorities are to turn it the right way up and stream the sea anchor (drogue) as soon as practical to prevent sea conditions causing another capsize. The procedure for one man is first to turn the raft into the wind and then climb over the gas bottle onto the base of the raft (Figure 8.12). It will be seen that a righting strap passes directly under the base of the raft and runs across the full diameter. The mariner should take hold of the righting strap and endeavour to climb up over the base of the raft. Once he has gained enough height, he should lean backwards, dragging the raft over by the strap virtually on top of himself, in the water. If the raft has been turned into the wind, then the wind will assist the motion of righting the raft, acting on the area of the raft as on a sail.

Should the raft initially inflate in this manner, it can normally be expected to lie on its side like a capsized cone, since the weight of the gas bottle causes the balance of the raft to be offset, and the heavy side of the raft to lie well down in the water. For this reason it is essential that the mariner attempts to right the raft from the low side, pushing the heavy gas bottle with his feet under the base of the raft as righting occurs.

Plate 82 Evacuation drill being conducted with the RFD Marine-Arc, Marine Evacuation System. Two chutes, two life rafts being deployed on the portside of a North Sea ferry.

Once the raft has been righted, the mariner may find that he is underneath it. At this stage there is no need for panic, because a saucer shape containing air is formed between the raft base and the sea surface, which will allow the individual to breathe for some considerable time. However, once the raft has been righted, the seafarer should endeavour to come out from under the raft, hand for hand, against the base of the raft. He should leave the righted raft from the opposite end to where the bottle is lying in the water. Once clear, the mariner should then board and commence normal survival techniques, bearing in mind that he is not a survivor until rescued.

Figure 8.12 Righting life raft.

Beaching a Life Raft

If it becomes necessary to beach a raft, the operation should preferably be carried out during the hours of daylight on a sandy beach having a gradual slope. Circumstances will dictate the conditions, however, and the mariner should base his tactics on the weather and the proposed beaching area.

Preparations to beach the raft should be made well outside the line of surf and should include the inflating of the double floor, the donning of lifejackets by all persons, the opening of entrances and the manning of paddles, and the streaming of one or two sea anchors.

The raft should be allowed to drift into the line of surf, while the weight on the sea anchors is checked. A sharp lookout should be kept for rocks and similar obstructions and the raft fended off from the dangers. On striking the shoreline, two men should endeavour to pull the raft ashore as far up the beach as possible. Other survivors should disembark as quickly as they can, then salvage equipment and the raft itself.

It should be borne in mind that the survival craft contains all the necessary life-support systems. The raft itself will provide shelter in its inflated condition even on dry land. The gas bottle could be removed from the underside in order to allow the raft to lie flat, but the survival craft should not be deflated until a rescue is evident. The bright colour of the canopy will attract the rescue services, and to this end the survival craft should not be pulled up under cover of trees or cliff overhangs.

Stowage of Life Rafts (Cargo Ships)

Every life raft, other than those required by Regulation 31.1.4, shall be stowed with its painter permanently attached to the ship and with a float-free arrangement which complies with the regulations. The arrangement should allow the life raft to float free and in the case of an inflatable raft, it should inflate automatically.

In the case of passenger ships the stowage shall be as far as practical the same as for cargo ships.

Regulation 31.1.4 (6 man life raft)

Cargo ships where the survival craft are stowed in a position which is more than 100 m from the stem or stern shall carry, in addition to the life rafts required by paragraphs 1.1.2 and 1.2.2 (of this regulation) a life raft stowed as far forward or aft, or one as far forward and another as far aft, as is reasonable and practicable. This life raft must permit manual release and need not be of a type which can be launched from an approved launching device.

Containers for Inflatable Life Rafts

Containers for life rafts shall be constructed to withstand hard wear under conditions encountered at sea. They should have inherent buoyancy when packed with the life raft and its equipment, to pull the painter from within and to operate the inflation mechanism should the ship sink. Containers should be watertight, except for drain holes in the container bottom.

Servicing of Life Rafts

- Every inflatable life raft shall be serviced at intervals not exceeding 12 months. However, in cases where it appears proper and reasonable, the authority may extend this period to 17 months.
- Servicing should take place at an approved service station which maintains proper service facilities and only by properly trained personnel.

Ro-Pax Vessels in Coastal Waters

Roll on–roll off/passenger vessels are now required to carry life rafts of either the automatically self-righting or canopied reversible types. Such rafts must be fitted with float-free facilities and a boarding ramp. In coastal and inland waters these vessels are permitted to be fitted with 'open reversible life rafts'.

Equipment Contained in Open Reversible Life Rafts

- One buoyant rescue quoit attached to not less than 30 m of buoyant line (breaking strength of the line of at least 1 kN).
- Two safety knives of the non-folding type, having a buoyant handle, should be fitted attached to the open reversible life raft by light lanyards. These should be stowed in pockets so that irrespective of the way in which the life raft inflates, one will be readily available on the top surface of the upper buoyancy tube in a suitable position to enable the painter to be readily cut.
- One buoyant bailer.
- Two sponges.

Plate 83 Lifejacket stowage arrangement on the embarkation deck of a passenger ferry as fitted with overhead gravity davits.

- One sea-anchor permanently attached to the life raft in such a way as to be readily deployable when the open reversible life raft inflates. The position of the sea anchor should be clearly marked on both buoyancy tubes.
- Two buoyant paddles.
- One first aid outfit in a waterproof container capable of being closed tightly after use.
- One whistle or equivalent sound signal.
- Two hand flares.
- One waterproof electric torch suitable for Morse signalling, together with one spare set of batteries and one spare bulb in a waterproof container.
- One repair outfit for repairing punctures in buoyancy compartments.
- One topping up pump or bellows.

The above equipment would be designated for high-speed craft operating in coastal waters.

> NB. Standard life rafts carry alternative equipment to the open reversible life rafts.

Lifejacket Requirements

Passenger ships

A lifejacket shall be provided for every person on board the ship and in addition:

- a number of lifejackets suitable for children equal to at least 10 per cent of the number of passengers on board shall be provided or such greater number as may be required to provide a lifejacket for each child;
- a sufficient number of lifejackets shall be carried for persons on watch and for use at remotely located survival craft stations;

- every passenger ship shall carry additional lifejackets for not less than 5 per cent of the total number of persons on board, these lifejackets being stowed in conspicuous places on deck or at muster stations;
- as from 1 July 1991, every lifejacket carried on a passenger ship which is engaged on international voyages shall be fitted with a light which complies with the regulations.

Cargo ships

A lifejacket shall be provided for every person on board the ship. Each lifejacket will be fitted with a light which complies with the regulations (applicable from 1 July 1991).

Lifejackets shall be so placed on every vessel as to be readily accessible and their position shall be plainly indicated. Where, due to the particular arrangements of the ship, the individual lifejackets may become inaccessible, alternative provisions shall be made to the satisfaction of the authority. This may include an increase in the number of lifejackets to be carried.

Every lifejacket should comply with the regulation specifications and may be of the automatic inflated type. They should be constructed to the general requirements and fitted with retro-reflective material, be rot-proof and corrosive-resistant and should not be unduly affected by sea water, sunlight, oil or fungal attack. They should be capable of satisfactory operation and be manufactured in a highly visible colour.

General Requirements for Lifejackets

1 A lifejacket shall not sustain burning or continue melting after being totally enveloped in fire for a period of two seconds.
2 A lifejacket shall be so constructed that:
   - after demonstration a person can correctly don it within a period of one minute without assistance;
   - it is capable of being worn inside out, or is clearly capable of being worn in only one way and, as far as possible, cannot be donned incorrectly;
   - it is comfortable to wear;
   - it allows the wearer to jump from a height of at least 4.5 m into the water without injury and without dislodging or damaging the lifejacket.
3 A lifejacket shall have sufficient buoyancy and stability in calm and fresh water to:
   - lift the mouth of an exhausted or unconscious person not less than 120 mm clear of the water with the body inclined backwards at an angle of not less than 20° and not more than 50° from the vertical position;
   - turn the body of an unconscious person in the water from any position to one where the mouth is clear of the water in not more than five seconds.

4 A lifejacket shall have buoyancy which is not reduced by more than 5 per cent after 24 hours' submersion in fresh water.
5 A lifejacket shall allow the person wearing it to swim a short distance and to board a survival craft.
6 Each lifejacket shall be fitted with a whistle securely fastened by a cord.

Lifejacket lights

Each lifejacket light shall:

- have a luminous intensity of not less than 0.75 cd (practical range 1.0 nautical mile);
- have a source of energy capable of providing a luminous intensity of 0.75 cd for a period of at least eight hours;
- be visible over as great a segment of the upper hemisphere as is practicable when attached to a lifejacket.

If the light of the lifejacket is a flashing light it shall in addition:

- be provided with a manually operated switch;
- not be fitted with a lens or reflector to concentrate the beam; and
- flash at a rate of not less than 50 flashes per minute with a luminous intensity of at least 0.75 cd.

Inflatable Lifejackets

A lifejacket which depends on inflation for buoyancy shall have not less than two separate compartments and comply with the regulations for fixed-buoyancy type lifejackets.
They should:

- inflate automatically on immersion, be provided with a device to permit inflation by a single manual motion and be capable of being inflated by the mouth;
- in the event of loss of buoyancy in any one compartment, the lifejacket should still be capable of maintaining the standards set by the regulations for standard lifejackets.

General Requirements and Specifications for Lifebuoys

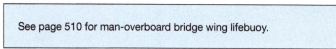

See page 510 for man-overboard bridge wing lifebuoy.

Passenger and cargo ships

Lifebuoys should be so distributed that they are readily available on both sides of the ship and as far as practicable on all open decks extending to the ship's sides. At least one should be placed in the vicinity of the stern.

They should be stowed so that they are capable of being cast loose and not permanently secured in any way. At least one lifebuoy on each side of the vessel should be fitted with a buoyant line, equal in length to not less than twice the height at which it is stowed above the waterline in the lightest seagoing condition, or 30 m, whichever is the greater.

Not less than one-half of the total number of lifebuoys should be provided with self-igniting lights and not less than two of these should also be provided with self-activating smoke signals and capable of quick release from the navigation bridge. These lifebuoys should be equally distributed on either side of the ship and should not be the lifebuoys previously stated with buoyant lines.

Each lifebuoy shall be marked in block capitals of the Roman alphabet with the name and port of registry of the ship on which it is carried.

Specifications

Every lifebuoy shall:

- have an outer diameter of not less than 800 mm and an inner diameter of not less than 400 mm;
- be constructed of inherently buoyant material – it shall not depend on rushes, cork shavings or granulated cork, any other

Table 8.1 Passenger ship's minimum number of lifebuoys

Length of ship (m)	Minimum number of lifebuoys
Under 60	8
60 and under 120	12
120 and under 180	18
180 and under 240	24
240 and over	30

Passenger ships of under 60 m in length shall carry not less than six lifebuoys provided with self-igniting lights.

Table 8.2 Cargo ship's minimum number of lifebuoys

Length of ship (m)	Minimum number of lifebuoys
Under 100	8
100 and under 150	10
150 and under 200	12
200 and over	14

loose granulated material or any air compartment which depends on inflation for buoyancy;
- be capable of supporting not less than 14.5 kg of iron in fresh water for a period of 24 hours;
- have a mass of not less than 2.5 kg;
- not sustain burning or melting after being totally enveloped in a fire for a period of two seconds;
- be constructed to withstand a drop into the water from a height at which it is stowed above the waterline in the lightest seagoing condition or 30 m, whichever is the greater without impairing either its operating capability or that of its attached components;
- if it is intended to operate the quick release arrangement provided for the self-activated smoke signals and self-igniting lights, have a mass sufficient to operate the quick release arrangement or 4 kg, whichever is the greater;
- be fitted with a grabline not less than 9.5 mm in diameter and not less than four times the outside diameter of the body of the buoy in length. The grabline shall be secured at four equidistant points around the circumference of the buoy to form four equal loops.

Lifebuoy self-igniting lights

As required by the regulations, self-igniting lifebuoy lights shall:

- be such that they cannot be extinguished by water;
- be capable of burning continuously with a luminous intensity of not less than 2 cd in all directions of the upper hemisphere or flashing at a rate of not less than 50 flashes per minute with at least the corresponding effective luminous intensity;
- be provided with a source of energy capable of meeting the stated requirements for a period of at least two hours;
- be capable of withstanding the drop test into water from the stowed position or from 30 m, whichever is the greater.

Self-igniting lights on lifebuoys of tankers shall be of an electric battery type.

Lifebuoy self-activating smoke signals

As required by the regulations, self-activating smoke signals shall:

- emit smoke of a highly visible colour at a uniform rate for a period of at least 15 minutes when floating in calm water;
- not ignite explosively or emit any flame during the entire smoke emission of the signal;
- not be swamped in a seaway;
- continue to emit smoke when fully submerged in water for a period of at least ten seconds;
- be capable of withstanding the drop test into the water from the stowed position or from 30 m, whichever is the greater.

Buoyant Lifelines (attached to lifebuoys)

The buoyant lifelines attached to lifebuoys should be of a nature and structure which is non-kinking and have a diameter of not less than 8 mm, with a breaking strength of not less than 5 kN.

Immersion Suits

Passenger ships shall carry for each lifeboat on the ship at least three immersion suits which comply with the regulations. In addition, a thermal protective aid must be provided for every other person who is to be accommodated in the lifeboat, who is not provided with an immersion suit.

These immersion suits and thermal protective aids need not be carried if:

- persons are to be accommodated in totally or partially enclosed lifeboats; or
- if the ship is constantly engaged on voyages in warm climates where in the opinion of the authority thermal protective aids are unnecessary.

With respect to the rescue boat of passenger and cargo ships, each person assigned to the crew of a rescue boat will be provided with an immersion suit of appropriate size which complies with the regulations.

Cargo ships shall carry one immersion suit for every person on board the ship.

However, in addition to immersion suits required for life rafts, lifeboats and rescue boats, the vessel shall carry thermal protective aids for persons not provided with immersion suits.

Plate 84 Lifejacket and immersion suit stowage locker, usually positioned close to launch stations.

These immersion suits and thermal protective aids need not be required if the ship:

- has totally enclosed boats on each side of the ship of such aggregate capacity as will accommodate the total number of persons aboard;
- has totally enclosed lifeboats capable of being launched by free fall over the stern of the ship of such aggregate capacity as will accommodate the total number of persons on board, launched directly from the stowed position, together with life rafts on each side of the ship of such aggregate capacity as will accommodate the total number of persons aboard; or
- is constantly engaged on voyages in warm climates where in the opinion of the authority immersion suits are unnecessary.

Cargo ships of less than 85 m length other than oil tankers, chemical tankers and gas carriers shall carry immersion suits which comply with the regulations for every person on board unless the ship:

- has davit-launched life rafts;
- has life rafts served by equivalent approved appliances, capable of being used on both sides of the ship and which do not require entry into the water to board the life raft; or
- is constantly engaged on voyages in warm climates where in the opinion of the authority immersion suits are unnecessary.

The immersion suits required to be carried by cargo vessels may be used to comply with the requirements for rescue boats.

Life rafts shall be provided with thermal protective aids which comply with the regulations, sufficient for 10 per cent of the number of persons the raft is permitted to carry or two, whichever is greater.

Lifeboats shall be provided with thermal protective aids which comply with the regulations, sufficient for 10 per cent of the number of persons the lifeboat is permitted to accommodate or two, whichever is the greater.

General Requirements for Immersion Suits

1 The immersion suit shall be constructed with waterproof material so that:
   - it can be unpacked and donned without assistance within two minutes, taking into account any associated clothing and a lifejacket if the immersion suit is to be worn in conjunction with a lifejacket;
   - it will not sustain burning or continue melting after being totally enveloped in a fire for a two-second period;
   - it will cover the whole body with the exception of the face – hands shall also be covered unless permanently attached gloves are provided;

Plate 85 Example of immersion suit and lifejacket in place on the wearer (source: Shutterstock).

- it is provided with arrangements to minimize or reduce free air in the legs of the suit;
- following a jump from a height of not less than 4.5 m into the water there is no ingress of water.

2   An immersion suit which complies with regulations concerning lifejackets may be classified as a lifejacket.

3   An immersion suit shall permit the person wearing it, and also wearing a lifejacket if the suit is to be worn in conjunction with a lifejacket, to:
- climb down a vertical ladder at least 5 m in length;
- perform normal duties during abandonment;
- jump from a height of not less than 4.5 m into the water without damaging or dislodging the immersion suit or being injured;
- swim a short distance through the water and board a survival craft.

4   An immersion suit which has buoyancy and which is designed to be worn without a lifejacket shall be fitted with a light and whistle that comply with the lifejacket regulations.

5   If the immersion suit is designed to be worn in conjunction with a lifejacket, the lifejacket should be worn over the immersion suit. The person wearing such an immersion suit shall be able to don a lifejacket.

Performance requirements

An immersion suit made of material which has no inherent insulation shall be:

- marked with instructions that it must be worn in conjunction with warm clothing;
- so constructed that when worn in conjunction with warm clothing and a lifejacket, if the suit is to be worn with a lifejacket, it shall continue to provide sufficient thermal protection following one jump, by the wearer into the water from a height of 4.5 m, to ensure that when it is worn for a period of one hour in calm, circulating water at a temperature of 5°C, the wearer's body core temperature does not fall more than 2°C.

An immersion suit made of material with inherent insulation, when worn either on its own or with a lifejacket, if the suit is to be worn with a lifejacket, shall provide the wearer with sufficient thermal insulation following one jump into the water from a height of 4.5 m to ensure that the wearer's body core temperature does not fall more than 2°C after a period of six hours in calm, circulating water at a temperature of between 0° and 2°C.

The immersion suit shall permit the person wearing it, with hands covered, to pick up a pencil and write after being immersed in water at 5°C for a period of one hour.

Buoyancy requirements of immersion suits

A person in fresh water wearing either an immersion suit complying with the lifejacket regulations or an immersion suit with a separate lifejacket shall be able to turn from face down to a face up position in not more than five seconds.

Thermal protective aids (not to be confused with immersion suits)

1. A thermal protective aid shall be made of waterproof material having a thermal conductivity of not more than 0.25 W/(m·K) and shall be so constructed that, when used to enclose a person, it shall reduce both the convective and evaporative heat loss from the wearer's body.
2. The thermal protective aid shall:
   - cover the whole body of a person wearing a lifejacket with the exception of the face – hands shall also be covered unless permanently attached gloves are provided;
   - be capable of being unpacked and easily donned without assistance in a survival craft or rescue boat;
   - permit the wearer to remove it in the water in not more than two minutes if it impairs the ability to swim.
3. The thermal protective aid shall function properly throughout an air temperature range of –30°C to +20°C.

Plate 86 Rescue boat stowed at the boat deck level of a large passenger vessel.

Rescue Boats

A rescue boat is defined as a boat designed to rescue persons in distress and to marshal survival craft. The 1983 amendments to the SOLAS convention of 1974 requires:

- Passenger ships of 500 tonnes gross and over shall carry at least one rescue boat which complies with the regulations, on either side of the ship.
- Passenger ships of less than 500 tonnes gross shall carry at least one rescue boat which complies with the regulations.
- Cargo ships shall carry at least one rescue boat which complies with the regulations. A lifeboat may be accepted as a rescue boat, provided that it complies with the requirements for rescue boats.

If the rescue boat carried is a lifeboat it may be included in the aggregate capacity for cargo ships less than 85 m in length. This is provided that the life raft capacity on either side of the vessel is at least 150 per cent of the total number of persons on board.

Similarly, for passenger ships of less than 500 tonnes gross, and where the total number of persons on board is less than 200, if the rescue boat is also a lifeboat, then it may be included in the aggregate capacity. This is provided that the life raft capacity on either side of the ship is at least 150 per cent of the total number of persons on board.

The regulations state:

- The number of lifeboats and rescue boats that are carried on passenger ships shall be sufficient to ensure that in providing for abandonment by the total number of persons on board not more than six life rafts need be marshalled by each lifeboat or rescue boat.

Plate 87 Semi-rigid fast rescue craft held by launching bridle under the davit fast launch system (source: Shutterstock).

- The number of lifeboats and rescue boats that are carried on passenger ships engaged on short international voyages shall be sufficient to ensure that in providing for abandonment by the total number of persons on board not more than nine life rafts need be marshalled by each lifeboat or rescue boat.

Launching Arrangements for Rescue Boats

Rescue boat arrangements shall be such that the rescue boat can be boarded and launched directly from the stowed position with the number of persons assigned to crew the rescue boat on board.

If the rescue boat is also a lifeboat, and other lifeboats can be boarded and launched from an embarkation deck, the arrangements shall be such that the rescue boat can also be boarded and launched from the embarkation deck.

Every rescue boat launching device shall be fitted with a power winch motor of such capacity that the rescue boat can be raised from the water with its full complement of persons and equipment. It should be kept at a continuous state of readiness for launching in not more than five minutes and be stowed in a suitable position to allow launch and recovery.

General requirements for rescue boats

1 Rescue boats may be either of rigid or inflated construction or a combination of both.
2 Rescue boats will be not less than 3.8 m and not more than 8.5 m in length. They will be capable of carrying at least five seated persons and a person lying down.

3   Rescue boats which are a combination of rigid and inflated construction shall comply with the appropriate requirements of the regulations affecting rescue boats to the satisfaction of the authority.

4   Unless the rescue boat has adequate sheer, it shall be provided with a bow cover extending for not less than 15 per cent of its length.

5   Rescue boats shall be capable of manoeuvring at speeds up to six knots and maintaining that speed for a period of at least four hours.

6   Rescue boats shall have sufficient mobility and manoeuvrability in a seaway to enable persons to be retrieved from the water, marshal life rafts and tow the largest life raft carried on the ship with its full complement of persons and equipment or its equivalent at a speed of at least two knots.

7   A rescue boat shall be fitted with an inboard engine or outboard motor. If it is fitted with an outboard motor, the rudder and the tiller may form part of the engine. Outboard engines with an approved fuel system may be fitted in rescue boats provided the fuel tanks are specially protected against fire and explosion.

8   Arrangements for towing shall be permanently fitted in rescue boats and shall be sufficiently strong to marshal or tow life rafts as required in item 6 above.

9   Rescue boats shall be fitted with weathertight stowage for small items of rescue equipment.

Plate 88 Single-arm launching system for a semi-rigid fast rescue craft. Operates up to 2.5 tons SWL. The boat is griped in by quick release web straps and provided with a crew boarding platform.

Rescue boats should be constructed in a manner to provide adequate stability and with inherent strength to withstand launching, its internal capacity being established by similar means as with lifeboats. Access should be such as to permit rapid boarding by individuals and also allow persons to be retrieved from the sea or brought aboard on stretchers. Acceptable propulsion and steering arrangements, together with a release launching mechanism, must be approved by the authority. There must be a method of release of the forward painter and skates for launching down the ship's sides if necessary to prevent damage to the boat.

Figure 8.13 Recovery by rescue boat. Recovery methods should endeavour to bring casualties aboard over the weather bow by means of a 'house recovery net' or other horizontal method.

Rescue Boat Equipment

All items of rescue boat equipment, with the exception of boat hooks, which shall be kept free for fending-off purposes, shall be secured within the rescue boat by lashings, storage in lockers or compartments, storage in brackets or similar mounting arrangements or other suitable means. The equipment shall be secured in such a manner as not to interfere with any launching or recovery procedures. All items of rescue boat equipment shall be small and of as little mass as possible and shall be packed in suitable and compact form. This equipment will include:

- sufficient buoyant oars or paddles to make headway in calm seas. Thole pins, crutches or equivalent arrangements shall be provided for each oar. Thole pins or crutches shall be attached to the boat by lanyards or chains;
- a buoyant bailer;
- a binnacle containing an efficient compass which is luminous or provided with suitable means of illumination;
- a sea anchor and tripping line with hawser of adequate strength, and not less than 10 m in length;
- a painter of sufficient length and strength, attached to the release device complying with the regulations, to enable the forward painter to be released under tension;
- one buoyant line, not less than 50 m in length, of sufficient strength to tow a life raft as required by the regulations;
- one waterproof electric torch suitable for Morse signalling, together with one spare set of batteries and one spare bulb in a waterproof container;
- a whistle or equivalent sound signal;
- a first aid outfit in a waterproof container capable of being closed tightly after use;
- two buoyant rescue quoits, attached to not less than 30 m of buoyant line;
- a searchlight capable of effectively illuminating a light-coloured object at night having a width of 18 m at a distance of 180 m for a total period of six hours and of working continuously for at least three hours;

Plate 89 A 6.4 m Avon Searider operating at speed is an example of a semi-rigid fast rescue craft now carried by many vessels.

- an efficient radar reflector;
- thermal protective aids complying with the regulations and sufficient for 10 per cent of the number of persons the rescue boat is permitted to carry (a minimum of two).

Additional equipment requirements for rescue boats

Every rigid rescue boat shall include with its normal equipment:

- a boat hook;
- a bucket;
- a knife or hatchet.

Every inflated rescue boat shall include with its normal equipment:

- a buoyant safety knife;
- two sponges;
- an efficient, manually operated bellows or pump;
- a repair kit in a suitable container for repairing punctures;
- a safety boat hook.

Additional requirements for inflated rescue boats

The mariner should also be aware of the following general particulars which affect rescue boats of the inflated type. Unlike hulls and rigid covers of lifeboats they do not have to be fire retardant, but they should be of sufficient strength and rigidity to withstand launch and recovery in the inflated condition when slung from its bridle or lifting hook (with full complement). The strength should be such as to withstand four times the load of the total mass of persons and equipment and capable of withstanding exposure on an open deck of a ship at sea or 30 days afloat in all sea conditions.

They should be marked as for an ordinary lifeboat but carry, in addition, a serial number, the maker's name or trademark and the date of manufacture. Underneath the bottom and on vulnerable places on the outside of the hull, in the inflated condition, rubbing strips shall also be provided to the satisfaction of the authority.

The buoyancy of inflated rescue boats shall be a single tube which is subdivided into at least five separate compartments of approximate equal volume, or two separate tubes neither of which exceed 60 per cent of the total volume. The tubes should be so arranged that in the event of any one of the compartments becoming damaged, the intact compartments shall be capable of supporting the full complement. The buoyancy tubes, when inflated, forming the boundary of the boat, shall provide a volume not less than 0.17 m^3 for each person the boat is permitted to carry. Each buoyancy compartment will be provided with a non-return valve for manual inflation and means should be provided for deflation. A safety relief valve will also be fitted if the authority considers this a necessary requirement.

If a transom stern is fitted, it should not be inset by more than 20 per cent of the boat's length. Suitable patches shall be provided for securing painters fore and aft and securing of the becketed lifeline inside and outside the boat. The boat itself should be maintained at all times in the inflated condition.

Emergency Equipment Lockers for Ro-Ro Passenger Ships

Since April 1989 (Ref. S.I. 1988 No. 2272) it has been a requirement under the regulations for all UK ro-ro passenger vessels to be fitted with an 'emergency equipment locker'. This locker should be constructed in steel or GRP and fitted to either side of the vessel and contain the equipment as listed (M 1359). These lockers must be clearly marked and located on high decks as near to the ship's side as possible so that at least one of the lockers will be accessible in all foreseen circumstances.

The need for the lockers to be instigated was generated after the loss of the *Herald of Free Enterprise*. The purpose of the equipment so contained is to assist passengers and crew to escape from enclosed parts of the vessel when normal escape routes may be found to be inaccessible due to an excessive list on the vessel or other unforeseen situations. The equipment is suitable for use by damage-control parties and should be considered useful in exercise and in maintaining the freedom of access in emergencies.

The locker contents are:

- one long-handled fire axe
- one fire axe, short handle
- one 7 lb pin maul
- one crowbar
- four torches or lamps
- one lightweight collapsible ladder (at least three metres)

Plate 90 Example of emergency equipment locker as required by UK registered ro-ro passenger vessels.

- one lightweight rope ladder (ten metres)
- one first aid kit
- six sealed thermal blankets or thermal protective aids
- four sets of waterproof clothing
- five padded lifting strops for adults
- two padded lifting strops for children
- three hand-powered lifting arrangements (tripod).

It is expected that all crew members would become familiar with equipment during exercise and drills.

The lockers and the contents would also be subject to inspection under Chapter III of the SOLAS Regulations by an Examining Surveyor prior to issuing the Passenger Ship Safety Certificate.

If the ship does not comply with this requirement, it is liable to be detained and the Master and owners are liable to a fine (level five of the standard scale) or indictment to imprisonment for a period not exceeding two years, or both, on summary conviction.

Emergency Communications (Applicable to Passenger and Cargo Ships)

General emergency alarm signal

The general emergency alarm signal shall be capable of sounding the general alarm signal, consisting of seven or more short blasts followed by one long blast on the ship's whistle or siren, and additionally on an electrically operated bell, klaxon or other equivalent warning system, which shall be powered from the ship's mains supply and the emergency source of electrical power required by the regulations.

The system shall be capable of operation from the navigation bridge and, except for the ship's whistle, also from other strategic points. The system shall be audible throughout all accommodation and normal crew working spaces, and supplemented by a public address or other suitable communication system.

An emergency means comprising either fixed or portable equipment, or both, shall be provided for two-way communications between emergency control stations, muster and embarkation stations and strategic positions on board.

Two-way radiotelephone apparatus

Two-way radiotelephone apparatus which complies with the regulations shall be provided for communication between survival craft, between survival craft and the parent ship and between the ship and rescue boat. An apparatus need not be provided for every survival craft, but at least three apparatus shall be provided on each ship. This requirement may be complied with by other apparatus used on board, provided such apparatus is not incompatible and is appropriate for emergency operations.

Muster lists

Clear instructions shall be provided for every person to follow in the event of an emergency. Muster lists which specify the requirements laid down by the regulations shall be exhibited in conspicuous places throughout the ship, including the navigation bridge, engine room and crew accommodation spaces. Illustrations and instructions in the appropriate language shall be posted in passenger cabins and displayed at muster stations and passenger spaces to inform passengers of:

NB. Passenger vessels with a rescue boat on either side require to be fitted with a fixed radio installation in each rescue boat.

- their muster station;
- the essential actions they should take in an emergency;
- the method of donning lifejackets.

Content of muster lists

The muster list shall specify detail of the general alarm signal and also the action to be taken by crew and passengers when the alarm is sounded. The list will specify how the order to abandon ship will be given.

The muster list shall show designated, qualified personnel in charge of survival craft and the duties assigned to different members of the crew including:

- closing of watertight doors, fire doors, valves, scuppers, sidescuttles, skylights, portholes and other similar openings in the ship;
- the equipping of survival craft and other life-saving appliances;

- the preparation and launching of survival craft;
- the general preparations of other life-saving appliances;
- the muster of passengers;
- the use of communication equipment;
- manning of fire parties to deal with fires;
- special duties assigned in respect of the use of fire-fighting equipment and installations.

The muster list shall specify which officers are assigned to ensure that life-saving appliances and fire appliances are maintained in good condition and ready for immediate use. Muster lists should also specify substitutes for key persons who may become disabled, taking into account that different emergencies may call for different actions.

The muster list shall show the duties assigned to crew members in relation to passengers in case of emergency. These duties shall include:

1   warning the passengers;
2   seeing that they are suitably clad and have donned their life-jackets correctly;
3   assembling passengers at muster stations;
4   keeping order in passageways and on stairways and generally controlling the movements of passengers;
5   ensuring that a supply of blankets is taken to the survival craft.

The muster lists shall be prepared before the ship proceeds to sea and be of an approved type in the case of passenger ships.

On Board: Passenger Ship Drills and Training

Drills

An abandon ship drill and fire drill shall take place weekly.

Applicable to all vessels

Each member of the crew shall participate in at least one abandon ship drill and one fire drill every month. The drills of the crew shall take place within 24 hours of the ship leaving a port if more than 25 per cent of the crew have not participated in abandon ship and fire drills on board that particular ship in the previous month. The authority may accept other arrangements that are at least equivalent for those classes of ship for which this is impracticable.

On a ship engaged on an international voyage, which is not a short international voyage, musters of the passengers shall take place within 24 hours after embarkation. Passengers should be instructed in the use of lifejackets and actions to take in the event of an emergency.

Content of abandon ship drill

Each abandon ship drill shall include:

1   summoning passengers and crew to muster stations with the alarm required by the regulations and ensuring that they are made aware of the order to abandon ship, as specified in the muster list;
2   reporting to stations and preparing for the duties described in the muster list;
3   checking that passengers and crew are suitably dressed;
4   checking of lifejackets to ensure they are correctly donned;
5   lowering of at least one lifeboat after any necessary preparation for launching;
6   starting and operating the lifeboat engine;
7   operation of davits used for launching life rafts.

The regulations specify that each lifeboat shall be launched with its assigned crew aboard and manoeuvred in the water at least once every three months during an abandon ship drill. However, ships operating on short international voyages may be relieved of this obligation if their berthing arrangements do not permit the launching, but all such boats should be lowered at least once every three months and launched at least annually. As far as practicable, rescue boats, if not lifeboats, should be launched with their assigned crew at least each month and in any event launched at least once in three months.

Drills should be conducted as if there were an actual emergency and different boats used at successive drills. If drills are carried out with the ship making headway, because of the dangers involved, practice should take place in sheltered waters, under the supervision of an experienced officer.

The regulations now require that on board training in the use of the ship's life-saving appliances be given as soon as possible, but not later than two weeks after a crew member joins the ship. Instruction in survival at sea should be given at the same intervals as drills, and coverage of all the ship's life-saving equipment should be covered within a period of two months.

Instruction should include specifically:

- instruction in the use of inflatable life rafts;
- treatment and associated problems of hypothermia and first aid procedures;
- special instructions in the use of gear when in severe weather and sea conditions.

On board training in the use of davit-launched life rafts shall take place at intervals of not more than four months on every ship fitted with such appliances. Whenever practicable this shall include the inflation and lowering of a life raft. This life raft may be a special raft, intended for training purposes only, and if so it should be conspicuously marked as such.

Training manual

A training manual will be provided for each crew messroom, recreation room or in each crew member's cabin. The contents of this manual shall include:

- the donning of lifejackets and immersion suits as appropriate;
- muster arrangements at the assigned stations;
- a method of launching from within the survival craft;
- the releasing method from the launching appliance;
- methods and use of devices for protection in launching areas, where appropriate;
- illumination in launching areas;
- use of all survival equipment;
- use of all detection equipment;
- with the assistance of illustrations, the use of radio life-saving appliances;
- use of drogues;
- use of the engine and accessories;
- recovery of survival craft and rescue boats, including stowage and securing;
- hazards of exposure and the need for warm clothing;
- methods of retrieval, including the use of helicopter rescue gear (slings, baskets, stretchers), and shore life-saving apparatus and ship's line-throwing apparatus;
- all other functions contained in the muster list and emergency instructions;
- instructions for emergency repair of the life-saving appliances.

The manual, which may comprise several volumes, shall contain the information in an easily understood form, illustrated as appropriate, and may be provided as an audio-visual aid in lieu of a book format.

Manning Requirements for Survival Craft (Applicable to All Ships)

The regulations specify that there shall be a sufficient number of trained persons on board a ship for the mustering and assisting of untrained persons. There will be a sufficient number of crew members, who may be Deck Officers or certificated persons, on board for the operation of survival craft and their respective launching arrangements required for an abandonment by the total number of persons on board.

A Deck Officer or certificated person shall be placed in charge of each survival craft. However, the authority, having due regard to the nature of the voyage and the characteristics of the ship, may permit persons practised in the handling and operation of life rafts to be placed in charge of life rafts instead of the persons qualified as above.

The Master shall ensure that persons qualified to carry out respective duties regarding the boatwork operations shall be allocated among all the ship's survival craft. Every motorized boat shall have a person assigned to it who can operate the engine and carry out minor adjustments. Every lifeboat, which carries a radio telegraph installation, shall also have a person assigned to it who is capable of operating the equipment.

The person in charge of a survival craft shall have a list of the survival craft crew and ensure that they are familiar with their duties. The second in command of the lifeboats will also have a similar list.

Standards of Training Certification and Watchkeeping for Seafarers (STCW)

New amendments will directly affect the training of seafarers effective from 2012 through to 2017. New levels of competence for deck officers in the use of ECDIS are required and engineering officers must be competent in the use of oil pollution prevention equipment.

Additionally, leadership and teamwork, alongside managerial skills, will be expected for both engineering and deck officers. Ratings in both departments will also be expected to demonstrate competence through a development of a 'Training Record Book'.

Security training will be required for all grades of shipboard personnel. Refresher training in fire-fighting, first aid and survival practices will be introduced to cover five-year periods in order to maintain levels of competence.

Specific 'tanker training' at both basic and advanced levels will be required for oil, chemical and gas tanker operations.

9
Communications

Introduction

Communications have many forms but are somewhat specialized within the maritime environment. Yes, we have the standard speech (in English language) and written communication just as virtually all other industries, but then not everyone uses the Morse code flashlight. Neither do all companies communicate with the International Code Flags.

These methods may seem outdated alongside the facsimile machine and sat-coms, but they do have their place and can often provide direct communication faster than many of the technological innovations of the current environment (albeit limited in range).

Communications have always been the key element of military campaigns. This is similar regarding the movement of a ship engaged on its voyage. The fact is that internal and external communications, in their every form, are continuously employed to execute the ship's safe navigation.

Terminology and Definitions for Communications

Bridge to bridge

This refers to safety communications between ships from the position from which the ships are navigating.

Digital selective calling (DSC)

This is a technique using digital codes which enables a radio station to establish contact with and transfer information to another station or group of stations complying with the International Radio Consultative Committee (CCIR).

INMARSAT

This is the organization established by the Convention on the International Maritime Satellite Organization adopted in September 1976.

International NAVTEX Service

This is the coordinated broadcast and automatic reception on 518 kHz of maritime safety information (MSI) by means of narrow-band, direct-printing telegraphy in the English language.

Maritime safety information (MSI)

This refers to the navigational and meteorological warnings, meteorological forecasts and other urgent safety-related messages broadcast to ships.

Methods Employed in the Marine Industry

GMDSS

Increased technology has caused communications to mushroom around the globe and nowhere more so has this occurred than in the maritime environment. Wireless telegraphy has been dispatched to the archives and since February 1999 the Global Maritime Distress and Safety System (GMDSS) has become compulsory for all passenger ships and cargo ships over 300 gross tonnes. The system has been developed by the IMO and has formed part of the amendments to SOLAS.

Plate 91 GMDSS communication console seen aboard a new ro-ro ferry vessel, operating out of the United Kingdom (source: Capt. David A. McNamee AFNI).

Requirements for the Carriage of Equipment

In order to comply with the regulations, ships will require specific items of equipment for operation in designated areas:

- Sea area A1: ships will carry VHF equipment and either a satellite EPIRB or a VHF EPIRB.
- Sea area A2: ships will carry VHF and MF equipment and a satellite EPIRB.
- Sea area A3: ships will carry VHF, MF, a satellite EPIRB and either HF or satellite communication equipment.
- Sea area A4: ships will carry VHF, MF and HF equipment and a satellite EPIRB.

Additionally, all ships will also be equipped to receive MSI broadcasts and have continuous scanning ability of all HF safety frequencies with automatic printing ability of MSI messages. A Navtex receiver will also be on board to continuously monitor 518 kHz with programming capability to be selective of message category.

Ships will also be required to have a 'radar transponder' (SART) and two-way radios for designated use with survival craft (a minimum of three walkie-talkies designated for survival craft use only).

The vessel must also have at least one radar set capable of operating in the 9 GHz band; terminals to provide this requirement are available. Inmarsat A offers the use of voice-, data-, facsimile- and telex-based communications; Inmarsat C is a smaller unit and offers text and data messaging at reduced speed. Both terminals provide world coverage with the exception of the extreme polar regions.

Sea Area Definition

- Sea area A1: an area within the radiotelephone coverage of at least one VHF coast station in which continuous DSC alerting is available, as may be defined by a contracting government. (The UK government has not defined a sea area A1, so all vessels departing UK are immediately into sea area A2.)
- Sea area A2: an area excluding A1 within the radiotelephone coverage of at least one coast station providing continuous watch on 2182 kHz and continuous DSC alerting on 2187.5 kHz, as may be defined by a contracting government. (The UK has provided six Coastguard stations with continuous DSC alerting capability which extends sea area A2 approximately 150 nm from the UK coastline.)
- Sea area A3: an area excluding A1 and A2 within the coverage of an INMARSAT geostationary satellite by which continuous distress alerting is available. (This is effectively everywhere between latitudes 70°N and 70°S.)
- Sea area A4: an area outside sea areas A1, A2 and A3 (Figure 9.1) (namely the polar regions).

Figure 9.1 Radio reaction to distress (key numbers in bold).

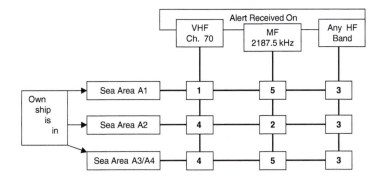

Radio reaction to distress

1 (a) Tune to RT Channel 16 and listen for distress communication.
   (b) Acknowledge receipt of the alert using RT on channel 16 and carry out distress communication.
   (c) If the alert is not responded to by a shore station, acknowledge by DSC on Channel 70 and relay the alert ashore by any means.
2 (a) Tune to 2182 kHz and listen for distress communication.
   (b) Acknowledge receipt of the alert using RT on 2182 kHz and carry out distress communication.
   (c) If the alert is not responded to by a shore station, acknowledge by DSC on 2187.5 kHz and relay the alert ashore by any means.
3 (a) Tune to RT distress frequency in the band on which the distress alert is received.
   (b) *Do not* acknowledge either by RT or DSC.
   (c) Wait at least three minutes for a shore station to send DSC acknowledgement.
   (d) If no shore station acknowledgement or RT distress communication is heard, relay the alert ashore by any means.
   (e) If within VHF or MF range of the distress position, try to establish RT contact on Channel 16 or on 2182 kHz.
4 (a) Tune to RT VHF Channel 16 and listen for distress.
   (b) Acknowledge receipt of the alert using RT Channel 16 and carry out distress communications.
   (c) If the alert continues, relay it ashore by any means.
   (d) Acknowledge the alert by DSC on Channel 70.
5 (a) Tune to RT 2182 kHz and listen for distress communication.
   (b) Acknowledge receipt of the alert using RT 2182 kHz and carry out distress communications.
   (c) If the alert continues, relay it ashore by any means.
   (d) Acknowledge the alert by DSC on 2187.5 kHz.

EPIRBS

It is an IMO requirement that all vessels that are GMDSS compliant will carry a float-free EPIRB. This is usually mounted in a bracket and fitted with a hydrostatic release unit (HRU) to enable float-free capability in the event that the vessel sinks. The HRU has limited endurance and must be serviced or replaced periodically as per type/manufacturer's recommendations. HRUs are activated at a depth of 1.5 to 4.0 metres.

Although it has a shelf life of six years (although it is normally changed after two years), the output capability of the battery of the EPIRB must be sufficient to provide transmission power for a 48-hour continuous operating period. EPIRBs must also be tested on a monthly basis as per manufacturer's instructions. The SARSAT/COSPAS polar orbiting satellites receive transmitted EPIRB signals on 406 MHz. INMARSAT satellites receive transmitted EPIRB signals on 1.6 GHz. The 406 MHz EPIRBs are fitted with a strobe light to aid visual location and also contain a homing signal operating on 121.5 MHz, or may operate on 243 MHz.

Operation of the 406 MHz

Once a signal is received on board the satellite, Doppler shift measurements are taken and time coded. This is then processed as digital data and added to the digital message transmitted from the beacon. The whole is then stored on board the satellite for future transmission and also transmitted immediately in real time to any current land user terminal (LUT) within view from the space craft.

One of the main differences between the operation of the alternative frequencies is that transmissions from 406 MHz beacons are acted on immediately, whereas transmissions on 121.5 MHz are not acted upon until two transmissions have been received. It should also be noted that Doppler shift measurement is more accurate and 406 MHz beacons can be located to within approximately three miles. In comparison, the location accuracy on 121.5 MHz is about 12 miles. The other main difference between the two is that the 406 MHz EPIRB transmits an identification code, whereas the 121.5 MHz homing beacon does not.

EPIRB Function and Purpose

The primary function of the EPIRB is to aid location of survivors in a distress situation. The signal transmitted is meant to indicate that a person or persons who may have been separated from their parent vessel and consequently may not have radio capability are in distress and require immediate assistance. Clearly, an EPIRB on board a distressed vessel could equally be used as a secondary means of transmitting a distress alert signal.

SARTs

The SART is a search and rescue radar transponder. It is portable and can be operated from the parent vessel or from a survival craft.

SART Radar Signatures

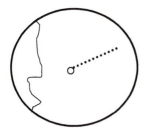

(i) SART showing 12 dot blip code
bearing approximately 060°.

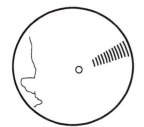

(ii) Range of search craft at approximately
1 nautical mile dots change to wide arcs.

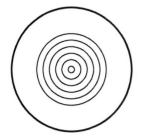

(iii) Wide arcs change to concentric
circles as the SART is closed.

Once operated, the SART is intended to indicate its position to search units of either the surface vessel or aircraft type.

Operation of the SART is on 9 GHz and can be interrogated by units operating 3 cm radar. Specification of the SART is such that it is identifiable at 5 nm from operation at one metre above surface. In comparison, an aircraft could detect it at about 40 miles because of its increased height.

SARTs are activated by the radar energy of an approaching search unit and start their own transmission. The radar signature produced is a radial line of 12 dots over an approximate eight-mile length on the search vehicle's radar, from a position just beyond the position of the distress. As the search unit closes the target to about 1 nm, the dots become small arcs. The signature will change again to concentric circles when the range closes to a few hundred metres.

For best results, search radar operators should:

- turn off the anti-rain control;
- place gain control to maximum;
- operate on 6 nm range or less;
- reduce clutter by down-tuning slightly;
- reduce range and gain control when closing in on the target.

> NB. SARTs are not designed to operate with 10 cm radar. The International Code of Signals is expected to be combined with Volume III of the IAMSAR manual in the near future.

International Code of Signal Flags

This is a method of communication over a short range, being limited by the need for transmitting and receiving stations to be in sight of each other. All British ships are obliged to carry a full set of international code flags, together with a copy of the code book, *International Code of Signals*. The flags cover the alphabet from A to Z, plus three substitutes, and the numerals from 0 to 9; and there is a code and answering pendant.

The system is extremely useful for communication if radio silence has to be maintained, as when a country is engaged in hostilities. The method proved ideal for signal transmission between ships in convoy and has shown itself to be a speedy communications system when employing single or even double flag hoists. Where code groups are used, it is time-consuming in transmission and in decoding the messages received.

During peaceful trading, single-letter hoists tend to play a major part in harbour or port control with regard to the pilot service, port health clearance, fishing boats, emergency signals and the like. It is a requirement of all Deck Officers taking a watch to be alert for the use of single-letter and important double-letter flag hoists.

Morse Code by Flashing Light

This is a very fast method of communication when transmitted by an experienced operator. It may be carried out by Aldis lamp, direct to the receiving station, or by an all-round lamp, either method being limited in range to the visual distance between transmitting and receiving stations. This method of communication is used ship-to-shore as well as ship-to-ship. Normal speed of transmission is 4–6 words per minute.

The Aldis lamp should be kept readily available for emergency use if required. It is normally provided with sun shades, usually red and green, to allow operation during the hours of daylight.

Morse Code by Wireless Telegraphy

With the full implementation of GMDSS from February 1999, Radio Officers and the use of wireless telegraphy (W/T) are virtually no longer used within the maritime communications field. Full world satellite coverage is now possible and all vessels over 300 gross tonnes are equipped to operate under the GMDSS regulations. It is a requirement of GMDSS compliance that a vessel must carry at least two qualified and certificated personnel capable of operating the communication equipment.

> NB. In the event of the vessel being requisitioned for a search and rescue operation, a designated communications operator must be identified within the bridge team.

Morse Code by Flag

This is a very slow and tiring method of communication. Transmission should be carried out slowly unless a person is very experienced in this style of signalling. Limited in range, it has the advantage that an injured party with only one arm available can use it.

Semaphore (Obsolete)

This method employs two hand flags which are positioned to indicate letters of the alphabet. It is limited in range to the visual distance between the transmitting and receiving stations. A good operator can transmit 15–18 words per minute. However, semaphore has been made obsolete in both the Royal Navy and the British Mercantile Marine, and it is no longer a requirement for Deck Officers to have detailed knowledge of it in order to qualify for a certificate of competency.

Very High Frequency (VHF)

Without doubt this is the most popular method of communication employed to date. It is limited in range with normal equipment to under 100 miles, though this distance will vary with weather and atmospheric conditions. It is a direct method of voice communication, which can be linked to land-based local telephone exchange systems or used between ships direct.

Miscellaneous Communications

Signals of distress are dealt with in Chapter 13. Small-boat landing signals are covered in Chapters 7 and 8 and signals used in conjunction with the coastguard may be found in Chapter 16.

Flag Signalling Terms

At the dip

This term is used to describe a signal which is hoisted to approximately half the extent of the halyard (often displayed as a mark of respect following a death).

Close up

This denotes that the flag hoist is flying at its maximum height, e.g. close up to the truck, at the mainmast.

Signal letters

These are letters allocated to vessels to signify the 'call sign' of that vessel for the purpose of communication and identification.

Tackline

This is a length of halyard about one fathom in length, the tackline is used to separate two distinctive hoists on the same halyard.

Single-Letter Meanings: International Code of Signals

A   I have a diver down; keep well clear at slow speed.
B   I am taking in, or discharging, or carrying dangerous goods.
C   Yes (affirmative or 'The significance of the previous group should be read in the affirmative').
D   Keep clear of me; I am manoeuvring with difficulty.
E   I am altering my course to starboard.
F   I am disabled; communicate with me.
G   I require a pilot. When made by fishing vessels operating in close proximity on the fishing grounds it means: 'I am hauling nets.'
H   I have a pilot on board.
I   I am altering my course to port.
J   I am on fire and have dangerous cargo on board: keep well clear of me.
K   I wish to communicate with you.
L   You should stop your vessel instantly.
M   My vessel is stopped and making no way through the water.
N   No (negative or 'The significance of the previous group should be read in the negative'). This signal may be given only visually

or by sound. For voice or radio transmission the signal should be 'No'.
O Man overboard.
P *In harbour*: All persons should report on board as the vessel is about to proceed to sea.
At sea. It may be used by fishing vessels to mean: 'My nets have come fast upon an obstruction.'
Q My vessel is 'healthy' and I request free pratique.
S I am operating astern propulsion.
T Keep clear of me; I am engaged in pair trawling.
U You are running into danger.
V I require assistance.
W I require medical assistance.
X Stop carrying out your intentions and watch for my signals.
Y I am dragging my anchor.
Z I require a tug. When made by fishing vessels operating in close proximity on the fishing grounds it means: 'I am shooting nets.'

Signalling by International Code Flags

How to Call

When calling a vessel which is known and identified, the transmitting station should hoist the signal letters of the vessel with which it wishes to communicate. When the identity of the vessel called is not known, then a general signal of:

VF = 'You should hoist your identity signal', or
CS = 'What is the name or identity signal of your vessel?'

should be made. The transmitting station should also display its own signal letters while calling up the other vessel.

Alternative Call Up

The transmitting station may hoist, instead of the signals listed above:

YQ = 'I wish to communicate by … (Complements Table 1; Section 1) with vessel bearing from me.'

If the signal had been YQ3 = 'I wish to communicate by Morse signalling lamp.' The numeral 3 is cross-referenced with complements table.

How to Answer a Signal

The receiving station, once called up, should hoist the answering pendant, at the dip. Once the signal is received and understood, the answering pendant should be closed up. The transmitting station will then haul down the signal while the receiving station returns the answering signal to the dip position and awaits the next signal.

How to End Transmission

The transmitting station will hoist the answering pendant as a single hoist to indicate that the communication is complete. This will be answered in the normal manner by the receiving station.

Signals Not Understood

Should confusion arise during transmission and the signal not be fully understood by the receiving station, then that station should not close up the answering pendant but leave it at the dip position. The receiving station may make the following signals when the signal is distinguished but the meaning of it is unclear:

ZQ = 'Your signal appears incorrectly coded. You should check and repeat the whole'; or
ZL = 'Your signal has been received but not understood'.

Names and Spelling

Should any signal contain names requiring full spelling, then these names should be spelled out with international code alphabetical flags. If it is felt necessary, then the hoist:

YZ = 'The words which follow are in plain language'

may be used for clarification.

General Notes on Transmission

Signals should always be hoisted in such a manner as to fly clear of obstructions and not be obscured by smoke, etc. It is normal practice to exhibit one hoist at any one time, but should it become necessary to exhibit more than the one hoist, two hoists may be flown from the same halyard, provided they are separated by a length of tackline. The upper of the two groups is the one to be read first.

Order of Transmission with Several Hoists

Should a number of flag hoists be made simultaneously, they should be read in the following order:

1   Signal at masthead.
2   Signal at triatic stay.
3   Signal at starboard yard arm.
4   Signal at port yard arm.

Should there be more than one group flown from the triatic stay, then they should be read from forward to aft. Should several

Alphabetical flags

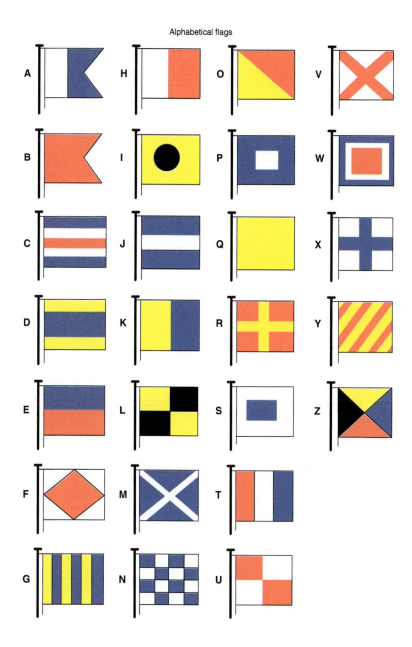

Figure 9.2 International code flags.

hoists be displayed from the same yard arm, then these should be read from outboard to inboard.

International Code: Use of Substitutes

The purpose of the three substitute flags is to enable a letter or numeral pendant to be repeated in the same hoist. A letter or

Figure 9.3 Substitutes, code and numeral flags.

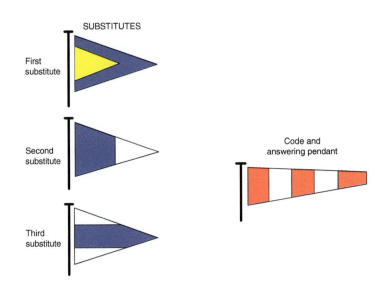

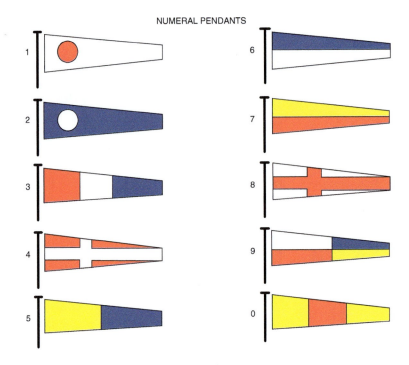

number may have to be repeated more than once within the same hoist so there are three substitute flags, not just one.

When a substitute flag is used, it refers only to the hoist in which it is contained. If the first substitute is used, it will mean a repeat of the first letter/numeral flag of the hoist. If the second substitute flag is used, it will mean that the second flag of the hoist, letter or numeral is repeated. The third substitute indicates that the third flag of the hoist is repeated.

Exceptions with the Use of Substitutes

Many hoists are significant in that they have a prefix letter, e.g. L3640N = latitude 36°40 north, the prefix of the hoist being the letter 'L' for latitude. Similarly, a hoist indicating a bearing/ azimuth is prefixed by the letter 'A', e.g. A200 = bearing 200°.

Whenever a hoist containing three or four numerals is made and there is a prefix letter in front of the hoist, substitute flags should *not* include the prefix letter. See the following examples:

Examples

1. *Bearing 200° = A200.* Signal made up as follows:
   'A' Flag.
   '2' Numeral.
   '0' Numeral.
   Second substitute.

2. *Longitude 22°32 W = G2232W.* Signal made up as follows:
   'G' Flag.
   '2' Numeral.
   First substitute.
   '3' Numeral.
   Second substitute.

Table 9.1 Examples of flag hoists, *International Code of Signals*

Flag hoist	Meaning	Reference
'K' '3'	I wish to communicate by Morse signalling lamp.	Complements Table 1
'A' '3' 1st sub. 2nd sub. Code flag 3rd sub.	Bearing 333.3°	Prefix letter 'A' to indicate bearing. Code flag represents Decimal point. Use of all substitutes.
'C' '1' 1st sub. '5'	True course 115°	Prefix letter 'C', general instructions. All courses are transmitted in three-figure notation. Always expressed as true, unless stated otherwise.
'D' '1' 1st sub. '0' '7' '8' '4'	11 July 1984	Prefix letter 'D', general instructions. First two numerals refer to days. Second two numerals refer to the month. Last two numerals refer to the year.

Important Two-Letter Signals

AL   I have a doctor on board.

AN   I need a doctor.

CB   I require immediate assistance.

FR   I am (or vessel indicated is) in charge of coordinating search (intended for use with SAR activity).

GU   It is not safe to fire a rocket.

GT   I will endeavour to connect with line-throwing apparatus.

LO   I am not in my correct position (to be used by a light vessel).

NC   I am in distress and require immediate assistance.

NE   You should proceed with great caution.

NO   Negative – No – or the significance of the previous group should be read in the negative.

OK   Acknowledging a correct repetition or 'It is correct.'

RA   My anchor is foul.

SP   Take the way off your vessel.

Table 9.2 Tables of complements (found in the *International Code of Signals*)

Table I: Methods of communication

1 Morse signalling by hand flags or arms

2 Loud hailer (megaphone)

3 Morse signalling lamp

4 Sound signals

Table II: Services

0 Water

1 Provisions

2 Fuel

3 Pumping equipment

4 Fire-fighting appliances

5 Medical assistance

6 Towing

7 Survival craft

8 Vessel to stand by

9 Ice-breaker.

Table III: Compass directions

0 Direction unknown (or calm)

1 Northeast

2 East

3 Southeast

4 South

5 Southwest

6 West

7 Northwest

8 North

9 All directions (or confused or variable).

TJ   You should navigate with caution. There are nets with a buoy in this area.

YG   You appear not to be complying with the traffic separation scheme.

Sample Messages Employing International Code of Signals

1   I am in distress in position 212°T, 15 miles from Morecambe Bay and require a vessel to stand by immediately:

    CC8   A212   MORECAMBE BAY   R15.

2   My ETA at FOLKESTONE is 0700 BST on 14 December 1980:

    UR   FOLKESTONE   T0700   D141280.

3   My position at 1200 GMT was latitude 50°15 N, longitude 0°20 W:

    ET   Z1200   L5015N   G0020W.

4   I require a helicopter urgently to pick up an injured person; the patient has a severe burn on central upper abdomen:

    BR3   MGH19.

5   I will try to proceed under my own power but request you to keep in contact with me by R/T on 2182 kHz:

    IJ   YZ   RT   2182.

6   A tropical storm on course 290°T at 15 knots is centred at latitude 26°30 N, longitude 60°16 W at 1800 GMT; NW storm force winds are expected:

    VM   Z1800   L2630N   G6016W   C290 S15   VK7.

Flag Maintenance

Ships' flags are normally stowed on the bridge or near the bridge area in an easily accessible flag locker. It is not unusual for modern vessels to carry two flag lockers, one each side of the wheelhouse. Flag halyards are normally stretched to port and starboard bridge wings and the monkey island from the triatic stay. The old-fashioned flag lockers were open to allow easy flag withdrawal, but nowadays open-cupboard design is more popular, with protective cupboard door arrangements. The 'pigeon-hole' design of one pocket per flag is still accepted.

Flags should be fitted with brass ingle field clips, one of the clips being fitted with a swivel fitment. Halyards should be similarly

Figure 9.4 Flag construction.

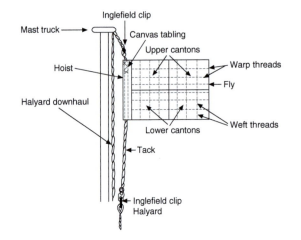

fitted with the clips to facilitate speedy attachment of the hoist (Figure 9.4).

Modern flags are manufactured in a brush nylon fabric; they are washable and light in weight, allowing the flag to fly clear in light winds. Previously, flags were manufactured in 'bunting', a wool-woven cloth which was and is a lightweight material that does not fray easily. Calico was also used in the past, but proved most unsatisfactory, especially when wet. It was a heavy material basically, but when wet absorbed the water and hung limp in all but heavy winds.

Flags are expensive, and should be treated with care. Inspection of the flag locker at periodic intervals gives the locker and the flags an 'airing', removing any musty smells that may have accumulated.

Flags should be thoroughly dried, preferably by natural rather than artificial means, before being stowed. Nylon fabric flags are washable and, provided they have not started to fray around the 'fly' edge, will last for many years. The older style 'bunting' can be washed in fresh water, but strong detergents should be avoided. Bunting is not as hard-wearing as the synthetic material and more liable to rot, especially if oil contaminates the fabric.

Wearing of Ensigns

White Ensign

The flag is flown by all ships of Her Majesty's Navy when in commission or when being launched, and by Royal Navy shore establishments when commanded by a commissioned officer.

Royal Standard

This is the personal flag of the sovereign and should be flown only when the sovereign is present aboard the vessel. It should be exhibited at the mainmast, and in a vessel without a mainmast the flag should be flown from the most conspicuous part of the vessel.

Personal Royal Standards

These are exhibited in a similar manner to the Royal Standard. Other members of the Royal Family than the Queen, e.g. the Prince of Wales, may appropriately use a personal standard when visiting aboard ships.

Jack

Often referred to as the pilot jack, this name may have been derived from King James I, who signed his name 'Jackques', and under whose instructions the flag was constructed. It was always flown from the 'jackstaff at the stem'. Ships of the Royal Navy fly a square or rectangular union flag, whereas the Merchant Navy tends to exhibit a small 'pilot jack'. A union flag, surrounded by a white border, is also flown from the jackstaff.

Originally the flying of the pilot jack meant that the ship required a pilot, but this requirement is now indicated by the International Code Flag 'G'. The flying of the 'jack' in the forepart is generally restricted to times when the vessel is in harbour, either alongside or at anchor.

House Flag

This flag is flown by merchant vessels to indicate their ownership or the company that has chartered the vessel. It will bear the company's insignia and will normally be flown when entering or leaving port, or when the vessel is at anchor. It is generally exhibited at the mast-head, but has in the past been flown at the stem on the jackstaff.

Courtesy Flag

It is general practice for merchant vessels to fly the colours of the country the vessel is visiting as a mark of respect to the host nation, and care should be taken not to fly the flag incorrectly. Nowadays the courtesy ensign is flown from the starboard yard arm. Previously it was broken out at the forward masthead.

Blue Ensign

British ships of the Mercantile Marine are allowed to fly the blue ensign, provided that the following conditions are complied with:

- The Master or officer in charge of the vessel is an officer on the retired or emergency list of the Royal Navy or the Royal Australian Navy, or an officer of the Royal Navy Reserve, the Royal Australian Naval Reserve (Seagoing), the Royal Canadian Naval Reserve or the Royal Naval Reserve (New Zealand Division).
- Crew members, in addition to the officer commanding, must include members and officers of the following: Royal Naval Reserve, Royal Australian Naval Reserve (Seagoing), Royal

Canadian Naval Reserve or Royal Naval Reserve (New Zealand Division). The total numbers are specified by the Admiralty at periodic intervals and may also include Royal Fleet Reservists of the British, Australian and New Zealand fleets, together with persons holding deferred pensions from the Royal Naval Reserve and the Royal Navy.

- Before hoisting the blue ensign the commanding officer must be in possession of an Admiralty warrant indicating this entitlement. Royal Naval Reserve officers wishing to apply for the Admiralty warrant should in the first instance make application to the local Mercantile Marine Office.
- The ship's articles of agreement should bear a statement to the fact that the commanding officer of the vessel is authorized to hoist the blue ensign, and holds the warrant from the Admiralty.

British merchant ships requisitioned by the Admiralty to assist in operations will also be allowed to fly the blue ensign, under the Admiralty warrant. Application to do so must be made by the owners of the vessel to the Admiralty through the Registrar General of Seamen.

Red Ensign (British Ships)

This is the national ensign of the United Kingdom, flown by all vessels owned by British subjects other than Her Majesty's ships. It is normally flown by all merchant vessels when in port or at anchor from the ensign staff at the after end of the vessel. When at sea it is flown at the peak of the 'gaff ' of the aftermast. If the vessel is not fitted with an aftermast or gaff, the ensign may be flown from the ensign staff aft.

The ensign should never be 'broken out' at the masthead. When hoisting, it should always be hauled 'close up' and maintained on a tight halyard. It should be exhibited when:

- entering and leaving any foreign port,
- entering or leaving any British port (if the vessel is over 50 tonnes), or
- when in contact or communication with a naval vessel.

National Colours (Non-British)

The national colour of a ship of any nation should be flown in port between 0800 hrs and sunset between 25 March and 20 September, and from 0900 hrs to sunset between 21 September and 24 March.

If a vessel is entering or leaving port, or coming into an anchorage, it is customary to leave the ensign flying until the operation of anchoring or berthing has been completed. Should this occur after the time of sunset, the colours are struck as soon as is practicable.

Dipping the Ensign

This is a salute made to naval vessels. Traditionally it was to indicate that the vessel dipping her ensign was engaged on peaceful and lawful trade. Dipping the ensign means to lower the colours from the 'close up' position to the 'dip' position. This signal should be acknowledged and answered by the receiving vessel by lowering her own ensign to the dip and immediately returning it to the 'close up' position. Once this action is seen by the saluting vessel, then she should return her own ensign to the close up position.

Morse Code Procedure: Signalling by Flashlight

The following are some procedural signals:

AA AA AA, etc.	Call for an unknown station or general call.
EEEEEEEEE, etc.	Erase signal.
AAA	A full stop or a decimal point.
TTTTTT etc.	The answering signal.
T	Word or group has been received.

Table 9.3 Phonetic alphabet and figure spelling

Letter	Word	Pronunciation	Number	Word	Pronunciation
A	Alfa	Al Fah	0	Nadazero	Nah Dah Zay Roh
B	Bravo	Brah Voh	1	Unaone	Oo Nah Wun
C	Charlie	Char Lee	2	Bissotwo	Bees Soh Too
D	Delta	Dell Tah	3	Terrathree	Tay Rah Tree
E	Echo	Eck Oh	4	Kartefour	Kartay Fower
F	Foxtrot	Foks Trot	5	Pantafive	Pan Tah Five
G	Golf	Golf	6	Soxisix	Sok See Six
H	Hotel	Hoh Tell	7	Setteseven	Say Tay Seven
I	India	In Dee Ah	8	Oktoeight	Ok Toh Ait
J	Juliett	Jew Lee Ett	9	Novenine	No Vay Niner
K	Kilo	Key Loh			
L	Lima	Lee Mah	Decimal point	Decimal	Day See Mal
M	Mike	Mike			
N	November	No Vem Ber			
O	Oscar	Oss Car			
P	Papa	Pah Pah			
Q	Quebec	Keh Beck			
R	Romeo	Row Me Oh			
S	Sierra	See Air Rah			
T	Tango	Tang Oh			
U	Uniform	You Nee Form			
V	Victor	Vik Tah			
W	Whiskey	Wiss Key			
X	X-Ray	Ecks Ray			
Y	Yankee	Yang Key			
Z	Zulu	Zoo Loo			

Table 9.4 International Morse Code

Letter	Symbol	Number	Symbol
A	• —	1	• — — — —
B	— • • •	2	• • — — —
C	— • — •	3	• • • — —
D	— • •	4	• • • • —
E	•	5	• • • • •
F	• • — •	6	— • • • •
G	— — •	7	— — • • •
H	• • • •	8	— — — • •
I	• •	9	— — — — •
J	• — — —	0	— — — — —
K	— • —		
L	• — • •		
M	— —		
N	— •		
O	— — —		
P	• — — •		
Q	— — • —		
R	• — •		
S	• • •		
T	—		
U	• • —		
V	• • • —		
W	• — —		
X	— • • —		
Y	— • — —		
Z	— — • •		

To Call an Unknown Station

The general call AA AA AA, etc. is made by the transmitting station to attract the attention of all other stations within visible range. This transmission is continued until the required station answers. The directional light of an Aldis lamp provides a more distinctive identification than, say, an all-round signalling lamp if several stations are in close proximity. Should the call sign or identity of the receiving station be known, then the identity signal of that station may be used as an alternative to the general call.

To Answer the Transmitting Station

The call-up transmission is answered by TTTTT, etc. (a series of 'Ts') and this answer should continue until the transmitting station stops transmitting the call-up signal.

Message Transmission

The receiving station should acknowledge receipt of each group or word of the message by the single transmission of the letter 'T'. The message should be concluded by the signal AR by the transmitting station. On receipt of the AR, the receiving station will signal 'R' to signify that the 'message has been received'.

Morse Code Regular Procedural Signals

AA	'All after...' (used after the 'repeat signal' (RPT)) means 'Repeat all after...'
AB	'All before...' (used after the 'repeat signal' (RPT)) means 'Repeat all before...'
AR	Ending signal or end of transmission or signal.
AS	Waiting signal or period.
BN	'All between ... and ...' (used after the 'repeat signal' (RPT)) means 'repeat all between ... and ...'
C	Affirmative – 'Yes' or 'The significance of the previous group should be read in the affirmative.'
CS	'What is the name or identity signal of your vessel (or station)?'
DE	'From ...' (used to precede the name or identity signal of the calling station).
K	'I wish to communicate with you' or 'Invitation to transmit.'
NO	Negative – NO or 'The significance of the previous group should be read in the negative'. When used in voice transmission, the pronunciation should be 'NO'.
OK	Acknowledging a correct repetition or 'It is correct.'
RQ	Interrogative, or 'The significance of the previous group should be read as a question.'
R	'Received' or 'I have received your last signal.'
RPT	Repeat signal, 'I repeat', or 'Repeat what you have sent' or 'Repeat what you have received.'
WA	'Word or group after ...' (used after the 'repeat signal' (RPT)) means 'Repeat word or group after ...'
WB	'Word or group before ...' (used after the 'repeat signal' (RPT)) means 'Repeat word or group before ...'

Morse Code Signalling by Hand Flags or Arms

Figure 9.5 illustrates this signalling method.

To Call Up a Station

The general call sign may be made by the transmitting station: AA AA AA. An alternative call-up signal may be the transmission of K1 by any means, meaning 'I wish to communicate with you by Morse signalling by handflags or arms.'

Figure 9.5 Morse signalling by hand-flags or arms.

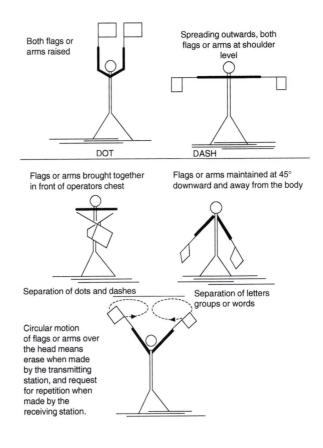

Both flags or arms raised

Spreading outwards, both flags or arms at shoulder level

DOT

DASH

Flags or arms brought together in front of operators chest

Flags or arms maintained at 45° downward and away from the body

Separation of dots and dashes

Separation of letters groups or words

Circular motion of flags or arms over the head means erase when made by the transmitting station, and request for repetition when made by the receiving station.

MORSE CODE SIGNALLING BY HAND FLAGS OR ARMS

A Receiving Station: in Answer

On receiving the call-up signal the receiver should answer the call by the answer signal: TTTTT. Should a receiving station be unable to communicate by this method, the signal YS1 should be displayed by any available means, meaning 'I am unable to communicate by Morse signalling by hand-flags or arms.'

Completing the Signal

All signals are completed by the transmission of the ending signal: AR. This method of signalling can be carried out by flags or just by the operator's arms, and if the operator has only one arm available, the system is still feasible. But it is tiring and time-consuming, and unpopular with mariners.

10
Watchkeeping Duties

Introduction

The shipping industry is currently passing through probably the most dramatic change it has seen in the last 200 years. History has shown one or two milestones in the development of the industry, not least the move from wood to steel hulls and from sail to steam. However, the most recent innovation, which is forcing all mariners to re-think basic seamanship, is the advance of electronic navigation systems and, in particular, the introduction of the electronic navigation chart and full ECDIS use.

Watchkeeping operations are changing, with an enforced one-man bridge system. Officers are being given additional tasks to carry out (e.g. communications) while some of the older operations (like chart correction, as known) are changing to becoming virtually obsolete. Hi-technology is here to stay, within the integrated bridge, where GPS has become the dominant position-fixing system.

Mariners may be reluctant to embrace the new changes, but to stay abreast of the industrial changes training and development of the individual need to be increased. Many basic elements remain untouched, like the principle of calling the Master, or keeping an efficient lookout. Overall, the functions of the watchkeeper have not disappeared, but have embraced the essential needs of security in all its many forms. However, key elements are changing with legislation, equipment styles, improved management and, of course, the age of computer literacy.

Master's Responsibilities*

It is the duty of every Master to ensure that the watchkeeping arrangements for his ship are adequate for maintaining a safe navigational watch. All watchkeeping

---

* Based on recommendations published by IMO.

STCW '95 requires that watch-keeping personnel must have ten hours off duty, of which six hours must be continuous.

Refer to the IMO publication, 'Guidelines on Fatigue'.

officers are the Master's representatives and are individually responsible for the safe navigation of the vessel throughout their period of duty.

It falls to the Master of the vessel that watch arrangements are such that:

- At no time shall the bridge (Plate 92) be left unattended.
- The composition of the watch is adequate for the prevailing circumstances and conditions.
- Watchkeepers are capable and fit for duty, and in no way fatigued in such a manner as to impair their efficiency.
- At all times a proper lookout is maintained by sight and hearing, as well as by all other available means.
- The Master is aware of all navigational hazards that may make it necessary for the Officer of the Watch to carry out additional navigational duties. During such periods ample personnel should be available to provide full coverage for all duties.
- The Master is satisfied that watchkeeping personnel are familiar with all navigational equipment at their disposal.
- The voyage is well planned beforehand and all courses laid down are checked in advance.
- Correct handover and relief watchkeeping procedures are in practice.
- The limitations of the vessel and its equipment are known to watchkeeping personnel.
- When navigating under the advice of a pilot, the presence of that pilot in no way relieves the Master or the Officer of the Watch of their duties and obligations for the safe navigation of the vessel.

Plate 92 Example bridge layout as seen looking forward through the bridge windows from the autopilot/steering position. Multifunctional ARPA and communication console are set either side of centre.

The Master should further be aware that the protection of the marine environment is a major consideration. He should take all necessary precautions to ensure that no operational or accidental pollution of the environment takes place, being guided in this matter by the existing international regulations.

The Integrated Bridge

Many vessels and virtually all new builds of all ship types are being constructed with the integrated navigational bridge. These lend themselves to one-man operations and centralize all the essential controls for the vessel, inclusive of: remote control of propulsion, alarm and monitoring systems, power management and supply, navigation controls, safety management, control of utilities, and all internal/external communications.

Specific vessels also incorporate CCTV. All functional elements tend to be supported by built-in redundancy in such a way that the failure of a unit can be isolated to ensure against the failure of other functions (Figures 10.1 and 10.2).

Functions of the Integrated Bridge

Many ships are now moving towards the integrated bridge with this major change in the navigation and control element of ship-handling. Each integrated bridge is customized to suit the needs of 'that ship', but general functions are usually common to meet the standard needs of every vessel.

Typical functions would be expected to include:

- remote control of propulsion systems (diesel or electric)
- alarm and monitoring systems;
- power management supply and monitoring

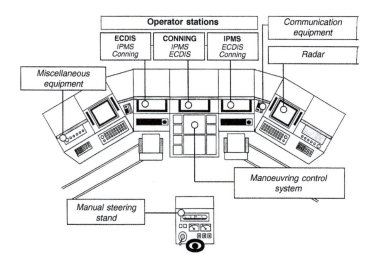

Figure 10.1 Example of an integrated bridge. ECDIS: Electronic Chart Display and Information System; IPMS: Integrated Platform Management System.

Figure 10.2 Operational aspects: the integrated bridge.

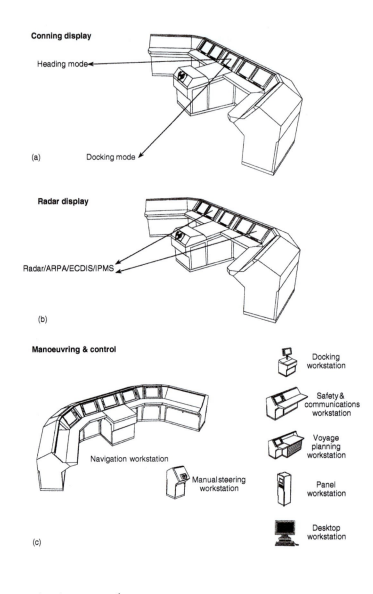

- navigation controls
- manoeuvring control (auto & manual systems)
- safety management
- utility control
- internal and external communications.

In the case of the passenger/ro-ro ferry, the ship's bridge would expect to have custom CCTV displays covering surveillance of all hull access and opening points, together with a light 'telltale' of watertight integrity. Docking, workstations with full manoeuvring and thruster control would usually be incorporated at prominent viewing positions in bridge wing situations, while safety management would ensure that fire protection systems are kept under positive observation at all times and alarm monitoring is continuous. Watertight and fire doors are readily activated from this central position.

Interfaces of speed log, gyro compass, autopilot, radar, echo-sounder, GPS, ENC, propellers and thrusters would all be alarm monitored for off-course, off-track, close-quarter situations, malfunction, power fluctuations, etc., so as to comply with IHO and IMO requirements.

The overall design would be compatible for one-man bridge operations and could expect additional inputs from any or all of the following:

- roll/pitch sensors
- engine performance parameters
- rate of turn
- Doppler stern radars – rate of approach
- sea temperature
- anemometer
- reefer/cargo temperatures
- damage stability data
- tank levels
- stabilizers/foil deployment.

Safety and Redundancy

The integrated navigation and ship-handling systems of today raise the obvious questions of 'What if ... an essential component or system fails?'

To counter this, built-in redundancy is part of each system, and all the essential units are often doubly represented. Design is also such that in the event of failure of one function then this is isolated to avoid leading to the failure of other functions. When classified for 'one-man bridge operation' (OMBO) it is usual to find two radars, two gyro compasses, two GPSs and two ECDIS-compliant units, each element being capable of operating independently and/or within the integrated system. In some cases the classification societies require automatic changeover in case of failure, or a manual selection as with autopilots, to provide an additional option.

NB. OMBO at night for UK ships is currently illegal and the carriage of navigation equipment requirements is based respective to tonnage. E.g. a 10,000 gross registered tonnage ship must have an automatic radar plotting aid (ARPA).

Many shipboard operations are now conducted under a label of 'bridge control' with a one-man bridge (OMB) concept. To ensure that such a routine is viable, the integrated bridge is designed to take account of:

- ergonomics
- automation of navigation
- monitoring
- safety and redundancy
- system support and maintenance.

Correct layout and unit position is usually determined between the shipowner and the classification regulations. Det Norske Veritas (DNV) were the first society to define how an 'integrated system' should look and subsequently provided a specific designation, DNV-W1, to signify OMB operational capability.

Such designation takes full account of ergonomics within the design, instrumentation, documentation and training.

Checklist of Items for Passage Appraisal

1   Select the largest scale-appropriate charts for the passage.
2   Check that all charts to be used have been brought up to date from the latest information available.
3   Check that all radio navigational warnings affecting the area have been received.
4   Check that sailing directions and relevant lists of lights have been brought up to date.
5   Estimate the draught of the ship during the various stages of the passage to ensure loadline zones are not compromised.
6   Study sailing directions for advice and recommendations on the route to be taken.
7   Consult a current atlas to obtain direction and rate of set.
8   Consult tide tables and a tidal atlas to obtain times, heights and direction and rate of set.
9   Study climatological information for weather characteristics of the area.
10   Study charted navigational aids and coastline characteristics for landfall and position monitoring purposes.
11   Check the requirements of traffic separation and routing schemes.
12   Consider volume and flow of traffic likely to be encountered.
13   Assess the coverage of radio aids to navigation in the area and the degree of accuracy of each.
14   Study the manoeuvring characteristics of the ship to decide upon safe speed and, where appropriate, allowance for turning circle at course alteration points (wheel over points).
15   If a pilot is to be embarked, make a careful study of the area at the pilot boarding point for pre-planning intended manoeuvres.
16   Where appropriate, study all available port information data and routing charts.
17   Check any additional items which may be required by the type of ship, the particular locality or the passage to be undertaken.

Watchkeeping: General Duties

Various duties are carried out by watchkeepers either individually or as a bridge team. Experience has shown that the bridge becomes the operational centre for the watch period, with all relevant information and orders processed through it.

The deck log book is maintained on the bridge by the Officer of the Watch, together with continual observation and supervision of the following items:

- Watertight integrity of the hull, together with the opening and closing of watertight doors.
- Fire watch, with continual observation of smoke detector systems.
- Special cargo surveillance, as and when required.
- Correct display of all lights and shapes.
- Weather conditions affecting the ship and its course.
- Routine working of the deck, inclusive of rigging pilot ladders, deployment of logs, organizing boat and fire drills, etc.
- All emergencies affecting the safety of the vessel.

Duties of the Officer of the Watch

The Officer of the Watch must supervise the efficient running of the watch and ensure the safe navigation of the vessel throughout the watch period; his or her main duty is to maintain a proper lookout whenever the vessel is at sea, regardless of other personnel engaged on a similar duty. Navigational duties include the regular checking of the ship's course and the comparison of the gyroscopic compasses with the magnetic compass. The former should be checked by obtaining the compass error at least once per watch or on every alteration of the vessel's course.

The position of the ship should be plotted at regular intervals. Depending on the circumstances, the time interval between separate positions will vary, especially when navigating in coastal waters.

Plate 93 Modern integrated bridge layout

Traffic avoidance is of prime importance in the safe naviga-tion of the vessel throughout the passage. The Officer of the Watch should use all available means at his or her disposal to ensure the safe passage of the vessel and should not hesitate to use any of the following if required: whistle, engines, radar, additional manpower, helmsman or anything else considered necessary for the safety of the vessel.*

The Officer of the Watch should make full use of navigational aids such as the echo-sounder whenever possible to check navi-gational accuracy. Radar plotting should be carried out in good weather as well as foul to ensure that the mariner becomes proficient in the correct plotting techniques.

Any Watch Officer should also be aware of the ship's capabil-ities in the way of 'turning circles' and emergency stop distances. He or she should not hesitate to summon the Master at any time, day or night, should assistance be required. In any event the Master should be kept informed by the Officer of the Watch of all the movements and events affecting the vessel's progress.

Calling the Master

The Officer of the Watch should notify the Master immediately in the following circumstances:

- If restricted visibility is encountered or suspected.
- If the traffic conditions or the movements of other vessels are causing concern.
- If difficulty is experienced in maintaining course.
- On failure to sight land or navigation mark, or to obtain sound-ings by the expected time.
- If land or navigation mark is sighted or a change of soundings occurs unexpectedly.
- On the breakdown of the engines, steering gear or any essential navigational equipment.
- In heavy weather, or if in any doubt about the possibility of weather damage.
- In any other emergency or situation in which there is doubt.

The requirement for the Officer of the Watch to call the Master in the above situations does not relieve the officer of taking any immediate action necessary for the safety of the vessel.

Duties of the Lookout

The principal duty of the lookout is to maintain a continuous watch for all hazards that may impair the safe navigation of the vessel (Figure 10.3).

---

* Further reading: *IMO Bridge Procedures Manual.*

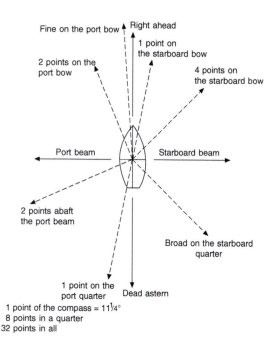

Figure 10.3 Field of view of the lookout.

Plate 94 Primary action of the Officer of the Watch is to maintain a continuous and effective lookout and not stand the vessel into danger (source: Shutterstock).

The lookout is obliged to give his full uninterrupted attention to this duty, reporting any of the following to the Officer of the Watch:

- All ships, irrespective of size or position in relation to the vessel on which the lookout is sailing.
- All navigation marks or lights.
- All floating objects.
- Any sightings of ice, no matter in what form.

- Sandbanks or prominent navigational features.
- Derelicts and any other hazard considered dangerous to navigation.
- The malfunction of the ship's lights, and their correct functioning at hourly intervals.

The lookout is also obliged to remain at his or her position until correctly relieved of duties. On being relieved, the relief should be given any relevant information concerning the items reported.

The lookout has a very responsible job and the duties must never be taken lightly. Rule 2 of the Regulations for Preventing Collisions at Sea states: 'Nothing in these Rules shall exonerate any vessel, or the owner, Master or crew thereof, from the consequence of any neglect to comply with these Rules or of the neglect of any precaution which may be required by the ordinary practice of seamen, or by the special circumstances of the case.' Rule 5 states: 'Every vessel shall at all times maintain a proper lookout by sight and hearing as well as by all available means appropriate in the prevailing circumstances and conditions so as to make a full appraisal of the situation and of the risk of collision.'

Duties of the Helmsman

The function of the 'helmsman' is to steer the vessel when it is not engaged on automatic pilot. On large passenger vessels the steering duty is normally carried out by the 'quartermaster' and the terms helmsman and quartermaster are both in common use in all types of vessels. The duty can be tedious and tiring and regular reliefs are employed to maintain efficiency, since the responsibility for the safe passage of the vessel lies in the hands of the man steering.

A helmsman taking the wheel should come on duty earlier rather than later. It is a duty in which punctuality is always expected. At the changeover the course being steered should always be repeated out loud, from one person to another, in order to allow all personnel on the bridge to be aware of the course being steered. Once the wheel has been relieved, the helmsman whose duty has finished should report the course to the Officer of the Watch, who will repeat it in acknowledgement.

It is common practice when handing over the wheel to a relief to advise on the amount of 'helm' that the vessel is carrying in order to maintain her course reasonably steadily. Such information can enable the relief to avoid excessive helm movement while keeping a steady course.

Although an old-fashioned custom which has largely dropped out of practice with the advent of the 'gyro compass' and 'reflector units', it was normal procedure for the helmsman to remove his knife and spike and any other metal about his person which might have affected the magnetic compass. He would at the same time remove any lanyard or loose-fitting clothing which could possibly catch the spokes of the wheel when turning.

Before sailing, the steering gear should be tested, under the supervision of the Duty Officer, by putting the helm hard over to both port and starboard and holding the pressure in both positions so that the rudder indicator will show the extent of movement in response to the helm.

Pilot Wheels

When a vessel is navigating in coastal waters a pilot is often employed and manual steering is used. The helmsman should take orders for the wheel movements from the pilot and repeat each order, word for word, back to the pilot before executing the movement. The man at the wheel should bear in mind that the pilot is an advisor to the Master and his representative (Officer of the Watch), and at any time the Master or the Officer of the Watch may countermand the orders of the pilot.

Duties of the Standby Man

These are numerous, depending on the time of day, the ship's position and possibly the weather. In whatever situation the vessel finds herself, the function of the standby watchman is to back up the efficient running of the watch at sea (see Table 10.2). During daylight hours it is normal practice for this standby duty to be carried out in or around the navigational bridge area, where the Officer of the Watch can easily contact the duty man if required. There are often plenty of odd maintenance jobs to be carried out in this area, and normal working within earshot of the bridge helps to fulfil both requirements.

However, the main duties of the standby man are as follows:

- To call the next duty watch in ample time to relieve watchkeepers.
- To double up as additional lookout in the event of poor visibility.
- To rig and display or strike flag signals as required.
- To carry out normal deck duties throughout the watch period, e.g. rig pilot ladders, check cargo lashings, renew faulty navigation lights, etc.

To summarize standby duty, one might say a 'jack of all trades', to be held in readiness for the unexpected incident. The standby is often in great demand to assist the Officer of the Watch when approaching a port after a sea passage or when leaving a port at the start of a deep-sea passage.

Table 10.1 Example of helm orders

Order (by Officer of the Watch, pilot or Master)	Helmsman's reply	Helmsman's actions	Helmsman's final report
Starboard twenty degrees	Starboard twenty, sir	Turns the wheel until twenty degrees to starboard is shown by the rudder indicator	Twenty degrees of starboard helm 'on', sir
Midships	Midships, sir	Returns the wheel to the midships position. Checks that rudder indicator shows midships	Wheel amidships, sir
Hard a-port	Hard a-port, sir	Turns the wheel as far to port as it will go. Checks that rudder indicator shows maximum port helm	Wheel hard a-port, sir
Ease the wheel to port ten degrees	Ease the wheel to port ten degrees, sir	Allows the wheel to return towards the midships position, but retains ten degrees of port helm as shown by rudder indicator	Wheel eased to port ten degrees, sir
Check her (should be understood to mean, check the swing of the vessel)	Check her, sir	Turns the wheel against the swing of the vessel, up to approx. ten degrees of opposing helm being applied to reduce the rate of swing. Eases the wheel back to the midships position once the vessel stops swinging	
Steady	Steady, sir	Observes compass heading, or land reference point, and steadies the ship's head in/on that heading. Applies helm as required in order to maintain a steady course	Steady on course, X,Y, Z (or whatever the heading happens to be)

* The supervising officer may, in addition, use the following phrases:

'*How is your head?*', a question as to the ship's compass heading.
'*Alter course to …*', ordering the helmsman to apply helm to change the ship's course to whatever is stated.
'*State when the vessel stops steering*', or when vessel no longer responds to helm movement because she has reduced her way, i.e. helm hard a-port and the vessel paying off to starboard.
'*Finished with the wheel*', when the helmsman is no longer required. The wheel is returned to midships and the helmsman can stand down.

Official Publications

British merchant vessels are obliged by law to carry certain marine publications of the following nature, in addition to sufficient navigational charts for the intended journey, together with the operating manuals for all the navigational aids carried aboard the vessel:

- *Weekly Notices to Mariners*
- *Annual Summary of Admiralty Notices to Mariners*
- *Marine Shipping Notices*
- *Marine Information Notices*
- *Marine Guidance Notices*

Table 10.2 Example of four-hour watch operated by three men under supervision of the Officer of the Watch

	1st hour	2nd hour	3rd hour	4th hour
Man A	First wheel (1st hour)	Standby	First wheel (2nd hour)	Lookout
Man B	Lookout	Second wheel (1st hour)	Standby	Second wheel (2nd hour)
Man C	Standby	Lookout	Lookout	Standby

The two men A and B each take two hours of the watch in steering duty. Each man has one hour of standby duty with one hour of lookout duty. Man C, known as the farmer, has no duty at the wheel but has two hours lookout duty and two hours of standby duty. In this manner the full four hours of wheel, lookout and standby are equally distributed between the three men. Rotation between first wheel (A), second wheel (B) and the farmer (C) ensures a fair allocation between all concerned.

- *Mariner's Handbook*
- *Admiralty List of Radio Signals*
- *Admiralty List of Lights*
- *Admiralty Sailing Directions*
- *Admiralty Tide Tables*
- *Nautical Almanac*
- nautical tables
- oil record book
- *International Code of Signals*
- *Tidal Stream Atlas*
- *Code of Safe Working Practice*
- *IMDG Code.*

In addition to the above, a well-found ship would normally carry such publications as:

- A copy of the *Regulations for the Prevention of Collision at Sea*
- A copy of *Chart Abbreviations* (No. 5011)
- *IAMSAR Manual*
- A copy of *Ship Routing and Traffic Separation Schemes*
- A *Ship Master's Medical Guide*
- *Stability Information Booklet*
- relevant statutory instruments

and, if relevant to the trade:

- *Code of Safe Working Practice for Bulk Cargoes*
- *Manual on the Avoidance of Oil Pollution*
- *Tanker Safety Guide*
- *Cargo Securing Manual* (ro-ro vessels).

Other countries, especially those which comply with IMO conventions, have similar rules for the carriage of official publications.

The only difference is that their titles may differ from the British ones.

Official Publications in Detail

Weekly Notices to Mariners

These are published by the Hydrographic Department of the Admiralty in booklet form. They contain an index of all the navigational corrections included in the booklet for that week, together with temporary (T) and preliminary (P) notices regarding navigational corrections. At the end of the booklet there are sections giving corrections to the *Admiralty List of Radio Signals* and the *Admiralty List of Lights*. Information regarding new charts, new editions and large and small corrections to charts are all included in the weekly list.

Annual Summary of Admiralty Notices to Mariners

This publication is issued annually by the Hydrographic Department of the Admiralty. The notices contain information on such items as tide table corrections, list of agents for charts, radio message procedures, search and rescue operational details, firing and practice areas, submarine information, coastal radio warning stations, mine-laying operations, etc., ocean weather ship details and navigational warnings. The *Annual Summary* and the *Weekly Notices to Mariners* are obtainable from any of the Mercantile Marine Offices, chart agents and custom offices (National Publication 247, published 2003).

Merchant Shipping Notices ('M' Notices)

These are published by the Marine Safety Agency and from 1997 known and titled as MSNs. They convey mandatory information which must be complied with under UK law. They amplify and expand on statutory instruments. In addition, Marine Guidance Notes (MGNs) will be issued regarding specific topic areas, e.g. SOLAS, MARPOL, etc., also Marine Information Notes (MINs) will be issued aimed at training establishments, equipment manufacturers, etc. These will be published periodically and will also carry a self-cancellation date.

NB. Each of the above notices will be pre-fixed by:

(M) for merchant vessels
(F) for fishing vessels
(M+F) for both merchant and fishing vessels.

Mariner's Handbook

The *Mariner's Handbook* was first published in 1962 by the Hydrographer of the Royal Navy. It has since been reissued nine times, the last time in 2009. The handbook is a reference book for mariners, giving information with regard to the following: charts and publications, the use of navigational aids, navigational hazards, natural conditions pertinent to weather, ice navigation and various tables of conversion factors (National Publication 100, published 2009).

Admiralty List of Radio Signals

These comprise six volumes published by the Admiralty which contain details of individual radio stations and the services they provide, and information with regard to radio procedures, together with extracts of regulations governing transmission and reception of radio signals. The volumes cover all areas and provide information to operators regarding navigational warnings, time signals, medical advice, weather bulletins, satellite information and distress procedures (volumes may have several parts).

Admiralty List of Lights

This is a publication in several volumes which lists the characteristics of all navigation lights and beacons. The lights are geographically listed to allow a prudent navigator to achieve continuity and easy comparison from the coastline against the listing. Information on each light includes the position of the light, the name of the light, its range and any flashing characteristic (published in both paper and digital format).

Admiralty Sailing Directions

Often referred to as 'pilot books', covering the whole world, they are published by the Admiralty in over 72 volumes. Each volume provides the mariner with general information and local knowledge of the area in which he intends to navigate. Items are covered in detail, including port facilities, navigational hazards regarding port entry, systems of buoyage, coastline views and chart information.

Admiralty Tide Tables (*Annual Publications*)

Published by the Hydrographic Department of the Navy, these are tidal predictions for all 'standard' and 'secondary' ports. They are published in four volumes:

- Volume I covers the United Kingdom and Ireland.
- Volume II covers Europe (excluding the United Kingdom and Ireland), the Mediterranean Sea and Atlantic Ocean.
- Volume III covers the Indian Ocean and China Sea.
- Volume IV covers the Pacific Ocean.

The tables provide the time and the heights of high and low waters for the 'standard' port on a daily basis. In the second part of the volumes the details for 'secondary' ports can be calculated by making minor adjustment to the heights and times of tides at the nearest 'standard' port.

Nautical Almanac (*Annual publication*)

This is a book of information relating to the sun, planets, moon and stars, published jointly by HM Nautical Almanac Office and the US Naval Observatory. The listing provides information with regard to Greenwich hour angle, meridian passage, sidereal hour angle, etc. of celestial bodies. The listed information is required in the calculations made to ascertain the ship's position by observing heavenly bodies.

There are a number of publishers of nautical almanacs at present in the commercial field. The almanacs are published annually and should be collected well in advance for vessels engaged on long voyages which may run into the following year.

Nautical Tables

Every vessel is obliged to carry a set of these tables in order to carry out the basic navigational functions. Each set will contain computation tables of logarithms, traverse tables, square roots, etc., together with tables for celestial navigation: A, B, C tables, amplitude corrections, total correction tables, etc. Additional tables cover coastal navigation, day's run, radar range, distance by vertical angle, etc. Several publishers are engaged in the distribution of nautical tables. The most popular are probably *Norie's Nautical Tables* and *Burtons Nautical Tables*.

Oil Record Book

This book's purpose is to record all oil movements aboard the vessel. It is carried by tanker vessels and vessels other than tankers. All entries should bear the signature of the Master of the vessel, together with those of officers concerned in the movement of oils or oily waste. A copy of the entries made into the oil record book for a non-tanker-type vessel is given in Table 10.3.

International Code of Signals

Published by Her Majesty's Stationery Office, it is used for coding and decoding signals made by international signal flags. It also contains information on methods of marine communication and procedures for their execution (see Chapter 9).

Table 10.3 Entries in oil record book for ships other than tankers

(a) Ballasting or cleaning of bunker fuel tanks

    1 Identity of tank(s) ballasted.....................................

    2 Whether cleaned since last containing oil and if not, type of oil previously carried.

    3 Date and position of ship at start of cleaning.........................

    4 Date and position of ship at start of ballasting........................

(b) Discharge of dirty ballast or cleaning water from tanks referred to under (a)

    5 Identity of tank(s)..............................

    6 Date and position of ship at start of discharge.........................

    7 Date and position of ship at finish of discharge........................

    8 Ship's speed(s) during discharge...................................

    9 Method of discharge (state whether separator was used).................

    10 Quantity discharged..

Signature of officer or officers in charge of the operations and date.
Signature of Master and date.

(c) Disposal of residues

    11 Quantity of residue retained on board..............................

    12 Methods of disposal of residue:...................................

        (a) reception facilities..

        (b) mixed with next bunkering...............................

        (c) transferred to another (other) tank(s).........................

    13 Date and port of disposal of residue*...............................

Signature of officer or officers in charge of the operations and date.
Signature of Master and date.

(d) Discharge of oily bilge water which has accumulated in machinery spaces while in port

    14 Port..

    15 Duration of stay................................

    16 Quantity disposed..............................

    17 Date and place of disposal........................

    18 Method of disposal (state whether separator was used)..................

(e) Routine discharge at sea of oily bilge water from machinery spaces*

    19. Frequency of discharge and method of disposal (state whether or not a separator was used)*†

Signature of officer or officers in charge of the operations and date.
Signature of Master and date.

(f) Accidental or other exceptional discharge of oil.

    20 Date and time of occurrence.....................................

    21 Place or position of ship at time of occurrence.........................

    22 Approximate quantity and type of oil..............................

    23 Circumstances of discharge or escape and general remarks................

Signature of officer or officers reporting the occurrence and date.
Signature of Master and date.

---

* In accordance with regulations such discharges need not be entered into this book if entered in the engine-room log book.

---

† Where pumps start automatically and discharge through a separator at all times, it will be sufficient to enter each day, 'Automatic discharge from bilges through separator'.

Tidal Stream Atlas

This contains detailed plans of the British Isles and other selected areas, showing the state of the tide at hourly intervals either side of high water. Indication of the tidal flow is shown by arrows referring to times and state of tide at a major port, e.g. 'Dover, High Water'. By comparing the ship's position and the time to the time of high water at a particular port, the navigator can ascertain the state of the tide in relation to the ship.

Code of Safe Working Practice

A recent publication by Her Majesty's Stationery Office on behalf of the Maritime and Coastguard Agency, this details safe working practice for seamen in all departments of the ship. It covers deck work, engine-room practice, electrical apparatus, etc., together with working practices in and around the galley and catering department.

IMDG Code

Mariners are guided by rules and recommendations regarding the carriage of dangerous goods at sea; these rules and recommendations are contained in what is known as *The International Maritime Dangerous Goods (IMDG) Code*. It is published by the IMO and is directly concerned with the recommendations relevant to the carriage of dangerous substances. It specifies in detail the method of stowage and packaging of dangerous cargoes.

It is contained in five volumes and provides relevant information on the required documentation and respective class of cargoes.

Shippers are required under the Merchant Shipping (Dangerous Goods) Rules to provide a certificate stating that the goods are properly marked, labelled and packaged in accordance with the rules. The goods are classified as follows:

Class 1   Explosives.

Class 2   Flammable gases, poisonous gases or compressed, liquefied or dissolved gases which are neither flammable nor poisonous.

Class 3   Flammable liquids subdivided into three categories:

  3.1   Low flashpoint group of liquids having a flashpoint below $-18°C$ (0°F) closed cup test, or having a low flashpoint in combination with some dangerous property other than flammability.

  3.2   Intermediate flashpoint group of liquids having a flashpoint of $-18°C$ (0°F) up to but not including 23°C (73°F), closed cup test.

  3.3   High flashpoint liquids having a flashpoint of 23°C (73°F) up to and including 61°C (141°F), closed cup test.

Class 4
    4.1   Flammable solids.
    4.2   Flammable solids or substances liable to spontaneous combustion.
    4.3   Flammable solids or substances which in contact with water emit flammable gases.

Class 5
    5.1   Oxidizing substances.
    5.2   Organic peroxides.

Class 6
    6.1   Poisonous (toxic) substances.
    6.2   Infectious substances.

Class 7   Radioactive substances.

Class 8   Corrosives.

Class 9   Miscellaneous dangerous substances, that is any other substance which experience has shown, or may show, to be of such a dangerous character that this class should apply to it.

Security

In July 2004 new security measures were set to be adopted by the International Shipping Community. Ships' personnel could expect to see the International Ship and Port Facility Security Code (ISPS) come into effect. This was to include regulations to enhance maritime security and confirm the Master in his authority to exercise his professional judgement over decisions necessary to maintain the security of the ship.

All ships were to be provided with a ship security alert system, probably by 2006. Once activated it was intended that this would initiate and transmit a ship to shore security alert. One must anticipate that the duties of Watch Officers will play a major role regarding on-board security.

Under the recommendations, ships will be required to have a 'Security Plan' and a 'Ship Security Officer'. The shipping company will also be required to have a designated 'Company Security Officer'. Increased vigilance by all personnel within the maritime industries, to the threat of acts of terrorism, is now considered an essential element of maintaining a safe environment.

A ship Security Plan will mean a plan developed to ensure the application of measures on board the ship designed to protect persons on board, cargo, cargo transport units, its stores or the ship from the risks of a security incident.

'Ship Security Officer' will mean that person on board the ship accountable to the Master, designated by the company as being responsible for the security of the ship. This person will be responsible for the implementation and the maintenance of the ship security plan and for liaison with the company security officer and port facility security officers.

'Company Security Officer' will mean the person designated by the company for ensuring that a ship security assessment is carried out: that a ship security plan is developed, submitted for approval and thereafter implemented and maintained. This person will, in addition, liaise with port facility security officers and ship security officers.

The new security code is meant to ensure confidence that adequate and proportionate maritime security measures are in place.

In May 2011, IMO approved interim measures to employ armed security personnel aboard ships.

Security code 1 The level for which minimum appropriate protective security measures shall be maintained at all times.

Security code 2 The level for which appropriate additional protective security measures shall be maintained for a period of time as a result of heightened risk of a security incident.

Security code 3 The level for which further specific protective measures shall be maintained for a limited period of time when a security incident is probable or imminent, although it may not be possible to identify the specific target.

Rigging Pilot Ladders

The Rigging and Use of Pilot Ladders

Each pilot ladder shall be suitable for the purpose of enabling a pilot to embark and disembark safely. Such ladders must be used only by officials and other persons while a ship is arriving or leaving a port and for the disembarkation and embarkation of pilots. Every pilot ladder must be secured in a position clear of any possible discharge from the ship and so that the steps rest firmly against the ship's side (Figure 10.4), providing the pilot convenient access to the vessel after climbing not less than 5 ft (1.5 m) nor more than 30 ft (9 m).

A single length of ladder should be used, and it should be capable of reaching the water from the point of access to the ship when the ship is in the unloaded condition and in normal trim with no list. Whenever the distance to the water exceeds 30 ft (9 m), then access to the vessel must be by means of an accommodation ladder or other equally safe means (Figure 10.5).

The treads of the ladder must be made of a hardwood such as ash, oak, elm or teak. Each step must be made from a piece free of knots, having a non-slip surface and must be not less than 480 mm long, 115 mm wide and 25 mm in depth. The steps should be spaced not less than 300 mm nor more than 380 mm apart, and individually secured in such a manner that they will remain horizontal.

Plate 95 Pilot boat alongside during pilot embarkation to vessel.

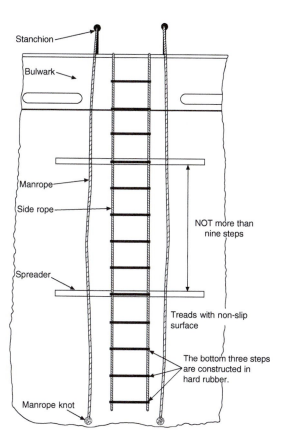

Stanchion

Bulwark

Manrope

Side rope

NOT more than nine steps

Spreader

Treads with non-slip surface

The bottom three steps are constructed in hard rubber.

Manrope knot

Figure 10.4 Pilot ladder.

Figure 10.5 Combined rigging of accommodation and pilot ladders.

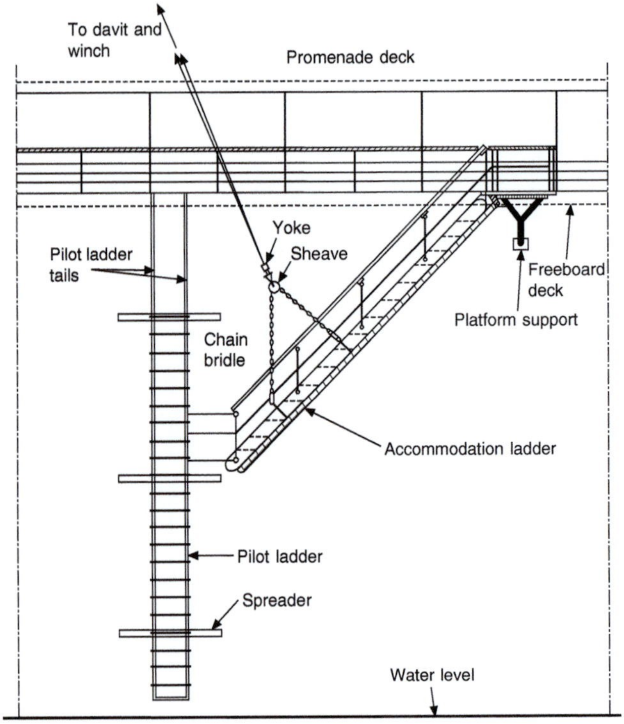

Accommodation ladders when rigged in combination with a pilot ladder should not exceed 45° from the horizontal.

The four lower steps may be constructed in rubber or other suitable material of sufficient strength and similar character. No pilot ladder must have more than two replacement steps secured in a different manner from the original method of securing, and these must be secured in place by the original method as soon as possible.

The side ropes of the ladder must consist of manila rope 18 mm in diameter. Each rope must be continuous without joins and should be left uncovered. Manropes of not less than 20 mm in diameter should be secured to the ship, and a safety line kept ready for use if required.

Hardwood battens (spreaders or anti-twist battens) between 1,800 and 2,000 mm long must be provided at such intervals as will prevent the pilot ladder from twisting. They must be so fixed that the lowest batten comes no lower than the fifth step from the bottom, and the interval between battens must not be more than nine steps. Each batten must be made of ash, oak or similar material, and free of knots.

Safe access must be provided between the pilot ladder and the vessel. Gateway or hand holds are suitable safety measures.

The bulwark ladder must be hooked onto the bulwark in a secure manner. It must be fitted with two handhold stanchions at the point of boarding. The stanchions should be not less than 700 mm nor more than 800 mm apart, be rigid in their construction, and extend not less than 1,200 mm above the top of the bulwark (Figure 10.6).

A light must be provided at night, and a lifebuoy equipped with a self-igniting light kept ready for immediate use.

Mechanical Pilot Hoists

Mechanical pilot hoists generally consist of three main parts:

1   A mechanical powered appliance together with means of safe passage from the hoist to the deck and vice versa.
2   Two separate falls (steel wire rope).
3   A ladder consisting of two parts:
    (a)   a rigid upper part for the transportation of the pilot upwards or downwards,
    (b)   a lower section consisting of a short length of pilot ladder which enables the pilot to climb from the pilot launch to the upper part of the hoist and vice versa.

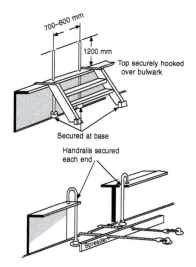

Figure 10.6 Access to the deck from pilot ladders.

Mechanical Powered Appliance

The source of power may be electrical, hydraulic or pneumatic. The winch arrangement should incorporate some form of braking system capable of supporting the working load in the event of power failure, and be equipped with crank handles for manual operation at a reasonable speed. These crank handles, when inserted, automatically cut off the power supply.

All hoists are fitted with emergency stop arrangements and automatic power cut-outs in the event of the hoist becoming jammed. Should the hoist be of a pneumatic type, the power cut-outs may be omitted, provided that the maximum torque available from the air motor cannot result in the overloading of the gear.

Pilot hoists should be securely attached to the ship's structure with safe means of access from the hoist to the ship and vice versa. The platform arrangement should be securely guarded by hand-rails. The wire falls must be wound evenly onto the winch drums and the winch controls clearly marked to indicate the 'hoist', 'stop' and 'lower' positions.

The powered appliance must be capable of hoisting or lowering at a speed of between 15 and 30 m per minute. If it is electrically powered, the voltage should not be in excess of 25 volts.

NB. Additional reference regarding pilot transfer arrangements can be found in Chapter 5 of SOLAS.

Falls

The two separate flexible steel wire rope falls should be resistant to corrosion in a salt-laden atmosphere. They should be securely attached to the winch drums and the ladder by fitments capable of withstanding a proof load of not less then 2.2 times the load on such fitments. The length of the falls should be sufficient to allow at least three turns to be retained on the winch drum when the hoist is in the lowest position, at all levels of freeboard.

Ladder Sections

The rigid section should be not less than 2.5 m in length and arranged so that the pilot may take up a safe position while being hoisted. It should be provided with adequate means of communication between the pilot and the hoist operator, together with an emergency stop control within easy reach of the pilot.

The section should be fitted at the lower end with a spreader not less than 1.8 m in length. The ends of the spreader should be provided with rollers which will allow the section to roll freely on the ship's side during the operation of embarking and disembarking pilots. A sufficient number of steps with non-skid surfaces should be included in the section to facilitate safe and easy access to the deck of the vessel.

The section should also be provided with suitable handholds, which will protect operators' hands from extreme temperatures and provide a safe and secure hold. In addition, it needs an effective guard ring, well padded to support the pilot without hampering his movements.

The flexible section consists of a pilot ladder length of eight steps. Manufactured in hardwood, it should be free of knots like the conventional pilot ladder, and of the same size.

Both the rigid and the flexible sections should be in the same vertical line, of the same width and placed as near to the ship's side as practicable. Both sections should be so secured that the handholds are also aligned as closely as possible.

Testing of New Hoists

All new pilot hoist systems are subjected to an overload test of 2.2 times the working load. During the test the load should be lowered a distance of not less than 5 m, the weight of each person being taken as 150 kg.

After installation has been completed, a 10 per cent overload test should be carried out to check securing attachments. Regular test rigging and inspection should be carried out by the ship's personnel at intervals not exceeding six months and a record of these checks maintained by the Master in the ship's log. Subsequent examinations of the hoists, under working conditions, should be made for each survey for the vessel's safety equipment certificate.

Rigging and Operational Aspects

Before use, the rigging of the hoist should be supervised by a responsible officer and the system tested to the satisfaction of the operator before embarking or disembarking pilots. The operational area should be adequately lit and a lifebuoy should be on hand for immediate use.

A conventional pilot ladder should be rigged adjacent to the hoist for use by the pilot should he so prefer. This pilot ladder should

Plate 96 Accommodation ladder seen in the turned out position and deployed with the vessel at anchor. Accommodation ladders should not be rigged more than 55° from the horizontal. (source: Shutterstock).

Plate 97 Department of Trade gangway, rigged at right angles to the fore and aft line, with gangway net rigged aboard the passenger vessel *Seawing* deployed from shell door. Angle from the horizontal should not exceed 30 degrees.

Plate 98 Alternative short gangway deployed from lower shell door from high freeboard vessels.

Plate 99 Accommodation ladder seen rigged with stanchions, manropes and gangway net landed to the quay with the bottom roller clear of obstructions.

be fully rigged during the operation of the hoist in such a manner as to be accessible to the pilot from any point on the mechanical hoist during the period of travel.

Protective stowage for pilot hoists is required, especially in cold weather, when ice formation may cause damage to the equipment. Bearing this fact in mind, the portable hoist should not be rigged until its use is imminent.

NB. For additional reference, see Chapter 5 of SOLAS regarding the rigging and operation of pilot ladders and hoists.

Masters/Pilot Exchange

When a ship takes a pilot on board either by launch or helicopter delivery, the initial meeting between the two professionals is usually one of cordiality. The Master would expect to see the identification and the administration documents from the pilotage authority. In return, the pilot would anticipate being given the ship's general particulars regarding its movements from which port and bound to which destination. Other details could expect to cover nature of cargo and, in particular, the ship's draught.

The Master would probably supply manoeuvring information and machinery status by means of a pilot information card and present the pilot with his current passage plan. It would be considered standard practice to ask the following questions:

1   Is the current 'passage plan' acceptable to the pilot or will it need to be amended?
2   With the ship's present draught, where are the lowest positions of underwater keel clearance (UKC)?
3   Are there any navigation hazards/warnings on the route, through the course of the pilotage, or special traffic congestion focal points?
4   Are there any changes to the local by-laws?
5   What state of tide and currents can be expected to affect the ship's progress and where will the effects become the most prominent?
6   What berthing details are known? Which side to? Tug assistance yes or no? Docking pilot yes or no? Also need the ETA.

Any communication criteria would also need to be clarified.

Comment

Pilots are taken in virtually all ports around the world, especially so in waters which are designated compulsory pilotage areas. The Master takes the services of the marine pilot because of his local knowledge. Pilots generally have a good reputation as being capable seamen, but they are not infallible, and the prudent Master would not take the competence of the pilot for granted.

Ship to Shore Transit

Once the ship has arrived in port, accommodation ladders and/or gangways must be rigged (see Plates 98 and 99).

11
Marine Instruments Monitoring and Measurement

Introduction

Instrumentation employed in the marine industry continues to go through radical change on the back of the microchip. The needs of the vessel have not changed; we still need the speed of the vessel and we still need to provide a heading to steer by, but the methods of fulfilling these needs have removed many of the seaman's old skills (e.g. streaming the patent rotator log).

The new hardware is more accurate and more convenient to operate, provided individuals have the computer and technological background; such attributes are not generally assumed to be in the armoury of the mariner of yesterday.

Miniaturization, fibre optics and visual display units have all become part of the modern ship's navigation bridge. We still have the magnetic compass, but also the transmitting magnetic compass. We still have radar, but now with ARPA, electronic navigation charts and real-time graphics. One might easily feel that the Officer of the Watch is becoming superfluous to shipboard operations.

This is not the case, in any shape or form. What is happening is that the capabilities of the Officer of the Watch are changing and the interpretation and monitoring skills have increased extensively. Part of this changing work pattern is the keeping of records. Again very often completed electronically, but with more accuracy and in greater quantity.

At the end of the day the many ships engaged at sea are not moving into the technological age at a uniform rate. The yachts, fishing boats, small craft under 400 gross registered tonnes, etc., tend to be on the whole left behind, through the problems of finance, the lack of legislation or just a preference by owners for the 'old ways'.

The end result is that the seaman will remain on the ship's bridge for the foreseeable future, but his job specification will be changed by the instrumentation around him.

Sextant

The sextant's purpose is to measure angles, either vertical or horizontal, to obtain the necessary data to check the vessel's position. Latitude and longitude may be determined by a combination of sextant, chronometer and nautical almanac readings.

This precision instrument is based on the principle, enunciated by the First Law of Light, that when a ray of light is reflected from a plane mirror, then 'The angle of incidence of the ray equals the angle of reflection'. In the sextant a ray of light is reflected twice by two mirrors, the index and horizon mirrors, in the same plane. When a ray of light is reflected in this way by two plane mirrors, then the angle between the direction of the original ray and the direction of the final reflected ray is twice the angle between the mirrors (see Figures 11.1–11.2 and Plate 100).

Principle of the Sextant

The principle of the sextant is based on the fact that twice the angle between the mirrors (HAI, Figure 11.2) must equal the angle between the initial and final directions of a ray of light which has undergone two reflections.

Proof

Let α represent the angle between the mirrors; let Ø represent the angle between the initial and final directions of a ray of light.

The required proof is:

$$2\alpha = \emptyset$$

Construction

Extend the ray of light from the object to intersect the reflected ray from the horizon mirror, H, at point L.

With modern developments in GPS and DGPS, celestial navigation methods and use of the marine sextant have become close to obsolete. However, marine examiners still expect candidates to know how to use the instrument.

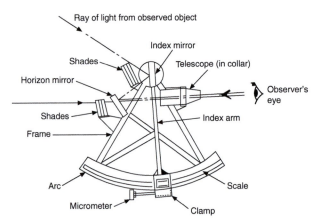

Figure 11.1 Sextant.

Proof of theory

1   The angle between the mirrors, α, is equal to the angle between the normals to the mirrors.

2   In triangle *HIK*:

$$\beta = \alpha + X$$

and $2\beta = 2\alpha + 2X$

3   In triangle *HIL*:

$$2\beta = \emptyset + 2X$$

Therefore from equations 2 and 3:

$$2\alpha + 2X = \emptyset + 2X$$

and $2\alpha = \emptyset$

i.e. twice the angle between the mirrors is equal to the angle between the initial and final directions of a ray of light which has undergone two reflections in the same plane, by two plane mirrors.

Errors of the Marine Sextant

There are three main errors which can quite easily be corrected by the mariner. A fourth error, for 'collimation', can also be corrected,

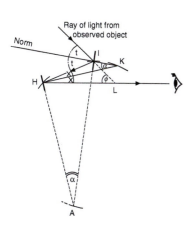

Figure 11.2 Principle of the sextant.

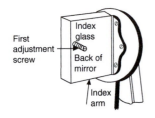

Figure 11.3 Adjustment screw on index mirror.

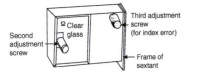

Figure 11.4 Adjustment screws on horizon mirror, seen from behind.

Figure 11.5 Images of true and reflected stars, showing side error.

with care and attention, but only to an older sextant where telescope collars are fitted with adjusting screws.

The first error, of *perpendicularity*, is caused by the index mirror not being perpendicular to the plane of the instrument. To check if this error is present, clamp the index arm between one-third and halfway along the arc, remove the telescope and look obliquely into the index mirror, observing the true and reflected arcs of the sextant. Hold the sextant horizontal, arc away from the body. If the true and reflected arcs are not in line with each other, then an error of perpendicularity must be considered to exist (Figure 11.3).

To correct the error, adjust the screw at the rear of the index mirror until the true and reflected arcs are brought together in line.

The second error, *side error*, is caused by the horizon mirror not being perpendicular to the plane of the instrument. There are two ways of checking if this error is present. The first is by observing a star. Hold the sextant in the vertical position with the index arm set at zero, and observe a second magnitude star through the telescope. If the true and reflected stars are side by side, then side error must be considered to exist (Figure 11.5). It is often the case when checking the instrument for side error that the true and reflected stars are coincident. If this is the case, a small amount of side error may exist, but a minor adjustment of the micrometer should cause the true star to appear below the reflected image. Should, however, the reflected image move to one side rather than move in a vertical motion, side error may be considered to exist.

The second way is by observing the horizon. Set the index arm at zero and hold the sextant just off the horizontal position. Look through the telescope at the true and reflected horizons. If they are misaligned, as indicated in Figure 11.6, then side error must be considered to exist.

To correct for side error, adjust the centre screw furthest from the plane of the instrument at the back of the horizon mirror to bring either the star and its image into coincidence or the true and reflected horizons into line.

The third error, *index error*, is caused by the index mirror and the horizon mirror not being parallel to each other when the index arm is set at zero. To check whether index error is present by observing a star, look through the telescope when the sextant is set at zero, and if the reflected image of the star is above or below the true image, then index error must be considered to exist. Should the true and reflected images be coincident, then no error will exist. To check by observing the horizon, set the index arm at zero, hold the sextant in the vertical position, and observe the line of the true and reflected horizons; if they are seen as one continuous line, then no error exists, but if the line between the true and reflected horizons is broken, an adjustment needs to be made to remove the error. This adjustment is made by turning the screw nearest to the plane of the instrument. Index error may also be checked by observing the sun. Fit the shaded eye piece to the telescope. Clamp the index arm at about 32′ off the arc and observe the true and reflected images to the position of limb upon limb. Repeat the observation with index arm set at about 32′ on the arc, and note the two readings of both observations. The numerical value of the index error is the difference between the two readings divided by two, and would be called 'on the arc' if the 'on the arc' reading were the greater of the two, and 'off the arc' if the 'off the arc' reading were the greater.

Let us consider an example:

Adjust the micrometer to bring the true sun into contact with the reflected sun. Note the reading; for example:

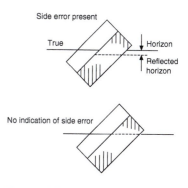

Figure 11.6 Indication of side error.

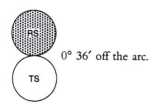

0° 36′ off the arc.

Repeat the observation, but with images the other way round. Note the reading; for example:

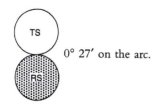

0° 27′ on the arc.

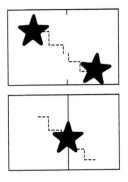

Figure 11.7 Dealing with combination of side and index error.

Take the difference of the two readings and divide by 2:

$$\text{Index error is } \frac{36-27}{2} = 4.5' \text{ off the arc.}$$

This error must be added to the future sextant readings.

The accuracy of the observations may be checked by adding the numerical values of both readings together and dividing the number by four. The resulting value should equal the semi-diameter of the sun for the period at which the observation was taken.

Sometimes an instrument suffers from side error and index error combined. Should this undesirable condition be apparent, the mariner can resolve the problem by removing each error a little at a time, as shown in Figure 11.7. The correction is made by turning the second and then the third adjustment screws alternately, by a small amount each time, until coincidence of image is achieved.

Collimation error

This is an error caused by the axis of the telescope not being parallel to the plane of the instrument. To check whether the error is present, insert the inverting telescope, setting the eyepiece so that one pair of the cross wires is parallel to the plane of the sextant.

To check by observation of two stars (selected about 90° apart), move the index arm to bring the two stars into exact contact with each other, resting on the wire nearest to the plane of the sextant. Now tilt the sextant upwards so as to bring them onto the wire which is furthest from the plane of the instrument. Should the images diverge or converge from the top intersections of the wires, it must be assumed that an error of collimation exists, and that the axis of the telescope is not parallel to the plane of the instrument.

This error can be corrected by adjustment of the two screws in the collar or telescope mounting. The screws are moved together, one being tightened, the other slackened, to align the stars on the top intersection which will bring the telescope back to parallel with the sextant frame. (Not all sextants, however, have adjustable collar screws.)

Non-adjustable errors

1 *Centering error.* This error could be caused by wearing of the pivot on which the index arm moves, perhaps because the index arm is not pivoted at the exact point of the centre of curvature of the arc.
2 *Prismatic error.* This error is caused by the two faces of the mirror not being parallel to each other.

3   *Shade error*. This is an error caused by the faces of shades not being parallel to each other. If it is known to exist, the telescope is used in conjunction with the dark eyepiece.

4   *Graduation error*. This error may be encountered on the arc itself or on the vernier or micrometer scales. If the micrometer drum is known to be correct, then the first and last graduations on the drum should always be aligned with graduation marks on the arc.

The manufacturer tables all the non-adjustable errors and issues the sextant with a certificate, usually secured inside the lid of the case. The combination of the above four errors is known as 'instrument error'.

Marine Chronometer

The chronometer represents a fine example of precision engineering. The instrument is manufactured and tested under stringent quality-control methods to comply with marine authorities' regulations. The mechanical movement of the timepiece is manufactured as near to perfection as is humanly possible.

It is used for the purpose of navigation and is generally the only instrument aboard which records GMT (Greenwich Mean Time), all other clocks tending to indicate local mean time or zone time. It is normal practice for two chronometers to be carried by modern vessels as a safeguard against mechanical failure or accident.

The chronometer is stowed if possible in a place free of vibration and maintained at a regular and even temperature. It must be accessible to the navigation officer but not so exposed as to allow irresponsible handling. By experience it has been found that the chartroom or wheelhouse area are ideal positions for this most important of ship's instruments.

The timepiece itself is slung in a gimbal arrangement which can be locked in position should the instrument have to be transported, the whole being encased in a strong wooden box fitted with a lock and binding strap. Most vessels are fitted with a glass-covered well which holds the ship's chronometers. These wells are often padded to reduce vibration effects, while the glass acts as a dust cover and permits observation of the clock.

Usually a brass bowl is made to encase the mechanism. The bowl is maintained in the horizontal position by the gimbal arrangement set on stainless steel pivot bearings. A sliding, spring-loaded dust cover set in the base of the bowl allows access for winding.

Regularity is achieved via a torque-equalizing chain to a fusee drum. The main spring is non-magnetic (of platinum, gold or palladium alloy), and is fully tested before the instrument is released.

The chronometer is fitted into an inner guard box fitted with a hinged, glazed lid. The outer wooden protective box is normally removed once the instrument has been transported to the vessel and secured in place.

> Modern chronometers are quartz/battery operated and tend not to require daily winding. However, they are still rated daily to ascertain any error.

Plate 102 Marine chronometer
(source: Shutterstock).

Two-day chronometers should be wound daily at the same time. The winding key, known as the 'tipsy key', is inserted into the base of the instrument after inverting the bowl in the gimbals and sliding the dust cover over the key orifice. Chronometers are manufactured so that they cannot be overwound, the majority being fully wound after 7.5 half-turns of the key anticlockwise. At this stage the person winding will encounter a butt stop which prevents further winding. A small indicating dial on the clock face also provides indication that the instrument is fully wound.

Should the chronometer have stopped through oversight or other reason, it may become necessary to reset the hands on the face before restarting the mechanism. If time permits, it is best to wait until the time indicated is arrived at 12 hours later, then just restart the instrument. However, this is not always practical, and if the hands need to be reset, they can be by means of the following method:

1   Unscrew the glass face plate of the chronometer.
2   Fit the 'tipsy key' over the centre spindle, holding the hands.
3   Carefully turn the key to move the hands in the normal clockwise direction.

Under no circumstances must the hands be turned anticlockwise as this will place excessive strain on the mechanism and may cause serious damage. Starting the chronometer should be done in conjunction with a radio time signal once the mechanism has been fully wound. It will be necessary for any person restarting a chronometer after it has stopped to give the timepiece a gentle circular twist in the horizontal plane. This effectively activates the balance and sets the mechanism in motion.

After starting, the chronometer should be rated on a daily basis against reliable time signals. Any error, either fast or slow, should be recorded in the chronometer error book, small errors being taken account of in navigation calculations.

Speed and Depth

Impeller Log

The impeller log may be considered an electric log, since its operation is all electrical, except for the mechanical rotation of the impeller. There are several designs in general use, but probably the most common is the 'Chernikeeff'.

The principle of operation is based on turning an impeller by a flow of water, the speed of rotation being proportional to the rate of flow past the impeller (turbine principle). As previously stated, designs vary, the two most popular being one with a ring magnet attached to the spindle and one with the magnet incorporated in the blades of the impeller. In either case a pick-up coil transmits the generated pulses via an amplifier to an electromagnetic counter. This signal is then displayed by a speed indicator and distance recorder.

Additional sensors will provide the opportunity for various repeaters to include a direct link to allow speed input into true motion radar. Operating power is normally 230/240 volts.

It is worth noting that the load on the impeller is negligible; consequently the slip, if any, on the impeller is minimal and can be ignored. The extended log, when in operation, projects approximately 14 in. (35 cm) below the ship's hull, usually from the engine-room position. The log shaft should be housed in the stowed position for shallow water, drydocking, etc. The sea valve sluice need only be closed if the log is to be removed for maintenance. However, it must be considered good seamanship practice to close the sluice each time the log is housed.

Performance of the log is in general considered to be very good, but obvious problems arise in dirty water areas with a muddy bottom and heavily polluted canals (see Figure 11.8).

Speed Logs

There are several systems of speed logs commercially available, but one of the most common is the electromagnetic speed log manufactured

Figure 11.8 Impeller log.

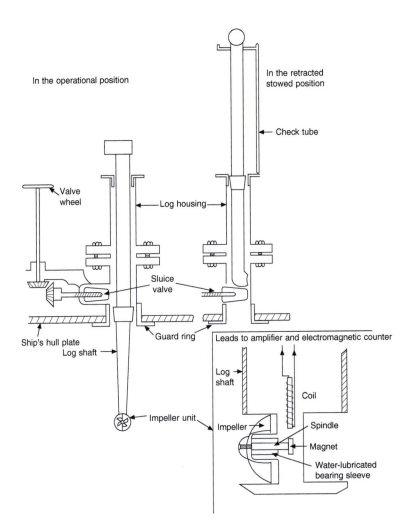

In the operational position

In the retracted stowed position

Check tube

Valve wheel

Log housing

Sluice valve

Ship's hull plate
Log shaft

Guard ring

Impeller unit

Leads to amplifier and electromagnetic counter

Log shaft

Coil

Impeller

Spindle

Magnet

Water-lubricated bearing sleeve

by C. Plath. The Naviknot III, flush-fitting speed sensor has the capability of measuring speed over a range from –5 to +80 knots (Figure 11.9). The fitting is suitable for steel and aluminium hull designs of any mono- or multi-hull vessel, inclusive of surface effect ships (SES) and the small water-plane area twin hull ships (SWATHS).

Speed log features

- Liquid crystal display (LCD) screen
- complies with IMO Resolution A-478 (XII)
- satisfies IMO requirements for ARPA
- measurement accuracy fulfils IMO Resolution A.824(19)
- built-in test facility
- high accuracy even at low speeds
- no moving parts in the sensor

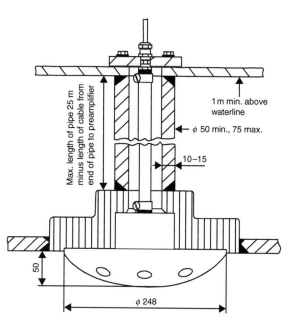

Figure 11.9 Naviknot III FNF (flush-fitted sensor) for external hulls in either steel or aluminium. Speed range from −5 to 35 knots.

- single system has the capability to drive a maximum of 20 remote control and display units
- resettable daily mile counters
- total mile counter
- microprocessor technology provides exceptional reliability
- operational data remains stored in the event of a power failure
- digital output and user friendly
- sensors can be replaced without drydocking
- integrated stopwatch facility
- water depth alarm facility
- facility for water temperature and salinity values to be stored
- electromagnetic and Doppler sensors
- each system can be operable with two sensors
- satisfies specifications of Classification Society Germanischer Lloyd.

Accuracy

- Speed relative to the water flow at the location of the sensor <±0.1 knot
- distance travelled, based on the measured speed <±0.1%.

Model variants for

- Vessel types
- manual deployment
- speed variations (inclusive of high-speed craft) – external fitment as opposed to internal

- extra additions to: navigation data display, digital repeaters and sensors.

Hand Lead

The normal length of the hand lead line is about 25 fathoms, and the line used is 9 mm untarred cable-laid hemp (left-hand lay). A rawhide becket attached to an eye splice in the end of the line secures the lead, the weight of which is 7–9 lb (3.2–4 kg) when operating from vessels moving at less than six knots.

From the eye splice, i.e. 'lead out', which has the extra safety factor of the length of the lead, or 'lead in', measured from the base of the lead, the markings are as follows:

- At 2 fathoms a piece of leather with two tails.
- At 3 fathoms a piece of leather with three tails.
- At 5 fathoms a piece of white linen.
- At 7 fathoms a piece of red bunting.
- At 10 fathoms a piece of leather with a hole in it (leather washer).
- At 13 fathoms a piece of blue serge.
- At 15 fathoms a piece of white linen.
- At 17 fathoms a piece of red bunting.
- At 20 fathoms a piece of cord with two knots.

Markings of metric hand lead line are as follows:

- 1 and 11 m – one strip of leather
- 2 and 12 m – two strips of leather
- 3 and 13 m – blue bunting
- 4 and 14 m – green and white bunting
- 5 and 15 m – white bunting
- 6 and 16 m – green bunting
- 7 and 17 m – red bunting
- 8 and 18 m – yellow bunting
- 9 and 19 m – red and white bunting
- 10 m – leather with a hole in it
- 20 m – leather with a hole and two strips of leather

The different materials indicating the various marks are distinctive to allow the leadsman to feel rather than see the difference during the hours of darkness. The intermediate whole-fathom values, i.e. 1, 4, 6, 8, 9, 11, 12, 14, 16, 18 and 19 fathoms, are known as deeps.

The leadsman used to stand in the 'chains', from where he would take the cast and call up the sounding to the Officer of the Watch. The lead line is not used in this manner today, but the soundings are still occasionally called in a traditional manner of stating the actual number of fathoms last. For example:

At 7 fathoms ... 'by the mark seven'.

At $7\frac{1}{4}$ fathoms ... 'and a quarter seven'.

At $7\frac{1}{2}$ fathoms ... 'and a half seven'.

At $7\frac{3}{4}$ fathoms ... 'a quarter less eight'.

At 8 fathoms ... 'by the deep eight'.

Should the bottom not be reached, then 'no bottom' is reported.

Constructing a new line

Splice the eye into one end of the line, then soak and stretch the line, possibly by towing astern. Mark the line off when wet from measured distances marked off on deck, and tuck the fabrics of the marks through the lay of the line.

Benefit of the lead

This is the term used to describe the length from the base of the lead to the eye splice. The actual distance is about 12 inches (30 cm) and is always 'beneficial' to the soundings, giving more water for the benefit of the ship.

Arming the lead

This describes the action of placing tallow into the 'arming recess', found at the base of the lead. The purpose of the soft tallow is to act as a glue to obtain the nature of the sea bottom. If tallow is not available, a soft soap will be equally good. The information is passed to the Officer of the Watch with the depth of sounding. It allows an additional comparison with the charted information.

Echo-Sounding

Principle of the echo-sounder

The echo-sounding depth recorder emits a pulse of sound energy from a transmitter, and the time this pulse takes to reach the sea bed and be reflected back to the vessel is directly related to the distance. Speed of sound through water is known to be 1,500 metres per second (see Figures 11.10–11.12).

However, that value will vary with water temperature and salt content (salinity).

Let us work out an example:

Let the velocity of sound in water = v metres per second. Let the time between transmission and reception of the pulse = t seconds. Let the distance to the sea bed and back = 2s metres.

But the distance = speed × time

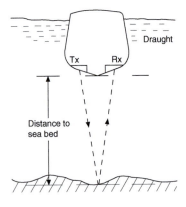

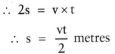

$$\therefore\ 2s\ =\ v \times t$$

$$\therefore\ s\ =\ \frac{vt}{2}\ \text{metres}$$

Therefore, s represents the depth of water under the vessel.

Figure 11.10 Principle of the echo-sounder.

Modern-day echo sounding

The majority of ships' sounders employ the 'Doppler effect' and are taken from what the military refer to as SONAR (sound navigation and ranging).

SONAR can be active or passive. Active SONAR emits pulses of sounds for acoustic location (still known as echo sounding). Doppler effect can be used to measure depth variations, and with advances in digital processing can interpret more accurately the distance between the transmitter and the sea bed.

Variations of the same principle are found as fish finders throughout the associated fishing industry.

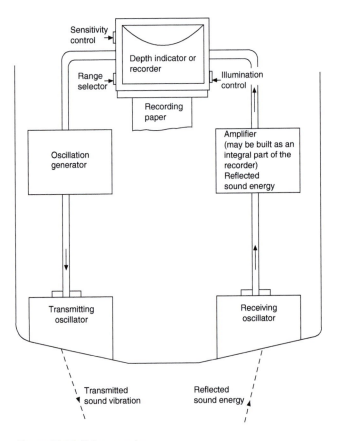

Figure 11.11 Echo-sounder.

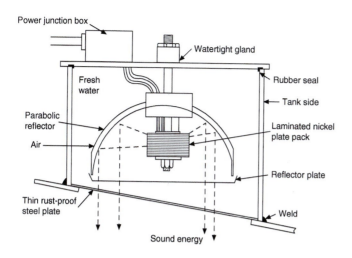

Figure 11.12 Echo-sounder's transmitting oscillator (magnetostriction type).

Possible errors of echo-sounding equipment

1 *Differences of the velocity of propagation.* Owing to the differences of salinity and temperature encountered in various parts of the world, adjustment tables are available, published by the Admiralty.

2 *Transmission line error.* This is caused by the misalignment of the reference 'zero' on the scale. Reference 'zero' sets the timer of the recorder unit, and if it is not set at 'zero' then a false time and recording will be obtained.

3 *Pythagorean error.* This error is encountered with separated transducers rather than with the combined transmit/receive unit. The error is caused by the measuring of the 'slant distance' as opposed to the vertical distance under the keel.

4 *Aeration.* The presence of air in the water will affect the speed at which sound travels through it, since the velocity of sound through air is much less than that in water (330 m/s compared with 1,500 m/s). The main causes of aeration are:

   • turbulence caused by having the rudder hard over;
   • having a light ship which is pitching heavily;
   • having sternway on the vessel;
   • having broken water over shoals;
   • entering an area where prevalent bad weather has left pockets of air bubbles over comparatively long periods.

Possible cures for the above include stopping or reducing the vessel's speed, and abrupt movement of the rudder either way to sweep away formed bubbles.

False echoes

1 *False bottom echo.* This may occur if the echo-sounder is incorrectly set in such a manner that in deep water a returning echo is received after the stylus has completed one revolution.

2 *Multiple echoes.* These are caused by the transmitted pulse being reflected several times between the sea bed and the water surface before its energy is dispersed. Such multiple reflection may cause multiple echoes to be recorded on the trace of the sounding machine. They can, however, be reduced in strength by decreasing the sensitivity control on the equipment.

3 *Double echo.* This type of echo is a double reflection of the transmitted pulse. It occurs when the energy is reflected from the sea bed and then reflected back from the surface of the water before being received by the transducer. A double echo is always weaker than the true echo, and can be expected to fade quickly with a reduction in the sensitivity of the equipment.

4 *Other causes.* Side echo may come from objects not directly under the keel of the vessel reflecting the sound energy, e.g. shoals of fish or concentrations of weed or kelp. There may be electrical faults or man-made noise in and around the hull. In addition, turbulence may be caused by the vessel herself, with or without interaction between the shore or other shipping.

5 *Deep scattering layer.* This is a level of several layers believed to consist of fish and plankton which will scatter and reflect sound energy. The layer has a tendency to move from as much as 450 m below the surface during daylight hours to very near the surface at night. It becomes more noticeable during the day when the cloud cover is sparse than when the sky is overcast.

The Gyro Compass

The Sperry, Anschutz and Brown are three well-known makes of conventional gyro compass, and one of them will often be found in most deep-sea ships. The compass provides a directional reference to true north and is unaffected by the earth's magnetism and that of the ship.

A brief description follows, but readers requiring more information on the theory and construction of the compass should consult more specialist literature.

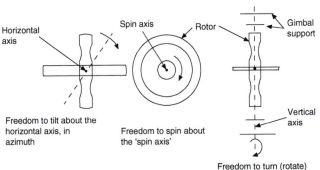

Figure 11.13 Degrees of freedom of rotor of free gyroscope.

Description and Application (the Three Degrees of Freedom)

The free gyroscope consists of a fast-spinning rotor, mounted to provide three degrees of freedom: freedom to spin; freedom to turn about a vertical axis; and freedom to tilt about a horizontal axis. As the rotor is constructed to have a high mass in relation to its dimensions, such a gyroscope displays two important properties:

1   Gyroscopic inertia (rigidity in space), whereby it will point in space to a fixed direction and thus follow the apparent motion of a fixed star.
2   Gyroscopic precession – the angular velocity acquired by the spin axis when torque is applied to the gyro in a plane perpendicular to the plane of the instrument.

These properties are made use of in the gyro compass, where a rotor spins at very high speed in nearly frictionless bearings, mounted with freedom to turn and tilt. The axle of the gyro is constrained by a system of weights producing a torque which causes the axle to precess (under the influence of gravity) in such a manner that it remains horizontal and in the meridian. The rate of precession of the gyro is equal to the rate at which the axle of the free gyroscope would appear to tilt and drift as the result of the earth's motion.

The properties of the free gyroscope

It is important that the mariner understands the properties of the free gyroscope in order to understand the gyro compass.

*   *Gyroscopic inertia.* This term is often referred to as 'rigidity in space', which better describes this property. It is the ability of the gyroscope to remain with its spin axis pointing in the same fixed direction in space regardless of how the gimbal support system may turn. The term may be illustrated by considering the direction of a star in space. If the free gyroscope is set spinning with the spin axis pointing to that star, then it will be seen that, as the earth turns, the spin axis will follow the apparent motion of that star.
*   *Precession.* If a torque is applied to the spin axis of the free gyroscope then it will be observed that the axis will turn in a direction at right angles to that applied torque. This movement by the rotor due to the applied force is known as precession.
*   *Torque.* Torque is defined as the moment of a couple or system of couples producing pure rotation. For a rotating body, torque is equal to the product of the moment of inertia and the angular acceleration.

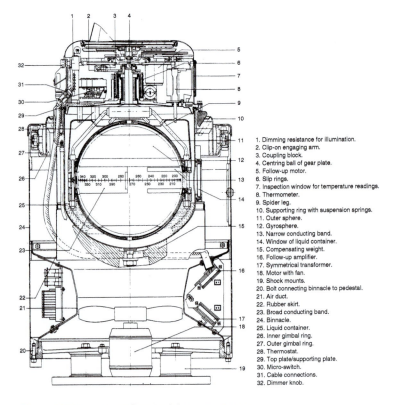

1. Dimming resistance for illumination.
2. Clip-on engaging arm.
3. Coupling block.
4. Centring ball of gear plate.
5. Follow-up motor.
6. Slip rings.
7. Inspection window for temperature readings.
8. Thermometer.
9. Spider leg.
10. Supporting ring with suspension springs.
11. Outer sphere.
12. Gyrosphere.
13. Narrow conducting band.
14. Window of liquid container.
15. Compensating weight.
16. Follow-up amplifier.
17. Symmetrical transformer.
18. Motor with fan.
19. Shock mounts.
20. Bolt connecting binnacle to pedestal.
21. Air duct.
22. Rubber skirt.
23. Broad conducting band.
24. Binnacle.
25. Liquid container.
26. Inner gimbal ring.
27. Outer gimbal ring.
28. Thermostat.
29. Top plate/supporting plate.
30. Micro-switch.
31. Cable connections.
32. Dimmer knob.

Figure 11.14 Anschutz Standard 4 gyro compass.

Changes in Ships' Compass Systems

The old sea dog of yesterday will tell today's hi-tech mariner that the compass was the most important instrument on board the vessel. This particular author would still agree with that old sea dog ... but the compass has changed somewhat from the sailing-ship days. The modern vessel would expect to be equipped with a 'digital gyro' – at least one and probably two, together with numerous repeaters to associated instruments. It could well have a transmitting magnetic compass (TMC) or 'flux gate' compass, which all tend to leave the somewhat old-fashioned magnetic compass in the shade.

NB. When a power failure occurs and the gyros go offline, that old-fashioned magnetic compass could well be reassuring to even the most modern of mariners; although a modern compass incorporates automatic changeover to emergency power and may not desynchronize because of a loss of power or supply interruption.

Modern Gyro Arrangements

The navigation requirements of modern shipping demand the highest standards of monitoring the ship's head. Several systems are commercially available which employ either a single or double unit operation. Twin gyro compasses satisfy redundancy needs for classification regulations.

Standard equipment will usually take the form of a gyroscopic compass unit being directly linked to a control element, operating on a direct current (DC) power supply of 24 volts. The controller will interact with the operator's input parameters to the gyro unit (and/or to a TMC unit, if fitted). A connection unit from the control element will allow numerous repeaters to function by way of: bridge wing repeaters, course printer, rate of turn (RoT) indicator, overhead full-view compass card (analogue or digital display), interfaced to radar/ECDIS operational units and with various alarm outputs.

The modern-day gyro compass will provide accurate heading with a built-in test facility. It will also have self-adjusting digital transmission to all repeater outputs, removing the need for manual adjustment to these units. Compass errors could be reduced by inclusion of a static and dynamic speed error correction, making the unit suitable for vessels operating at high speeds (up to 70 knots). A high rate of follow-up of 75° per second is not unusual. Standard construction would resist shock and vibration and the unit would reflect compliance with IEC 945 and IMO Resolution A 424 (XI).

TMC Unit

Where a TMC unit is engaged, values of variation and deviation can be set. Subsequently, all output users could be switched over to the magnetic compass sensor and the corrected magnetic compass course could be digitally displayed and be alarm-protected.

Fibre-Optic Gyroscope

This is a solid state, fully electronic gyro compass suitable for all marine applications, including high-speed craft. The unit, which is strapped down to the vessel, eliminates the use of a gimbal system and provides heading information, roll and pitch, together with rate of turn about three axes.

The fundamental principal of the fibre-optic gyro is the invariance of the speed of light. (The C. Plath model employs what is known as the 'Sagnac Effect'.) A fibre-optic coil is used as a sensitive rate sensor which is capable of measuring the speed of rotation of the earth. A combination of three such coils (gyroscopes) together with two electronic level sensors are able to determine the direction of true north.

The three RoT signals, with information from the level sensors, compute the direction of the earth's rotation from which geographical north can be derived. The unit has a short settling time

(30 minutes) and provides high dynamic accuracy without course and speed error.

The unit is without moving parts, very reliable and maintenance free. In the event of a malfunction a secondary redundancy unit can be exchanged. Various outputs inclusive of repeaters to radars and autopilots can be interfaced with displays in digital and analogue variants.

The main features include:

- solid-state technology with no moving parts
- short settling time
- meets IMO recommendations
- compact unit of low weight
- all repeaters are self-aligning
- built-in test function
- low power consumption
- high dynamic accuracy
- no maintenance
- allowance for second gyro compass input
- allowance for magnetic compass input
- automatic emergency power changeover
- LCD display
- alarm monitored.

Magnetic Compass

This is without doubt the most important of all instruments aboard even the most modern vessel, and it is probably the most reliable. Its origins go back as far as 2300 BC, but the Chinese development of the compass card dates to the fourteenth century, and the sophisticated instrument we know today became established with the advent of steel ships in the nineteenth century.

The compass bowl is supported in a binnacle usually constructed of wood, but increasingly many binnacles are being made in fibreglass (Plates 103–104). The natural resilience of fibreglass absorbs vibration from machinery and requires little maintenance.

The main function of the binnacle is to provide support and protection for the compass bowl. However, the structure also provides the ideal support for the standard correction elements, namely the quadrantal correctors, the flinders bar and the fore and aft and athwartships permanent magnets. Heeling error magnets are placed in a 'bucket' arrangement on the centre line of the binnacle, directly under the central position of the compass bowl (see Figure 11.15). The effect of heeling error magnets can be increased or decreased by respective adjustment of the chain raising or lowering the bucket.

Liquid Magnetic Compass

This compass is illustrated in Figure 11.17.

NB. The 'Sagnac Effect' is based on two light waves travelling in opposite directions around a circular light path, resulting in a phase shift. This phase difference is directly calibrated to rotation rate.

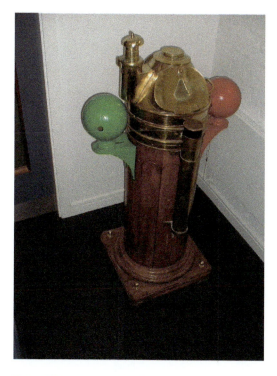

Plate 103a Binnacle of hardwood finish.

Plate 103b The forward face of a hardwood finish binnacle with 'helmet' in place over the compass bowl position.

Plate 104a Modern binnacle manufactured in glass reinforced plastic (GRP) seen *in situ* on the monkey island. A compass reflector unit would pass through the deck into the wheelhouse. The voyage data recorder (VDR) is also sighted on the deck area, providing the ship's black box technology.

Plate 104b Modern binnacle shown without 'helmet' in position but with azimuth circle in position on the top of the compass bowl.

Figure 11.15 Magnetic compass. The chain adjustment for the heeling error bucket can be reached via the panel under the compass bowl into the light chamber.

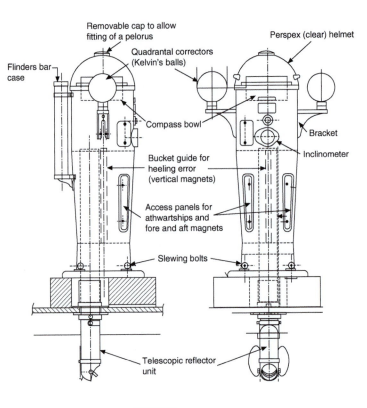

Removable cap to allow fitting of a pelorus

Quadrantal correctors (Kelvin's balls)

Perspex (clear) helmet

Flinders bar case

Compass bowl

Bracket

Inclinometer

Bucket guide for heeling error (vertical magnets)

Access panels for athwartships and fore and aft magnets

Slewing bolts

Telescopic reflector unit

Figure 11.16 Siting of magnetic compass's correctors.

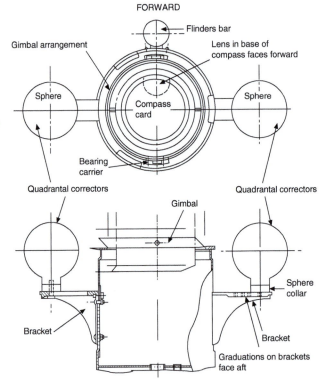

FORWARD

Flinders bar

Gimbal arrangement

Lens in base of compass faces forward

Sphere

Sphere

Compass card

Bearing carrier

Quadrantal correctors

Quadrantal correctors

Gimbal

Sphere collar

Bracket

Bracket

Graduations on brackets face aft

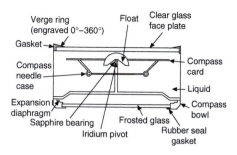

Figure 11.17 Liquid magnetic compass.

Compass Bowl

Manufactured in high-quality, non-magnetic brass, this has a clear glass face and a frosted glass base to diffuse the underside lighting. The older designs were fitted with chambers to allow for the expansion and contraction of the fluid, but the modern compass is fitted with a corrugated diaphragm (elastic membrane) at the base of the bowl for the same purpose.

Most compass manufacturers include a graduated verge ring around the clear glass face plate. Both the face plate and the frosted glass base are secured via rubber gaskets to prevent leaks from the bowl. Special paints, used both internally and externally, are 'stove baked' onto the compass body so that they will not cause discolouration of the fluid, and they last for many years. Many paints are magnetic, especially blacks, and their use on and around the binnacle should be limited. Brackets are fixed to the outside to connect to the gimbal system and a support is secured on the inside of the bowl to accommodate the pivot. The 'lubber line' is marked on the inside of the bowl in alignment with the ship's fore and aft line.

Compass Card

Usually made of glass melamine or mica, this must be the correct size for the bowl to which it is fitted. Should another sized card be used, then fluid disturbances could make the compass unsteady. The diameter of cards varies, but a 10 in. (254 mm) compass bowl one could expect a 7.5 in. (191 mm) diameter card, unless a specially reduced card is provided.

In the course of manufacture the card is corrected for its magnetic moment to limit its speed of movement, then checked for friction. Cards are normally screen-printed to indicate three-figure notation in degrees, and have the cardinal and half-cardinal points identified. Intermediate and by-points are indicated but not individually lettered.

A single circular magnet is secured beneath the card to produce the directive force required of the compass. The magnet system may consist of two parallel circular magnets disposed on either side of the central line.

Pivot Point

This is made of polished iridium, a member of the platinum family, and care should be taken in manufacture that the iridium element is neither too hard – or it may shatter – nor too soft – or it may collapse. The bearing point is an industrial jewel, usually a sapphire, fitted into the base of a float set into the centre of the compass card.

The pivot point effectively lowers the centre of gravity (C of G) of the card below the point of suspension. This arrangement is achieved by the inclusion of the dome-shaped float in the centre of the card. The magnetic system may be in the form shown, with a needle arrangement slung beneath the card, or a single circular magnet may be encased below the float. The casing is generally made of brass to prevent rusting and loss of magnetic effect.

One of the main advantages of the liquid compass over the dry card compass is that it is not as sensitive. Consequently, it makes an excellent steering compass. Oscillations of the card are greatly reduced by the dense liquid within the bowl and any induced movement is practically eliminated.

The term 'dead beat' applied to the liquid compass means slow moving, with a steady card. Undesirable oscillations of the card are kept to a minimum by the liquid.

Liquid

The older style of liquid magnetic compass contained a mixture of two parts distilled water to one part ethyl alcohol, providing a fluid with low viscosity and a small coefficient of expansion. The idea behind the mix was that the alcohol would reduce the freezing point of the mixture in cold climates and the water would reduce evaporation in the warm tropical climates. The modern liquid compass employs an oily liquid derived from 'Bayol', which not only provides additional flotation for the card but also lubricates the pivot and reduces motion on the card.

In the manufacture of modern compasses the bowl, once assembled, is passed through a vacuum before being filled. The actual filling is carried out at an ambient temperature, and any final air bubbles are removed by manual joggling of the instrument (see Plate 105).

Gimballing

This means slinging the compass in such a manner that it remains horizontal at all times, even in a heavy sea. Keeping the compass card horizontal at all times may be achieved in two ways:

1   Raising the point of support of the compass card above the C of G of the card.
2   Maintaining the compass bowl in the horizontal position by two axes/gimbal rings, one in the fore–aft line and the other in the athwartships line.

Plate 105 Liquid compass bowl in gimbal arrangement

It is usual to have the fore and aft axis secured to the outer gimbal ring rather than the inner ring, as this reduces the possibility of the lubber's line mark travelling to port or starboard when the vessel is rolling heavily.

Order of Placing Correctors

Mariners are advised that to attempt to cover compass adjustment within the bounds of this text would be impractical. The reduction of deviation affecting the magnetic compass is complex and should be studied in depth. Use of correctors to compensate for permanent and induced magnetic effects must be carried out in a correct and orderly procedure.

A method of adjustment within the Mercantile Marine employs coefficients A, B, C, D, E and J for heeling error. These coefficients are types of deviation which vary in accordance with some ratio of the compass course. For example, coefficient B is a deviation which varies as the sine of the compass course.

A thorough knowledge of the use of the coefficients together with a sound background of general magnetism must be considered essential to any mariner attempting the adjustment of the marine magnetic compass.

Marine students seeking further information should refer to: *The Ship's Compass* by G.A.A. Grant and J. Klinkert (Routledge and Kegan Paul).

Flinders bar

This bar usually comes in lengths of 12 inches (30.48 cm), 6 inches (15.24 cm), 3 inches (7.62 cm), and 1.5 inches (3. cm); all are of 3 inches (7.62 cm) diameter, with similar sized wood blocks to raise

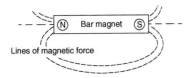

Figure 11.18 Lines of magnetic force.

the level of the bar and bring the pole of the bar level with the magnets of the card. The pole is assumed to be one-twelfth the length from the end. This is explained by the fact that the pole of a bar magnet is never at the very end of the bar (Figure 11.18).

Spheres

Employed in various sizes from 2 inches to 10 inches (5.08–25.4 cm) diameter, they may be of a solid or hollow construction. They are placed with their centres on a level with the magnets of the card, but not closer than 1.25 times the length of the longest needle in the card.

Heeling error magnets

Hard iron magnets, 9 inches in length (22.86 cm) by $^3/_8$ inch (0.93 cm) diameter, these compensate for heeling error due to field 'R' and vertical soft iron. They also induce magnetism into the flinders bar and spheres, which helps the heeling error correction.

Horizontal magnets

These are 8 inches in length (19.32 cm) and either $^3/_8$ or $\frac{3}{16}$ inch (0.93 lines of magnetic force or 0.46 cm) in diameter. They compensate for the effects of the fore and aft and athwartships components of semi-permanent magnetism.

Telescopic Reflector Unit

The majority of compass manufacturers will now supply ships' binnacles with or without telescopic reflector units (Plate 106), depending on the requirements of the shipowner. The reflector unit was an acceptable advance within the industry, since it obviated the need and the cost of providing a steering compass.

The idea of achieving a through-deck repeater from the standard compass on the 'monkey island' is based on the development of the submarine periscope. The unit is fitted under the forepart of the compass in order to reflect the lubber's line and the foremost section of the compass card within the standard compass. The reflector unit is not centrally positioned, as the operation of the bucket containing the heeling error magnets would obstruct its use.

A typical reflector unit would be manufactured in PVC, with brass fittings. It usually incorporates moisture-free, sealed-in mirrors, which are adjustable at eye level inside the wheelhouse, together with a detachable anti-glare shield.

Dry Card Magnetic Compass

This very sensitive instrument (Figure 11.19) has now been largely superseded by the more popular liquid magnetic compass. It is a

Plate 106 Compass reflector unit seen *in situ*, under the standard compass (not shown), suspended from the deckhead inside the wheelhouse. Rudder indicators are also seen at the deckhead either side of the reflector unit.

very light card constructed of 'rice paper', this being segmented to allow an even flow of air above and below the card to obtain an even balance of pressure around the whole. The weight of the card is usually around 15 grams, but it is given some rigidity by an aluminium ring beneath it which also supports the magnetic needle arrangement.

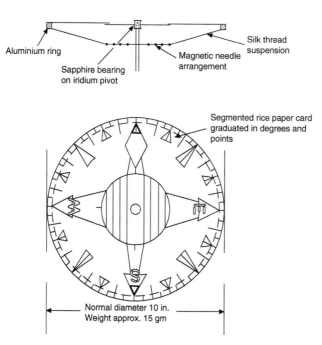

Figure 11.19 Dry card magnetic compass.

Azimuth Mirror

This instrument (Figure 11.20 and Plate 107) fits on the compass bowl of either the standard magnetic or the gyroscopic repeaters, and allows the navigator to obtain accurate compass bearings of either terrestrial or celestial objects. The instrument has a stand, either triangular or round and usually manufactured in anodized aluminium and brass, designed to grip the verge of the compass bowl and provide a firm support for a reflecting prism. Most designs incorporate neutral tinted shades to allow observation of the sun, together with a shadow pin to provide a reverse bearing of the sun. Some more modern styles include two shadow pins for transit sighting.

For bearings to be taken accurately, the axis of the prism must at all times remain horizontal. To this end, the prism is held in its frame by two small screws, which may be adjusted if the prism loses its alignment. For operation of the azimuth mirror, see Figure 11.21.

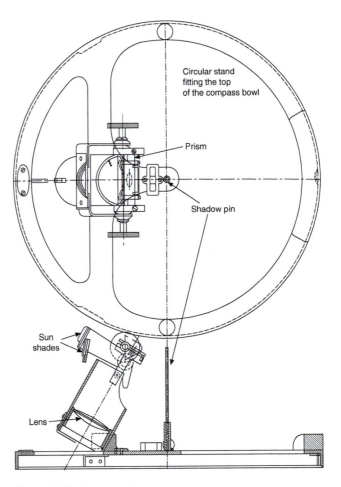

Figure 11.20 Azimuth mirror.

Plate 107 Azimuth mirror.

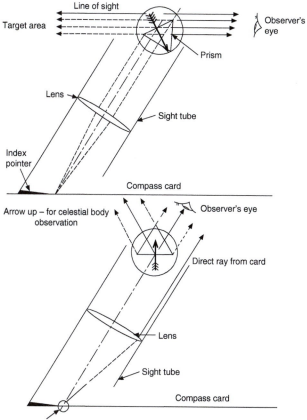

Arrow down – for surface/land observations

Line of sight

Target area

Observer's eye

Prism

Lens

Sight tube

Index pointer

Compass card

Arrow up – for celestial body observation

Observer's eye

Direct ray from card

Lens

Sight tube

Compass card

Image of sun or star

Figure 11.21 Operation of azimuth mirror.

Should the navigator at any time have cause to suspect the accuracy of the azimuth mirror, he or she may check by taking two bearings of the same object, one with the arrow in the up position and the second with the arrow in the down position. The results of both bearings – arrow up, arrow down – should be the same. Should a discrepancy exist between the results, then the prism should be adjusted.

Pelorus

The pelorus, which enables the navigator to obtain bearings of shore objects, is an alternative to the azimuth mirror. It is particularly useful when the line of sight of the azimuth mirror on the standard compass is obscured. Being a portable instrument, it can be transferred from bridge wing to bridge wing, so that the line of sight need never be impeded by such obstructions as the funnel. See Figure 11.22 and Plate 108.

In operation, the graduated bearing plate can be manually turned so as to be aligned with the ship's head, then clamped into position; and by having an observer watching the ship's head and noting when the vessel is exactly 'on' course, the navigator can observe the true bearing by means of the sight vanes aligned with the shore object. Relative bearings may also be obtained by having the lubber line indicator set at 000°.

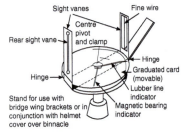

Figure 11.22 Pelorus.

Plate 108a Modern pelorus azimath bearing circle.

Automatic Identification System

In 1997 the draft performance standards for a universal shipborne automatic identification system (AIS) were finalized. Working parties were subsequently set up by IALA and the ITU to establish the technical characteristics for an operational system, and these were

Plate 108b Bridge wing control console showing a prominent gyro repeater which can be fitted with a pelorus bearing vane.

adopted by the IMO Maritime Safety Committee at its 69th session in 1998. Proposed amendments to Chapter V of the SOLAS convention were made in 1999 and a schedule for fitting AIS to new and existing ships was set.

The current identification of target ships at sea from radar is a problem. It is not possible to establish which echo is which ship or determine with complete certainty whether a collision situation exists or not. Where VTS is involved, a similar lack of target information makes the task of monitoring traffic that much more difficult without positive information.

One of the prime functions of the AIS operation will permit positive identification of targets and allow continuous tracking of vessels over lengthy voyages. If this is considered against continuous radar coverage an increased number of tracking stations would have to be established and this would be extremely costly to set up, operate and maintain.

Transponder Information

Each transponder will:

- provide the ship's identity, type, position, speed and navigational status;
- receive similar data from other ships;
- monitor and track ships;
- exchange data with shore bases.

AIS Unit Components

Each transponder is expected to contain four modular units:

1 GPS
2 VHF Tx/Rx

3   controller
4   power unit.

The ITU specify that the receiver is dual frequency.

The units are expected to function automatically with little or no human operator intervention. The controller assembles all positional data and message format for transmission and reception, respectively, over the various modes of operation.

Message Format

The IMO has specified three standard message types:

1   *Static message*. This contains the ship's IMO number, call sign and name, together with dimensions of ship's length and beam, type of ship and the location of the position-fixing antenna aboard the vessel.
2   *Dynamic message*. This contains the ship's position with indicated accuracy and integrity status, time of report, course over the ground, speed over the ground, heading, rate of turn, navigational status and optionally angles of heel, pitch and roll.
3   *Voyage-related messages*. These contain ship's draught, types of hazardous cargo carried, destination and ETA (at Master's discretion) and optionally the route plan in way-point format.

Operational AIS

Once operational, the AIS will be compatible with radar, ARPA and ECDIS for the multiple areas for which the system is designed, namely:

* *Identification of targets*. Each message will contain the MMSI number to effect identification from a database. It is probable that whichever on-board system is in use, be it radar, ARPA or ECDIS, an information window will open to display a complete ship's data profile.
  Targets acquired in a similar manner from a VTS shore base would detail target information in much the same way to a VTS operator.
* *Tracking*. Targets will have the capability of being tracked themselves by other ships or a VTS operation, and at the same time have the ability to track other ships themselves. This would effectively be achieved by frequently received AIS position reports from the intended target. The ITU has specified the update intervals for dynamic reports as noted in Tables 11.1 and 11.2.

Additional navigation/motion data

The AIS position reports already include notation for RoT and 'ship's heading'. Such information is relevant for anti-collision because it allows the determination of 'aspect', which to date has

NB. At the time of writing it is not yet known whether a separate GPS system will be a requirement or whether the ship's existing positioning system will be sufficient.

Table 11.1 AIS update intervals of message

Message type	Interval time/period
Static information	Every six minutes and on request
Dynamic information	Depends on speed/course alteration (see Table 11.2)
Voyage-related information	Every six minutes, when data has been amended and on request
Safety-related messages	As required

Table 11.2 AIS dynamic information – update intervals

Ship dynamics	Reporting interval
Ship at anchor	3 minutes
Ship at 0–14 knots	12 seconds
Ship at 0–14 knots and changing course	4 seconds
Ship at 14–23 knots	6 seconds
Ship at 14–23 knots and changing course	2 seconds
Ship at over 23 knots	3 seconds
Ship at over 23 knots and changing course	2 seconds

not been acquired accurately or confirmed without a visual contact or manual plot. This information could also be enhanced by a key code to specify the navigational status of the vessel, whether she is at anchor, underway, NUC or restricted.

Such detailed information on any target could determine the responsibility between 'give way vessels' and 'stand on' vessels when in a 'rules of the road' collision scenario.

Traffic movement-related messages

The benefits of monitoring traffic movement and providing continuous updated information to vessels in transit through narrow or congested waters is of potential benefit to pilots and ship's Masters alike. Canals, rivers, narrow fairways or shoal regions could all have tight monitoring systems linked directly to the VTS work station. All traffic-related signals and text messages can be displayed on board and in the VTS control environment without causing disruption to the normal functions of the AIS.

Additional uses

Interested parties in addition to VTS operations, such as coastguards, SAR personnel, ship reporting systems and pilotage services all have everything to gain from an effective operational AIS.

Summary

The AIS will at last identify the radar echo to the mariner. Since radar was developed over 50 years ago, and electronic detection of targets became possible, all that has been available is the knowledge of the changing position of the target. This mandatory system will shortly provide an essential profile on respective targets, allowing all participants access to all target information and anti-collision analysis. What it will not do is change actual weather conditions or install transponders on the smaller fishing boat or yacht under 300 gross tonnes. Neither will AIS replace radar and remove the need for good seamanship.

Its benefits will clearly be readily acceptable to the next generation, geared to a hi-tech future. However, the need to keep pace with innovation and train with new equipment will become the essential by-product for our personnel, both ashore and afloat.

Voyage Data Recorders

It is now a requirement that all passenger ships and all vessels over 3,000 gross tonnes, which are built after July 2002, be fitted with a voyage data recording (VDR) unit.

The Maritime Safety Committee (MSC) has also agreed that VDRs are to be fitted to all existing ro-ro passenger vessels and high-speed craft already in operation. The new regulations agreed by IMO in 2004 have introduced S-VDR (simplified VDR), which will be allowed as an alternative for ships over 3,000 grt from July 2006.

The principle of marine 'black boxes' has come about because of the transport relationship with the aviation industry, which has had black box technology on all passenger aircraft for many years. The monitoring of all principal elements within the mode of transport has shown itself indispensable in resolving aircraft accidents and subsequently improving long-term industrial safety. The question would now tend to be: why has the marine industry taken so long to introduce what many would now consider as a basic monitoring device?

The revised Chapter V of SOLAS will make carriage mandatory for certain types of vessels. IMO has made recommendations on the data that VDRs are expected to record and include the following:

- date and time
- ship's position and speed
- course/heading
- bridge audio – one or more microphones situated on the navigation bridge to record conversations near the conning position and at relevant operational stations like radar, chart tables, communication consoles, etc.
- main alarms and P/A systems
- engine orders and responses
- rudder orders and responses
- echo-sounder recordings
- status of watertight and fire doors

- status of hull openings
- acceleration and hull stress levels (only required where a vessel is fitted with response monitoring equipment).

The information should be stored 24 hours per day over a seven-day week period. It should be contained in a crash-proof box, painted orange and fitted with an acoustic device to aid recovery after an accident. The system will be a fully automatic memory unit which will be 'tamper free' and always watching, even when the vessel is tied up alongside. VDRs are expected to provide continuous operation for at least two hours following a power failure and are also alarm protected in the event of malfunction of any of the VDR elements.

Hydrometer

This instrument (Figure 11.23) is used by Ships' Officers to determine the relative density of fluids, including the dock water in which a vessel is lying while loading. It is necessary to obtain the amount that the loadline mark may be submerged when in dock water of less than 1,025 kg/m³, and then, by use of the dock water allowance formula, define the draught limits so that once the vessel is at sea, she is correctly loaded to her loadline marks.

Ships' hydrometers are usually made in polished steel or brass, but they may also be made of glass. A bulb weighted with lead shot or mercury acts to keep the graduated stem in the vertical position. The operational principle of the hydrometer is based on the laws of flotation of Archimedes. The mass of the stem, float chamber and shot is a constant. If the hydrometer floats in a fluid, then the weight of fluid displaced will equal the volume of hydrometer times the density of fluid. The density can be determined by practical use of the instrument.

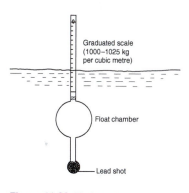

Figure 11.23 Hydrometer.

Operation

The following procedure is recommended for determining the density of the dock water.

Use a clean two-gallon bucket with a light line attached and obtain a sample of the dock water from about the midships point of the vessel. The bucket should be allowed to sink to the mean draught level and remain there for a short period to ensure that the temperature of the bucket will agree with that of the water at the mean level. The sample of water should not be taken from the upper surface, as this will probably be polluted in some way and will also be at a different temperature to the water around the submerged hull fluids. By the same reasoning the bucket should be cast clear of overboard discharges.

Float the hydrometer in the bucket of water once the liquid is still. Give the instrument a slight twist to break surface tension and allow it to settle. Read off the scale the level of the surface at which the instrument is floating.

The dock water allowance (DWA) formula is now applied. When a vessel is loading in a dock which is not a salt-water dock, the ship may submerge her appropriate loadline by an amount equal to that value obtained by the dock water allowance formula. This statement is only correct for vessels proceeding into salt water of 1025 kg/m³, and should a vessel be entering water of a different density this would have to be calculated accordingly.

$$\text{The amount a vessel may submerge her loadline mark}$$

$$\text{DWA} = \frac{1025 - \text{Density shown on hydrometer}}{25} \times \text{FWA}$$

where FWA represents fresh water allowance (the amount by which a ship may submerge her seasonal loadline when loading in fresh water of density 1000 kg/m³).

Hygrometer

This is an instrument for measuring relative humidity. Marine hygrometers are normally used in conjunction with a 'Stevenson's screen', which allows the air to circulate freely inside but protects the hygrometer from the direct force of the wind and the chill factor.

The hygrometer consists of two thermometers secured side by side. The mercury bulb of one is kept dry; this is known as the 'dry bulb thermometer'. The other thermometer has a muslin wick covering the mercury bulb, and the end of the wick is dipped into a small distilled water reservoir; this is known as the 'wet bulb thermometer'. The whole is often referred to as the Mason's hygrometer or just the 'wet and dry bulb thermometer'.

The hygrometer is used in conjunction with calibrated tables to obtain not only the relative humidity, but also the dewpoint. These values are indicated by the difference between the wet and dry bulb temperatures.

The process of evaporation requires heat, and this heat is drawn from the wet bulb thermometer. Evaporation of the distilled water in the reservoir and more directly from the wick takes place, leaving the wet bulb thermometer generally at a temperature below that of the dry bulb. Should the air be saturated at the time of observation, then the temperatures indicated by both wet and dry bulb thermometers will in fact be the same.

The readings are useful to Ships' Officers in predicting the condensation of moisture in the atmosphere. This fact is particularly relevant to vessels whose cargoes would be at risk from 'cargo sweat' owing to improper ventilation. It should also be noted that

Plate 109 Stevenson screen seen open to reveal the wet and dry bulb thermometers.

high levels of moisture in the air may indicate the approach of a tropical storm or depression.

Precision Aneroid Barometer

Precision aneroid barometers (Figure 11.24) have been widely used at sea since about 1963. Their compact, robust construction has proved ideal for the marine environment, especially when compared with the mercury barometers. They have proved easy to transport and are less liable to damage than mercury barometers, as well as having a simple method of reading from a digital display.

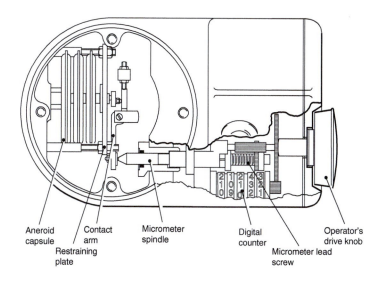

Aneroid capsule
Contact arm
Restraining plate
Micrometer spindle
Digital counter
Micrometer lead screw
Operator's drive knob

Figure 11.24 Precision aneroid barometer.

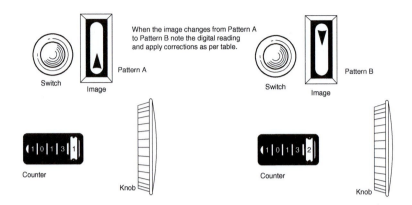

When the image changes from Pattern A to Pattern B note the digital reading and apply corrections as per table.

Figure 11.25 Operating procedures for precision aneroid barometer.

Plate 110 Conventional aneroid barometer (source: Shutterstock).

Plate 111 Precision aneroid barometer.

Plate 112 Marine baragraph (source: Shutterstock).

Displacement of the aneroid capsule is measured by means of a micrometer spindle which is connected via gears to the digital read-out counter. A contact arm positioned between the spindle and the capsule acts as a simple lever with a ratio of 3:1 for adjustment purposes and as a safeguard against damage. Barometric pressure is indicated on the digital counter after adjustment of the operator's drive knob. This adjustment causes the micrometer spindle to break contact with the contact arm. Contact between the arm and the spindle is detected by an electronic circuit powered by dry batteries having a service life of three to nine months, depending on use (see Figures 11.24 and 11.25).

The conventional and precision aneroid barometer are shown respectively in Plates 110 and 111, and the marine barograph in Plate 112.

Calibration correction

These corrections must always be applied. When the sign is +, add the correction to the MSL reading. When the sign is −, it should be subtracted from the MSL reading.

Table 11.3 Correction of millibar barometers to mean sea level (these corrections are to be added to the barometer readings)

Height in metres	Air temperature (dry-bulb in screen) °C										Heights in	
	−15	−10	−5	0	5	10	15	20	25	30	ft	ins
2	0.3	0.3	0.3	0.2	0.2	0.2	0.2	0.2	0.2	0.2	6	7
4	0.5	0.5	0.5	0.5	0.5	0.5	0.5	0.5	0.5	0.5	13	1
6	0.8	0.8	0.8	0.7	0.7	0.7	0.7	0.7	0.7	0.7	19	8
8	1.1	1.0	1.0	1.0	1.0	1.0	0.9	0.9	0.9	0.9	26	3
10	1.3	1.3	1.3	1.2	1.2	1.2	1.2	1.2	1.1	1.1	32	10
12	1.6	1.6	1.5	1.5	1.5	1.4	1.4	1.4	1.4	1.4	39	4
14	1.9	1.8	1.8	1.7	1.7	1.7	1.7	1.6	1.6	1.6	45	11
16	2.1	2.1	2.0	2.0	2.0	1.9	1.9	1.9	1.8	1.8	52	6
18	2.4	2.3	2.3	2.3	2.2	2.2	2.1	2.1	2.1	2.0	59	1
20	2.6	2.6	2.5	2.5	2.5	2.4	2.4	2.3	2.3	2.3	65	7
22	2.9	2.9	2.8	2.8	2.7	2.7	2.6	2.6	2.5	2.5	72	2
24	3.2	3.1	3.1	3.0	2.9	2.9	2.8	2.8	2.8	2.7	78	9
26	3.4	3.4	3.3	3.3	3.2	3.1	3.1	3.0	3.0	2.9	85	4
28	3.7	3.6	3.6	3.5	3.4	3.4	3.3	3.3	3.2	3.2	91	10
30	4.0	3.9	3.8	3.8	3.7	3.6	3.6	3.5	3.4	3.4	98	5
32	4.2	4.2	4.1	4.0	3.9	3.9	3.8	3.7	3.7	3.6	105	0
34	4.5	4.4	4.3	4.3	4.2	4.1	4.0	4.0	3.9	3.8	111	7
36	4.8	4.7	4.6	4.5	4.4	4.3	4.3	4.2	4.1	4.1	118	1
38	5.0	4.9	4.8	4.8	4.7	4.6	4.5	4.4	4.4	4.3	124	8

Table 11.3 (Continued)

Height in metres	Air temperature (dry-bulb in screen) °C										Heights in	
	−15	−10	−5	0	5	10	15	20	25	30	ft	ins
42	5.6	5.5	5.4	5.3	5.2	5.1	5.0	4.9	4.8	4.7	137	10
44	5.8	5.7	5.6	5.5	5.4	5.3	5.2	5.1	5.0	5.0	144	4
46	6.1	6.0	5.9	5.8	5.7	5.6	5.5	5.4	5.3	5.2	150	11
48	6.4	6.2	6.1	6.0	5.9	5.8	5.7	5.6	5.5	5.4	157	6
50	6.6	6.5	6.4	6.3	6.2	6.0	5.9	5.8	5.7	5.6	164	0

Serial No.		Serial No.		Serial No.	
Reading	Correction	Reading	Correction	Reading	Correction
1050	......	1050	......	1050	......
1020	......	1020	......	1020	......
1000	......	1000	......	1000	......
980	......	980	......	980	......
950	......	950	......	950	......
920	......	920	......	920	......

12
Meteorology

Introduction

The mariner that neglects the weather factor will undoubtedly pay the price, sooner or later. Whether in ship-handling or in the protection of cargo, meteorology has always been an essential consideration to the prosecution of the voyage. In the days of sail the wind was all-important. Favourable winds and currents still remain basic elements of passage planning and the successful execution of the ship's endeavours.

Bearing the seafarers' needs in mind, no seamanship text would be complete without an element of meteorology. The extremes are, of course, in the avoidance of tropical revolving storms (TRSs) and ice navigation, both of which are topics of serious concern to ships' crews.

To know the subject is to avoid heavy weather and all that goes with it. Heavy rolling or heavy pitching is, at the very least, uncomfortable. If and when it causes damage to ship or cargo then it should have been avoided, if at all possible. Such avoidance cannot take place unless seafarers know their enemy and respect it across the waters of the planet.

Meteorological Terms

Anabatic

The upward movement of air due to convection is anabatic; an anabatic wind ascends a hillside or blows up a valley.

Anemometer

This is an instrument used to register and determine the velocity of the wind.

Aneroid barometer

This is a dry mechanical instrument for measuring changes of pressure in the atmosphere.

Plate 113 Precision aneroid barometer and barograph shelf mounted against a bulkhead.

Anticyclone

An area of high pressure with clockwise circulation of air in the northern hemisphere, and anticlockwise in the southern hemisphere, defines an anticyclone. Winds are generally light to moderate.

Aurora

This shimmering area of light is caused by an electrical discharge in the atmosphere over high northern and southern latitudes. The Northern Lights are called the Aurora Borealis, and the Southern Lights the Aurora Australis.

Backing

This is a change in the direction of the wind in an anticlockwise sense, e.g. from north through west to south and then east; this is the opposite of veering, which occurs when the wind direction changes in a clockwise direction.

Bar

This is an international unit of atmospheric pressure; at sea level a bar is equal to the pressure of a column of mercury 29.53 in. high at a temperature of 32°F at latitude 45°.

Barograph

This instrument provides a permanent record, in graphical form, of the continuous changes in atmospheric pressure; it may be described as a continuous recording aneroid barometer.

Barometer

This is an instrument for measuring barometric pressure. Corrections are made to the readings for latitude, temperature and height above sea level. The instrument will also carry an index error, which may be found on its certificate.

Barometric tendency

This is the change in barometric pressure indicated during the three hours preceding observation. It shows the rise or fall of atmospheric pressure.

Cold front

Cold air travelling over the earth's surface can sometimes lodge itself under warmer air. A sloping separation between the layers of warm and cold air is defined as a 'cold front'.

Condensation

This is the process of converting a vapour into liquid.

Conduction

This is heat transfer through a body from places of higher temperature to those having a lower temperature.

Convection

This is a process of heat transfer carried out in a fluid, when the heat is carried by the motion of the hot fluid itself.

Corona

A corona is faint blue rings about the moon, brought about by diffraction of and interference with light by water droplets in the atmosphere.

Cyclone

Often referred to as a depression or just as simply a low, this is an area of low pressure about which the air moves in an anticlockwise direction in the northern hemisphere and vice versa in the southern hemisphere. It is also the term given to violent TRSs.

Dew

This is an accumulation of water droplets on objects cooler than the temperature of the air.

Dewpoint

This is the temperature to which air can be cooled without condensation taking place.

Diffraction

Light is diffracted when light waves pass through narrow apertures or between bodies forming narrow apertures. *See* 'corona'.

Doldrums

This area of calm, variable winds lies between the NE and SE Trades. Occasional squalls and torrential rain may be encountered within the area.

Etesian

This is a northerly wind encountered among the Greek islands, the Etesian is of katabatic origin. *See* 'Katabatic wind'.

Evaporation

This is the process by which water or ice is converted into an aqueous vapour.

Fog

Fog is defined as visible water vapour at the earth's surface. Mist may be similarly defined, except that mist tends not to impede navigation to the same degree as fog. A state of fog exists when visibility is less than 1000 yd (914.4 m).

Gale

A strong wind in excess of 40 knots and represented by forces 8 and 9 on the Beaufort Wind Scale constitutes a gale. Cone-shaped signals exhibited by coastal stations give warning of the approach and direction of a gale.

Gulf Stream

This warm-water current flows from the Gulf of Mexico up the east coast of the United States and then moves in an easterly direction, as the North Atlantic drift current, towards the European continent.

Hail

A hard ice pellet which generally falls from cumulonimbus cloud, hail is usually associated with thunderstorms. Hailstones vary in size.

They are built up by concentric layers of ice forming on top of each other. One theory is that the nucleus is a particle of dust which attracts moisture, and the moisture subsequently freezes.

Halo

This is a circle of light caused by refraction, which forms about the sun or moon.

Haze

A reduction of visibility caused by dust or smoke in the atmosphere, limiting the range to about 1.25 miles (2 km), haze is not to be confused with mist, which is brought about by condensed water particles.

Horse latitudes

The area of calm and light variable winds between the 30th and 40th parallels. In general, they lie between the trade winds and the prevailing westerly winds.

Humidity

This is the amount of moisture in the air.

Hurricane

An exceptionally strong wind, measuring force 12 on the Beaufort Wind Scale. A tropical cyclone, not uncommon in the North Atlantic and the Caribbean Sea, is a hurricane. Its counterpart in the Indian Ocean and the Far East is known as a typhoon, from the Chinese word *Tai-fung*.

Hydrometer

This is an instrument for obtaining the relative density of a fluid, it is used extensively to obtain the dock water allowance and test boiler water.

Hygrometer

This is an instrument for obtaining the relative humidity of the air; it comprises two thermometers, one a wet bulb and the other a dry bulb. The thermometers are usually contained in a box known as a Stevenson's screen, which allows the passage of air currents.

Ice

Ice is frozen water. For types of ice, see 'Ice terminology', p. 412.

Isobars

Isobars are lines drawn on a weather map to connect areas of the same barometric pressure.

Isotherms

These are lines drawn on a weather map to connect areas of the same temperature; they may also be used to express the sea or air temperatures.

Katabatic wind

This is the name given to a wind produced by a downward current, which is especially prevalent in high coastal areas. The wind 'runs' down the hillside, its velocity increasing with gravity, and it can expect to meet the sea often with great violence.

Land and sea breezes

Evening temperatures over land and sea tend to be reasonably equal, but at night the temperature over the land falls and the pressure increases, the state of equilibrium is upset and a current of air moves towards the sea. The opposite phenomenon takes place in the morning.

Lightning

A discharge of electricity between two clouds, or between a cloud and the earth.

Mirage

This is abnormal refraction and reflection of light rays may cause a false horizon in the lower layers of the atmosphere because of the differing densities of the layers. When a mirage is seen over water, distant ships may appear, sometimes upside down.

Monsoon

This seasonal wind blows over much of Southeast Asia, sometimes from the land and sometimes from the sea. In fact, it may be compared to the definition for land and sea breezes above, except that the occurrence is seasonal rather than daily, and over a much larger area.

Phosphorescence

This luminous effect on the surface of the water, showing bluish points of light, has never been explained satisfactorily.

Polar front

This is the line of demarcation between a cold polar air mass and warmer air from more temperate latitudes.

Precipitation

This is the conversion of water vapour into visible rain, snow, sleet, hail, dew etc.

Radiation

This is the process of heat being transferred by wave energy.

Rain

This comprises water droplets, formed by the condensation of water vapour. The maximum size of each droplet will not exceed 5.5 mm, and its maximum velocity, depending on size, when falling will not exceed 17.9 mph (29 kph).

Rainbow

This is an arc formed by refracted and reflected light from water droplets in the atmosphere; it can only be seen when the observer is looking into a rain cloud or shower of rain with the sun at his back.

Recurvature of storm

Often referred to as the vertex of the path of the storm, the recurvature represents that point which is as far west as the centre of the tropical storm will reach. Also known as the 'cod'.

Refraction

This is the bending of a ray of light when passing from one medium to another of different density.

Ridge

This term may be applied to a 'ridge of high pressure', indicating a bulge or extension of a high-pressure area between two lows.

Sleet

This is a mixture of rain and snow, or partially melted snow becomes sleet.

Snow

Light ice crystals fall as snow.

Squall

This is a sudden change in wind velocity, often increasing considerably over a short period of time, with little warning. It can consequently cause serious damage, especially to small craft.

Stratosphere

This is the region of the atmosphere above the troposphere in which the lapse rate is about zero and in which the phenomena comprising 'weather' do not occur. The stratosphere begins at a height of some 11 miles at the equator.

Temperature

A condition which determines heat transfer from a hot to a colder body. Temperature may be expressed in degrees Fahrenheit (°F), Celsius (°C), Kelvin (°K) or Absolute (°A).

Thunder

A violent report caused by the expansion of air as it becomes heated along the path of a lightning flash. Rumbling thunder is experienced at a distance from the lightning, and may be accentuated by echoes. As sound travels through the air at 1,100 ft per second, and light travels at the rate of 186,000 miles per second, there is always a delay after a lightning flash before the observer hears the sound of thunder.

Tornado

A violent whirlwind about an area of low pressure, the tornado is most common in the United States, where they have been known to cause considerable damage. The diameter of the whirlwind area is small, usually 50–200 m, but wind speeds may be in excess of 200 knots about the centre. Actual wind speed in the centre is zero, but updraft may lift objects into the air.

Trade winds

Permanent winds which blow towards the equator, trade winds usually measure between 3 and 5 on the Beaufort Scale. They are generally referred to as NE Trades when they blow over the North Atlantic and North Pacific from below latitude 30°N towards the equator, and SE Trades when they blow from latitude 30°S towards the equator over areas of the South Atlantic and the South Pacific.

Trough

This is an extension of low pressure from a low-pressure centre. It is the opposite of a ridge, which is the outward extension from a high-pressure centre.

Twilight

This is a period of reduced light which occurs after the sun dips below the horizon; it is caused by the rays of sunlight being refracted in the atmosphere towards the earth.

Veering

See 'Backing'.

Vertex

The turning point in the path of a TRS, the vertex is the position in which the path of the storm moves to an easterly from a westerly direction in the northern hemisphere.

Visibility

This is the maximum range at which an object is discernible. The state of visibility may be assessed by using the length of the ship when in dense or thick fog conditions. It may similarly be assessed when in poor visibility by noting the time taken for an approaching vessel to become visible, making due allowance for the respective speeds of the two ships. When assessing good visibility, it is not good practice to use the range of the visible horizon, owing to the possibility of distortion by refraction, especially in misty or hazy conditions. Excellent visibility may be ascertained when heavenly bodies are seen to be coming over or dropping under the horizon when rising and setting.

Warm front

This is a line of demarcation between advancing warm air and a mass of cold air, over which the warm air is rising.

Waterspout

This is caused by an extension, usually from a nimbus cloud, it will extend to the surface of the sea, causing agitation of the water, which effectively turns to a spout. The result is a column of water vapour which may last for up to half an hour. Shipping is advised to give it a wide berth.

Wave

A disturbance of the surface of the sea, a wave is caused by the wind. Waves will vary in size and height. When a wave breaks on the coastline it is referred to as a 'breaker'.

Wedge

This is a ridge of relatively high pressure situated between two low-pressure areas; it is often roughly wedge-shaped.

Wind

The movement of air parallel or nearly parallel to the surface of the earth, the wind is named after the direction from which it comes.

Figure 12.1 shows a synoptic chart of the North Atlantic region with the European continent 48 hours after issue on 14th May, 2013. Figure 12.3 shows the North Atlantic areas.

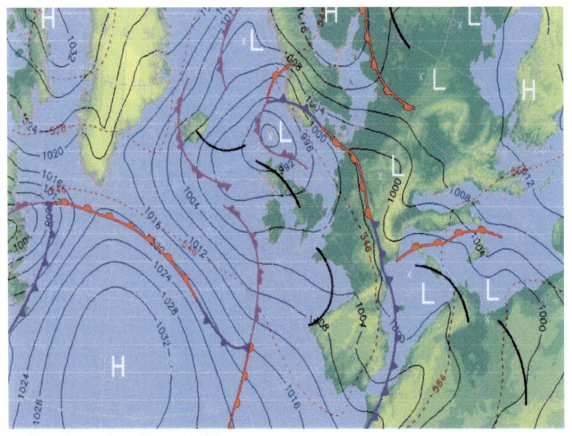

Surface pressure chart - Forecast T+48
Issued at: 0100 on Tue 14 May 2013

Figure 12.1 Synoptic chart (source: adapted from Met Office).

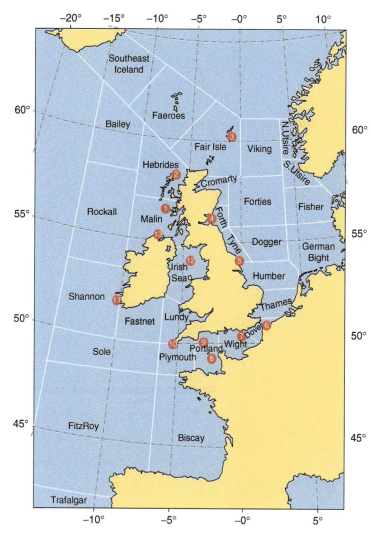

Figure 12.2 Present weather: UK coastal stations. Reports from these coastal stations and automatic weather logging stations in the British Isles are included in the extended Shipping Forecasts on BBC Radio 4.

The stations are listed in the order they are read in the forecast, the numbers in brackets refer to the map. Weather reports included in the forecasts are issued at 23:00 local time for the late broadcast and 04:00 for the early one, although reports issued at other times may be included if for some reason the most recent did not arrive.

- Tiree Automatic (1)
- Stornoway (2)
- Lerwick (3)
- Wick Automatic (00:48 only)
- Aberdeen (00:48 only)
- Leuchars (4)
- Boulmer (00:48 only)
- Bridlington (5)
- Sandettie Light Vessel Automatic (6)
- Greenwich Light Vessel Automatic (7)
- St. Catherine's Point Automatic (00:48 only)
- Jersey (8)
- Channel Light Vessel Automatic (9)
- Scilly Automatic (10)
- Milford Haven (00:48 only)
- Aberporth (00:48 only)
- Valley (00:48 only)
- Liverpool Crosby (00:48 only)
- Valentia (11)
- Ronaldsway (12)
- Malin Head (13)
- Machrihanish Automatic (00:48 only)

Weather Scales

Tables 12.1–12.3 cover the Beaufort Wind Scale and weather notation, fog and visibility scale and wave scale.

Construction and Interpretation of Synoptic Chart

Meteorological offices around the world in many participating countries collect and collate weather reports and related information for the benefit of safe navigation. Weather reporting vessels, together with aircraft and satellite pictures, provide reasonable forecasts for all major shipping areas.

Figure 12.3 Sea areas and associated marine communication areas effective under the GMDSS operation.

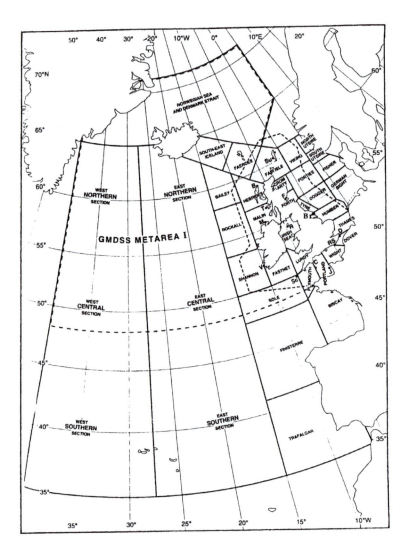

The reports from all sources allow a comprehensive weather chart to be produced. Symbols used are shown in Figure 12.4.

The following information is typical of the normal weather report:

- Position of reporting station, latitude and longitude.
- Speed of reporting station (knots).
- Course of vessel.
- Barometric pressure, corrected for sea level.
- Indication of movement.
- Weather description in Beaufort Scale notation (letter).
- Air temperature.
- Wind force and direction.

Table 12.1 Beaufort Wind Scale

Beaufort Scale number	Wind description	Wave description	Height of sea in feet	Knots in nautical mph
0	Calm	Flat calm, mirror smooth	—	0–1
1	Light airs	Small wavelets, without crests	0.25	1–3
2	Light breeze	Small wavelets, crests glassy but not breaking	0.5	4–6
3	Light breeze	Large wavelets, crests beginning to break	2.0	7–10
4	Moderate breeze	Small waves, becoming longer, crests breaking frequently	3.5	11–16
5	Fresh breeze	Moderate waves, longer with crests breaking	6.0	17–21
6	Strong breeze	Large waves forming, crests breaking more frequently	9.5	22–27
7	Strong wind	Large waves, streaky foam	13.5	28–33
8	Gale	High waves, increasing in length, continuous streaking of crests	18.0	34–40
9	Strong gale	High waves, crests rolling over, dense streaking	23.0	41–47
10	Storm	Very high waves, overhanging crests, surface white with foam	29.0	48–55
11	Violent storm	Exceptionally high waves, surface completely covered with foam	37.0	56–65
12	Hurricane	Air filled with spray, visibility impaired	—	over 65

Table 12.2 Beaufort weather notation

Symbol	Meaning
b	Blue sky with clear or hazy atmosphere, with less than one-quarter of the sky area clouded
c	Cloudy with detached opening cloud, where more than three-quarters of the sky area is clouded
bc	Sky area clouded over between one-quarter and three-quarters of the total area
d	Drizzle or fine rain
e	Wet air with no rain falling
f	Fog
fe	Wet fog
g	Gloomy
h	Hail
kq	Line squall
l	Lightning
m	Mist
o	Overcast sky
p	Passing showers
q	Squalls
r	Rain
rs	Sleet
s	Snow
t	Thunder
tl	Thunderstorm
u	Ugly threatening sky
v	Unusual visibility
w	Dew
z	Dust haze

Table 12.3 Wave scale	
State of sea	**Height in metres**
Calm – glassy	0
Calm – rippled	0–0.1
Smooth wavelets	0.1–0.5
Slight	0.5–1.25
Moderate	1.25–2.5
Rough	2.5–4.0
Very rough	4.0–6.0
High	6.0–9.0
Very high	9.0–14.0
Phenomenal	Over 14.0
Length of swell	**Length in metres**
Short	0–100
Average	100–200
Long	Over 200
Height of swell	**Height in metres**
Low	0–2.0
Moderate	2.0–4.0
Heavy	Over 4.0

- Sea state.
- Description of any swell.
- Ice accretion.
- Cloud cover and description.
- Date and time of observation.

For ease of transmission, reports are coded by use of the *Code and DeCode Booklet*, issued by the Meteorological Office and obtainable from Her Majesty's Stationery Office.

Once all the coded reports from stations in the area have been received, decoding takes place and the lowest barometric pressure is marked on the weather chart at its point of observation. Due allowance is made for the station's course and speed from the time of observation to the moment of reception. The term LOW is then recorded on the chart, and isobars, joining places of equal barometric pressure, are sketched in lightly.

Arrows are then added to indicate wind direction. The mariner should bear in mind that the arrows will generally cross the isobars in the direction of the LOW. Speed of the wind in knots is indicated, together with barometric pressure in numerical form. Wind speed used to be indicated by the number of feathers attached to the drawn arrows, to represent wind speed under the Beaufort Scale, but this practice is no longer as popular as in the past.

The letters of the Beaufort notation are added to describe the apparent weather condition around the observer's area, together with any relevant information regarding storms, ice, fog, etc. (see Figure 12.5).

Heavy Weather Precautions (General Cargo Vessel) Open Water Conditions

Stability

- Improve the 'GM' of the vessel (if appropriate).
- Remove free surface elements if possible.
- Ballast the vessel down.
- Pump out any swimming pool.
- Inspect and check the freeboard deck seal.
- Close all watertight doors.

Navigation

- Consider re-routing.
- Verify the vessel's position.
- Update weather reports.
- Plot storm position on a regular basis.
- Engage manual steering in ample time.
- Reduce speed if required and revise ETA.
- Secure the bridge against heavy rolling.

Type of front	Symbol as used on printed charts
Quasi-stationary	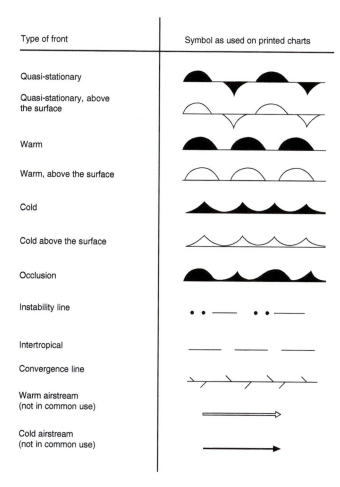
Quasi-stationary, above the surface	
Warm	
Warm, above the surface	
Cold	
Cold above the surface	
Occlusion	
Instability line	
Intertropical	
Convergence line	
Warm airstream (not in common use)	
Cold airstream (not in common use)	

Figure 12.4 Symbols for fronts, as plotted on a synoptic weather chart.

Deck

- Ensure lifelines are rigged to give access fore and aft.
- Tighten all cargo lashings, especially deck cargo securings.
- Close up ventilation as necessary.
- Check the securings on:
  - accommodation ladder
  - survival craft
  - anchors
  - derricks/cranes
  - hatches.
- Reduce manpower on deck and commence heavy weather work routine. Close up all weather deck doors.
- Clear decks of all surplus gear.
- Slack off whistle and signal halyards.
- Warn all heads of departments of impending heavy weather.
- Note preparations in the deck log book.
- Ideally the ship should be kept not too tender and not too stiff.

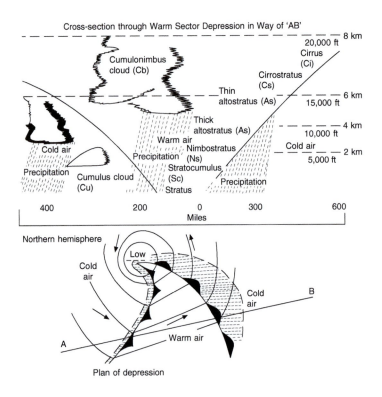

Figure 12.5 Interpretation of a synoptic weather chart.

- Masters/Chief Officers of vessels other than cargo ships should take account of their cargo, e.g. containers, oil, bulk products, etc., and act accordingly to keep their vessels secure. Long vessels like the large ore carriers or the VLCC can expect torsional stresses through their length in addition to bending and shear force stresses.
- Re-routing to avoid heavy weather should always be the preferred option whenever possible. If unavoidable, reduce speed in ample time to prevent pounding and structural damage.

Effects of Heavy Weather on Vessels at Sea

To describe the behaviour of any vessel in a heavy sea, the mariner should first be aware that every vessel, depending on her build, GM, state of loading, etc. will perform differently.

Stiff and Tender

A large GM will render a vessel stiff, i.e. give her a short period of roll, and subsequent damage may be sustained by rapid rolling. A small GM will render the vessel tender, i.e. she will have a long, slow, roll motion. These two conditions, usually brought about by incorrect loading or ballasting, should be avoided so that unnecessary stress in the structure of the vessel when in a seaway is avoided.

NB. When a ship has a large GM she will have a tendency to roll quickly and possibly violently (stiff ship). Raise 'G' to reduce GM.

When the ship has a small GM she will be easier to incline and not easily returned to the initial position (tender ship). Increase GM by lowering 'G'.

Periods of Roll and Encounter

Period of roll may be defined as that time taken by a ship to roll from port to starboard, or vice versa, and back again. The 'period of roll' will be to a great extent controlled by the GM of the vessel and by the disposition of weights away from the fore and aft line.

The period of encounter may be defined as that time between the passage of two successive wave crests under the ship.

If we consider the behaviour of a vessel with a short period of roll compared to the period of encounter, then the vessel will tend to lie with her decks parallel to the water surface or wave slope. The ship will probably suffer violent and heavy rolling and may suffer damage because of this. However, she will not generally ship a lot of water in this condition (see Figure 12.6).

If we consider the behaviour of a vessel with a long period of roll compared to the period of encounter, then the vessel may be expected to roll somewhat slowly and independently of the waves. The vessel will probably experience only moderate angles of roll, and the waves may be expected to break near the ship's side (see Figure 12.7).

Synchronism

Synchronism is most dangerous and a highly undesirable condition for a vessel to experience. It occurs when the period of roll is equal, or nearly equal, to the half-period of the waves. Successive waves tend to increase the angle of roll of the vessel, producing the possible danger of capsize. It is imperative that the Watch Officer should recognize the condition immediately, especially in a small vessel or when the range of stability is small. An immediate alteration of the vessel's course will effectively change the period of encounter and eliminate the condition, which is probably at its most dangerous when a beam sea is experienced and the ship reaches a greater maximum inclination at each crest and hollow.

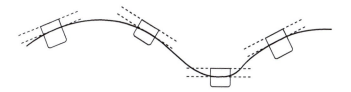

Figure 12.6 Vessel with short period of roll compared to period of encounter.

Figure 12.7 Vessel with long period of roll compared to period of encounter.

Cargo is liable to shift and the vessel will most certainly be damaged if the condition is left uncorrected for any length of time.

Synchronized pitching – when the period of encounter is similar to the vessel's period of pitch – may also occur. This situation can be alleviated by an alteration of speed, preferably a reduction as an increase may cause the vessel to 'pound'. A vessel which has suffered engine failure is most vulnerable to synchronized rolling and efforts to bring the vessel's head into the wind should be made while she still has headway (headreach).

General Behaviour of Vessels in Heavy Weather

The options available to a vessel running into heavy weather can be restricted to five main categories:

1 Head to sea, or with wind and sea fine on the bow, running at reduced speed.
2 Stern to sea, at reduced speed, running before the wind.
3 Heaving to, preferably in the lee of a land mass, to allow the weather to pass.
4 Anchoring, depending on depth of water and shelter afforded.
5 Altering course in plenty of time to take evasive action away from adverse weather conditions.

It has been pointed out that the successful handling of any ship will be dependent on the circumstances at the time and the characteristics of the ship in question. Draught, state of loading, superstructure, turning circles, etc. will all influence decisions taken for the safety of the vessel.

Head to Sea (or with Wind and Sea Fine on the Bow)

This is probably the most favoured position for a deep-draughted vessel. Leeward drift is minimized, but the vessel is liable to sustain considerable punishment, owing to continual pounding. Should a vessel be designed with increased scantlings, as for ice navigation, the concern might not be as great as in, say, a vessel with no additional strengthening built in.

The object is to head the vessel into the weather, with the idea of letting the weather pass over her. To this end, the speed of the vessel is considerably reduced, which will affect the period of encounter of the oncoming wave formations and subsequently reduce any pounding that the vessel is experiencing. Courses and speed should be altered to remove the possibility of hogging or sagging, and to prevent synchronism.

This situation can be a most uncomfortable one, with the vessel pitching violently at times. Violent pitching may result in 'racing propellers', which in turn puts excessive stress on engines. Absolute control of rudder and power is essential. As a rule of thumb, power should be reduced to the minimum necessary to maintain steerage way and avoid undue stress on machinery. Two steering motors should be operational, if fitted, and any zone of critical revolutions should be avoided.

Stern to Sea

Bad weather may often overtake the vessel at sea and she will effectively find herself running before the wind. It is usual to take up a course with the wind on the quarter rather than dead astern, this action tending to make things more comfortable on board for all concerned. If the wind and sea are acting directly from astern, then a vessel will run the risk of a surf effect, as waves build up under the stern. In addition, vessels with a low freeboard will run the risk of 'pooping'.

Pooping occurs when a vessel falls into the trough of a wave and does not rise with the wave, or if the vessel falls as the wave is rising and allows the wave to break over her stern or poop deck area – hence the name 'pooping' – which may cause considerable damage in the stern area.

The mariner should consider the speed of the vessel in all conditions of heavy weather, and what the effects of an increase or decrease would be on the periods of encounter and the effective wave impact; but generally the vessel's speed should be eased down until she is handling comfortably.

Generally speaking, the vessel with a following sea will not move as violently as a vessel head to sea. Trial and error will determine an optimum speed and minimize adverse motions of the ship. Speed adjustment, together with the long period of encounter, will probably reduce wave impact without any great delay to a ship's schedule.

The main concern for a vessel with the wind and sea abaft the beam arises if and when the vessel is required to turn. A distinct danger of attempting to turn across the wave front is that the vessel may 'broach to'.

A following sea reduces the flow of water past the rudder so that steering may become difficult and prevent the vessel's head coming up to wind. With reduced rudder effect, the vessel may be caught in an undesirable beam sea and may 'broach to', being unable to come into the wind and sea.

Heaving to

The prudent Master, after due consideration of all the circumstances, might be well advised to take what may at first appear to be the easy option. This may prove to be just that, with the wind blowing itself out in a very short time. However, this is not always the case, and a Master may encounter problems associated with crew fatigue or the spoiling of cargo through heaving to for a lengthy period of time.

Obviously, circumstances must dictate the actions in every case, but if it is possible to take advantage of a lee caused by some land mass, then this can often be the answer to the immediate problem. This practice is employed frequently in the coastal trades, especially with vessels carrying cargoes liable to shift, e.g. roll on–roll off, grain, etc.

If general heavy weather is encountered at sea, well away from coastlines, the action taken by the Master will depend on the type and form of the vessel. A reduction of speed will probably be one of the early actions to reduce the motions of the vessel and eliminate the possibility of cargo shift. Such reductions in the vessel's speed should be limited, to permit correct steerage under the adverse weather conditions. Power should not be reduced to such an extent that stalling of the main machinery occurs, nor should revolutions be allowed to oscillate about any critical zone of revolutions for that type of main engine.

Another alternative under the heading of heaving to is when it is decided to stop main engines altogether. This action could result in considerable drifting of the ship and sufficient sea room should be available before the operation is begun. Heavy rolling can be expected, with the ever-present risk of synchronism and the real problem of shifting cargoes.

For this alternative to be successful, a vessel needs to have good watertight integrity, together with an adequate GM. The use of storm oil may become a necessity once the vessel has taken up her own position. Oil should only be used to maintain the safety of the vessel and/or life. It should be distributed on the windward side of the vessel in an amount sufficient to reduce the immediate hazards.

Spreading vegetable or animal oil on heavy seas will prevent wave crests from breaking over the vessel but will have little or no effect on the swell waves about the hull area. The use of mineral oils should be avoided, especially if people are in the water. Lubricating oils are a possible alternative but heavy fuel oils should be avoided at all costs.

Use of Anchors

One of the greatest fears of any Master is that of being blown down onto a lee shore. Many shipwrecks caused in this way could have been avoided by anchoring in deep water, say 25–50 fathoms.

If the vessel is in shallow water, consideration should be given to the use of two anchors, and the expected strain on cable(s). Many vessels founder on a lee shore because they become disabled, loss of power resulting in subsequent grounding, or insufficient power preventing them from 'beating out' to seaward. The process of anchoring with or without engine power will reduce the rate of the vessel's drift to leeward. The possibility of the anchors holding is a real one. Even if grounding is not prevented, then refloating may very well be assisted by heaving on cables.

Use of Sea Anchors

The idea of rigging an efficient sea anchor to keep the vessel head to wind is feasible for a small vessel, if a sea anchor can be constructed easily, but it is doubtful if any Master of a supertanker or even just a large vessel would consider the idea. To be effective, the sea anchor would have to be of an unmanageable size, even if the ship were equipped with the necessary lifting gear and materials to make one, which is highly unlikely.

For small craft such as coasters and large yachts, a sea anchor will reduce the lee drift, and keep the boat head to wind, but for the majority of vessels it is a non-starter and they should consider other possibilities. Any floating object that will offer reasonable resistance to the drift of the ship will behave as a sea anchor, and mooring lines paid out over the bow will sometimes be useful. Large ships, especially those having high freeboard, would probably need outside help, such as a tug, in dangerous situations.

The situation may be more appreciated if the mariner considers a VLCC or ULCC ship, with few crew and little in the way of suitable equipment for jury rigging. In an emergency, walking back on the ship's bow anchors to just below the surface may have the desired effect of keeping the vessel head to sea.

Abnormal Waves

A lot of research has been carried out regarding the existence of abnormal waves. The *Mariner's Handbook* acknowledges that an area off the South African coast at the southern-most end of the 'Agulhas Current' experiences periods of abnormal wave sizes.

Storm waves are built up by the wind passing over the surface, like causing wrinkles in the skin of water. These can build up to an exceptional height, coupled with very deep troughs. There is seemingly no pattern and they have been reported in the North Sea, the Antarctic region and the North and South Atlantic Oceans.

Swell waves have been reported at over 30 metres, again with obvious dangers to shipping. The difference between swell and sea waves and the tsunami is that the swell and sea wave will exist for a brief interlude, while the wall of water of the tsunami can take several minutes to pass over.

Other reports and research indicate that, in fact, abnormal waves may be experienced in any part of the world's oceans. The findings also reflect that 'tsunamis' are generated by underwater volcanic eruptions, earthquakes and landslips on the ocean floor. This seismic energy displaces an immense volume of water. Over a vast expanse of ocean this may be only several centimetres high, but the wave is long and fast-moving (up to 700 kph) and is forced into a higher feature when it approaches shallow water. By the time the wave comes ashore it may be a wall of water up to about 15 metres high which can, and has, caused major disasters (Chilean Coast 1960, a 25-metre water wall was reported and 2,500 persons lost their lives).

'Tsunami' is from a Japanese word meaning 'harbour waves'. They are nearly always formulated from an earthquake or similar ocean-floor disturbance. The effect for shipping in coastal regions can sometimes be catastrophic and if known about in advance Masters are advised to move to deep water prior to the wave reaching the shore, time permitting. Countries around the Pacific Rim are particularly vulnerable to the effect of tsunamis.

Tropical Revolving Storm

The TRS normally forms in low latitudes, usually between 5° and 10° north or south of the equator. It cannot form in very low latitudes, or for that matter near land masses. These storms are often called hurricanes, typhoons or cyclones, but to seafarers they are all TRSs (see Figure 12.8).

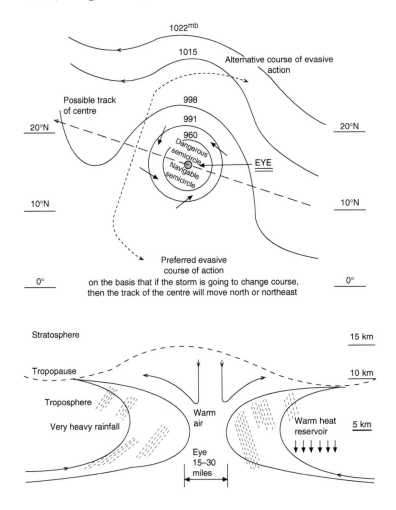

Figure 12.8 Tropical storms.

A storm will develop over open seas where the temperature and humidity are high, and some form of trigger action is available to set off the violent convection. The storm is made up of intense asymmetrical line squalls, which spiral inwards towards a central point known as the 'eye'. The eye of the storm will vary in size but is generally 15–30 miles in diameter. This area is known to be a comparatively calm area of warm air from the sea surface, right up to the stratosphere.

Wind speeds are in excess of 64 knots and may reach as high as 200 knots, with gale force winds extending from the eye up to a range of 300 miles. The barometric pressure will be exceptionally low and may fall below 900 mbs. A very high humidity level will be experienced, together with very heavy rainfall in the area.

A tropical storm may take up to five days to form and reach a mature stage, but once this stage is reached it may take several weeks before it dissipates. However, the normal period for a tropical storm's life is between one and two weeks. Should the storm move inland, the associated violent weather can be expected to diminish within 48 hours of crossing the coastline.

The general movement of a tropical storm will probably be in a westerly direction, either in the northern or southern hemispheres. Speed of movement will vary, but the average is about ten knots. Once clear of the tropical latitudes, it is not uncommon for the track to move north or northeast, or south or southeast, in the northern and southern hemispheres, respectively.

Evasive Action

It is the duty of the Master of any vessel to report the position and movement of any tropical storm if it has not already come to the attention of the authorities. The eye of the storm should be plotted, together with its rate of movement and probable path. Other dangers to the vessel's navigation should also be plotted in relation to the storm's path.

The Master of any vessel should ascertain at the earliest possible moment his own vessel's position and which 'semi-circle' he is in or entering. By full consideration of all the facts, a course of evasive action should be taken as quickly as possible to avoid crossing the path of the storm. Prudent use of the vessel's full speed should be quickly made, as it may become necessary to reduce speed later to avoid pounding and damage to the vessel.

Obviously, any Master faced with a tropical storm must make his decisions on the particular case. It may become a practical proposition to take up a satisfactory position and ride the storm out, letting the bad weather move past the vessel. When making the decision on the type of evasive action to take, Masters should bear in mind that tropical storms have a general tendency to move towards the pole of whichever hemisphere the vessel is navigating in. This, of course, is not inevitable, and a storm has been known to double back on itself more than once.

Setting an evasive course towards the equator when on the predicted track of the storm would seem to be the soundest action, provided that the storm maintains its predicted movements.

Ice Terminology

Anchor Ice

Submerged ice attached or secured to the bottom of the sea bed is known as anchor ice.

Bare Ice

This is ice without any snow covering.

Bergy Bit

A large piece of floating ice, between 1 m and 5 m above the surface of the water.

Brash Ice

An accumulation of broken, floating ice; this contains pieces up to approximately 2 m across.

Compact Pack Ice

This is a heavy concentration of pack ice, when no water is visible.

Compacted Ice Edge

A clear-cut ice edge, this is generally found on the windward side of an area of pack ice, compacted by the action of wind or current.

Concentration

A ratio expressing the density of ice accumulation; concentration is expressed in tenths of the total area.

Consolidated Pack Ice

A concentration of 10/10, where the ice floes are frozen together.

Crack

This is a split or fracture in the ice surface, which has not parted.

Difficult Area

This is a general term used to describe the area as difficult for purposes of navigation.

Easy Area

This is a general term used to describe the area as not too difficult for the purposes of navigation.

Fast Ice

This is sea ice which has become 'fast' to the shore, ice wall or other similar surface. It may be formed by the freezing of sea water close inshore or by pack ice freezing to the shore or other surfaces. Should its height extend more than 2 m, it would be referred to as an 'ice shelf'.

First Year Ice

A term derived from young ice, being sea ice of not more than one winter's growth, this ice is between 30 cm and 2 m thick.

Flaw

A narrow dividing section between the pack ice and fast ice, a flaw is formed by the shearing of the former from the latter.

Floating Ice

This general term is also used with regard to grounded or stranded ice.

Floe

This is a flat piece of ice more than 20 m across. Floes are subdivided according to size as giant, vast, big, medium and small.

Floeberg

A massive piece of sea ice, a floeberg is made up of one or more hummocks frozen together, the whole being separated from any other surrounding ice.

Fracture

- A general term used to describe any fracture/break of unspecified length. The width of the break is called:
- large when over 500 m,
- medium when 200–500 m,
- small when 50–200 m, and
- very small when less than 50 m.

Glacier

A continuously moving mass of snow and ice, a glacier moves from high to lower ground or, when afloat, its mass is continuously spreading.

Plate 114 Navigation at the front edge of a glacier (source: Shutterstock).

Glacierberg

An iceberg of irregular shape.

Grey Ice

Young ice up to 15 cm in thickness, grey ice has a tendency to break up in a swell and will be seen to 'raft' under pressure.

Grounded ice

These are large or small pieces of ice gone aground/ashore in shoal water.

Plate 115 Small tabular iceberg in Antarctic waters (source: Shutterstock).

Growler

A growler is a piece of ice that shows less than 1 m above the surface of the water. Its volume is less than that of a 'bergy bit', and it usually has an area of approximately 20 m². As a growler makes a very poor radar target, it is often very dangerous to navigation.

Hummock

A build-up of ice forced up by pressure is called a hummock, and a similar build-up of broken ice forced downwards by pressure is referred to as a 'bummock'.

Ice Belt

A long pack ice feature, an ice belt is longer than it is wide. Length will vary from about half a mile (1 km approx.) to more than 62 miles (100 km).

Iceberg

An enormous piece of ice more than 5 m in height above the surface of the water, an iceberg originates from a glacier and may be afloat or aground. When afloat, the greatest volume of the iceberg is beneath the surface.

Ice Bound

When navigation in or out of a harbour is restricted by an accumulation of ice the harbour is said to be 'ice bound'.

Ice Cake

A flat piece or cake of sea ice less than 20 m across.

Ice Edge

The dividing line between the open sea and the limit of sea ice (ice boundary).

Ice Field

This is pack ice composed of various sized floes in close proximity over an unspecified distance greater than 10 km across (6.2 miles).

Ice-Free

Open water, clear of any ice.

Plate 116 *Bransfield* research vessel secured alongside an ice front in Antarctica.

Ice Patch

A quantity of pack ice, less than 10 km across (6.2 miles).

Ice Shelf

A very thick layer of ice. An ice shelf could be up to 50 m above the surface of the water, and of any length. The seaward edge is termed an ice front.

Ice Tongue

A major ice projection from the coastline, this comprises several icebergs joined by 'fast ice'. Some or all of the icebergs may be grounded.

Lead

A visible fracture or passage which is navigable by surface craft through the ice regions.

Level Ice

Flat sea ice unaffected by deformation.

Multi-Year Ice

This is ice which has survived for more than two summers without melting. Its thickness is variable, but generally up to about 3 m. It is also practically salt-free.

New Ice

Newly formed ice.

Plate 117 The passenger ship *Star Princess* navigating close to glacial ice in the Tracy Arm Fiord of Alaska.

Nilas

A crust of thin ice approximately 10 cm in thickness, this often bends with the swell and wave motion on the surface. It may be subdivided into dark nilas and light nilas.

Nip

The vessel is said to be nipped when ice under pressure is pressed into the ship's side; she is sometimes damaged in the process.

Open Pack Ice

A concentration of pack ice of between four- and six-tenths coverage with extensive leads and floes not in contact with each other.

Open Water

Clear water free of obstruction ice and navigable to surface craft, with ice concentration not exceeding one-tenth.

Pack Ice

This is a general term to include areas of sea ice; it does not include 'fast ice'.

Pancake Ice

Circular pieces of ice up to 3 m in diameter and about 10 cm in thickness, pancake ice curls up at the edges when pieces crash into each other.

Rafted Ice

Deformed ice caused by layers riding on top of each other. Pressure changes cause the overriding, which is more often found in young ice.

Rotten Ice

Ice in an advanced state of decomposition, usually consisting of light, small pieces breaking up continuously.

Sea Ice

Ice formed from freezing sea water; found at sea.

Stranded Ice

This is ice left ashore by a falling tide.

Tabular Berg

A flat-topped iceberg in the southern hemisphere.

Very Close Pack Ice

A concentration of pack ice between nine- and ten-tenths coverage.

Ice Navigation

In general, when a vessel has to advance through ice areas, the progress of the ship will be dependent on:

- the nature of the ice;
- the qualities of the vessel, scantlings, ice-breaker bow construction and motive power of machinery;
- expertise and experience of the Master;
- operational qualities of navigational instruments;
- assistance of tugs or ice-breaker vessels;
- ice convoy facilities.

The Master of any vessel coming up to or approaching dangerous ice is obliged by the International Convention for the Safety of Life at Sea, 1960, to report any dangerous ice formation sighted on or near his course. His ice report should contain the following information: type of ice encountered; position of this ice; and GMT and date of sighting the same. The Master is further obliged to proceed at a moderate speed or alter his course to pass clear of ice dangers.

Ice reports are despatched to the International Ice Patrol, operated by the United States Coast Guard throughout the ice season, usually beginning about February and ending about June/July.

The prime function of the Ice Patrol is to warn shipping of the extent of sea ice and icebergs which may affect vessels on the main shipping routes.

Ice reports from shipping, together with weather reports from shipping, assist the Ice Patrol to piece together any movement of ice, and allow the construction of a facsimile chart of conditions for general broadcast to all shipping within the area. Reports are made by the Ice Patrol twice daily, together with the despatch of the facsimile chart. Additional reports of ice sightings are broadcast whenever considered necessary. Transmitting stations, together with frequencies and channels, are as described in the *Admiralty List of Radio Signals* (Vol. 3, Part 2). Mariners should be aware that this service is provided for them, and is greatly enhanced by their own cooperation. Reports of actual sightings help the flight planning of Ice Patrol aircraft.

Operating in Ice

Mariners entering ice regions should take early action to seek up-to-date ice reports from the Ice Patrol, as distributed by the US Naval Oceanographic Office. Ice limits should then be marked on navigational charts, and any particular hazards, such as single icebergs, plotted. Course and speed of the vessel can then be adjusted accordingly, circumstances dictating the safest route. A lookout is essential during daylight hours, even in so-called good visibility.

Vessels without operational radar should be prepared to stop during the hours of darkness if the concentration of dangerous ice warrants such action, and should at any time proceed only at a safe speed. Ice reports should be continually obtained and charts updated in accordance with the vessel's progress. A combination of fog and ice is not only a dangerous combination, but unfortunately a common occurrence.

Vessels attempting to negotiate ice regions should be equipped with reliable engines and steering gear. It is an advantage if the ship is ice-strengthened or longitudinally framed with an ice-breaker bow.

It has been found by experience that ship-handling in ice can be achieved by observing a few basic principles:

1 The vessel must endeavour to keep moving into the ice and making headway. Even if the movement is only very slight, it must be maintained.
2 It is best for the vessel to move with the ice, not against it.
3 Maintain freedom to move, bearing in mind that excessive speed lends itself to ice damage.
4 The mariner will require a great deal of patience.

Should the vessel become trapped in the ice and held, bear in mind that freedom of movement is lost and the ship will then only move with the ice, going wherever the ice is going. Should the forward motion of the vessel be impeded, a movement astern should be considered as an option, while searching for another 'lead' through the pack ice or ice field. Continuous movement astern should be avoided because of the very real danger to rudder and propeller. Continuous plotting of the ship's position in confined waters is essential at this stage in order to keep the vessel clear of shoals and to prevent disorientation. Regular checks on compasses and prominent landmarks must be considered essential.

The alternatives open to the Master are limited in the event of his vessel becoming 'ice bound'. Owners may decide to re-route the vessel to another port, but, failing this, one or more of the following actions are advised:

1 Assemble with other ships for movement in an ice convoy, usually escorted by ice-breaker vessels.
2 Follow the track of an ice-breaker vessel or ice-strengthened vessel towards the destination.
3 If equipped with an ice-breaker bow and also ice-strengthened, attempt passage independently.
4 Before leaving port, add ice-strengthening to the forepart of your own vessel. (This can be done relatively quickly by building a framework in the forepeak tanks out of pit prop beams, and covering or filling the whole tank area with concrete. This construction may later become permanent ballast for the vessel, as it is unlikely that it could be easily removed without drydocking and cutting into the shell plate.)

Single-Letter Signals Between Ice-Breaker and Assisted Vessels

The following single-letter signals, when made between an ice-breaker and assisted vessels, have only the significations given in this table and are only to be made by sound, visual or radiotelephony signals.

WM Ice-breaker support is now commencing. Use special ice-breaker support signals and keep continuous watch for sound, visual or radiotelephony signals.

WO Ice-breaker support is finished. Proceed to your destination.

Plate 118 An ice-breaker engages in close-proximity towing to assist a vessel in the Baltic Sea.

Table 12.4 Signals between ice-breaker and assisted vessels

Code letters or figures	Ice-breaker	Assisted vessel(s)
A • —	Go ahead (proceed along the ice channel)	I am going ahead (I am proceeding along the ice channel)
G — — •	I am going ahead; follow me	I am going ahead; I am following you
J • — — —	Do not follow me (proceed along the ice channel)	I will not follow you (I will proceed along the ice channel)
P • — — •	Slow down	I am slowing down
N — •	Stop your engines	I am stopping my engines
H • • • •	Reverse your engines	I am reversing my engines
L • — • •	You should stop your vessel instantly	I am stopping my vessel
4 • • • • —	Stop. I am ice-bound	Stop. I am ice-bound
Q — — • —	Shorten the distance between vessels	I am shortening the distance
B — • • •	Increase the distance between vessels	I am increasing the distance
5 • • • • •	Attention	Attention
Y — • — —	Be ready to take (or cast off) the towline	I am ready to take (or cast off) the towline

Ice Damage

The extent of any damage will depend on the condition of the ice the vessel is passing through. The mariner should be prepared to accept some damage to the vessel, while limiting the amount as much as possible.

Severe wear of the outer shell plating will be experienced at the waterline level, and for some height and depth above this level, according to the thickness of the ice the ship is passing through. All paint work on superstructures can be expected to flake and bare steelwork become badly pitted, especially if the temperature is continuously below freezing for any period.

Denting of shell plates in the bow area must be anticipated. The stem will be stripped clean of all paint and protective covering.

Plate 119 Ten-tenths ice coverage. Pancake ice; total coverage in the Baltic Sea, as seen from the bridge of a container/vehicle ferry.

The rudder and propeller area is extremely susceptible to ice damage from large floes passing down the ship's side and colliding with the upper area of the rudder and the rudder securing to the stock.

Lifeboat water tanks should be part-emptied to avoid fracture. Steam lines should be drained. Lagging on pipes should be regularly checked for expected deterioration. Ballast tanks and fresh water tanks should be inspected daily to prevent freezing over.

13

Preventing Collisions at Sea

Introduction

The International Regulations for the Prevention of Collision at Sea (COLREGS) originated from the Trinity Masters as far back as the 1840s. The Steam Navigation Act was passed in 1846 and gave the regulations, as was, statutory credibility. From those early beginnings the present-day COLREGS, incorporating the latest amendments, effective November 2003, stand to serve present-day shipping needs.

Under the auspices of the IMO, the regulations have been amended on numerous occasions and it is expected that individual seafarers make themselves aware of the content, interpretation and ramifications of adherence to the rules of the road.

Marine students are advised that the following content has been supported by numerous diagrams. These should, where appropriate, be considered as a self-teaching aid which it is hoped will facilitate learning.

International Regulations for Preventing Collisions at Sea, 1972 (as Amended by Resolution A464(XII)) A626(15) A678(16) A736(18) and A.910(22)

Part A: General

Rule 1: application

(a) These Rules shall apply to all vessels upon the high seas and in all waters connected therewith navigable by seagoing vessels.
(b) Nothing in these Rules shall interfere with the operation of special rules made by an appropriate authority for roadsteads, harbours, rivers, lakes or inland waterways connected with the high seas and navigable by seagoing vessels. Such special rules shall conform as closely as possible to these Rules.

(c) Nothing in these Rules shall interfere with the operation of any special rules made by the Government of any State with respect to additional station or signal lights, shapes or whistle signals for ships of war and vessels proceeding under convoy, or with respect to additional station or signal lights or shapes for fishing vessels engaged in fishing as a fleet. These additional station or signal lights, shapes or whistle signals shall, so far as possible, be such that they cannot be mistaken for any light, shape or signal authorized elsewhere under these Rules.

(d) Traffic separation schemes may be adopted by the Organization for the purpose of these Rules.

(e) Whenever the Government concerned shall determine that a vessel of special construction or purpose cannot comply fully with the provisions of any of these Rules with respect to the number, position, range or arc of visibility of lights or shapes, as well as to the disposition and characteristics of sound-signalling appliances, such vessels shall comply with such other provisions in regard to the number, position, range or arc of visibility of lights or shapes, as well as to the disposition and characteristics of sound-signalling appliances, as her Government shall have determined to be the closest possible compliance with these Rules in respect of that vessel.

Rule 2: responsibility

(a) Nothing in these Rules shall exonerate any vessel, or the owner, Master or crew thereof, from the consequences of any neglect to comply with these Rules or of the neglect of any precaution which may be required by the ordinary practice of seamen, or by the special circumstances of the case.

(b) In construing and complying with these Rules due regard shall be had to all dangers of navigation and collision and to any special circumstances, including the limitations of the vessels involved, which may make a departure from these Rules necessary to avoid immediate danger.

Rule 3: general definitions

For the purpose of these Rules, except where the context otherwise requires:

(a) The word 'vessel' includes every description of water craft, including nondisplacement craft, WIG craft and seaplanes, used or capable of being used as a means of transportation on water.

(b) The term 'power-driven vessel' means any vessel propelled by machinery.

(c) The term 'sailing vessel' means any vessel under sail provided that propelling machinery, if fitted, is not being used.

(d) The term 'vessel engaged in fishing' means any vessel fishing with nets, lines, trawls or other fishing apparatus which restrict manoeuvrability, but does not include a vessel fishing with trolling lines or other fishing apparatus which do not restrict manoeuvrability.

(e) The word 'seaplane' includes any aircraft designed to manoeuvre on the water.

(f) The term 'vessel not under command' means a vessel which through some exceptional circumstances is unable to manoeuvre as required by these Rules and is therefore unable to keep out of the way of another vessel.

(g) The term 'vessel restricted in her ability to manoeuvre' means a vessel which from the nature of her work is restricted in her ability to manoeuvre as required by these Rules and is therefore unable to keep out of the way of another vessel. The term 'vessels restricted in their ability to manoeuvre' shall include but not be limited to:

  (i) a vessel engaged in laying, servicing or picking up a navigation mark, submarine cable or pipeline;

  (ii) a vessel engaged in dredging, surveying or underwater operations;

  (iii) a vessel engaged in replenishment or transferring persons, provisions or cargo while underway;

  (iv) a vessel engaged in the launching or recovery of aircraft;

  (v) a vessel engaged in mine clearance operations;

  (vi) a vessel engaged in a towing operation such as severely restricts the towing vessel and her tow in their ability to deviate from their course.

(h) The term 'vessel constrained by her draught' means a power-driven vessel which, because of her draught in relation to the available depth and width of navigable water, is severely restricted in her ability to deviate from the course she is following.

(i) The word 'underway' means that a vessel is not at anchor, or made fast to the shore, or aground.

(j) The words 'length' and 'breadth' of a vessel mean her length overall and greatest breadth.

(k) Vessels shall be deemed to be in sight of one another only when one can be observed visually from the other.

(l) The term 'restricted visibility' means any condition in which visibility is restricted by fog, mist, falling snow, heavy rain-storms, sandstorms or any other similar causes.

(m) The term 'wing-in-ground (WIG) craft' means a multimodal craft which, in its main operational mode, flies in close proximity to the surface by utilizing surface-effect action.

Part B: Steering and Sailing Rules

Section I. conduct of vessels in any condition of visibility

Rule 4: application

Rules in this Section apply in any condition of visibility.

Figure 13.1 WIG craft design.

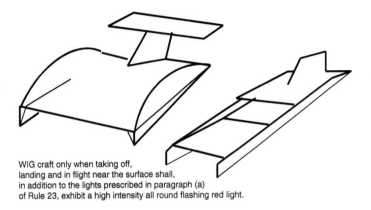

WIG craft only when taking off,
landing and in flight near the surface shall,
in addition to the lights prescribed in paragraph (a)
of Rule 23, exhibit a high intensity all round flashing red light.

Rule 5: look-out

Every vessel shall at all times maintain a proper look-out by sight and hearing as well as by all available means appropriate in the prevailing circumstances and conditions so as to make a full appraisal of the situation and of the risk of collision.

Rule 6: safe speed

Every vessel shall at all times proceed at a safe speed so that she can take proper and effective action to avoid collision and be stopped within a distance appropriate to the prevailing circumstances and conditions.

In determining a safe speed the following factors shall be among those taken into account:

(*a*) By all vessels:
    (i) the state of visibility;
    (ii) the traffic density including concentrations of fishing vessels or any other vessels;
    (iii) the manoeuvrability of the vessel with special reference to stopping distance and turning ability in the prevailing conditions;
    (iv) at night the presence of background light such as from shore lights or from back scatter of her own lights;
    (v) the state of wind, sea and current, and the proximity of navigational hazards;
    (vi) the draught in relation to the available depth of water.
(*b*) Additionally, by vessels with operational radar:
    (i) the characteristics, efficiency and limitations of the radar equipment;
    (ii) any constraints imposed by the radar range scale in use;
    (iii) the effect on radar detection of the sea state, weather and other sources of interference;
    (iv) the possibility that small vessels, ice and other floating objects may not be detected by radar at an adequate range;

(v) the number, location and movement of vessels detected by radar;

(vi) the more exact assessment of the visibility that may be possible when radar is used to determine the range of vessels or other objects in the vicinity.

Rule 7: risk of collision

(a) Every vessel shall use all available means appropriate to the prevailing circumstances and conditions to determine if risk of collision exists. If there is any doubt such risk shall be deemed to exist.

(b) Proper use shall be made of radar equipment if fitted and operational, including long-range scanning to obtain early warning of risk of collision and radar plotting or equivalent systematic observation of detected objects.

(c) Assumption shall not be made on the basis of scanty information, especially scanty radar information.

(d) In determining if risk of collision exists, the following considerations shall be among those taken into account:

    (i) such risk shall be deemed to exist if the compass bearing of an approaching vessel does not appreciably change;

    (ii) such risk may sometimes exist even when an appreciable bearing change is evident, particularly when approaching a very large vessel or a tow or when approaching a vessel at close range.

Rule 8: action to avoid collision

(a) Any action taken to avoid collision shall be taken in accordance with the rules of this Part and, if the circumstances of the case admit, be positive, made in ample time and with due regard to the observance of good seamanship.

(b) Any alteration of course and/or speed to avoid collision shall, if the circumstances of the case admit, be large enough to be readily apparent to another vessel observing visually or by radar; a succession of small alterations of course and/or speed should be avoided.

(c) If there is sufficient sea room, alteration of course alone may be the most effective action to avoid a close-quarters situation, provided that it is made in good time, is substantial and does not result in another close-quarters situation.

(d) Action taken to avoid collision with another vessel shall be such as to result in passing at a safe distance. The effectiveness of the action shall be carefully checked until the other vessel is finally past and clear.

(e) If necessary to avoid collision or allow more time to assess the situation, a vessel shall slacken her speed or take all way off by stopping or reversing her means of propulsion.

(f) (i) A vessel which, by any of these Rules, is required not to impede the passage or safe passage of another vessel shall, when required by the circumstances of the case, take early action to allow sufficient sea room for the safe passage of the other vessel.

(ii) A vessel required not to impede the passage or safe passage of another vessel is not relieved of this obligation if approaching the other vessel so as to involve risk of collision and shall, when taking action, have full regard to the action which may be required by the Rules of this Part.

(iii) A vessel the passage of which is not to be impeded remains fully obliged to comply with the Rules of this Part when the two vessels are approaching one another so as to involve risk of collision.

Rule 9: narrow channels

(a) A vessel proceeding along the course of a narrow channel or fairway shall keep as near to the outer limit of the channel or fairway which lies on her starboard side as is safe and practicable.

(b) A vessel of less than 20 metres in length or a sailing vessel shall not impede the passage of a vessel which can safely navigate only within a narrow channel or fairway.

(c) A vessel engaged in fishing shall not impede the passage of any other vessel navigating within a narrow channel or fairway.

(d) A vessel shall not cross a narrow channel or fairway if such crossing impedes the passage of a vessel which can safely navigate only within such channel or fairway. The latter vessel may use the sound signal prescribed in Rule 34(d) if in doubt as to the intention of the crossing vessel.

(e) (i) In a narrow channel or fairway when overtaking can take place only if the vessel to be overtaken has to take action to permit safe passing, the vessel intending to overtake shall indicate her intention by sounding the appropriate signal prescribed in Rule 34(c)(i). The vessel to be overtaken shall, if in agreement, sound the appropriate signal prescribed in Rule 34(c)(ii) and take steps to permit safe passing. If in doubt she may sound the signals prescribed in Rule 34(d).

(ii) This Rule does not relieve the overtaking vessel of her obligation under Rule 13.

(f) A vessel nearing a bend or an area of a narrow channel or fairway where other vessels may be obscured by an intervening obstruction shall navigate with particular alertness and caution and shall sound the appropriate signal prescribed in Rule 34(e).

(g) Any vessel shall, if the circumstances of the case admit, avoid anchoring in a narrow channel.

Author's Comments

The intention of 'X' is to overtake vessel 'Y'. It is the obligation of 'X' to keep out of the way of the vessel being overtaken, under Rule 13. Where an obvious clear passage exists, the overtaking operation may take place. No sound signals need be given by either vessel, because there is no requirement for the vessel being overtaken to alter her course or speed or change her intended actions in any way whatsoever.

Under Rule 9 the overtaking vessel need only give the appropriate sound signals prescribed by Rules 34(*c*) and 34(*d*), when vessel 'Y' is required to alter her position to allow safe passage. Rule 9 does not apply to lanes of traffic separation schemes, even though such lanes may be relatively narrow (see Rule 10).

Sources of reference: Rules 9, 13, 17 and 34.

The intention of vessel 'X' is to overtake and pass vessel 'Y' on the starboard side of vessel 'Y'. Vessel 'X' should indicate her intention by two prolonged blasts, followed by one short blast to mean 'I intend to overtake you on your starboard side'. Vessel 'Y' should answer only if she agrees to the operation taking place by sounding one prolonged, one short, one prolonged and one short blast in succession. If vessel 'Y' does not agree with the operation, she should sound no signal at all at this stage. Should doubt arise on vessel 'Y' after some movement by vessel 'X', vessel 'Y' may indicate such doubt by sounding at least five short and rapid blasts.

Sources of reference: Rules 9, 13, 17 and 34.

The intention of vessel 'X' is to overtake vessel 'Y' in a narrow channel by passing down the port side of vessel 'Y'. Vessel 'X' should indicate her intentions by two prolonged blasts, followed by two short blasts, to mean 'I intend to overtake you on your port side'. Vessel 'Y' should answer only if she agrees to the operation taking place by sounding one prolonged, one short, one prolonged and one short blast in succession. If vessel 'Y' does not agree with the operation, she should sound no signal at all at this stage. Should doubt arise on vessel 'Y' after some movement by vessel 'X', vessel 'Y' may indicate such doubt by sounding at least five short and rapid blasts.

Sources of reference: Rules 9, 13, 17 and 34.

Comment

Interpretation of the sound signals prescribed by Rule 34(*c*), and 34(*d*) being made in conjunction with vessels navigating in narrow channels (Rule 9) is open to considerable debate. In the author's view, doubt will only be created by the overtaking vessel if she starts to overtake without the agreement of the vessel to be overtaken.

There would appear to be a strong case for the vessel being overtaken to be allowed to sound one prolonged followed by one short blast (N in Morse code) to indicate 'Negative, do not attempt to overtake', or even X in Morse code to signify 'Stop carrying out

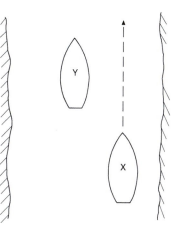

Figure 13.2 Overtaking vessel in a narrow channel.

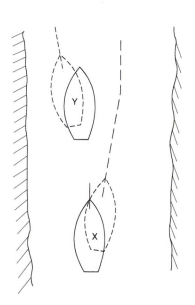

Figure 13.3 Overtaking vessels within the confines of a narrow channel.

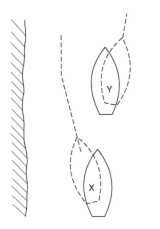

Figure 13.4

your intentions and watch for my signals'. However, additional signals may only add confusion to what could become a period saturated by sound signals, causing unnecessary concern all round.

Some mariners may feel that the sounding of five short and rapid blasts should be interpreted by other vessels as the negative signal, but the author feels that this is an incorrect assumption. The signal should only be sounded if and when an incorrect movement by the overtaking vessel creates doubt on board the vessel being overtaken (see Figures 13.2–13.4).

Rule 10: traffic separation schemes

(a) This Rule applies to traffic separation schemes adopted by the Organization and does not relieve any vessel of her obligation under any other Rule.

(b) A vessel using a traffic separation scheme shall:
   (i) proceed in the appropriate traffic lane in the general direction of traffic flow for that lane;
   (ii) so far as practicable keep clear of a traffic separation line or separation zone,
   (iii) normally join or leave a traffic lane at the termination of the lane, but when joining or leaving from either side shall do so at as small an angle to the general direction of traffic flow as practicable.

(c) A vessel shall so far as practicable avoid crossing traffic lanes, but if obliged to do so shall cross on a heading as nearly as practicable at right angles to the general direction of traffic flow.

(d) (i) a vessel shall not use an inshore traffic zone when she can safely use the appropriate traffic lane within the adjacent traffic separation scheme. However, vessels of less than 20 metres in length, sailing vessels and vessels engaged in fishing may use the inshore traffic zone.
   (ii) Notwithstanding sub-paragraph (d)(i), a vessel may use an inshore traffic zone when en route to or from a port, offshore installation or structure, pilot station or any other place situated within the inshore traffic zone; or to avoid immediate danger.

(e) A vessel other than a crossing vessel or a vessel joining or leaving a lane shall not normally enter a separation zone or cross a separation line except:
   (i) in cases of emergency to avoid immediate danger;
   (ii) to engage in fishing within a separation zone.

(f) A vessel navigating in areas near the terminations of traffic separation schemes shall do so with particular caution.

(g) A vessel shall so far as practicable avoid anchoring in a traffic separation scheme or in areas near its terminations.

(h) A vessel not using a traffic separation scheme shall avoid it by as wide a margin as is practicable.

(i) A vessel engaged in fishing shall not impede the passage of any vessel following a traffic lane.

(*j*) A vessel of less than 20 metres in length or a sailing vessel shall not impede the safe passage of a power-driven vessel following a traffic lane.

(*k*) A vessel restricted in her ability to manoeuvre when engaged in an operation for the maintenance of safety of navigation in a traffic separation scheme is exempted from complying with this Rule to the extent necessary to carry out the operation.

(*l*) A vessel restricted in her ability to manoeuvre when engaged in an operation for the laying, servicing or picking up of a submarine cable, within a traffic separation scheme, is exempted from complying with this Rule to the extent necessary to carry out the operation.

Section II: Conduct of vessels in sight of one another

Rule 11: application

Rules in this Section apply to vessels in sight of one another.

Rule 12: sailing vessels

(*a*) When two sailing vessels are approaching one another, so as to involve risk of collision, one of them shall keep out of the way of the other as follows:
  (i)   when each has the wind on a different side, the vessel which has the wind on the port side shall keep out of the way of the other;
  (ii) when both have the wind on the same side, the vessel which is to windward shall keep out of the way of the vessel which is to leeward;
  (iii) if a vessel with the wind on the port side sees a vessel to windward and cannot determine with certainty whether the other vessel has the wind on the port or on the starboard side, she shall keep out of the way of the other.

(*b*) For the purposes of this Rule the windward side shall be deemed to be the side opposite to that on which the mainsail is carried or, in the case of a square-rigged vessel, the side opposite to that on which the largest fore-and-aft sail is carried.

Rule 13: overtaking

(*a*) Notwithstanding anything contained in the Rules of Part B, Sections I and II, any vessel overtaking any other shall keep out of the way of the vessel being overtaken.

(*b*) A vessel shall be deemed to be overtaking when coming up with another vessel from a direction more than 22.5 degrees abaft her beam, that is, in such a position with reference to the vessel she is overtaking, that at night she would be able to see only the sternlight of that vessel but neither of her sidelights.

(c) When a vessel is in any doubt as to whether she is overtaking another she shall assume that this is the case and act accordingly.

(d) Any subsequent alteration of the bearing between the two vessels shall not make the overtaking vessel a crossing vessel within the meaning of these Rules or relieve her of the duty of keeping clear of the overtaken vessel until she is finally past and clear.

Comment

The target vessel is bearing exactly two points abaft the beam. The observing vessel must assume that it is a crossing situation, because the target is not, in the words of Rule 13, 'more than 22.5 degrees abaft her beam'. Vessel 'A' would normally be expected to give way to vessel 'B'. (To be an overtaking vessel, the target must be more than two points abaft the beam of your own vessel.)

If the observing vessel, namely your own ship, is in doubt as to the bearing of the other vessel and whether that vessel is an overtaking vessel, she must assume under Rule 7 that when risk of collision *may* exist, it *does* exist, and take appropriate action to avoid collision (Rule 13c).

The target vessel is 'more' than two points abaft the beam of the vessel to be overtaken. The deciding factor making the target vessel an overtaking vessel is the word 'more' written into Rule 13. Should any vessel be in doubt as to whether she is an overtaking vessel or a crossing vessel, she must assume herself to be an overtaking vessel and keep out of the way by acting accordingly.

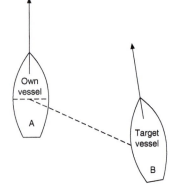

Figure 13.5 Illustration of Rule 13.

Rule 14: head-on situation

(a) When two power-driven vessels are meeting on reciprocal or nearly reciprocal courses so as to involve risk of collision each shall alter her course to starboard so that each shall pass on the port side of the other.

(b) Such a situation shall be deemed to exist when a vessel sees the other ahead or nearly ahead and by night she could see the mast head lights of the other in a line or nearly in a line and/or both sidelights and by day she observes the corresponding aspect of the other vessel.

(c) When a vessel is in any doubt as to whether such a situation exists she shall assume that it does exist and act accordingly.

Rule 15: crossing situation

When two power-driven vessels are crossing so as to involve risk of collision, the vessel which has the other on her own starboard side shall keep out of the way and shall, if the circumstances of the case admit, avoid crossing ahead of the other vessel.

Rule 16: action by give-way vessel

Every vessel which is directed to keep out of the way of another vessel shall, so far as possible, take early and substantial action to keep well clear.

Rule 17: action by stand-on vessel

(*a*) (i) Where one of two vessels is to keep out of the way the other shall keep her course and speed.

(ii) The latter vessel may however take action to avoid collision by her manoeuvre alone, as soon as it becomes apparent to her that the vessel required to keep out of the way is not taking appropriate action in compliance with these Rules.

(*b*) When, from any cause, the vessel required to keep her course and speed finds herself so close that collision cannot be avoided by the action of the give-way vessel alone, she shall take such action as will best aid to avoid collision.

(*c*) A power-driven vessel which takes action in a crossing situation in accordance with sub-paragraph (*a*)(ii) of this Rule to avoid collision with another power-driven vessel shall, if the circumstances of the case admit, not alter course to port for a vessel on her own port side.

(*d*) This Rule does not relieve the give-way vessel of her obligation to keep out of the way.

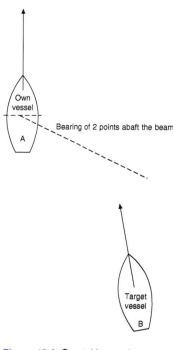

Figure 13.6 Overtaking under Rule 13.

Rule 18: responsibilities between vessels

Except where Rules 9, 10 and 13 otherwise require:

(*a*) A power-driven vessel underway shall keep out of the way of:
(i) a vessel not under command;
(ii) a vessel restricted in her ability to manoeuvre;
(iii) a vessel engaged in fishing;
(iv) a sailing vessel.

(*b*) A sailing vessel underway shall keep out of the way of:
(i) a vessel not under command;
(ii) a vessel restricted in her ability to manoeuvre;
(iii) a vessel engaged in fishing.

(*c*) A vessel engaged in fishing when underway shall, so far as possible, keep out of the way of:
(i) a vessel not under command;
(ii) a vessel restricted in her ability to manoeuvre.

(*d*) (i) Any vessel other than a vessel not under command or a vessel restricted in her ability to manoeuvre shall, if the circumstances of the case admit, avoid impeding the safe passage of a vessel constrained by her draught, exhibiting the signals in Rule 28;

    (ii) A vessel constrained by her draught shall navigate with particular caution having full regard to her special condition.

(e) A seaplane on the water shall, in general, keep well clear of all vessels and avoid impeding their navigation. In circumstances, however, where risk of collision exists, she shall comply with the Rules of this Part.

(f) (i) A WIG craft when taking-off, landing and in flight near the surface, shall keep well clear of all other vessels and avoid impeding their navigation;

    (ii) a WIG craft operating on the water surface shall comply with the Rules of this Part as a power-driven vessel.

Section III: Conduct of Vessels in Restricted Visibility

Rule 19: conduct of vessels in restricted visibility

(a) This Rule applies to vessels not in sight of one another when navigating in or near an area of restricted visibility.

(b) Every vessel shall proceed at a safe speed adapted to the prevailing circumstances and conditions of restricted visibility. A power-driven vessel shall have her engines ready for immediate manoeuvre.

(c) Every vessel shall have due regard to the prevailing circumstances and conditions of restricted visibility when complying with the Rules of Section I of this Part.

(d) A vessel which detects by radar alone the presence of another vessel shall determine if a close-quarters situation is developing and/or risk of collision exists. If so, she shall take avoiding action in ample time, provided that when such action consists of an alteration of course, so far as possible the following shall be avoided:

    (i) an alteration of course to port for a vessel forward of the beam, other than for a vessel being overtaken;

    (ii) an alteration of course towards a vessel abeam or abaft the beam.

(e) Except where it has been determined that a risk of collision does not exist, every vessel which hears apparently forward of her beam the fog signal of another vessel, or which cannot avoid a close-quarters situation with another vessel forward of her beam, shall reduce her speed to the minimum at which she can be kept on her course. She shall if necessary take all her way off and in any event navigate with extreme caution until danger of collision is over.

Part C: Lights and Shapes

Rule 20: application

(a) Rules in this Part shall be complied with in all weathers.

(b) The Rules concerning lights shall be complied with from sunset to sunrise, and during such times no other lights shall be exhibited, except such lights as cannot be mistaken for the lights specified in these Rules or do not impair their visibility or distinctive character, or interfere with the keeping of a proper look-out.

(c) The lights prescribed by these Rules shall, if carried, also be exhibited from sunrise to sunset in restricted visibility and may be exhibited in all other circumstances when it is deemed necessary.

(d) The Rules concerning shapes shall be complied with by day.

(e) The lights and shapes specified in these Rules shall comply with the provisions of Annex I to these Regulations.

Rule 21: definitions

(a) 'Masthead light' means a white light placed over the fore and aft centreline of the vessel showing an unbroken light over an arc of the horizon of 225 degrees and so fixed as to show the light from right ahead to 22.5 degrees abaft the beam on either side of the vessel.

(b) 'Sidelights' means a green light on the starboard side and a red light on the port side each showing an unbroken light over an arc of the horizon of 112.5 degrees and so fixed as to show the light from right ahead to 22.5 degrees abaft the beam on its respective side. In a vessel of less than 20 metres in length the sidelights may be combined in one lantern carried on the fore and aft centreline of the vessel.

(c) 'Sternlight' means a white light placed as nearly as practicable at the stern showing an unbroken light over an arc of the horizon of 135 degrees and so fixed as to show the light 67.5 degrees from right aft on each side of the vessel.

(d) 'Towing light' means a yellow light having the same characteristics as the 'sternlight' defined in paragraph (c) of this Rule.

(e) 'All-round light' means a light showing an unbroken light over an arc of the horizon of 360 degrees.

(f) 'Flashing light' means a light flashing at regular intervals at a frequency of 120 flashes or more per minute.

Sidelights	
Range	3 miles
Arc	112.5°
Mainmast/foremast	
Range	6 miles
Arc	225°
Sternlight	
Range	3 miles
Arc	135°

Rule 22: visibility of lights

The lights prescribed in these Rules shall have an intensity as specified in Section 8 of Annex I to these Regulations so as to be visible at the following minimum ranges:

(a) In vessels of 50 metres or more in length:
- a masthead light, 6 miles;
- a sidelight, 3 miles;
- a sternlight, 3 miles;

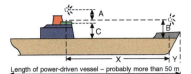

Length of power-driven vessel – probably more than 50 m

Figure 13.7 Navigation lights under Rule 23.

A   Second light at least 4.5 m higher than foremast light.
B   Forward light not less than 6 m high. If the breadth of the vessel exceeds 6 m, it should be at a height not less than such breadth, but need not be over a height of 12 m.
C   Sidelight at a height not greater than three-quarters of that of the foremast light.
X   Second light not less than half of the vessel's length away from foremast light, but need not be more than 100 m.
Y   Foremast light not more than a quarter of the vessel's length from the stem.

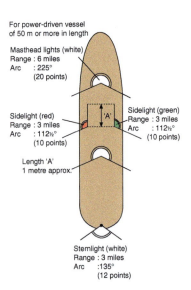

For power-driven vessel of 50 m or more in length

Masthead lights (white)
Range : 6 miles
Arc   : 225°
         (20 points)

Sidelight (red)
Range : 3 miles
Arc   : 112½°
         (10 points)

Sidelight (green)
Range : 3 miles
Arc   : 112½°
         (10 points)

Length 'A'
1 metre approx.

Sternlight (white)
Range : 3 miles
Arc   :135°
        (12 points)

Figure 13.8 Arcs and ranges of lights for power-driven vessel. Sources of reference: Rules 21 and 22, Annex 1, and Surveyors' Instructions.

- a towing light, 3 miles;
- a white, red, green or yellow all-round light, 3 miles.

(*b*)   In vessels of 12 metres or more in length but less than 50 metres in length:
- a masthead light, 5 miles; except that where the length of the vessel is less than 20 metres, 3 miles;
- a sidelight, 2 miles;
- a sternlight, 2 miles;
- a towing light, 2 miles;
- a white, red, green or yellow all-round light, 2 miles.

(*c*)   In vessels of less than 12 metres in length:
- a masthead light, 2 miles;
- a sidelight, 1 mile;
- a sternlight, 2 miles;
- a towing light, 2 miles;
- a white, red, green or yellow all-round light, 2 miles.

(*d*)   In inconspicuous partly submerged vessels or objects being towed:
- a white all-round light, 3 miles.

A power-driven vessel less than 50 m in length need not show the second masthead light but may do so.

Source of references: Rules 21, 22, 23, Annex 1, Statutory Instrument 1983 No. 708.

Rule 23: power-driven vessels underway

(*a*)   A power-driven vessel underway shall exhibit:
   (i)   a masthead light forward;
   (ii)   a second masthead light abaft of and higher than the forward one; except that a vessel of less than 50 metres in length shall not be obliged to exhibit such light but may do so;
   (iii)   sidelights;
   (iv)   a sternlight.
(*b*)   An air-cushion vessel when operating in the non-displacement mode shall, in addition to the lights prescribed in paragraph (*a*) of this Rule, exhibit an all-round flashing yellow light.
(*c*)   A WIG craft only when taking-off, landing and in flight near the surface shall, in addition to the lights prescribed in paragraph (*a*) of this Rule, exhibit a high intensity all-round flashing red light.
(*d*)   (i)   A power-driven vessel of less than 12 metres in length may in lieu of the lights prescribed in paragraph (*a*) of this Rule exhibit an all-round white light and sidelights.
   (ii)   a power-driven vessel of less than 7 metres in length whose maximum speed does not exceed 7 knots may in lieu of the lights prescribed in paragraph (*a*) of this Rule exhibit an all-round white light and shall, if practicable, also exhibit sidelights;

(iii) the masthead light or all-round white light on a power-driven vessel of less than 12 metres in length may be displaced from the fore and aft centreline of the vessel if centreline fitting is not practicable, provided that the sidelights are combined in one lantern which shall be carried on the fore and aft centreline of the vessel or located as nearly as practicable in the same fore and aft line as the masthead light or the all-round white light.

Small-boat owners should be aware that if practical their boats should also exhibit sidelights. Screens need not be fitted to the combined lantern when complying with the specifications of Annex 1.

Sources of reference: Rules 20, 21, 22, 23, and Annex 1.

Rule 24: towing and pushing

(a) A power-driven vessel when towing shall exhibit:
   (i)   instead of the light prescribed in Rule 23(a)(i) or (a)(ii), two masthead lights in a vertical line. When the length of the tow, measuring from the stern of the towing vessel to the after end of the tow exceeds 200 metres, three such lights in a vertical line;
   (ii) sidelights;
   (iii) a sternlight;
   (iv) a towing light in a vertical line above the sternlight;
   (v)   when the length of the tow exceeds 200 metres, a diamond shape where it can best be seen.
(b) When a pushing vessel and a vessel being pushed ahead are rigidly connected in a composite unit they shall be regarded as a power-driven vessel and exhibit the lights prescribed in Rule 23.
(c) A power-driven vessel when pushing ahead or towing alongside, except in the case of a composite unit, shall exhibit:
   (i)   instead of the light prescribed in Rule 23(a)(i) or (a)(ii), two masthead lights in a vertical line;
   (ii) sidelights;
   (iii) a sternlight.

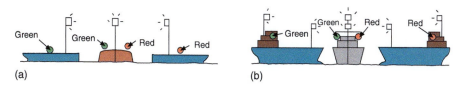

Figure 13.9 Power-driven vessel's lights as seen at night, or in any condition of restricted or poor visibility or whenever it is deemed necessary. *(a)* More than 12 m in length but less than 50 m, masthead light five miles (may be three miles if vessel's length less than 20 m), sidelights two miles, sternlight two miles (not shown in diagram). *(b)* Underway, but not necessarily making way, masthead light forward and second masthead light higher and abaft this forward light. (A vessel of less than 50 m in length need not show this second masthead light, but may do so if desired.) Range of masthead lights six or five miles, sidelights three or two miles, sternlight three or two miles (not shown in this diagram).

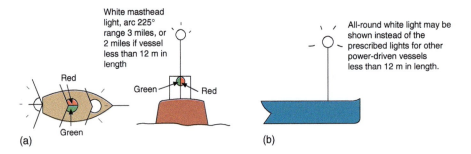

Figure 13.10 Small power-driven vessels: combined lantern. Each arc of the combined lantern must show from right ahead to two points abaft the vessels beam (112.5°). Range two miles or, if vessel less than 12 m, one mile.
(a) Power-driven vessel less than 20 m in length showing the sidelights in a combined lantern on the fore and aft line. Sternlight arc of visibility over 12 points of the compass, (135°), range two miles. *(b)* Power-driven vessel less than 7 m in length whose maximum speed does not exceed 7 knots.

(d) A power-driven vessel to which paragraph (a) or (c) of this Rule apply shall also comply with Rule 23(a)(ii).

(e) A vessel or object being towed, other than those mentioned in paragraph (g) of this Rule, shall exhibit:
 (i) sidelights;
 (ii) a sternlight;
 (iii) when the length of the tow exceeds 200 metres, a diamond shape where it can best be seen.

(f) Provided that any number of vessels being towed alongside or pushed in a group shall be lighted as one vessel:
 (i) a vessel being pushed ahead, not being part of a composite unit, shall exhibit at the forward end, sidelights;
 (ii) a vessel being towed alongside shall exhibit a sternlight and at the forward end, sidelights.

(g) An inconspicuous, partly submerged vessel or object, or combination of such vessels or objects being towed, shall exhibit:
 (i) if it is less than 25 metres in breadth, one all-round white light at or near the forward end and one at or near the after end except that dracones need not exhibit a light at or near the forward end;
 (ii) if it is 25 metres or more in breadth, two additional all-round white lights at or near the extremities of its breadth;
 (iii) if it exceeds 100 metres in length, additional all-round white lights between the lights prescribed in sub-paragraphs (i) and (ii) so that the distance between the lights shall not exceed 100 metres;
 (iv) a diamond shape at or near the aftermost extremity of the last vessel or object being towed and if the length of the tow exceeds 200 metres an additional diamond shape where it can best be seen and located as far forward as is practicable.

(h) Where from any sufficient cause it is impracticable for a vessel or object being towed to exhibit the lights or shapes prescribed in paragraph (e) or (g) of this Rule, all possible measures shall be taken to light the vessel or object towed or at least to indicate the presence of such vessel or object.

(*i*) Where from any sufficient cause it is impracticable for a vessel not normally engaged in towing operations to display the lights prescribed in paragraph (*a*) or (*c*) of this Rule, such vessel shall not be required to exhibit those lights when engaged in towing another vessel in distress or otherwise in need of assistance. All possible measures shall be taken to indicate the nature of the relationship between the towing vessel and the vessel being towed as authorized by Rule 36, in particular by illuminating the towline. A vessel engaged in towing operations that render her unable to deviate from her course shall, in addition to the black diamond day signal or the towing lights at night, exhibit the lights or shapes for a vessel restricted in her ability to manoeuvre.

Sources of reference: Rules 3, 24, 27 and Annex 1.

Mariners should exercise extreme caution when identifying a tug and tow, bearing in mind that a tug less than 50 m in length with a tow exceeding 200 m in length will show similar lights to a larger tug with a tow of less than 200 m when seen end on.

Sources of reference: Rules 21, 22, 24, and Annex 1.

The towing light is to provide a reference point to assist the tow to steer by, and provide identification for other vessels approaching from astern. It must be shown by all power-driven vessels when engaged in towing.

Sources of reference: Rules 21, 22, 24 and Annex 1.

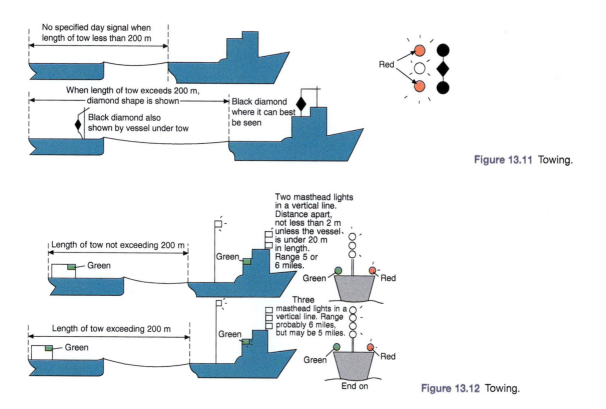

Figure 13.11 Towing.

Figure 13.12 Towing.

Figure 13.13 Towing.

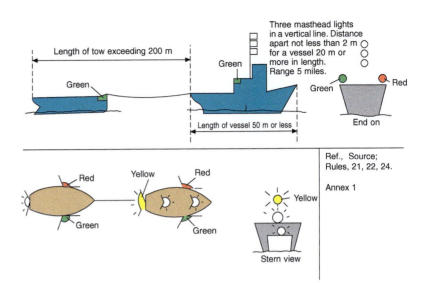

Length of tow exceeding 200 m

Green

Green

Three masthead lights in a vertical line. Distance apart not less than 2 m for a vessel 20 m or more in length. Range 5 miles.

Green

Red

Length of vessel 50 m or less

End on

Red

Yellow

Red

Green

Green

Yellow

Stern view

Ref., Source; Rules, 21, 22, 24.

Annex 1

Rule 25: sailing vessels underway and vessels under oars

(*a*) A sailing vessel underway shall exhibit:
(i) sidelights;
(ii) a sternlight.
(*b*) In a sailing vessel of less than 20 metres in length the lights prescribed in paragraph (*a*) of this Rule may be combined in one lantern carried at or near the top of the mast where it can best be seen.
(*c*) A sailing vessel underway may, in addition to the lights prescribed in paragraph (*a*) of this Rule, exhibit at or near the top

If the object being towed is less than 25 m in breadth, one all-round white light aft, one all-round white light forward.

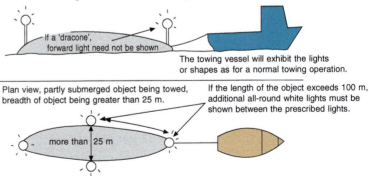

If a 'dracone', forward light need not be shown

The towing vessel will exhibit the lights or shapes as for a normal towing operation.

Plan view, partly submerged object being towed, breadth of object being greater than 25 m.

If the length of the object exceeds 100 m, additional all-round white lights must be shown between the prescribed lights.

more than 25 m

Figure 13.14 Vessels towing inconspicuous, partly submerged objects.

The object must exhibit, at or near its extremities, two additional all-round white lights By day any partially submerged object should exhibit at or near the aftermost extremity a black diamond and should the length of two exceed 200 m, an additional diamond located as far forward as practical.

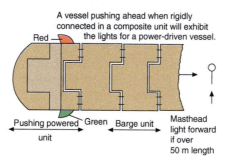

The pushing vessel will exhibit two masthead lights in a vertical line in the forward part, together with sidelights and sternlight.

The vessel or unit being pushed will exhibit sidelights forward.

Range of masthead lights 2, 3, or 5, 6 miles, depending on vessel's length

Green   Green

A vessel pushing ahead when rigidly connected in a composite unit will exhibit the lights for a power-driven vessel.

Red

Pushing powered unit   Green   Barge unit   Masthead light forward if over 50 m length

Figure 13.15 *Above*, power-driven vessel engaged in pushing operations when not rigidly connected. *Below*, power-driven vessel pushing ahead when rigidly connected.

of the mast, where they can best be seen, two all-round lights in a vertical line, the upper being red and the lower green, but these lights shall not be exhibited in conjunction with the combined lantern permitted by paragraph (*b*) of this Rule.

(*d*) (i)   A sailing vessel of less than 7 metres in length shall, if practicable, exhibit the lights prescribed in paragraph (*a*) or (*b*) of this Rule, but if she does not, she shall have ready at hand an electric torch or lighted lantern showing a white light which shall be exhibited in sufficient time to prevent collision.

(ii)   A vessel under oars may exhibit the lights prescribed in this Rule for sailing vessels, but if she does not, she shall have ready at hand an electric torch or lighted lantern showing a white light which shall be exhibited in sufficient time to prevent collision.

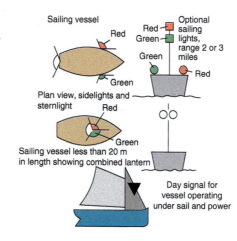

Sailing vessel

Red   Red   Optional sailing lights, range 2 or 3 miles
Green

Green   Red

Green

Plan view, sidelights and sternlight   Red

Sailing vessel less than 20 m in length showing combined lantern

Day signal for vessel operating under sail and power

Figure 13.16 Distinguishing lights and shape for vessel under sail and under combined sail and power.

(e) A vessel proceeding under sail when also being propelled by machinery shall exhibit forward where it can best be seen a conical shape, apex downwards.

Rule 26: fishing vessels

(a) A vessel engaged in fishing, whether underway or at anchor, shall exhibit only the lights and shapes prescribed in this Rule.

(b) A vessel when engaged in trawling, by which is meant the dragging through the water of a dredge net or other apparatus used as a fishing appliance, shall exhibit:
  (i)   two all-round lights in a vertical line, the upper being green and the lower white, or a shape consisting of two cones with their apexes together in a vertical line one above the other;
  (ii) a masthead light abaft of and higher than the all-round green light; a vessel of less than 50 metres in length shall not be obliged to exhibit such a light but may do so;
  (iii) when making way through the water, in addition to the lights prescribed in this paragraph, sidelights and a sternlight.

(c) A vessel engaged in fishing, other than trawling, shall exhibit;
  (i)   two all-round lights in a vertical line, the upper being red and the lower white, or a shape consisting of two cones with apexes together in a vertical line one above the other;
  (ii) when there is outlying gear extending more than 150 metres horizontally from the vessel, an all-round white light or a cone apex upwards in the direction of the gear;
  (iii) when making way through the water, in addition to the lights prescribed in this paragraph, sidelights and a sternlight.

(d) The additional signals described in Annex II to these Regulations apply to a vessel engaged in fishing in close proximity to other vessels engaged in fishing.

(e) A vessel when not engaged in fishing shall not exhibit the lights or shapes prescribed in this Rule, but only those prescribed for a vessel for her length.

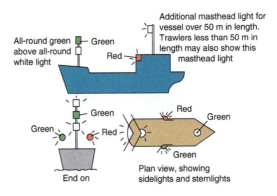

Figure 13.17 Lights on trawler engaged in fishing at night.

Rule 27: vessels not under command or restricted in their ability to manoeuvre

(a) A vessel not under command shall exhibit:
   (i) two all-round red lights in a vertical line where they can best be seen;
   (ii) two balls of similar shapes in a vertical line where they can best be seen;
   (iii) when making way through the water, in addition to the lights prescribed in this paragraph, sidelights and a sternlight.

(b) A vessel restricted in her ability to manoeuvre, except a vessel engaged in mine clearance operations, shall exhibit:
   (i) three all-round lights in a vertical line where they can best be seen. The highest and lowest of these lights shall be red and the middle light shall be white;
   (ii) three shapes in a vertical line where they can best be seen. The highest and lowest of these shapes shall be balls and the middle one a diamond;
   (iii) when making way through the water, a masthead light or lights, sidelights and a sternlight, in addition to the lights prescribed in subparagraph (i);
   (iv) when at anchor, in addition to the lights or shapes prescribed in sub-paragraphs (i) and (ii), the light, lights or shape prescribed in Rule 30.

(c) A power-driven vessel engaged in a towing operation such as severely restricts the towing vessel and her tow in their ability to deviate from their course shall, in addition to the lights or shapes prescribed in Rule 24(a), exhibit the lights or shapes prescribed in sub-paragraphs (b)(i) and (ii) of this Rule.

(d) A vessel engaged in dredging or underwater operations, when restricted in her ability to manoeuvre, shall exhibit the lights and shapes prescribed in subparagraphs (b)(i), (ii) and (iii) of this Rule and shall in addition, when an obstruction exists, exhibit:
   (i) two all-round red lights or two balls in a vertical line to indicate the side on which the obstruction exists;
   (ii) two all-round green lights or two diamonds in a vertical line to indicate the side on which another vessel may pass;
   (iii) when at anchor, the lights or shapes prescribed in this paragraph instead of the lights or shape prescribed in Rule 30.

(e) Whenever the size of a vessel engaged in diving operations makes it impracticable to exhibit all lights and shapes prescribed in paragraph (d) of this Rule, the following shall be exhibited:
   (i) three all-round lights in a vertical line where they can best be seen. The highest and lowest of these lights shall be red and the middle light shall be white;

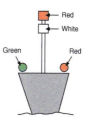

A vessel engaged in fishing, but not trawling, where the nets do not extend more than 150 m into the sea.

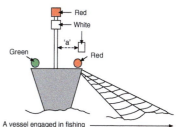

A vessel engaged in fishing with nets or lines extending over the port side more than 150 m into the sea. Distance 'a' not less than 2 m and not more than 6 m.

Figure 13.18 Fishing vessel other than a trawler at night. Sources of reference: Rule 26 and Annex 1 Section 4a.

A vessel of unspecified length engaged in fishing.

A vessel engaged in fishing, other than a trawler, with nets or lines extending more than 150 metres into the sea, with starboard side exhibiting a single cone apex up in the direction of the extending gear.

Figure 13.19 Fishing vessel's day signals. Sources of reference: Rule 26 and Annex 1, Section 6.

(ii) a rigid replica of the International Code flag 'A' not less than 1 metre in height. Measures shall be taken to ensure its all-round visibility.

(*f*) A vessel engaged in mine clearance operations shall in addition to the lights prescribed for a power-driven vessel in Rule 23 or to the lights or shape prescribed for a vessel at anchor in Rule 30 as appropriate, exhibit three all-round green lights or three balls. One of these lights or shapes shall be exhibited near the foremast head and one at each end of the fore yard. These lights or shapes indicate that it is dangerous for another vessel to approach within 1000 metres of the mine clearance vessel.

(*g*) Vessels of less than 12 metres in length, except those engaged in diving operations, shall not be required to exhibit the lights and shapes prescribed in this Rule.

(*h*) The signals prescribed in this Rule are not signals of vessels in distress and requiring assistance. Such signals are contained in Annex IV to these Regulations.

Vessels not under command

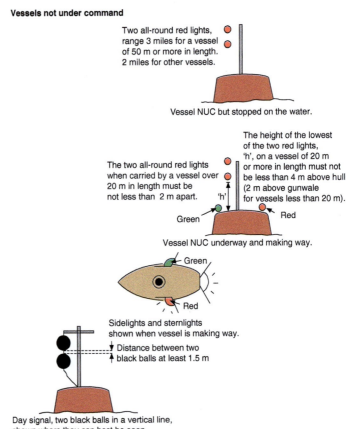

Two all-round red lights, range 3 miles for a vessel of 50 m or more in length. 2 miles for other vessels.

Vessel NUC but stopped on the water.

The two all-round red lights when carried by a vessel over 20 m in length must be not less than 2 m apart.

The height of the lowest of the two red lights, 'h', on a vessel of 20 m or more in length must not be less than 4 m above hull (2 m above gunwale for vessels less than 20 m).

Green   'h'   Red

Vessel NUC underway and making way.

Green

Red

Sidelights and sternlights shown when vessel is making way.

Distance between two black balls at least 1.5 m

Day signal, two black balls in a vertical line, shown where they can best be seen.

Figure 13.20 Vessel not under command (NUC). Sources of reference: Rules 22, 27 and Annex 1, Section 6.

Vessels restricted in ability to manoeuvre

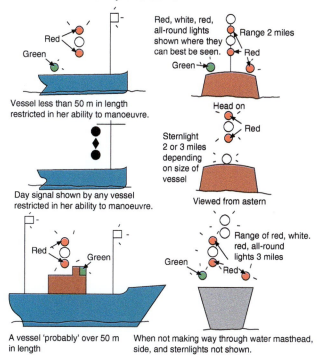

Figure 13.21 Vessel restricted in her ability to manoeuvre. Sources of reference: Rules 3, 21, 22, 27 and Annex 1.

A vessel restricted in her ability to manoeuvre, when at anchor

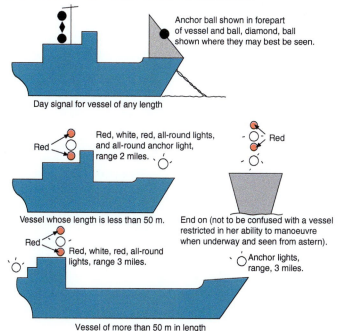

Figure 13.22 Vessel restricted in her ability to manoeuvre, when at anchor. Sources of reference: Rules 22, 27, 30 and Annex 1.

Figure 13.23 Vessels engaged in replenishment at sea, restricted in their ability to manoeuvre. Sources of reference: Rules 3, 27, 36 and Annex 1.

Replenishment at sea

Two vessels in close proximity, day signal black ball, diamond, ball, shown where they can best be seen.

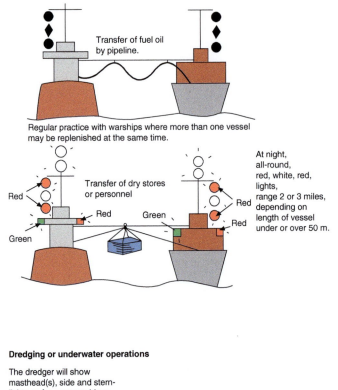

Transfer of fuel oil by pipeline.

Regular practice with warships where more than one vessel may be replenished at the same time.

Red

Green

Transfer of dry stores or personnel

Red

Green

Red

Red

At night, all-round, red, white, red, lights, range 2 or 3 miles, depending on length of vessel under or over 50 m.

Dredging or underwater operations

The dredger will show masthead(s), side and stern-lights as for a power-driven vessel underway.
In addition, she shall show the lights for a vessel restricted in her ability to manoeuvre, namely red, white, red, all-round lights, range 2 or 3 miles

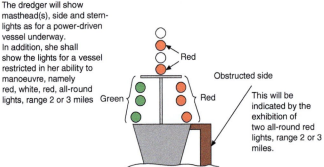

Red

Green

Obstructed side

Red

This will be indicated by the exhibition of two all-round red lights, range 2 or 3 miles.

Dredger underway and making way through the water.

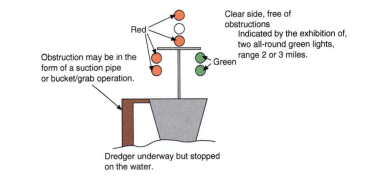

Red

Clear side, free of obstructions
Indicated by the exhibition of, two all-round green lights, range 2 or 3 miles.

Green

Obstruction may be in the form of a suction pipe or bucket/grab operation.

Dredger underway but stopped on the water.

Figure 13.24 Vessel engaged in dredging or underwater operations. Sources of reference: Rules 3, 22 and 27.

Dredging operations by day

Clear side free of obstructions

Obstructed side

Indicated by the exhibition of two black diamond shapes, at least 1.5 m apart.

Indicated by the exhibition of two black balls in a vertical line, not less than 1.5 m apart.

A vessel engaged in dredging or underwater operations underway and making way, or stopped or at anchor – day signal as seen from right ahead or right astern.

Figure 13.25 Vessel engaged in dredging operations by day. Sources of reference: Rules 3, 27 and Annex 1.

Vessels engaged in mine clearance

In addition to the lights for a power-driven vessel or for a vessel at anchor, vessels engaged in mine clearance shall exhibit three all-round green lights, one at the foremast head and one at each yard.

By day three black balls in place of the all-round green lights.

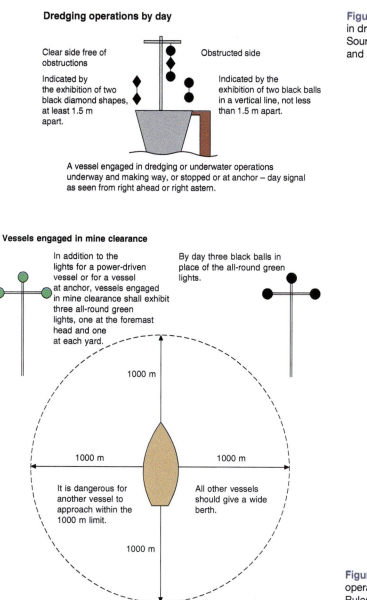

1000 m

1000 m

1000 m

1000 m

It is dangerous for another vessel to approach within the 1000 m limit.

All other vessels should give a wide berth.

Figure 13.26 Mine clearance operations. Sources of reference: Rules 3, 22, 27(f), 30 and Annex 1.

Rule 28: vessels constrained by their draught

A vessel constrained by her draught may, in addition to the lights prescribed for power-driven vessels in Rule 23, exhibit where they can best be seen three all-round red lights in a vertical line, or a cylinder.

Figure 13.27 Vessel constrained by her draught. Sources of reference; Rules 3, 22, 28 and Annex 1.

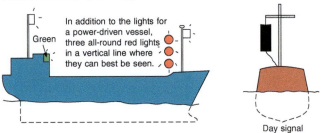

Vessel constrained by her draught

Green

In addition to the lights for a power-driven vessel, three all-round red lights in a vertical line where they can best be seen.

Day signal

Rule 29: pilot vessels

(a) A vessel engaged on pilotage duty shall exhibit:
   (i) at or near the masthead, two all-round lights in a vertical line, the upper being white and the lower red;
   (ii) when underway, in addition, sidelights and a sternlight;
   (iii) when at anchor, in addition to the lights prescribed in sub-paragraph (i), the light, lights or shape prescribed in Rule 30 for vessels at anchor.

(b) A pilot vessel when not engaged on pilotage duty shall exhibit the lights or shapes prescribed for a similar vessel of her length.

Rule 30: anchored vessels and vessels aground

(a) A vessel at anchor shall exhibit where it can best be seen:
   (i) in the fore part, an all-round white light or one ball;
   (ii) at or near the stern and at a lower level than the light prescribed in subparagraph (i), an all-round white light.

(b) A vessel of less than 50 metres in length may exhibit an all-round white light where it can best be seen instead of the lights prescribed in paragraph (a) of this Rule.

Plate 120 Example of pilot boat out of Lisbon, Portugal, seen at high speed at sea.

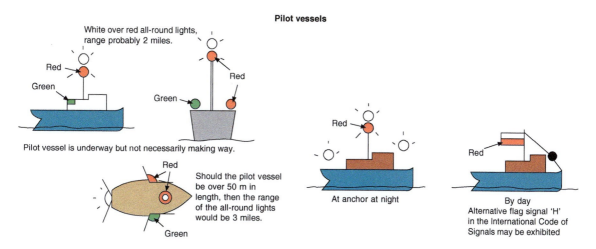

Figure 13.28 Pilot vessel engaged on pilotage duty. Sources of reference: Rules 21, 22, 29, 30, 35(i) and Annex 1.

(c) A vessel at anchor may, and a vessel of 100 metres and more in length shall, also use the available working or equivalent lights to illuminate her decks.

(d) A vessel aground shall exhibit the lights prescribed in paragraph (a) or (b) of this Rule and in addition, where they can best be seen:

   (i) two all-round red lights in a vertical line;

   (ii) three balls in a vertical line.

(e) A vessel of less than 7 metres in length, when at anchor, not in or near a narrow channel, fairway or anchorage, or where other vessels normally navigate, shall not be required to exhibit the lights or shape prescribed in paragraphs (a) and (b) of this Rule.

(f) A vessel of less than 12 metres in length, when aground, shall not be required to exhibit the lights or shapes prescribed in sub-paragraphs (d)(i) and (ii) of this Rule.

Rule 31: seaplanes and WIG craft

Where it is impracticable for a seaplane or a WIG craft to exhibit lights and shapes of the characteristics or in the positions prescribed in the Rules of this Part she shall exhibit lights and shapes as closely similar in characteristics and position as is possible.

Part D: Sound and Light Signals

Rule 32: definitions

(a) The word 'whistle' means any sound-signalling appliance capable of producing the prescribed blasts and which complies with the specifications in Annex III to these Regulations.

Figure 13.29 Vessel at anchor. Sources of reference: Rules 22, 30, and Annex 1, Sections 2(*k*) and 6(*a*).

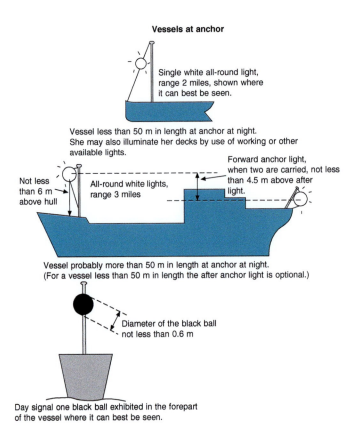

Vessels at anchor

Single white all-round light, range 2 miles, shown where it can best be seen.

Vessel less than 50 m in length at anchor at night. She may also illuminate her decks by use of working or other available lights.

Not less than 6 m above hull

All-round white lights, range 3 miles

Forward anchor light, when two are carried, not less than 4.5 m above after light.

Vessel probably more than 50 m in length at anchor at night. (For a vessel less than 50 m in length the after anchor light is optional.)

Diameter of the black ball not less than 0.6 m

Day signal one black ball exhibited in the forepart of the vessel where it can best be seen.

(*b*) The term 'short blast' means a blast of about one second's duration.

(*c*) The term 'prolonged blast' means a blast of from four to six seconds' duration.

Rule 33: equipment for sound signals

(*a*) A vessel of 12 metres or more in length shall be provided with a whistle, a vessel of 20 metres or more in length shall be provided with a bell in addition to a whistle, and a vessel of 100 metres or more in length shall, in addition, be provided with a gong, the tone and sound of which cannot be confused with that of the bell. The whistle, bell and gong shall comply with the specifications in Annex III to these Regulations. The bell or gong or both may be replaced by other equipment having the same respective sound characteristics, provided that manual sounding of the prescribed signals shall always be possible.

(*b*) A vessel of less than 12 metres in length shall not be obliged to carry the sound-signalling appliances prescribed in paragraph (*a*) of this Rule but if she does not, she shall be provided with some other means of making an efficient sound signal.

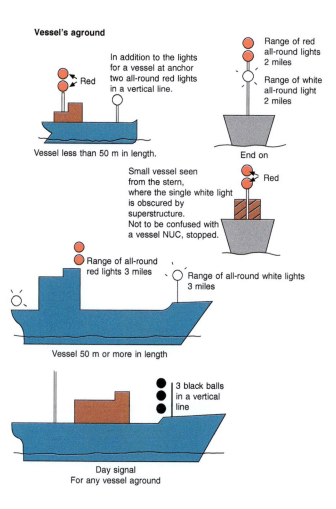

Vessel's aground

In addition to the lights for a vessel at anchor two all-round red lights in a vertical line.

Red

Vessel less than 50 m in length.

Range of red all-round lights 2 miles

Range of white all-round light 2 miles

End on

Small vessel seen from the stern, where the single white light is obscured by superstructure. Not to be confused with a vessel NUC, stopped.

Red

Range of all-round red lights 3 miles

Range of all-round white lights 3 miles

Vessel 50 m or more in length

3 black balls in a vertical line

Day signal
For any vessel aground

Figure 13.30 Vessel aground. Sources of reference: Rules 21, 22, 30, and Annex 1.

Rule 34: manoeuvring and warning signals

(*a*) When vessels are in sight of one another, a power-driven vessel underway, when manoeuvring as authorized or required by these Rules, shall indicate that manoeuvre by the following signals on her whistle:

- one short blast to mean 'I am altering my course to starboard';
- two short blasts to mean 'I am altering my course to port';
- three short blasts to mean 'I am operating astern propulsion'.

(*b*) Any vessel may supplement the whistle signals prescribed in paragraph (*a*) of this Rule by light signals, repeated as appropriate, whilst the manoeuvre is being carried out:

(i) these light signals shall have the following significance:

- one flash to mean 'I am altering my course to starboard';
- two flashes to mean 'I am altering my course to port';
- three flashes to mean 'I am operating astern propulsion';

(ii) the duration of each flash shall be about one second, the interval between flashes shall be about one second, and the

interval between successive signals shall be not less than
ten seconds;

(iii) the light used for this signal shall, if fitted, be an all-round
white light, visible at a minimum range of 5 miles, and shall
comply with the provisions of Annex I to these Regulations.

(c) When in sight of one another in a narrow channel or fairway:

(i) a vessel intending to overtake another shall in compliance
with Rule 9(e)(i) indicate her intention by the following
signals on her whistle:

• two prolonged blasts followed by one short blast to
mean 'I intend to overtake you on your starboard side';

• two prolonged blasts followed by two short blasts to
mean 'I intend to overtake you on your port side';

(ii) the vessel about to be overtaken when acting in accordance
with Rule 9(e)(i) shall indicate her agreement by the following signal on her whistle:

• one prolonged, one short, one prolonged and one short
blast, in that order.

(d) When vessels in sight of one another are approaching each
other and from any cause either vessel fails to understand the
intentions or actions of the other, or is in doubt whether sufficient action is being taken by the other to avoid collision, the
vessel in doubt shall immediately indicate such doubt by giving
at least five short and rapid blasts on the whistle. Such signal
may be supplemented by a light signal of at least five short and
rapid flashes.

(e) A vessel nearing a bend or an area of a channel or fairway
where other vessels may be obscured by an intervening obstruction shall sound one prolonged blast. Such signal shall be
answered with a prolonged blast by any approaching vessel
that may be within hearing around the bend or behind the
intervening obstruction.

(f) If whistles are fitted on a vessel at a distance apart of more than
100 metres, one whistle only shall be used for giving manoeuvring and warning signals.

Both vessels 'X' and 'Y' are keeping to their respective starboard
sides of the channel, as laid down by Rule 9(a). On approaching a
bend or obstruction where the intended path is obscured, a vessel
should sound one prolonged blast. Should any other vessel be
approaching the same bend or obstruction and be within hearing,
she should answer such a signal with a prolonged blast of her own.

Rule 35: sound signals in restricted visibility

In or near an area of restricted visibility, whether by day or night,
the signals prescribed in this Rule shall be used as follows:

(a) A power-driven vessel making way through the water shall sound
at intervals of not more than 2 minutes one prolonged blast.

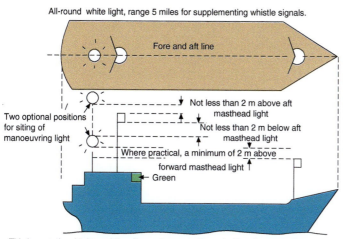

All-round white light, range 5 miles for supplementing whistle signals.

Fore and aft line

Two optional positions for siting of manoeuvring light

Not less than 2 m above aft masthead light

Not less than 2 m below aft masthead light

Where practical, a minimum of 2 m above forward masthead light

Green

This is an optional light, and therefore not necessarily carried by all vessels. When fitted, it should be in the same plane as the masthead lights and conform as near as is practicable to specifications in Annex 1.

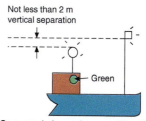

Figure 13.31 Manoeuvring light. Sources of reference: Rules 34, 36, and Annex 1, Section 12.

Not less than 2 m vertical separation

Green

On a vessel where only one masthead light is carried, the manoeuvring light, if fitted, should be carried where it can best be seen, not less than 2 m vertically apart from the masthead light.

(b) A power-driven vessel underway but stopped and making no way through the water shall sound at intervals of not more than 2 minutes two prolonged blasts in succession with an interval of about 2 seconds between them.

(c) A vessel not under command, a vessel restricted in her ability to manoeuvre, a vessel constrained by her draught, a sailing vessel, a vessel engaged in fishing and a vessel engaged in towing or pushing another vessel shall, instead of the signals prescribed in paragraphs (a) or (b) of this Rule, sound at intervals of not more than 2 minutes three blasts in succession, namely one prolonged followed by two short blasts.

(d) A vessel engaged in fishing, when at anchor, and a vessel restricted in her ability to manoeuvre when carrying out her

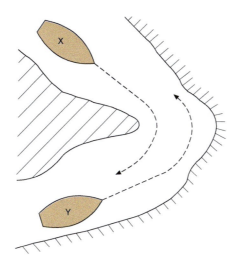

Figure 13.32 Two vessels approaching a bend or an obstruction which obscures the vision of each from the other. Sources of reference: Rules 7, 8, 9, 35 and 36.

work at anchor, shall instead of the signals prescribed in paragraph (g) of this Rule sound the signal prescribed in paragraph (c) of this Rule.

(e) A vessel towed or if more than one vessel is towed the last vessel of the tow, if manned, shall at intervals of not more than 2 minutes sound four blasts in succession, namely one prolonged followed by three short blasts. When practicable, this signal shall be made immediately after the signal made by the towing vessel.

(f) When a pushing vessel and a vessel being pushed ahead are rigidly connected in a composite unit they shall be regarded as a power-driven vessel and shall give the signals prescribed in paragraphs (a) or (b) of this Rule.

(g) A vessel at anchor shall at intervals of not more than one minute ring the bell rapidly for about 5 seconds. In a vessel of 100 metres or more in length the bell shall be sounded in the forepart of the vessel and immediately after the ringing of the bell the gong shall be sounded rapidly for about 5 seconds in the after part of the vessel. A vessel at anchor may in addition sound three blasts in succession, namely one short, one prolonged and one short blast, to give warning of her position and of the possibility of collision to an approaching vessel.

(h) A vessel aground shall give the bell signal and if required the gong signal prescribed in paragraph (g) of this Rule and shall, in addition, give three separate and distinct strokes on the bell immediately before and after the rapid ringing of the bell. A vessel aground may in addition sound an appropriate whistle signal.

(i) A vessel of 12 metres or more but less than 20 metres in length shall not be obliged to give the bell signals prescribed in paragraphs (g) and (h) of this Rule. However, if she does not, she shall make some other efficient sound signal at intervals of not more than 2 minutes.

(j) A vessel of less than 12 metres in length shall not be obliged to give the above-mentioned signals but, if she does not, shall make some other efficient sound signal at intervals of not more than 2 minutes.

(k) A pilot vessel when engaged on pilotage duty may in addition to the signals prescribed in paragraphs (a), (b) or (g) of this Rule sound an identity signal consisting of four short blasts.

Rule 36: signals to attract attention

If necessary to attract the attention of another vessel, any vessel may make light or sound signals that cannot be mistaken for any signal authorized elsewhere in these Rules, or may direct the beam of her searchlight in the direction of the danger, in such a way as not to embarrass any vessel. Any light to attract the attention of another vessel shall be such that it cannot be mistaken for any aid to navigation. For the purpose of this Rule the use of high intermittent or revolving lights, such as strobe lights, shall be avoided.

Rule 37: distress signals

When a vessel is in distress and requires assistance she shall use or exhibit the signals described in Annex IV to these Regulations.

Part E: Exemptions

Rule 38: exemptions

Any vessel (or class of vessels) provided that she complies with the requirements of the International Regulations for Preventing Collisions at Sea, 1960(a), the keel of which is laid or which is at a corresponding stage of construction before the entry into force of these Regulations may be exempted from compliance therewith as follows:

(a) The installation of lights with ranges prescribed in Rule 22, until four years after the date of entry into force of these Regulations.

(b) The installation of lights with colour specifications as prescribed in Section 7 of Annex I to these Regulations, until four years after the date of entry into force of these Regulations.

(c) The repositioning of lights as a result of conversion from Imperial to metric units and rounding off measurement figures, permanent exemption.

(d) (i) The repositioning of masthead lights on vessels of less than 150 metres in length, resulting from the prescriptions of Section 3(a) of Annex I to these Regulations, permanent exemption.

(ii) The repositioning of masthead lights on vessels of 150 metres or more in length, resulting from the prescriptions of Section 3(a) of Annex I to these Regulations, until nine years after the date of entry into force of these Regulations.

(e) The repositioning of masthead lights resulting from the prescriptions of Section 2(b) of Annex I to these Regulations, until nine years after the date of entry into force of these Regulations.

(f) The repositioning of sidelights resulting from the prescriptions of Sections 2(g) and 3(b) of Annex I to these Regulations, until nine years after the date of entry into force of these Regulations.

(g) The requirements for sound signal appliances prescribed in Annex III to these Regulations, until nine years after the date of entry into force of these Regulations.

(h) The repositioning of all-round lights resulting from the prescription of Section 9(b) of Annex I to these Regulations, permanent exemption.

Annex I: positioning and technical details of lights and shapes

1 Definition

The term 'height above the hull' means height above the uppermost continuous deck. This height shall be measured from the position vertically beneath the location of the light.

2 Vertical positioning and spacing of lights

(a) On a power-driven vessel of 20 metres or more in length the masthead lights shall be placed as follows:
  (i) the forward masthead light, or if only one masthead light is carried, then that light, at a height above the hull of not less than 6 metres, and, if the breadth of the vessel exceeds 6 metres, then at a height above the hull not less than such breadth, so however that the light need not be placed at a greater height above the hull than 12 metres;
  (ii) when two masthead lights are carried the after one shall be at least 4.5 metres vertically higher than the forward one.

(b) The vertical separation of masthead lights of power-driven vessels shall be such that in all normal conditions of trim the after light will be seen over and separate from the forward light at a distance of 1,000 metres from the stem when viewed from sea level.

(c) The masthead light of a power-driven vessel of 12 metres but less than 20 metres in length shall be placed at a height above the gunwale of not less than 2.5 metres.

(d) A power-driven vessel of less than 12 metres in length may carry the uppermost light at a height of less than 2.5 metres above the gunwale. When however a masthead light is carried in addition to sidelights and a sternlight, then such masthead light shall be carried at least 1 metre higher than the sidelights.

(e) One of the two or three masthead lights prescribed for a power-driven vessel when engaged in towing or pushing another vessel shall be placed in the same position as either the forward masthead light or the after masthead light: provided that, if carried on the aftermast, the lowest after masthead light shall be at least 4.5 metres vertically higher than the forward masthead light.

(f) (i) The masthead light or lights prescribed in Rule 23(a) shall be so placed as to be above and clear of all other lights and obstructions except as described in sub-paragraph (ii).
  (ii) When it is impracticable to carry the all-round lights prescribed by Rule 27(b)(i) or Rule 28 below the masthead lights, they may be carried above the after masthead light(s) or vertically in between the forward masthead light(s) and after masthead light(s), provided that in the latter case the

requirement of Section 3(c) of this Annex shall be complied with.

(g) The sidelights of a power-driven vessel shall be placed at a height above the hull not greater than three-quarters of that of the forward masthead light. They shall not be so low as to be interfered with by deck lights.

(h) The sidelights, if in a combined lantern and carried on a power-driven vessel of less than 20 metres in length, shall be placed not less than 1 metre below the masthead light.

(i) When the Rules prescribe two or three lights to be carried in a vertical line, they shall be spaced as follows:
  (i) on a vessel of 20 metres in length or more such lights shall be spaced not less than 2 metres apart, and the lowest of these lights shall, except where a towing light is required, be placed at a height of not less than 4 metres above the hull;
  (ii) on a vessel of less than 20 metres in length such lights shall be spaced not less than 1 metre apart and the lowest of these lights shall, except where a towing light is required, be placed at a height of not less than 2 metres above the hull;
  (iii) when three lights are carried they shall be equally spaced.

(j) The lower of the two all-round lights prescribed for a vessel when engaged in fishing shall be at a height above the sidelights not less than twice the distance between the two vertical lights.

(k) The forward anchor light prescribed in Rule 30(a)(i), when two are carried, shall not be less than 4.5 metres above the after one. On a vessel of 50 metres or more in length this forward anchor light shall be placed at a height of not less than 6 metres above the hull.

3 Horizontal positioning and spacing of lights

(a) When two masthead lights are prescribed for a power-driven vessel, the horizontal distance between them shall not be less than one-half of the length of vessel but need not be more than 100 metres. The forward light shall be placed not more than one-quarter of the length of the vessel from the stem.

(b) On a power-driven vessel of 20 metres or more in length the sidelights shall not be placed in front of the forward masthead lights. They shall be placed at or near the side of the vessel.

(c) When the lights prescribed in Rule 27(b)(i) or Rule 28 are placed vertically between the forward masthead lights(s) and the after masthead light(s) these all-round lights shall be placed at a horizontal distance of not less than 2 metres from the fore and aft centreline of the vessel in the athwartship direction.

4 Details of location of direction-indicating lights for fishing vessels, dredgers and vessels engaged in underwater operations

(a) The light indicating the direction of the outlying gear from a vessel engaged in fishing as prescribed in Rule 26(c)(ii) shall be placed at a horizontal distance of not less than 2 metres and not more than 6 metres away from the two all-round red and white lights. This light shall be placed not higher than the all-round white light prescribed in Rule 26(c)(i) and not lower than the sidelights.

(b) The lights and shapes on a vessel engaged in dredging or underwater operations to indicate the obstructed side and/or the side on which it is safe to pass, as prescribed in Rule 27(d)(i) and (ii), shall be placed at the maximum practical horizontal distance, but in no case less than 2 metres, from the lights or shapes prescribed in Rule 27(b)(i) and (ii). In no case shall the upper of these lights or shapes be at a greater height than the lower of the three lights or shapes prescribed in Rule 27(b)(i) and (ii).

5 Screens for sidelights

The sidelights of vessels of 20 metres or more in length shall be fitted with inboard screens painted matt black, and meeting the requirements of Section 9 of this Annex. On vessels of less than 20 metres in length the sidelights, if necessary to meet the requirements of Section 9 of this Annex, shall be fitted with inboard matt black screens. With a combined lantern, using a single vertical filament and a very narrow division between the green and red sections, external screens need not be fitted.

6 Shapes

Shapes shall be black and of the following sizes:

(a) (i) a ball shall have a diameter of not less than 0.6 metre;
    (ii) a cone shall have a base diameter of not less than 0.6 metre and a height equal to its diameter;
    (iii) a cylinder shall have a diameter of at least 0.6 metre and a height of twice its diameter;
    (iv) a diamond shape shall consist of two cones as defined in (ii) above having a common base.

(b) The vertical distance between shapes shall be at least 1.5 metres.

(c) In a vessel of less than 20 metres in length shapes of lesser dimensions but commensurate with the size of the vessel may be used and the distance apart may be correspondingly reduced.

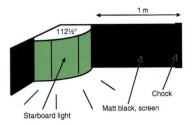

Figure 13.33 Starboard navigation sidelight.

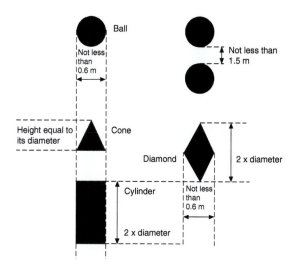

Figure 13.34 Shape specifications for vessel of 20 m or more in length. All shapes are black.

7 Colour specification of lights

The chromaticity of all navigation lights shall conform to the following standards, which lie within the boundaries of the area of the diagram specified for each colour by the International Commission on Illumination (CIE).

In a vessel less than 20 m in length the shapes' dimensions may be less than those given, but must be commensurate with the size of the vessel. Distance between shapes may also be correspondingly reduced.

The boundaries of the area for each colour are given by indicating the corner coordinates, which are as follows:

Table 13.1

(i)	White						
	x	0.525	0.525	0.452	0.310	0.310	0.443
	y	0.382	0.440	0.440	0.348	0.283	0.382
(ii)	Green						
	x	0.028	0.009	0.300	0.203		
	y	0.385	0.723	0.511	0.356		
(iii)	Red						
	x	0.680	0.660	0.735	0.721		
	y	0.320	0.320	0.265	0.259		
(iv)	Yellow						
	x	0.612	0.618	0.575	0.575		
	y	0.382	0.382	0.425	0.406		

8 Intensity of lights

(a) The minimum luminous intensity of lights shall be calculated by using the formula:

$$I = 3.43 \times 10^6 \times T \times D^2 \times K^{-D}$$

where I is luminous intensity in candelas under service conditions,
T is threshold factor 2×10^{-7} lux,
D is range of visibility (luminous range) of the light in nautical miles,
K is atmospheric transmissivity.

For prescribed lights the value of K shall be 0.8, corresponding to a meteorological visibility of approximately 13 nautical miles.

(b) A selection of figures derived from the formula is given in the following table:

Table 13.2

Range of visibility (luminous range) of light in nautical miles	Luminous intensity of light in candelas for K = 0.8
D	I
1	0.9
2	4.3
3	12
4	27
5	52
6	94

Note: The maximum luminous intensity of navigation lights should be limited to avoid undue glare. This shall not be achieved by a variable control of the luminous intensity.

9 Horizontal sectors

(a) (i) In the forward direction, sidelights as fitted on the vessel shall show the minimum required intensities. The intensities shall decrease to reach practical cut-off between 1 degree and 3 degrees outside the prescribed sectors.

(ii) For sternlights and masthead lights and at 22.5 degrees abaft the beam for sidelights, the minimum required intensities shall be maintained over the arc of the horizon up to 5 degrees within the limits of the sectors prescribed in Rule 21. From 5 degrees within the prescribed sectors the intensity may decrease by 50 per cent up to the prescribed limits; it shall decrease steadily to reach practical cut-off at not more than 5 degrees outside the prescribed sectors.

(b) (i) All-round lights shall be so located as not to be obscured by masts, topmasts or structures within angular sectors of more than 6 degrees except anchor lights prescribed in Rule 30, which need not be placed at an impracticable height above the hull.

(ii) If it is impracticable to comply with paragraph (b)(i) of this section by exhibiting only one all-round light, two all-round lights shall be used suitably positioned or screened so that they appear, as far as practicable, as one light at a distance of one mile.

10 Vertical sectors

(a) The vertical sectors of electric lights as fitted, with the exception of lights on sailing vessels shall ensure that:

(i) at least the required minimum intensity is maintained at all angles from 5 degrees above to 5 degrees below the horizontal;

(ii) at least 60 per cent of the required minimum intensity is maintained from 7.5 degrees above to 7.5 degrees below the horizontal.

(b) In the case of sailing vessels the vertical sectors of electric lights as fitted shall ensure that:

(i) at least the required minimum intensity is maintained at all angles from 5 degrees above to 5 degrees below the horizontal;

(ii) at least 50 per cent of the required minimum intensity is maintained from 25 degrees above to 25 degrees below the horizontal.

(c) In the case of lights other than electric these specifications shall be met as closely as possible.

11 Intensity of non-electric lights

Non-electric lights shall so far as practicable comply with the minimum intensities, as specified in the table given in Section 8 of this Annex.

12 Manoeuvring light

Notwithstanding the provisions of paragraph 2(f) of this Annex the manoeuvring light described in Rule 34(b) shall be placed in the same fore and aft vertical plane as the masthead light or lights and, where practicable, at a minimum height of 2 metres vertically above the forward masthead light, provided that it shall be carried not less than 2 metres vertically above or below the after masthead light. On a vessel where only one masthead light is carried the manoeuvring light, if fitted, shall be carried where it can best be seen, not less than 2 metres vertically apart from the masthead light.

13 High-speed craft

(a) The masthead light of high-speed craft may be placed at a height related to the breadth of the craft lower than that

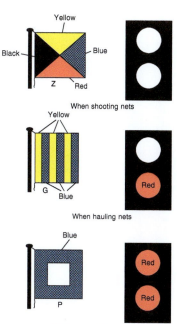

Figure 13.35 Additional signals for vessels fishing.

prescribed in paragraph 2(a)(i) of this Annex, provided that the base angle of the isosceles triangles formed by the sidelights and masthead light, when seen in end elevation, is not less than 27 degrees.

(b) On high-speed craft of 50 metres or more in length, the vertical separation between foremast and mainmast light of 4.5 metres required by paragraph 2(a)(ii) of this Annex may be modified provided that such distance shall not be less than the value determined by the following formula:

$$y = \frac{(a+17\psi)C}{1000} + 2$$

where: y is the height of the mainmast light above the foremast light in metres;

a is the height of the foremast light above the water surface in service condition in metres;

ψ is the trim in service condition in degrees;

C is the horizontal separation of masthead lights in metres.

14 Approval

The construction of lights and shapes and the installation of lights on board the vessel shall be to the satisfaction of the appropriate authority of the State.

Annex II: Additional Signals for Fishing Vessels Fishing in Close Proximity

1 General

The lights mentioned herein shall, if exhibited in pursuance of Rule 26(d), be placed where they can best be seen. They shall be at

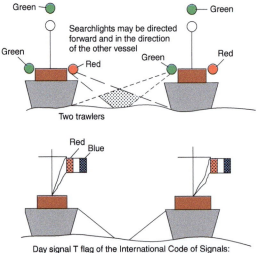

Figure 13.36 Fishing vessels engaged in pair trawling. Sources of reference: Rules 3, 9(c), 10, 18, 26, 36, Annex 1.

Day signal T flag of the International Code of Signals: 'Keep clear of me. I am engaged in pair trawling'.

least 0.9 metre apart but at a lower level than lights prescribed in Rule 26(*b*)(i) and (*c*)(i). The lights shall be visible all round the horizon at a distance of at least 1 mile but at a lesser distance than the lights prescribed by these Rules for fishing vessels.

2 Signals for trawlers

(*a*) Vessels when engaged in trawling, whether using demersal or pelagic gear, may exhibit:
  (i) when shooting their nets: two white lights in a vertical line;
  (ii) when hauling their nets: one white light over one red light in a vertical line;
  (iii) when the net has come fast upon an obstruction: two red lights in a vertical line.
(*b*) Each vessel engaged in pair trawling may exhibit:
  (i) by night, a searchlight directed forward and in the direction of the other vessel of the pair.
  (ii) when shooting or hauling their nets or when their nets have come fast upon an obstruction, the lights prescribed in 2(*a*) above.
(*c*) A vessel of less than 20 metres in length engaged in trawling, whether using demersal or pelagic gear or engaged in pair trawling, may exhibit the lights prescribed in paragraphs (a) or (b) of this Section, as appropriate.

3 Signals for purse seiners

Vessels engaged in fishing with purse seine gear may exhibit two yellow lights in a vertical line. These lights shall flash alternately every second and with equal light and occultation duration. These lights may be exhibited only when the vessel is hampered by its fishing gear.

Annex III: Technical Details of Sound Signal Appliances

1 Whistles

(*a*) *Frequencies and range of audibility*
  The fundamental frequency of the signal shall lie within the range 70–700 Hz.
  The range of audibility of the signal from a whistle shall be determined by those frequencies, which may include the fundamental and/or one or more higher frequencies, which lie within the range 180–700 Hz (±1 per cent) for a vessel of 20 metres or more in length, or 180–2100 Hz (±1 per cent) for a vessel of less than 20 metres in length and which provide the sound pressure levels specified in paragraph 1(*c*) below.

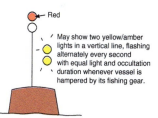

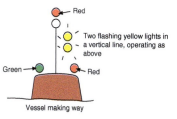

Figure 13.37 Special signals for purse seiners. Sources of reference: Annex II and M 587, July 1970.

(b) *Limits of fundamental frequencies*

To ensure a wide variety of whistle characteristics, the fundamental frequency of a whistle shall be between the following limits:

(i) 70–200 Hz, for a vessel 200 metres or more in length;

(ii) 130–350 Hz, for a vessel 75 metres but less than 200 metres in length;

(iii) 250–700 Hz, for a vessel less than 75 metres in length.

(c) *Sound signal intensity and range of audibility*

A whistle fitted in a vessel shall provide, in the direction of maximum intensity of the whistle and at a distance of 1 metre from it, a sound pressure level in at least one 1/3rd octave band within the range of frequencies 180–700 Hz (±1 per cent) for a vessel of 20 metres or more in length, or 180–2100 Hz (±1 per cent) for a vessel of less than 20 metres in length, of not less than the appropriate figure given in the table below.

Table 13.3

Length of vessel in metres	1/3 octave band level at 1 metre in dB referred to 2×10^{-5} N/m²	Audibility range in nautical miles
200 or more	143	2
75 but less than 200	138	1.5
20 but less than 75	130	1
Less than 20	120[1]	0.5
	115[2]	
	111[3]	

1 When the measured frequencies lie within the range 180–450 Hz.
2 When the measured frequencies lie within the range 450–800 Hz.
3 When the measured frequencies lie within the range 800–2100 Hz.

The range of audibility in the table above is for information and is approximately the range at which a whistle may be heard on its forward axis with 90 per cent probability in conditions of still air on board a vessel having average background noise level at the listening posts (taken to be 68 dB in the octave band centred on 250 Hz and 63 dB in the octave band centred on 500 Hz).

In practice the range at which a whistle may be heard is extremely variable and depends critically on weather conditions; the values given can be regarded as typical but under conditions of strong wind or high ambient noise level at the listening post the range may be much reduced.

(d) *Direction properties*

The sound pressure level of a directional whistle shall be not more than 4 dB below the prescribed sound pressure level on the axis at any direction in the horizontal plane within

±45 degrees of the axis. The sound pressure level at any other direction in the horizontal plane shall be not more than 10 dB below the prescribed sound pressure level on the axis, so that the range in any direction will be at least half the range on the forward axis. The sound pressure level shall be measured in that 1/3rd octave band which determines the audibility range.

(e) *Positioning of whistles*

When a directional whistle is to be used as the only whistle on a vessel, it shall be installed with its maximum intensity directed straight ahead.

A whistle shall be placed as high as practicable on a vessel, in order to reduce interception of the emitted sound by obstructions and also to minimize hearing damage risk to personnel. The sound pressure level of the vessel's own signal at listening posts shall not exceed 110 dB (A) and so far as practicable should not exceed 100 dB (A).

(f) *Fitting of more than one whistle*

If whistles are fitted at a distance apart of more than 100 metres, it shall be so arranged that they are not sounded simultaneously.

(g) *Combined whistle systems*

If due to the pressure of obstructions the sound field of a single whistle or of one of the whistles referred to in paragraph 1(f) above is likely to have a zone of greatly reduced signal level, it is recommended that a combined whistle system be fitted so as to overcome this reduction. For the purposes of the Rules a combined whistle system is to be regarded as a single whistle. The whistles of a combined system shall be located at a distance apart of not more than 100 metres and arranged to be sounded simultaneously. The frequency of any one whistle shall differ from those of the others by at least 10 Hz.

2 Bell or gong

(a) *Intensity of signal*

A bell or gong, or other device having similar sound characteristics shall produce a sound pressure level of not less than 100 dB at a distance of 1 metre from it.

(b) *Construction*

Bells and gongs shall be made of corrosion-resistant material and designed to give a clear tone. The diameter of the mouth of the bell shall be not less than 300 mm for vessels of 20 metres or more in length.

Where practicable, a power-driven bell striker is recommended to ensure constant force but manual operation shall be possible. The mass of the striker shall be not less than 3 per cent of the mass of the bell.

3 Approval

The construction of sound signal appliances, their performance and their installation on board the vessel shall be to the satisfaction of the appropriate authority of the State whose flag the vessel is entitled to fly.

Annex IV: distress signals

1   The following signals, used or exhibited either together or separately, indicate distress and need of assistance:
    (*a*) a gun or other explosive signal fired at intervals of about a minute;
    (*b*) a continuous sounding with any fog-signalling apparatus;
    (*c*) rockets or shells, throwing red stars fired one at a time at short intervals;
    (*d*) a signal made by radiotelegraphy or by any other signalling method consisting of the group ... − − − ... (SOS) in the Morse Code;
    (*e*) a signal sent by radiotelephony consisting of the spoken word 'Mayday';
    (*f*) the International Code Signal of distress indicated by N.C.;
    (*g*) a signal consisting of a square flag having above or below it a ball or anything resembling a ball;
    (*h*) flames on the vessel (as from a burning tar barrel, oil barrel, etc.);
    (*i*) a rocket parachute flare or a hand flare showing a red light;
    (*j*) a smoke signal giving off orange-coloured smoke;
    (*k*) slowly and repeatedly raising and lowering arms outstretched to each side;
    (*l*) the radiotelegraph alarm signal;
    (*m*) the radiotelephone alarm signal;
    (*n*) signals transmitted by emergency position-indicating radio beacons;
    (*o*) approved signals transmitted by radiocommunication systems, including survival craft radar responders.
2   The use or exhibition of any of the foregoing signals except for the purpose of indicating distress and need of assistance and the use of other signals which may be confused with any of the above signals is prohibited.
3   Attention is drawn to the relevant sections of the *International Code of Signals*, the *Merchant Ship Search and Rescue Manual* and the following signals:
    (*a*) a piece of orange-coloured canvas with either a black square and circle or other appropriate symbol (for identification from the air);
    (*b*) a dye marker.

Judging Another Vessel's Heading at Night

We know that the arc of a sidelight shows over 112.5°, ten points of the compass. Therefore we can assume that to an observer of a sidelight the direction of an observed ship's head must be related to the limits of that arc, otherwise the observer would see either the opposing sidelight or the sternlight. Refer to Figure 13.38.

The observer has narrowed the other vessel's heading from four points off the starboard quarter to two points off the starboard bow. By relating these limits to your own vessel's course, you may define the heading of the other vessel by points of the compass. If it is assumed in Figure 13.38, for example, that your own vessel's course is north, then the other vessel is heading between southeast and north-northeast.

The direction in which a sailing vessel is proceeding will depend on the wind direction. Let us assume your own vessel is moving north and a single red light is seen at three points off the starboard bow. Let us assume it is a sailing vessel, and consider her heading and direction of sailing. Refer to Figure 13.39.

Special Cases to Rules of the Road

Warships

Warships cannot always comply with the regulations concerning the disposition of navigation lights. Bearing this in mind, navigators should take extreme caution when navigating in their vicinity.

The aspect of an approaching warship may often be deceiving, especially when the separation between lights cannot be met by the

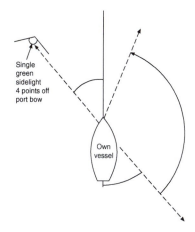

Figure 13.38 Judging heading from another vessel's green light.

First stage. Reverse the bearing of the observed light and you will get a line four points off the starboard quarter. This line gives one limit to the direction in which the other vessel is heading.

Second stage. From the reversed bearing traverse ten points to the left, which gives two points off the starboard bow. This is the second and final limit to the direction in which the other vessel is heading, as defined by the arc of visibility of the sidelight.

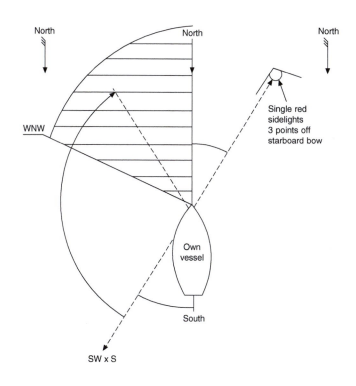

Figure 13.39 Judging heading of sailing vessel from red light.

First stage. Reverse the bearing of the observed light and you will get a line three points off the port quarter. This line gives one limit to the direction in which the other ship is heading.

Second stage. From the reversed bearing traverse ten points to the right, which gives three points off the port bow. This is the second limit to the direction in which the other vessel is heading, as defined by the arc of visibility of the sidelight.

Third stage. Assuming that the other vessel will not be able to sail closer to the wind than six points of the compass, a 'no go' area may be established six points either side of the wind direction. If the wind is blowing from the north, say, the other vessel is sailing between SW by S and WNW.

Figure 13.40 Submarine lights.

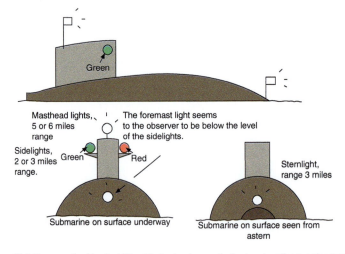

Green

Masthead lights,
5 or 6 miles
range

The foremast light seems
to the observer to be below the level
of the sidelights.

Sidelights,
2 or 3 miles
range.

Green　Red

Sternlight,
range 3 miles

Submarine on surface underway

Submarine on surface seen from
astern

Watchkeepers should set additional lookouts when navigating in submarine exercise areas or in the vicinity of ports servicing and accommodating submarines. Other warships often operate in conjunction with submarines and a wide berth is advised, if circumstances permit. Submarines at anchor exhibit normal anchor lights as for vessels of their length, but in addition will exhibit, midships, an all-round white light.

Some submarines are fitted with a quick-flashing yellow light for the purpose of identification in narrow channels and in dense traffic areas. This light, normally fitted above the after steaming light, flashes at a rate of 90 or more per minute.

very nature of the construction of the vessel. This separation between lights may lead to the misconception that there are two ships when in fact there is only one. Submarines also, when operating on or near the surface, may have the sternlight, which is carried close to the surface, obscured by wash or spray (see Figure 13.40).

Some warships of over 50 m in length, because of their construction, cannot be fitted with the second masthead light. This applies especially to certain frigates, minesweepers and boom defence vessels.

Aircraft carriers, although they generally have their sidelights positioned either side of the flight deck, sometimes have sidelights on either side of the island structure. Their red, white, red, all-round restricted lights should only be exhibited when they are engaged in the launching or recovery of aircraft, fixed wing or rotary wing (see Figure 13.41).

Collision Avoidance: High-Speed Craft

Despite all the new advances in ship design, including high-speed craft, the ColRegs still apply to all vessels on the high sea and it is stressed on all waters connected there to, navigable by seagoing vessels. This latter phrase is particularly important to high-speed craft operations, many of which extend into inland waterways.

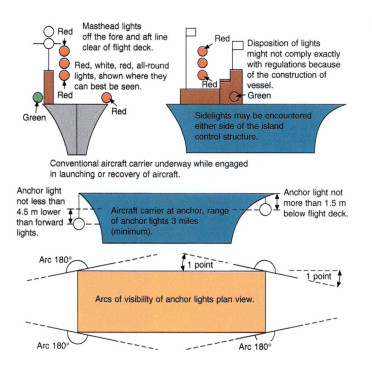

Figure 13.41 Aircraft carrier lights.

The profile of new tonnage and the light build features, often in aluminium, can influence the observed perspective from the point of view of other vessels. The modern ferry, as a result, may have an extremely low freeboard, and be unable to display navigation lights and signals in what is considered a normal configuration. (Ref., Regulation 1, part 'e', of the ColRegs allows for the closest possible compliance of the display of navigation lights and signals.)

Special features of certain types of craft also bring with them inherent problems. Hovercraft, for instance, generate considerable noise and as such operators of these craft may not readily hear sound manoeuvring signals made by other vessels. They also exhibit a flashing yellow light at 120 flashes or more per minute. A similar flashing yellow light, at a slower frequency, is relevant to some submarine types when navigating on the surface. Little wonder, then, that watch officers may be confused with what type of vessel they are dealing with.

The rate of flashing of the yellow light, is defined by Regulation 21(f) at 120 flashes or more per minute.

The ColRegs consider that hovercraft, hydrofoils and WIG craft are not classed as 'seaplanes', even when operating in the non-displacement mode, but are recognized as power-driven vessels. The fact remains that when such vessels are engaged at high speed, it may seem a prudent act of 'good seamanship' (Ref., Regulation 2a & 2b, Responsibilities) for these craft to take early action to avoid general shipping.

NB. It is pointed out that the yellow flashing light is only exhibited when the hovercraft is operating in the non-displacement mode (Ref., Regulation 23, part 'b' of the ColRegs). The purpose of the light is to indicate to mariners the type of craft and that her navigation lights may give a false indication of her direction of movement.

Plate 121 High-speed passenger
ferry operating in relatively calm
waters (source: Shutterstock).

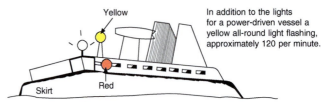

Figure 13.42 Hovercraft lights.

Yellow

In addition to the lights
for a power-driven vessel a
yellow all-round light flashing,
approximately 120 per minute.

Skirt   Red

High-Speed Craft: Features

The masthead light of high-speed craft with a length to breadth
ratio of less than 3.0 may be placed at a height related to the
breadth of the craft lower than that prescribed in paragraph $2(a)(i)$
of the Annex, provided that the base of the isosceles triangle formed
by the sidelights and masthead light when seen in end elevation is
not less than 27° (Additional Ref., Paragraph 13, of Annex 1, of the
ColRegs).

Rule $18(f)$ (i) specifically states that WIG craft, when taking off,
landing and in flight near the surface, shall keep well clear of *all*
other vessels and avoid impeding their navigation; while part (ii) of
the same Regulation states that WIG craft on the water surface
shall comply with the Rules of that part as a power-driven vessel.

High-Speed Craft: Watchkeeping
Practice

As with any other type of vessel, the watch arrangements for high-
speed craft evolve around a 'bridge team' which operates to recognized
routines. It should be realized that many of these craft operate on
precise routes, often with little deviation. Communication schedules

within VTS operations or other similar traffic control systems become regular activities during the navigation of the craft. The schedules also become well-known by local traffic and they tend to give a wide berth to specific routes. However, this does not relieve lookout duties and this prime function should be exercised with increased diligence.

It should be noted that one of the greatest hazards where the possibility of collision may exist is from other high-speed craft on the same or close intersecting routes. Various places around the world have recognized this danger and have organized specific tracks and fairway routes which contribute to overall safety. The English Channel and the Hong Kong areas immediately come to mind as falling into this category.

Bridge Officers are conscious of the fact that their increased speed creates a problem of 'wash'. This may not be as critical in open water as in confined waters, but recently a fisherman was swept overboard to his death from the deck of his own vessel. The observation of small craft, yachts, fishing craft, etc., is considered difficult enough on conventional vessels, but at speed from a high-speed craft detection needs to be sooner rather than later. Where fishing fleets or yacht flotillas are detected, a wider berth than usual is expected, coupled with a reduction of speed to reduce wash effects.

NB. The bridge of high-speed vessels are generally totally enclosed as it would be considered unwise to have personnel exposed when moving at 50-plus knots. To complement radar and lookout duties during the hours of darkness, many vessels are also fitted with night vision cameras to detect debris or floating hazards.

Oil Rig/Production Platform

The difference between a rig and a production platform is that a rig is of a temporary nature and moves from position to position, and a production platform is of a fixed nature and permanently attached to the sea bed. The safety zones for oil rigs and production platforms may vary from country to country, and in the absence of relevant information they must be assumed to exist.

Usually rigs and production platforms make extensive use of all available deck lighting as an additional aid to navigation as well as working illumination (see Figure 13.43).

For safety and profitability, production platforms need to fulfil the following conditions (by the very nature of their occupation and changing technology these criteria must be accepted as being variable):

- Height from sea bed to top of drilling derrick: 690 ft (210 m).
- Height of top deck from sea bed: 550 ft (167.6 m).
- Drilling radius covered from platform 9,000 ft (2,743 m).
- Directional drilling, maximum deviation, 55°.
- Nominal production per day: 125,000 barrels (variable).
- Total number of piles: 44.
- Number of piles in each leg: 11.
- Maximum depth of piles: 250 ft (76.2 m).
- Diameter of main flotation legs: 30 ft (9.1 m).
- Clearance of lowest deck from mean sea level: 77 ft (23.5 m).

Figure 13.43 Lights on oil rigs and production platforms.

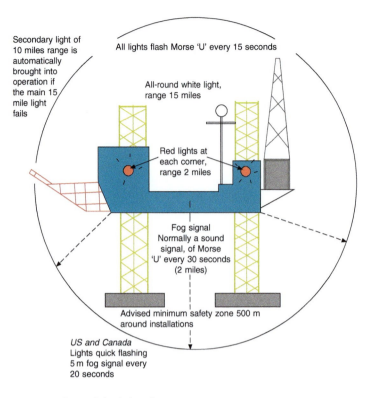

Secondary light of 10 miles range is automatically brought into operation if the main 15 mile light fails

All lights flash Morse 'U' every 15 seconds

All-round white light, range 15 miles

Red lights at each corner, range 2 miles

Fog signal
Normally a sound signal, of Morse 'U' every 30 seconds (2 miles)

Advised minimum safety zone 500 m around installations

US and Canada
Lights quick flashing
5 m fog signal every 20 seconds

- Number of deck levels: 3.
- Deck dimensions 175 ft (53.3 m) × 170 ft (51.8 m).

Accommodation on each platform will vary, but 200 men at any one time is not uncommon. Life-saving appliances generally include either Survival Systems International capsules or totally enclosed lifeboats together with life rafts.

Plate 122 Operational offshore installation (source: Shutterstock).

Plate 123 A jack-up mobile drilling rig under tow towards an operational site. Once in position the supporting structure legs are lowered to the sea bed.

Plate 124 Offshore fixed installation (source: Shutterstock).

Seaplane

The word 'seaplane' includes any aircraft designed to manoeuvre on the water. Non-displacement craft and seaplanes should be considered as vessels, but non-displacement craft are not to be considered as seaplanes.

Figure 13.44 Seaplane lights.

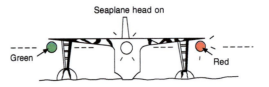

This seaplane on the water will exhibit the lights as for a power-driven vessel of her length. By the very nature of her construction the difference in height between white masthead lights and sidelights will be minimal, and may even appear in one horizontal line (see Figure 13.44).

Remote-Controlled Craft

These craft, normally 19.2 m in length, exhibit the usual navigation lights and shapes, together with 'not under command' signals. They are usually engaged with a controlling craft, which operates in or near the area and keeps a visual and radar watch for up to about 8 miles. Additional air cover is provided to cover greater ranges and ensure shipping safety.

Warning red fixed or red flashing lights may be exhibited at night, or red flags by day, to indicate target practice by aircraft, shore battery or ships at sea. Targets may be illuminated at night by bright red or orange flares. Mariners should be aware that the absence of such signals does not mean that a practice area does not exist (see Figure 13.45).

Additional information is given in the *Mariner's Handbook* and *Annual Summary of Notices to Mariners*. Questions and answers on the subject of rules of the road are contained in the Self Examiner Annex.

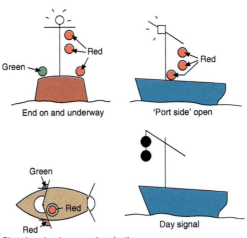

Figure 13.45 Lights and shape designating remote-controlled craft.

Plan view showing normal navigation lights together with NUC lights.

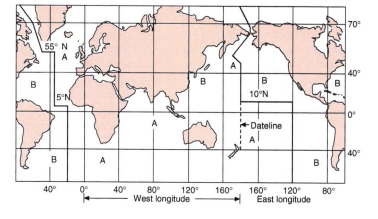

Figure 13.46 IALA maritime buoyage: areas covered.

IALA Maritime Buoyage Systems 'A' and 'B'

An acceptable worldwide buoyage system is not a new idea, and in 1973 the Technical Committee of the International Association of Lighthouse Authorities (IALA) attempted to provide such a system. The results showed that agreement on a single system could not be immediately achieved, but in conclusion found that an alternative double system 'A' and 'B' was in fact a practical proposition (Figure 13.46).

The rules governing the two systems being very similar, it was possible to combine them to form the IALA Maritime Buoyage System as we now know it. The system applies to all fixed or floating marks, other than lighthouses, sector lights, leading lights and marks, light vessels and lanbys. It serves to indicate the centrelines of channels and their sides, natural dangers such as sandbanks, as well as wrecks (described as new dangers when newly discovered), and also areas where navigation is subject to regulation.

Lateral marks employed in system 'B' are similar-shaped buoys but opposite in colour, namely green for port, red for starboard.

How the Systems Work

Each of the 'A' and 'B' systems comprises simplified lateral marks (port and starboard channel marks) and cardinal marks to be used in conjunction with the mariner's compass. The shapes of buoys, of which there are five, are common to both systems: can, conical, spherical, pillar and spar. In the case of can-, conical- and spherical-shaped buoys, these shapes indicate the side on which the buoy should be passed. The pillar and spar buoys provide indication to the mariner by colour and topmark, not by shape (see Figures 13.47 to 13.50).

Lateral marks are generally used for defining the navigable channel in and out of harbour. Port and starboard buoys are used in conjunction with the conventional direction of buoyage.

Cardinal marks are to be used by the mariner as guides or as an indication where the best navigable water may be encountered (Figure 13.51). They are used in conjunction with the mariner's

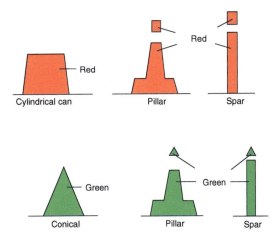

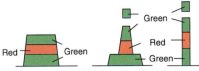

Figure 13.47 Buoys, System 'A'. Lights, when fitted, red for port and green for starboard, have any rhythm other than the composite group (2 + 1) as used for preferred channel marks.

Preferred channel to starboard, green with red stripe, and green light flashing 2 + 1 (group flashing).

Figure 13.49 Buoys, System 'B'. Lights, when fitted, green for port and red for starboard, have any rhythm other than the composite group (2 + 1) as used for preferred channel marks.

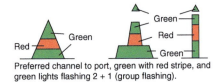

Preferred channel to starboard, red with green stripe, and red light flashing 2 + 1 (group flashing).

Preferred channel to port, green with red stripe, and green lights flashing 2 + 1 (group flashing).

Figure 13.48 Preferred channel marks, System 'A'.

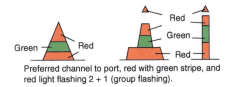

Preferred channel to port, red with green stripe, and red light flashing 2 + 1 (group flashing).

Figure 13.50 Preferred channel marks, System 'B'.

compass, e.g. the mariner would be expected to pass northward of a north cardinal mark. Topmarks and a colour scheme of black and yellow go to indicate the distinction and intention of the buoy during the hours of daylight. Lights are always white, and the rhythm distinctive to separate north, south, east and west.

Safe-water marks, isolated-danger marks and special marks are illustrated in Figures 13.52 and 13.53, and UK buoys in Figure 13.54.

Local direction of buoyage is the direction to be taken by the mariner when approaching a river estuary, harbour or other water-way from a seaward direction.

General direction of buoyage is a direction about a land mass, as predetermined by the buoyage authorities (Figure 13.55). The general direction for any given area may be obtained from the sailing directions or ascertained from the navigational chart.

Emergency Wreck-Marking Buoy

The emergency wreck-marking buoy is designed to provide a high visual and radio aid to navigation recognition. It is expected to be

New dangers are marked by double lateral or double cardinal marks. One of the cardinal marks would carry a 'racon' showing 'D' in Morse code on radar screens. (The purpose of the long flash in the rhythm of the south mark is for identification from the other cardinal marks.)

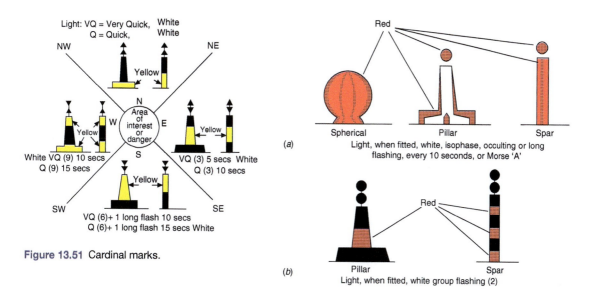

Figure 13.51 Cardinal marks.

Figure 13.52 (a) Safe water and (b) isolated danger marks.

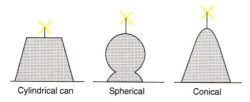

If can, spherical or conical shapes are used, it must be an indication of the side on which the buoys are to be passed:

Can shape, to be left on the mariner's port side.
Conical shape, to be left on the mariner's starboard side when travelling with the direction of buoyage
Spherical shaped buoys, in general, are passed either side. Mariners should navigate all buoys with direct consultation to the chart affecting the area and with regard to the observance of good seamanship.

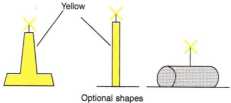

Light, when fitted, yellow having any rhythm not used for white lights.

Figure 13.53 Special marks.

placed as close as possible or in a pattern around a wreck and within any other marks that may be subsequently deployed, i.e. double cardinals.

It is expected to be retained in position until the wreck is fully known and has promulgated in nautical publications; it has been fully surveyed and the exact position and depth over the wreck are

Figure 13.54 Types of buoy in use about UK coast.

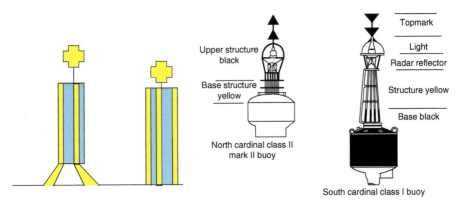

Upper structure black

Base structure yellow

North cardinal class II mark II buoy

Topmark

Light

Radar reflector

Structure yellow

Base black

South cardinal class I buoy

Figure 13.55 Direction of buoyage about UK.

Chart symbol for direction of buoyage

known; or until a permanent form of marking the wreck has been established.

The buoy will be in the form of a pillar or spar buoy carrying a yellow St. George's upright cross, if fitted. The buoy will be coloured in an equal number of blue and yellow stripes and will carry an alternating blue and yellow flashing light of a nominal range of 4.0 miles. The flash sequence will be: *Blue 1.0 s + 0.5 s + Yellow 1.0 s + 0.5 s = 3.0 seconds*. It may also be fitted with a racon-emitting Morse 'D'.

14
Emergencies

Introduction

With the establishment of the International Safety Management (ISM) system all shipboard emergencies have been investigated and subsequently documented. This action has provided ships' Masters with guidelines to support their decision-making in the event of a real-time emergency at sea. The system has also incorporated the shoreside back-up support, deemed necessary in the age of the desired and legally required green environment.

Emergency Contingency Planning

Amendments and Resolutions to the SOLAS convention now expect vessels to carry appropriate harmonized 'emergency plans' to provide guidelines for shipboard contingency planning for various types of emergency. A typical example of such emergencies are as follows, but should not be limited to:

- fire
- damage to the ship
- pollution
- unlawful acts threatening the safety of the ship and the security of its passengers and crew
- cargo-related accidents
- personnel accidents
- emergency assistance to other vessels.

These emergency plans should directly include the aspects of reporting the incident and should clarify:

1 when to report an incident;
2 how to report the incident;
3 whom to contact about the incident;
4 what to report regarding the incident.

It would be expected that the ship's Chief Officer would conduct an initial damage assessment to check on four elements:

1 Watertight integrity of the hull
2 Engine room wet or dry
3 Casualty report
4 Pollution yes/no.

Once the Master is aware of these points he can open up communication links.

Collision

Whenever a collision occurs, the vessel's sound watertight integrity is likely to suffer, and personnel may experience considerable shock, whether the collision is with another vessel, land mass or ice floe. No precise set of actions can be laid down for this situation, though certain general rules are applicable. A suggested line of action is:

1 Sound general emergency stations.
2 Stop main engines in most cases. Circumstances may dictate that if one vessel is embedded in another, it is desirable to maintain a few revolutions on the engines, since a rapid withdrawal

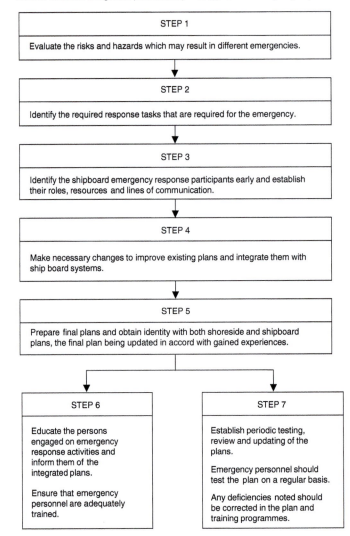

The flow charts are designed to provide a step-wise approach to emergency response.

STEP 1
Evaluate the risks and hazards which may result in different emergencies.

STEP 2
Identify the required response tasks that are required for the emergency.

STEP 3
Identify the shipboard emergency response participants early and establish their roles, resources and lines of communication.

STEP 4
Make necessary changes to improve existing plans and integrate them with ship board systems.

STEP 5
Prepare final plans and obtain identity with both shoreside and shipboard plans, the final plan being updated in accord with gained experiences.

STEP 6	STEP 7
Educate the persons engaged on emergency response activities and inform them of the integrated plans. Ensure that emergency personnel are adequately trained.	Establish periodic testing, review and updating of the plans. Emergency personnel should test the plan on a regular basis. Any deficiencies noted should be corrected in the plan and training programmes.

Figure 14.1 Emergency plan(s) implementation flow chart.

from a gashed hull could leave a massive hole, which would allow the passage of considerable water. This could result in the immediate sinking of the stricken vessel. While one ship is embedded in the other, it is acting as a plug and slowing the ingress of water through the area of damage.

3 Close all watertight and fire doors, and start bilge pumps on affected areas.

4 Assess the extent of the damage, as follows:
   (i) By visual inspection of the area of impact. Estimate tonnage of water inside the vessel and rate of water entering the vessel because of the damage.
   (ii) By estimating the size of the damaged area and its position in relation to the waterline. There may be temporary solutions, e.g. collision patch.
   (iii) By stability assessment. If cargo is affected, check the permeability and the subsequent loss of buoyancy, together with change in trim. Assess the immediate effect on GM and the continued effect on GM.
   (iv) Consider additional factors such as the risk of fire, gas or toxic fumes in the vicinity of the two vessels; or possibility of explosion resulting from withdrawal from the other vessel.
   (v) Check for casualties or missing persons aboard your own vessel

5 Establish communications and see that emergency procedures are operated as follows:
   (i) Damage-control party mustered at the incident scene.
   (ii) Emergency generator activated if required.
   (iii) Pumping arrangements set to gain maximum efficiency, with possible use of ballast pump instead of bilge pump.
   (iv) Communications officer to stand by.
   (v) Engine-room facilities kept on standby.
   (vi) Position on chart established and safe port options investigated.

6 The Master of every vessel colliding with another vessel is legally obliged to provide the other vessel with the following particulars:
   • the name of his ship;
   • the port of registry of his ship;
   • the port of departure;
   • the port of destination.
   The Master is further obliged to render all possible assistance to the other vessel.

7 Depending on the amount of damage and general circumstances, it must be assumed that at some stage a decision to abandon or not to abandon will be taken. In either case the communications officer will probably be ordered to despatch an 'urgency' signal, which may or may not be followed by a 'distress' message.

8   Many variable factors will come into play over the period from impact up to the time that any decision is taken to abandon the vessel, and these should be given full consideration when deciding any course of action:
   • The weather conditions at the time and in the future.
   • The expected time that vessels will stay afloat when taking in water.
   • The risk to personnel from fire or explosion by remaining aboard.
   • The odds of saving the vessel by beaching or steaming into shallow waters.
   • The question of pollution, especially in coastal waters.

9   The Master should report the casualty as soon as practicable, in any event within 24 hours of the ship's arrival at the next port. This report should be made to the Marine Accident Investigation Branch (MAIB) and should include a brief description of the incident, stating the time and place of its occurrence, together with the name and official number of the vessel, the next port of call and the position of the ship at the time of making the report. An entry should also be made in the official deck log book, describing the sequence of events surrounding the collision.

   Masters would under normal circumstances inform their owners at the earliest possible time after the collision, bearing in mind that repairs will probably have to be carried out immediately to render the vessel seaworthy.

10 Assuming that the vessel remains afloat, efforts should be made to prevent any increase in damage or further flooding. Should the vessel still be capable of manoeuvring, then she should do so under the correct signals, either NUC or International Code Flags.

   Any oil leakage should be reported in accordance with the pollution convention requirements regarding notification to the Marine Pollution Control Unit (MPCU), and a relevant entry made in the oil record book.

Flooding

The function of the outer shell, the hull, of a ship is to keep the water out and provide the displacement of water, and hence the upthrust, to keep the ship afloat. Exceptional circumstances may make this practical application of the Archimedes' Principle difficult to sustain. These circumstances may be caused from one or more of the following:

• collision with another vessel, or ice
• collision with the land or some object secured to the land
• collision with an underwater object
• explosion aboard the vessel
• stranding and/or beaching (grounding)
• military engagement
• fire.

To meet the above dangers, shipbuilders have incorporated in the ship's structure such items as the collision bulkhead, situated well forward, the cellular double bottom system over the keel area, together with increased scantlings where considered necessary for the trade. However, no company or shipbuilder can foresee the future and guard against the unexpected such as internal explosions or collision with underwater objects.

Main Dangers

The first and foremost hazard is the loss of watertight integrity in one or more compartments and the subsequent loss of internal buoyancy from the damaged areas. The immediate action of closing all watertight doors throughout the ship's length will reduce further loss of buoyancy.

The development of consecutive flooding throughout the vessel's length is a principal cause of foundering. Should this flooding occur, the total loss of internal buoyancy must be accepted as a probability, and the only course open is to delay the inevitable in order to get survival craft clear.

If total flooding of the vessel cannot be avoided, the decision to abandon the sinking vessel should be made in ample time for survival craft to clear the area. The dangers of lifeboats and rafts being dragged under by a sinking ship are obvious. Equally dangerous is the mast rigging and high superstructures fouling survival craft close to a capsizing hull.

Possible Solutions

For a ship holed above the waterline, build a collision patch (Figure 14.2) over the damaged area and investigate safe port options with repair facilities. Note any deviation of the ship, and note times and position in the ship's log. Pump out the bilges and take regular soundings. Avoid bad weather if at all possible. Inform the owner as soon as possible.

For a ship holed or cracked on or near the waterline, start the bilge pumps in the area affected, and pump out ballast water fore and aft to raise the vessel bodily. This should bring the damaged area above the waterline. If not, provide the vessel with a list on the opposite side to the damage.

Construct a collision patch on the outside, and/or a cement box on the inside, of the damaged plates. (This will depend on the size of the damaged area.) Have any temporary repairs kept under constant watch, with any adverse change being immediately reported. Check continually on soundings about the damaged area, especially in adjacent compartments.

Look at stability data for the compartment and check means of containment in adjoining spaces. Obtain the charted position and investigate safe port options. Inform the owner of the need for immediate repairs, together with an assessment of the damage.

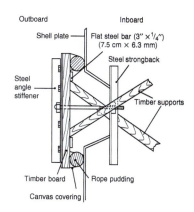

Figure 14.2 Collision patch.

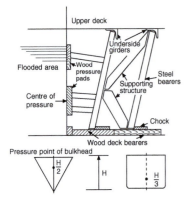

Figure 14.3 Shoring up a bulkhead. Assuming a space aboard the vessel to be flooded, then the decision to 'shore up' may be made regarding the reverse side of bulkheads, affecting the flooded area.

   The positions of shores will be determined by the pressure points calculated approximately by $\dfrac{H}{2}$ for ship-shape bulkheads; or by $\dfrac{H}{3}$ for box-shaped bulkheads where H is the vertical height.

Instigate temporary repairs if possible and be prepared to despatch an 'urgency' signal should the situation worsen. Keep a close watch on weather reports and inform passengers and crew of the situation.

If the damage is severe and the ingress of water is of massive proportion, loss of that compartment must be accepted. Attempts should be made to contain the flooded area. The shoring up of the bulkheads (Figure 14.3) must be considered, or deliberate part flooding of the adjoining compartments, provided the buoyancy of the vessel is adequate.

The weight of water entering the compartment will depend on the permeability of the cargo in that compartment. This fact alone may be the saving grace of the damaged vessel. However, in any case the normal procedures for investigation of safe port options must be made at the earliest possible moment, together with the despatch of an 'urgency' signal, which may be followed by a distress signal if the situation becomes beyond control.

The damaged stability data should be investigated, and a speedy conclusion reached. If the vessel is sinking, then this fact should be determined without delay to provide maximum time for abandoning ship. The amount of water entering the vessel can be found approximately from the following formula:

$$4.3A\sqrt{D} \text{ tonnes/per second}$$

where A represents area of damage and D represents the depth of the damage below the waterline. This formula can be used as a guide, but should not be considered as being totally accurate for every case.

Damage Control

Damage limitation is not commenced following the collision or the grounding incident. Damage limitation starts well before the incident takes place in the form of training, emergency planning, drills and acquiring the equipment and skills to perform damage-control action, as and when it is ever needed.

A ship's muster lists designate personnel into damage-control parties, but very often the ship is not equipped to carry the task through. Examples of this can be seen when a collision patch or the shoring of a bulkhead is required. How many vessels carry purpose-built collision patch material? Very few. How many carry shores to support a 20-metre cargo hold bulkhead, with water pressure building up behind it? Even fewer than few. Fortunately, some tankers are now carrying anti-pollution boom/barrier equipment, but again, the majority are relying on shoreside support only for such equipment.

The marine environment does not lend itself to early support; mariners cannot dial in the fire brigade when in open seas. It is therefore necessary that improvisation of all available means is utilized on board. As an example, the use of engine-room bottom

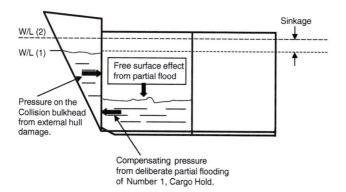

W/L (2)
W/L (1)
Sinkage

Free surface effect
from partial flood

Pressure on the
Collision bulkhead
from external hull
damage.

Compensating pressure
from deliberate partial flooding
of Number 1, Cargo Hold.

Figure 14.4 Adverse effects of deliberate, partial flooding.

plates for collision patches, where steel sheet is not carried. Partial flooding of an adjacent compartment to equal bulkhead pressure was an option discussed in the collision scenario. However, it should be realized that the effect of free surface is doubled and sinkage would be increased.

Some ships like bulk carriers may lend themselves more easily to such drastic measures when faced with the likelihood of a collapsed bulkhead. Without shoring equipment and the manpower to get it into place any method of bringing compensatory pressure to bear must be considered. Ideally, weather conditions, the ship's head, the stability criteria and permeability of cargo in the hold will favour and permit such extreme action.

Clearly, improvisation is not the ideal situation, but it is not practical for every ship to carry damage-control equipment to cover every task, which in any event hopefully will never be needed. Energy is far better deployed in the most likely areas of emergency: fire-fighting, casualty handling and ensuring the damage-control plans function correctly.

In support of this, regular drills where parties can be developed as a team can often deliver the desired rewards in the event of the unexpected emergency. Crowd control exercises for passenger vessels, crew instruction on equipment and rotation of key personnel will all affect how well the real-time incident could evolve.

It is far better not to have to action damage-control activity, on the basis that prevention is better than the cure. To this end good housekeeping, or in this case good shipboard practice, will go a long way to resolving incidents because the infrastructure is already in place. Keeping the ship well maintained, with emergency equipment correctly stowed, effective communications established and the crew with a positive mental attitude towards safety, will provide the foundation to tackle any incident.

Abandonment from the Vessel

This situation should only be considered in two circumstances: when all measures to keep the vessel afloat have failed and the ship is in a sinking condition; or when, although the vessel is floating, it is no longer possible to remain aboard because loss of life could result from heat, smoke or some other adverse condition. The decision to abandon ship must be made by the Master or officer in charge after assessing all the facts of the situation. The decision will be influenced by some of the following points:

- weather conditions
- stability of the vessel in distress
- geographic position
- life-saving appliances available and their condition
- casualties present, and expected, by remaining aboard or abandoning the vessel
- numbers of persons to consider and their experience
- range of search and rescue craft and the time factor
- communications with rescue services
- the danger of explosion
- the danger of the situation worsening, and the time when abandonment can be safely carried out being missed.

The order to abandon ship must be given by 'word of mouth', by the Master or the officer in charge, or a designated deputy (a responsible officer if such a person is still available). Loud hailers, public address systems, walkie-talkie radios, etc. may be used to ensure everyone hears the order. Certain equipment, such as PA systems, need a source of power, and persons finding themselves in this situation should be aware that in an emergency power may not be readily available.

Training and Drills

Warships carry out regular exercises in the methods of abandoning ship, but this is unfortunately not the case with many merchant vessels. Officers should be alert to the fact that simple briefings at the regular boat and fire drills could very well help in an emergency. Such briefings could include:

- the stowage positions of spare lifejackets;
- the duties of the boats' crews while awaiting the orders to abandon, e.g. gathering extra blankets, water and food;
- the need for any rescue boats launched to collect other survivors and marshal survival craft together;
- the closing of watertight doors before leaving one's post, so buying additional buoyancy time by common-sense actions;
- instruction of uncertified lifeboat men in the launching methods of survival craft.

Cause and Consequence

Experience has shown that ships have had to be abandoned after one or more of the following accidents: collision, explosion or fire.

Fire will often follow a collision or explosion, and in either case the loss of the ship may be imminent.

Once survival craft are launched, it is the duty of the person in charge of such craft to put sufficient distance between the craft and the sinking vessel so that masts, funnels and rigging, etc. will not fall on the craft and it will not be affected by the suction from the ship as she goes down. Survival craft are expected to wait at the scene of the sinking, if this is a practical proposition, for the purpose of picking up other survivors and to be picked up themselves by rescue craft, especially if the communications officer has been able to transmit a distress message and have it acknowledged before the sinking.

Tankers

The obvious danger with this type of vessel is from oil on the water, and the greater risk is burning oil on the surface. In either case the danger occurs to individuals jumping into the water. Swallowing oil and sea water causes choking and eventually drowning. For those who remain afloat there is the further danger of exhaustion and suffocation.

Possible solutions to these hazards are limited, but swimming underwater to clear smaller patches of oil on the surface should not be ruled out. Should the person have to come to the surface for air in the middle of oil patches, he may find that the breast stroke in swimming is better suited to clear away the oil film than, say, the crawl stroke.

It is a far better course of action to abandon ship by means of the survival craft if at all possible, especially if such craft include the totally enclosed type of lifeboat. Protection by the enclosed canopy from burning oil is part of their design, and more sophisticated types of survival craft have built-in cooling water sprays and self-righting capabilities.

Passenger/Ferry Vessels

With this type of vessel the real danger of panic setting in among passengers must be the Master's biggest worry once the decision to abandon has been made. Regular drills will play a major part in quelling any panic before it begins, and also provide the passengers with the necessary reassurance that the ship's personnel are trained and prepared to handle such emergencies.

Experience has shown that extensive loss of life may well occur if people abandon ship individually, and also if passengers without the required skills attempt to launch survival craft. In many cases the simple rules of warm clothing, donning the lifejacket correctly, and knowing where to go have shown themselves to be essential to survival.

With large numbers of passengers, the control of whom can easily get out of hand, assembly points like public rooms will be allocated at the beginning of the voyage. These will allow the relatively safe

Plate 125 A typical modern-day passenger/vehicle ferry catamaran hull of a type operated worldwide.

and calm assembly of passengers, who can be despatched at controlled intervals to the embarkation deck. This system also allows crew to turn out the boats and make survival craft ready for launch without being hindered by frightened people.

Roll On–Roll Off Vessels

The essential feature of this type of vessel is the large amount of open space on the vehicle decks, and the main danger, if the stowage area is holed to any extent, is the speed with which water may flood the vessel. The time in which to take the decision to abandon the vessel and launch survival craft will be limited.

Plate 126 A ro-ro passenger/vehicle ferry seen in a Mediterranean moor at Rhodes, in the Greek islands. Loading and discharge take place through the stern door/ramp arrangement.

The design of these vessels is generally such that they have a very large GM compared with the conventional type of vessel. They also tend to have large freeboards, with vehicle decks situated higher than the waterline. All these features reduce the risk of being holed below the waterline and in the stowage spaces, but the possibility always exists. This is especially the case if a vessel has grounded and the double bottom tanks and the tank tops have been pierced, allowing water to penetrate direct into the stowage areas.

A speedy assessment of damage would be essential with a ro-ro vessel, and once the decision to abandon has been taken, then swift, positive action would be required from all personnel. Abandoning the vessel by bow and stern doors, as well as by shell doors, should not be ruled out if the circumstances of the case admit, and if a source of power to operate them is still available. Individual abandonment should be avoided in view of the high freeboard, unless no alternative is left. Use should be made of disembarkation ladders, lifelines to boats, etc. as a realistic alternative to direct entry into the water from the freeboard deck (upper deck).

Scrambling Nets

Nets could be usefully employed in many types of rescue, especially on high freeboard vessels and where there are large numbers of people to consider. Nets may not always be available at short notice and improvisation in the way of gangway or cargo nets may be a useful alternative. When nets or other similar rescue equipment is to be used for the recovery of survivors, the physical condition of persons to be rescued should be considered, e.g. the injured and stretcher cases.

> NB. Nets require a self-help capability – that is, the survivor must be conscious in order to climb.

Rescue and Recovery of Survivors

Circumstances affecting the rescue of survivors will vary considerably, but might be categorized into three groups:

1   Recovery from survival craft or wreckage.
2   Recovery from the water.
3   Recovery from parent vessel before she sinks.

Recovery from Survival Craft

1   Prepare hospital and other reception areas to receive casualties. Provide medical aid for burns, oil cleansing and treatment of minor injuries with bandages, adhesive dressings and splints. Expect to treat for shock and hypothermia – blankets, warm clothing, hot drinks and stretchers should be made ready.
2   Rescue apparatus in the way of scrambling nets and boarding ladders should be rigged overside, together with a guest warp. Derricks and/or deck cranes may be swung overside to recover

survival craft, provided the safe working load (SWL) of the lifting gear is adequate. These may be used with or without cargo nets secured to the end of cargo runners. Cargo baskets may be useful for lifting injured people from boats.

3   Try to manoeuvre the rescue vessel to windward of the survival craft to create a lee, to aid recovery.

4   Establish communications with the survival craft as soon as is practical. Acknowledge distress signal flares by sound or light signals.

5   Have plenty of long heaving lines available, and also the rocket line-throwing gear.

6   Maintain normal bridge watch, checking navigation hazards in the vicinity. Display correct flag signals and keep other shipping, as well as the coastal radio station, informed of movements and the situation.

Recovery from the Water

1   Preparation should be as in (1) and (3) above.

2   Depending on the weather conditions prevailing at the time, the best method of recovering person(s) from the water would be by means of the rescue boat; launching the boat on the 'lee side' of the parent vessel and recovering the casualties into the rescue craft over its weather side.

3   Injured parties should be hoisted aboard the rescue boat carefully and in the horizontal aspect, preferably with a 'House Recovery Net' or similar device (see p. 289).

4   The condition of persons in the water, especially after a lengthy immersion period, will be generally poor. Such circumstances could expect to reduce the survivors' 'self-help capability' and as such they may require considerable assistance by the boat's crew. Where appropriate, boat's personnel should wear harnesses and lifeline/static lines as well as lifejackets.

5   It is not recommended that rocket lines are delivered in the direction of personnel or where oil is on the surface, unless it can be guaranteed to drop the line well over the top of persons and oil without incurring additional injuries. (As a general rule, a rocket line would not be used where a boat could carry out the task more safely.)

6   It should be realized that persons in exposed, cold water, without flotation aids, cannot be expected to remain afloat for long periods. Prudent use of lifebuoys should be considered as a serious option where a lot of casualties are involved.

NB. Where the launch of rescue boats is not possible, the alternative of using the ship's boat as an elevator up and down the high freeboard is offered as a practical option. The falls are moused onto the boat's lifting hooks to prevent the boat from releasing itself.

The parent vessel manoeuvres close in towards the casualties to allow them to gain access to the boat on the falls at the water surface. Once on board, the boat can be recovered. (Further reading on this method is found in House, D.J. (2000) *The Command Companion of Seamanship Techniques*, Volume III.)

Recovery from the Parent Vessel

1 Should an order to abandon ship be given while a rescue vessel is on the scene, it is an obvious move to attempt to recover personnel direct from the stricken vessel.

2 This operation could be carried out basically in two ways: by bringing the rescue vessel alongside the ship in distress or by use of the rescue ship's boats. Each case has its merits. A Master in a recovery operation would probably not endanger a tanker full of aviation spirit by drawing alongside another vessel on fire. In this case he would probably use his ship's boats. But say there were two vessels of different freeboards – the rescue vessel could manoeuvre her forecastle head into contact with that of the distressed vessel, and allow those being rescued to cross via the two forecastle head areas

Stranding/Grounding

This is physically the same action as beaching, but with the significant difference that beaching the vessel is an intentional action and under comparatively controlled conditions, whereas stranding is accidental. Circumstances will vary with different ships, but selecting a convenient position to 'set down' will in all probability never arise. In consequence, the double bottom area of the vessel will probably suffer considerable damage, especially if the ground is rocky.

The method of procedure to follow on stranding can only be an outline, considering how circumstances may vary. Here are some suggestions:

1 Stop engines.
2 Sound emergency stations.
3 Close all watertight and fire doors.
4 Damage-control party to assess damage. This must include sounding around the outside of the hull and checking the available depth of water. All of the vessel's tanks, especially double bottoms and bilges, should also be sounded and visually inspected wherever possible, air pipe and sounding pipe caps being well secured after the soundings have been obtained. This will prevent oil pollution as water pressure forces oil upwards through the outlet pipes above deck. An initial damage assessment would be made following any incidence of grounding/stranding/beaching.
5 Check the position on the chart and observe depths of water around the vessel.
6 The Master should consider refloating, though that depends on the extent of the damage, especially to tank tops. The tides should be assessed, and ballast tanks, together with additional weight (including fresh water), viewed for dumping in order to lighten the ship. Damage stability data should also be consulted. There may be value in dropping an anchor underfoot to prevent a damaged ship from sliding off into deep water.

7   Consider whether assistance is required in the form of tugs to drag the vessel astern clear of the beach into deeper water. In any event a standby vessel should be readily available prior to attempting any refloat operation.

8   As soon as practical, enter a statement into the deck log book and inform the owner and the Marine Authority (on state of seaworthiness).

Beaching Procedure

Beaching is defined as taking the ground intentionally, as opposed to accidental stranding. It is normally carried out for either or both the following reasons:

1   To prevent imminent collision.
2   To prevent loss of the vessel when damaged and in danger of sinking, damage having occurred below the waterline causing loss of watertight integrity. The intention is to carry out repairs in order to refloat at a later time.

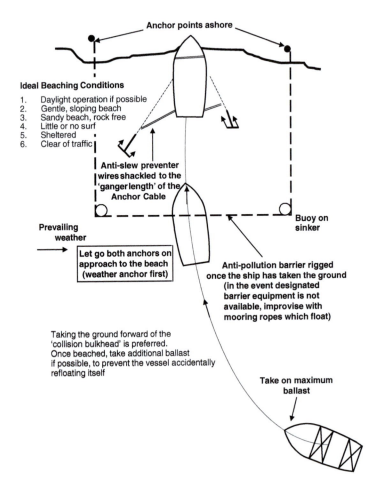

Figure 14.5 Beaching diagram.

Should time and options be available, the mariner should attempt to beach the vessel on a gentle sloping beach, which is rock free and ideally with little or no current. If possible it should be sheltered from the weather, free of surf action and any scouring effects.

Advantages and Disadvantages for 'Bow' Approach

When approaching bow-on, the obvious advantage is that a clear observation of the approach can be made and the vessel will probably have a favourable trim. The propeller and rudder will favour the deeper water at the stern, while the strengthened bow would cushion any pounding effects. The disadvantages of this approach are that the vessel is more likely to slew and the need for anti-slew wires used in conjunction with anchors may become necessary. Also it is difficult to lay ground tackle from this position, to assist with the refloating. In the majority of cases stern power would be used for refloating the ship and the average vessel normally operates with only 60 per cent of the ahead power when navigating stern first.

> Vessels would not want to beach stern first which would be likely to incur damage to rudders and propellers.

Actions Prior to Beaching

Provided that time and circumstances allow, the vessel to be beached should take on full ballast. This will make the operation of refloating that much easier.

Both anchors should be cleared away and made ready to let go. Care should be taken to lay anchors and cables clear of the position that the vessel is expected to come to rest, so minimizing the bottom damage if this is possible. Additional use of a stern anchor, if the ship is so equipped, would become extremely beneficial on the approach, with the view to refloating later.

Plate 127 A ro-ro vehicle ferry blown ashore and aground (source: Shutterstock).

On Taking the Ground

Drive the vessel further on and reduce the possibility of pounding. Take on additional ballast and secure the hull against movement from weather and sea/tide. Take precautions to prevent oil pollution. This can be achieved by discharge into oil barges, or transfer within the vessel into oil-tight tanks. Another alternative would be encircling the vessel with an oil pollution barrier, if one can be obtained quickly enough and positioned effectively. Damage reports should be made to the MAIB, together with a 'general declaration', the Mercantile Marine Office being informed and entries made into the official log book.

Grounding/Beaching: Summary

- Carry out a damage assessment following the action of the ship taking the ground. Damage assessment should initially cover:
    - watertight integrity of the hull
    - engine-room check, as to wet or dry
    - casualty report for injuries.
    - pollution assessment.

Subsequent actions are:

- Sound round all internal ship's tanks.
- Take full external soundings with particular attention to the forward and after end regions.
- Display aground signals as appropriate.
- Seal the uppermost continuous deck.
- Maintain a deck patrol for fire and security.

Plate 128 A general cargo vessel well aground in rocky shallows (source: Shutterstock).

- Calculate the next high water/low water times and heights.
- Investigate stability and refloating details following the instigation of repairs.
- Prior to attempting to refloat, call in a 'standby vessel'.
- Ensure log book accounts are entered of all events.

Master's advice:

- Order a position to be placed on the chart.
- Following the damage assessment results, open up communications with relevant authorities, inclusive of the coastguard.
- Engage tug assistance if appropriate.
- Investigate damage and stability criteria as soon as practical.
- Make a report to the MAIB.
- Investigate the local drydock capacity/availability/facilities with the owner's assistance and/or instigate diver inspection.

> The initial damage assessment would expect to include examination of the collision bulkhead and the condition of tank tops.

Deck Department Checklist for Watertight Integrity of Hull Following Grounding or Beaching

1 Check for casualties.
2 Assess internal damage by visual inspection where possible. (Special attention being given to the collision bulkhead and the tank tops.)
3 Look for signs of pollution from possible fractured oil tanks.
4 Make internal sounding of all double bottom and lower tanks, followed by a complete set of tank soundings at the earliest possible time.
5 Sound for available depth of water about the vessel, especially around stern and propeller area.
6 Check position of grounding on the chart. Determine the nature of the bottom and expected depth of water.
7 Obtain damage reports from all departments.
8 Determine state of tide on grounding, together with heights and times of the immediate high and low waters.
9 Order the Communications Officer to stand by.
10 Check condition of stability if the vessel has suffered an ingress of water.
11 Instigate temporary repairs to reduce the intake of any water, and order pumps to be activated on any affected areas.
12 Cause a statement to be entered into the deck log book, with a more detailed account to follow.

Engine Room Department Checklist for Machinery Spaces Following Grounding or Beaching

1 Check for casualties.
2 Assess damage inside the engine room and pump room and report to the Master.

3   Make ready fire-fighting equipment in case of fire outbreak.
4   Prepare pumps to pump out water from engine-room spaces.
5   Inspect all fuel and steam pipes for signs of fracture. A build-up of oil represents a fire hazard and must be located and corrected as soon as possible. Regular checks on bilge bays must be continued for a minimum period of three days after taking ground.
6   Inspect all piping, valves and auxiliary equipment before reporting to the Master on conditions.
7   Should water be entering the engine room, instigate immediate temporary repairs to reduce the ingress of water, and start the pumps on the affected areas.

The general alarm should be sounded before grounding or beaching, but if this has not been done, it would become the first action in the above lists.

Figure 14.6 Watertight door construction.

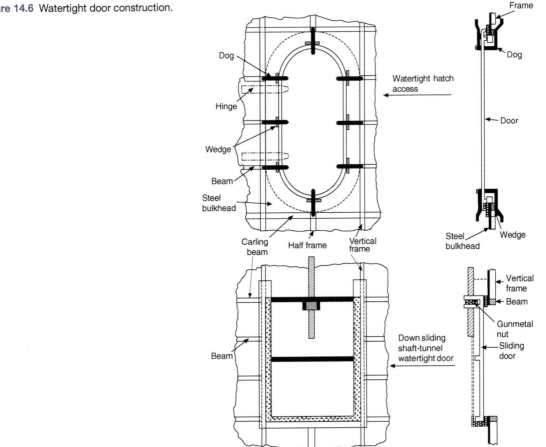

Perils of the Sea

Fortunately, accidents and loss of cargo and/or ship happen on rare occasions, but because of the inherent hazards associated with the maritime world they cannot be ruled out completely. Hence insurance of both ships and cargo has become in itself a major sector of the maritime environment.

Where a ship founders, it may not necessarily end in a total constructive loss. Salvage operations often recover much of the ship's cargo and hopefully the ship as well. This can then be repaired if damage is limited and it is practical to do so. Alternatively, the ship itself could be cut up into disposable sections for scrap value following the declaration of a total constructive loss.

Case Study

In 2008 the *Riverdance*, a ro-ro ferry, sailed from Warren Point in Ireland on a passage of less than 200 miles, towards Heysham, in England.

It was reported that the vessel encountered a very large freak wave, in bad weather which caused a loss of the ship's engines. Without power, listing heavily and drifting towards the northwest English coastline the ship's Master evacuated all the crew by helicopter, resulting in the vessel running aground close inshore, north of the town of Blackpool.

Tides and weather allowed a limited salvage operation of cargo, but the ship itself sank so deep into the sands that she became a total constructive loss and was cut into 12 movable sections for scrap.

Watertight Doors

There are many designs of watertight door and watertight hatch, the most common being those closed manually by means of 'butterfly clips' or 'double clips and wedges' (dogs), which are operable from either side of the door (Figure 14.6). The disadvantage of this type of closure is that it takes considerable time to secure. In an emergency it may even prove impossible to secure against water pressure on one side.

Regular and extensive maintenance is required on the clips to ensure they are free in movement and can be easily operated. Oiling and greasing of moving parts, especially of weather-deck hatches and doors, becomes an essential part of any planned maintenance operation. Regular inspection and periodic renewal of the hard rubber seal around the perimeter of the access door will ensure watertight integrity.

Where electrical, hydraulic (Figure 14.7) or pneumatic systems are installed, as in passenger vessels, each watertight door should be equipped with audible and visual alarms effective on both sides of the door, a local emergency stop control, a manual operation system located close to the door and emergency worming gear operative from

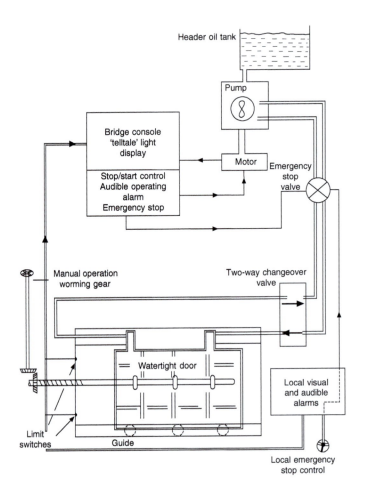

Figure 14.7 Control of hydraulic sliding watertight door.

an external point on the deck. The obvious advantage of a bridge controlling point which operates all doors simultaneously is that it increases the speed at which watertight integrity is achieved throughout a vessel, on all decks. The bridge control is fitted with a light 'telltale system', which tells the operator at a glance which doors are closed and which are open. When operating in a guide system under power, they are also effective, even against an inflow of water pressure. Should loss of power occur, then similar results may be achieved by external manual operation.

Drydocks and Docking Procedures

All ships during their working life will find themselves entering a drydock for survey work, routine maintenance or from the effects of an unscheduled accident. In any such case the choice and type of drydock will often be dictated by the physical dimensions of the vessel, especially the draught and the geography reflecting the location of the ship in relation to the position of the dockyard.

Plate 129 The drydock complex in Barcelona, Spain. A vessel is seen in the floating dock next to the empty graving dock with prominent cranes to service dock areas.

Clearly, owners will also be influenced by the overall costs of one dock compared with another for the intended repairs. To this end the economic factors could well win the day on the eventual choice of dock.

The fact that there are now several types of drydocking operations possible may also influence the choice, depending on the nature of the required services.

Types of Docking

Docking operations can take place in any of the following scenarios.

The Graving Dock

Usually stone built, with stepped sides known as alters, the graving dock has an access from the seaward, navigational channel. This is closed off once the ship has entered by a dock gate known as a caisson. Once aligned the water is pumped from the dock, allowing the vessel to take the pre-arranged blocks aligned on the floor of the drydock.

Docking Slip

This is usually an exposed slipway with a cradle arrangement that can transport the smaller vessel up the slip by mechanical means, effectively dragging the ship onto the shore line. Once the tide drops the vessel is left exposed, high and dry on the slipway. Docking work and repairs can be carried out and the vessel can be walked back down the slip into the water following completion of docking.

Floating Dock

An effective tanking system that is flooded and allowed to sink. This allows the waterline to be increased, permitting the vessel to

Plate 130 The bulbous bow of a ship seen on the keel blocks of a typical graving dock. Attention is drawn to the port and starboard anchor cables being ranged on the bottom of the dock.

be docked to float in and over the docking platform. The tanks are then emptied and the buoyancy allows the dock to rise, bringing the vessel with the platform above the waterline. Floating docks are particularly useful in the case of a damaged vessel because the dock itself can be listed to accommodate any adverse list that a damaged vessel might have acquired.

Synchro-Lift

A mechanical system capable of docking vessels up to about 10,000 gross registered tonnes. This system is a mechanical platform which is physically hoisted by a series of dockside winches. Once heaved up clear of the water, the vessel is pushed forward and heaved sideways on railed trackways into a docking bay.

Such a system allows numerous ships to be docked with a single docking system, a distinct advantage over the single-entry graving dock (Figure 14.8).

Hydrolift Docking Systems

In December 2000, the Lisnave Shipyard at Setúbal, in Portugal, opened a platform hydrolift docking system for ships. This new concept allowed for the docking of three Panamax-sized ships simultaneously.

The system works in conjunction with a wet basin which is entered via a caisson at the seaward end. Once the inward-bound ship is established in the basin, the outer caisson is closed. The water level is then increased in both the basin and the designated platform dock by pump operation. The platform dock can alternatively be filled by gravity from the basin area.

Once the basin is full and level with the dock space, the ship will have been elevated sufficiently to clear the sill of the platform dock.

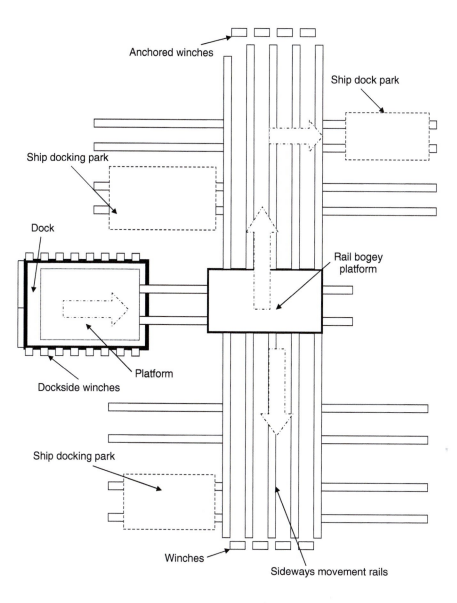

Figure 14.8 Open Docking – Synchro-Lift Operation.

This position allows the dock caisson to open and permit the transfer of the vessel from the basin to the dock platform. The dock caisson is subsequently closed and draining of the dock is then allowed by gravity (Figure 14.9).

Inward and Outward Procedures for Hydrolift Docking Systems

The docking method by means of the hydrolift system operates in a similar fashion to a vessel passing through locks. A prime example of this is seen in the Panama Canal passage, when vessels pass from the Caribbean Sea through to the Pacific Ocean.

Plate 131 Vessel on the tracks in a dock park after being raised on the synchro-lift. Such a docking principle allows easy access to all parts of the hull for maintenance purposes.

Plate 132 Vessel docked inside a floating dock (source: Shutterstock).

Log Book Entries: Entering Drydock (Assumed Graving Dock)

A typical, routine docking operation would expect to include log accounts of all activities leading up to and including the docking procedure:

1   Tugs engaged at rendezvous position.
2   Vessel proceeding towards open lock (usually under pilotage).

Plate 133 Graving dock
showing bottom blocks (source:
Shutterstock).

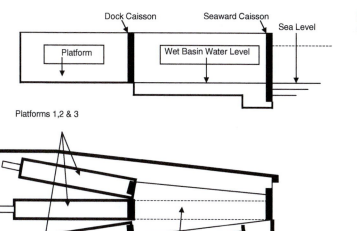

Figure 14.9 Hydrolift system. Top –
side elevation. Bottom – plan view.

3 Line ashore forward.
4 Line ashore aft.
5 Tugs dismissed.
6 Moorings carried up port/starboard.
7 Stern clears gates.
8 Vessel stopped making headway inside the dock.
9 Dock gates closed.
10 Moorings adjusted to align ship fore and aft.
11 Moorings checked to hold vessel.
12 Dock pumps commenced pumping out dock water.
13 Block contact made and vessel enters critical period.
14 Vessel sewn on blocks fore and aft.
15 Side shores passed to port and starboard.

16   Residual water cleared from dock.
17   Gangway access landed between shore and ship's side.
18   Gangway walkable.
19   Pilot dismissed.
20   Pumping of the dock complete and dock floor walkable.

> NB. Although the gangway access is provided by the dock authority, the responsibility of rigging and ensuring safe access remains the prerogative of the ship's Master.

Any additional tasks that are undertaken while the dock water is being pumped out would also be noted in the log book, e.g. cleaning and scrubbing around the hull by work punts in the area of the waterline as the level falls.

A separate record of tank soundings would be recorded in the tank sounding book for a complete set of 'on the block soundings'. These would be taken as soon as the vessel is 'sewn' on the blocks.

Chief Officer's Duties

Preparation and precautions for entry

1   All hatches and beams should be in the stowed position to ensure continuity of strength throughout the ship's length.
2   All derricks and cranes should be down and secured, not flying.
3   Any free surface in tanks should be removed or reduced to as little as possible, either by emptying the tank or pressing it up to the full condition.
4   Stability calculations should be made to ensure adequate GM to take into account the rise of 'G' when the vessel takes the blocks.
5   Consult dock authorities on draught of vessel and trim required. Generally a small trim by the stern is preferred in normal circumstances.
6   Inform dock authorities in plenty of time of any projections from the hull of the vessel, as indicated by the drydock plan.
7   Sound round all ship's tanks before entering the dock to be aware of quantities aboard. Note all soundings in sounding book.
8   Sound round all tanks once the vessel has taken the blocks to ensure a similar stability state when leaving the drydock.
9   Lock up ship's lavatories before entering the dock.
10   Ensure adequate fenders are rigged for entry into the dock and that dock shores are correctly placed against strength members once the vessel is positioned. If it is the custom in the graving dock, arrange for the forecastle head party to position wale shores on one side and the stern party to deal with the other side.
11   If required, endeavour to have the vessel cleaned and scrubbed as the dock water is pumped out.

When drydocking with cargo aboard

12   Inform dock authorities where to position extra shores or blocks to take account of additional stresses caused by the weight of cargo aboard.
13   Give cargo areas a lock-up stow whenever possible.

Plate 134 Blades of a controllable pitch propeller being removed (source: Shutterstock).

When in dock

14 Obtain telephone, electricity and water pressure, fire line, garbage and sanitation facilities as soon as possible.

15 Have documentation ready, inclusive of repair list, for dock personnel.

16 Should tank plugs need to be removed, sight their removal and retain the plugs for safe-keeping. Ensure that plugs are labelled after removal.

Draught and Trim

The vessel's required draught and trim will be decided by the dry-dock manager and the declivity of the drydock bottom. A small trim of between 12 in. (30 cm) and 18 in. (45 cm) is considered normal, but will be dictated by circumstances. If a floating drydock is to be engaged, the drydock itself can be trimmed to suit the vessel, especially if the vessel has sustained shell damage.

Drydock Plan

This is a plan carried aboard the vessel which shows recommended positions for keel blocks and shores. Normally the frames are numerically indicated from aft to forward, and the strakes lettered from the centreline out and upwards. Indicated on this plan will also be the position of any external projections from the hull, namely echo-sounder units, stabilizers, scoops for condensers, etc. Either a separate plug plan will be carried or the tank drain plugs will be indicated on the drydock plan.

Stability of Vessel

This is the responsibility of the vessel, and should be adequate to cope with the virtual rise of G as the vessel takes the blocks.

Plate 135 On the blocks in drydock the hull lines are clearly defined with both anchors walked back either side of the bow.

Plate 136 The passenger vessel *Miniver* in the Dubai floating drydock.

The vessel should not be listed. Should damage be such that the vessel cannot counter an acquired list, then shoreside weights should be taken aboard to bring the vessel to an even keel.

Positioning of Shores and Associated Docking Stresses

The docking procedure incurs many stresses on the hull from the shores placed in accordance with the docking plan. Incorrect placing of shores can and does cause damage to the vessel when she takes the blocks. This is especially so in the case of specialized vessels fitted with additional appendages like azimuth thrusters, long bilge keels or prominent condenser scoops. To this end, before dock pumping is commenced, many dock authorities are now employing the services of divers to ensure that correct line-up has been achieved and the ship will not incur additional damage during the critical period.

Shore positions should be placed with care and should ideally be placed in way of strength members like the intersection of deck stringers and frames. Bottom hull blocks, set for wide-beam vessels especially, should be placed to coincide with intercostals and other similar longitudinal members to avoid 'soft spots' which could lead to hull indentation of the shell plate.

Drydock stresses occur because of the loss of support, which is normally gained from the all-round water pressure. The vessel will become subject to an upward thrust from below the keel position caused through the lower blocks on the floor of the dock. There will also be a tendency for the ship's weight to cause a downward and outward stress action to the vessel's sides while in the dock. Provided the ship is only docked for a short period of time, any permanent or extensive distortion through stress is unlikely and the ship should revert back to her normal lines once refloating occurs.

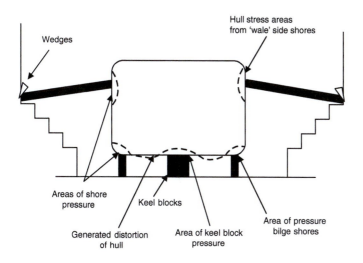

Wedges

Hull stress areas from 'wale' side shores

Areas of shore pressure

Keel blocks

Generated distortion of hull

Area of keel block pressure

Area of pressure bilge shores

Figure 14.10 Positioning of shores and associated docking stresses.

Plate 137 Typical drydock operation where the rudder post is withdrawn for inspection.

Plate 137 Typical drydock operation where the rudder post is withdrawn for inspection.

It is also possible to drydock a vessel with cargo on board for a short period of time. This can be done successfully without incurring undue stresses in the hull, provided the position of the cargo is known and additional shores are deployed to prevent undue deformity caused by the cargo weight. The possibility of being able to complete the drydock specification without pumping the dock completely dry is also an option to relieve stresses on the hull. Though this option is not always possible and restricts working on the hull, it remains an option, especially if the vessel is loaded or part loaded.

Repair Lists

It is normal practice to carry out repairs when entering drydock. These repairs may be expedited by detailed work lists covering expenditure limits, work monitoring, state of survey, maintenance of classification and protection of the owner's interests.

To Calculate the Virtual Loss of GM

There are two methods for ascertaining the virtual loss of GM. In each of the two methods the force, P, must be known. P represents the upthrust at the stern at the moment the vessel touches the keel blocks. The time the keel first touches the blocks until the vessel has taken the blocks overall is considered to be the critical period (Figure 14.11).

$$\text{Force P (tonnes)} = \frac{MCTC \times t}{L}$$

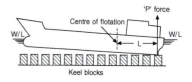

Figure 14.11 Drydocking: start of critical period.

where MCTC represents the moment to change trim one centimetre; t represents the trim in centimetres on entering the drydock; L represents the distance between the centre of flotation and the vertical line of action of the P force, in metres.

The first method considers the movement of the metacentre (M):

$$\text{Virtual loss of GM} = \frac{P \times KM}{W}$$

The second method considers the movement of the centre of gravity (G):

$$\text{Virtual loss of GM} = \frac{P \times KG}{W - P}$$

Either of the two methods are acceptable when W represents displacement of the vessel; KM represents the distance between the keel and the metacentre; KG represents the distance between the keel and the centre of gravity of the vessel.

Man Overboard

The actions of the Officer of the Watch and the ship will depend on the circumstances of each individual case. To take account of every eventuality would be impossible, but a general sequence of actions to take might be the following:

1 The alarm should be raised as soon as possible, once details of what has happened are known.
2 The Officer of the Watch should take the following immediate action:
   (i)   Order helm hard over towards the side on which the man fell, before commencing a Williamson turn (Figure 14.12).
   (ii) Release the bridge wing lifebuoy and combined smoke/light (Plate 138).
   (iii) Standby main engines to manoeuvre the vessel. (Do not stop engines unless the man in the water is in danger from the propeller.)

Point 'A' – man overboard

1 Rudder hard over to swing stern away from man.
2 Release lifebuoy.
3 Sound 'emergency stations'.
4 Engines on standby and reduce speed.

Point 'B' (60° off original course), reverse rudder to same angle in opposite direction to reduce speed and return vessel to reciprocal course.

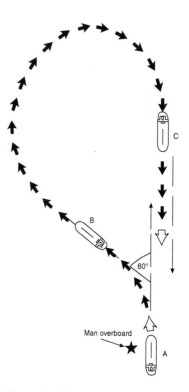

Figure 14.12 Williamson turn.

Plate 138 Emergency bridge lifebuoy for use in 'man overboard' emergencies, with combined light and smoke float attached. This one is secured in the stowage bracket on the after side of the starboard bridge wing.

Point 'C' – ship on reciprocal course

1 'Steady'.
2 Slow ship, launch rescue boat to pick up man.
3 Subsequently the Officer of the Watch should sound general emergency stations as soon as it is practical and the Master should be informed of the situation. Lookouts should be posted at strategic points while the turning manoeuvre is proceeding, and a VHF radio warning sent to nearby shipping and coastal radio station. If they see him, lookouts should point at the person in the water until he or she is picked up or lost to sight. The Officer of the Watch should delegate personnel to display 'O' flag, and have a man ready to act as helmsman if navigating on automatic pilot.
4 The emergency boat's crew should standby and be ready to launch the rescue boat to effect recovery, weather permitting. The Communications Officer should be ordered to standby and be prepared to transmit an urgency signal.
5 An efficient and effective watch should be maintained at all times throughout the manoeuvre, especially if other shipping is near. A position should be noted on the chart as soon as possible after the alarm is raised.
6 The Chief Steward/Catering Officer or Medical Officer aboard should be ordered to prepare the hospital reception space. Advice may be needed for the treatment of shock and hypothermia.
7 The vessel's speed should be reduced as the Williamson turn is completed, and the following points considered:
   • Whether to start a search pattern, from which point and at what time, and what type of search pattern to use (probably sector search).
   • Whether to let go a second bridge lifebuoy and combined smoke/light. If the second lifebuoy is released when the vessel is on a reciprocal course, a reference line of search can be established between the first and second lifebuoys. This would be of considerable help to a search vessel and

would provide an initial rate of drift over a greater area. However, when a second lifebuoy is released, the man in the water may assume that this is the first lifebuoy to be released and swim towards it. In so doing, he ignores the first buoy released and in confusion may drown through exhaustion while heading towards the second. (The combined smoke/light emits dense orange smoke for 15 minutes, and a light of 3.5 candela for 45 minutes.)

- The sounding of 'O', 'man overboard', on the ship's whistle to alert other shipping and reassure the man in the water that his predicament is known.

Factors influencing a successful recovery comprise weather conditions, sea water temperature, day or night operation, experience of crew members, geographic location, number of search units, time delay in the alarm being raised and condition of the man when falling.

Delayed Turn

The advantages of this optional alternative to the Williamson turn are that the man falling overboard is allowed to fall astern of the vessel, so clearing the propeller area. Turbulence in the vicinity of the man in the water is reduced by no sudden rudder movements, and the delay period, if used wisely, can become beneficial in a successful recovery (see Figure 14.13).

In this particular turn there is a very good case for releasing the second bridge lifebuoy. A line of direction is achieved for the vessel as the ship returns either between the two buoys (the first being released at point 'A' the second at point 'B' in Figure 14.13) or with the two buoys in transit.

The period of delay may vary according to the length of the vessel and the speed at which she was moving through the water, together with the perimeter of the turning circle. It is generally accepted that for most circumstances a delay period of approximately one minute would suit the majority of situations. The main disadvantage is that at some point in the turn the lookouts may lose sight of the man in the water.

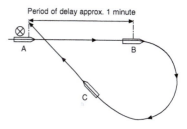

Period of delay approx. 1 minute

Figure 14.13 Delayed turn.

Double Turn

This has the distinct advantage that the man in the water remains on the same side of the ship throughout the manoeuvre, so that the job of lookouts to remain in visual contact is more likely to be successful. The turn begins when the helm is ordered hard over to the side on which the man fell once the alarm is raised, which will have the double effect of keeping the propeller clear of the man in the water, and bringing the vessel through 180° onto an opposite course. When the man is seen to be approximately three points abaft the beam, the vessel should complete the second, double, part of the turn in order to return to the position of the incident.

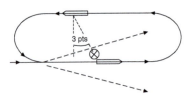

Figure 14.14 Double turn.

The approach to the man in the water should always be made to windward of him, so allowing the ship to drift down towards him. Recovery is achieved by boat or the use of scrambling nets down the ship's side and voluntary helpers wearing lifejackets with safety lines attached.

The success of the operation will depend on keeping the man in the water in sight, and if this is not possible because of fog or such other obstruction, the Williamson turn should be used.

Man Overboard: When Not Located

In the event that a Williamson turn, or other tactical turn, is completed and the man is not immediately located, the advice in the IAMSAR manual should be taken and a search pattern adopted. The recommendations from the manual suggest that where the position of the object is known with some accuracy, and the area of intended search is small, then a 'sector search pattern' should be adopted.

Although a table of suggested track spaces is recommended in the IAMSAR manual, factors like sea temperature can expect to be influential where a man overboard is concerned. In such cases a track space of 10 minutes might seem more realistic with regard to developing a successful outcome.

It will be seen that the alteration of course by the vessel is 120° on each occasion when completing this type of pattern. In the event of location still not being achieved after pattern completion, or in the event of two search units being involved, an intermediate track could be followed (see overlay in Figure 14.15).

> NB. Even at 10-minute track space intervals, at a search speed of three knots it would still take 90 minutes to complete a single sector search pattern.

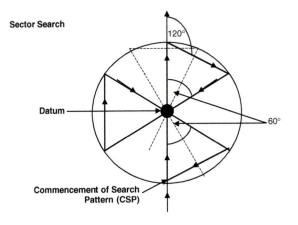

Figure 14.15 Sector search. The suggested construction circle to commence the search pattern as the vessel crosses the circumference: track space = radius of circle.

Helicopter Procedure

Any vessel which is to engage in helicopter operations should ensure that the crew are fully aware that the aircraft will need to overcome certain initial hazards. Any contact by the rotor blades of the aircraft with shrouds, stays masts, etc. could cause the helicopter to crash on or around the vessel itself. In order to receive the aircraft in a safe manner, the deck should be prepared and the following actions taken well in advance of the helicopter's arrival on the scene.

1 Clear away any small gear from the deck area, together with any rubbish which might be lifted by the down-draught into the rotor or engine of the helicopter.
2 Lower all radio aerials between masts and in the vicinity of the reception area.
3 Clear the reception area of derricks and other lifting gear, and mark the operational point with a 'white' letter 'H'.
4 Provide a visual indication, by use of flags or smoke, of the direction of the wind. If smoke is to be used, then this should be used with discretion for a limited amount of time, so as not to interfere with the pilot's visibility. The international code pendant should be used if flags are to be the method of wind indication.
5 An emergency party should be standing by, with hoses connected with spray nozzles and foam extinguishers available.
6 Adequate lighting of the area must be used if the operation is to be carried out at night (Figure 14.16).
7 Communication must be established between the vessel and the helicopter crew.
8 The ship should be kept on a steady course, with the wind about 30° on the port bow. The speed of the ship should be adjusted to produce minimal movement of the vessel from her heading.

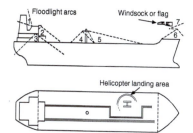

Figure 14.16 Deck lighting and landing areas for helicopters.

Extreme care should be taken not to direct floodlights into the helicopter cockpit, while providing maximum illumination to the operations area. The following lights should be available:

1 Floodlights illuminating funnel and high points.
2 Bridge front light illuminating after end of foredeck.
3 Overside illumination, to port and starboard.
4 Midships illumination, highlighting obstructions such as derrick posts, etc.
5 Low, direct floodlights over the operations area.
6 Floodlights illuminating danger areas and operations limits, forward.
7 Illumination of the wind direction indicator.

A flashing red light in the operations area will indicate to the pilot that the operation is curtailed, and the aircraft will immediately clear the area of the vessel.

Plate 139 Sikorsky 76 engaged in pilotage duty with a merchant vessel (source: Shutterstock).

Plate 140 French Allouette III helicopter engages in a personnel transfer offshore.

Landing and Evacuation of Personnel

1 All personnel should obey the instructions given by the helicopter crew when embarking or disembarking. There is a distinct danger of inadvertently walking into the tail rotor of the aircraft, which, especially at night, is sometimes difficult to see.

2 When engaged on hoist operations on no account must the winch wire be allowed to foul any part of the ship's rigging, and the end should not be secured in any manner whatsoever. Should the wire cable become caught in an obstruction, then the helicopter crew will cut the cable free.

3 No attempt should be made to handle the end of the hoist wire at deck level until any static electricity, which could have built up in the wire, has been discharged. Helicopters build up a charge of static electricity which could kill or cause severe injury, and pilots normally lower the cable into the sea before starting the operation, or let the cable touch the deck. Ship's personnel, when handling the cable after the static has been discharged, should wear rubber gloves, for the static could build up again during the operation. The static rod (hook) used throughout the operation should only be handled by one person (acting as the hook handler).

4 If the deck area cannot be adequately cleared, owing to permanent fittings, an alternative pick-up point could be established, e.g. a lifeboat towed astern.

5 Although emergency parties should be on standby, and firefighting gear on hand, hoses and loose gear could be drawn into the rotor by the down-draught. Essential equipment, therefore, should be ready for use but under cover to prevent accident.

6 Helicopters are limited in range and flight time, so that undue delay on site by personnel trying to save personal possessions could severely hamper the success of a rescue operation. In a distress situation transfers are restricted to personnel only.

7 When injured parties are to be transferred, helicopter crew may descend to the ship's deck with or without a stretcher. Time may be saved by having the patient already in the Neil-Robertson-type stretcher, which could then be lifted off directly or secured in the rigid frame stretcher of the aircraft.

8 In the majority of cases of personnel transfer, a strop is used on the end of the cable to accommodate the body (Figure 14.17).

Helicopter Evacuation Checklist (MEDIVAC)

Patient's condition

- Possible diagnosis, if known.
- Current condition of patient.
- Current medication patient is on and dosage taken.
- Pulse rate.
- Blood pressure.
- Past medical history, which may be significant.
- Any medication that patient may be allergic to.
- State: is the patient ambulatory or incapacitated.

Documentation (secured and in waterproof cover)

- Passport/discharge 'A' documents for identification.
- Record of medication administered, quantity and time of administration.
- Shoreside contact address and telephone numbers if known.

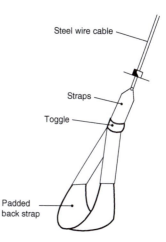

Figure 14.17 Helicopter lifting strop.

1 Take the strop and put both the head and the arms through the loop.
2 Secure the strop under the armpits, with the padded part positioned as high as possible across the back.
3 Tighten up the strop by pulling down the toggle as far as possible.
4 When secure inside the strop, extend one arm and give the thumbs-up sign to the winch man in the helicopter.
5 Put both arms down by the sides of the body.
6 On reaching the helicopter level do nothing until instructed by the helicopter crew.

Plate 141 Sikorsky Seahawk helicopter engaged in underslung load activity (source: Shutterstock).

Weather conditions

- Sky condition (clear, overcast, broken).
- Estimated cloud ceiling.
- Precipitation (rain, fog, drizzle, snow).
- Wind direction and speed.
- Sea state condition.

General information

- Ship's name and type of vessel.
- Vessel's present position.
- Intended rendezvous position (await confirmation).
- Vessel's course and speed.
- Departure port and next arrival port.
- ETA at next arrival port.

Helicopter Ship Operations (Non-Military)

In all combined ship–helicopter operations, marine personnel are advised to carry out operations to the safety standards as published by the International Civil Aviation Organization (ICAO) and in accordance with the International Chamber of Shipping's Guide to Helicopter/Ship Operations.

Plate 143 Wessex Mk4 SAR helicopter in operation with the RAF engages in stretcher casualty rescue from a totally enclosed lifeboat at the surface.

Designated responsibilities of involved parties:

- *The Master of the vessel* is responsible for the overall safety of the ship.
- *The pilot of the aircraft* is responsible for the safety of the helicopter at all times.
- *The offshore installation manager* is responsible for the appointment of a competent person to act as the Helicopter Landing Officer (HLO).
- *The owner of an offshore installation* is responsible for providing a safe landing area inclusive of all the necessary fire-fighting facilities, as required by the regulations.
- *The employer/user of the helicopter* is responsible for selecting a suitable helicopter to complete a safe operation, bearing in mind the constraints of the ship/installation and the limitations of the aircraft.
- *The Communications Officer* (where appropriate) is responsible for communications between the ship/installation and the aircraft, prior to landing, landing on or taking off, and at any time during the operation.
- *The HLO* is responsible for the safe control of all helicopter operations and movements which affect the ship, installation or base.
- *Aircrew* are responsible to the pilot for passengers entering or leaving the helicopter.

Plate 144 Sea King helicopter operating with HM Coastguard around the UK coastline (source: Shutterstock).

Plate 145 Sikorsky S61N engaged in lift of an underslung load.

- *The Winchman/observer or cabin attendant* is responsible for the supervision of the loading or unloading of the aircraft and for the completion of relevant documentation, e.g. manifest, customs declaration, etc. Also for observation of the pilot's blind spots.

Type and Capacity

Table 14.1 lists the operating ranges, carrying capacity and the type of helicopter which may be encountered in marine operations.

Table 14.1 Helicopters

Helio – Type	Employment	Capacity	Ops/range	Maximum range	Remarks
Dauphin 365	Offshore (O/S) ferry	12	350	506	All purpose
Super Puma	O/S ferry, search and rescue (SAR)	24	340 aux tanks	1080 ferry tanks	Civilian and military
Bell 214 ST	O/S transport	18	250	505	External lift capability
Westland 30	O/S passenger	19	185	481 aux tanks	Medivac alt. config.
MBB Bk 117	Inter-rig transport	8/12	135 standard tanks	500 aux tanks	Utility aircraft O/S
Sikorsky S61N	O/S ferry, SAR	24	215	495 30′ reserve transport	Amphibious O/S
Sea King	Military – SAR, ASW (anti-submarine warfare)	22 max	270	500	5.5 hrs endurance
Sikorsky 76	O/S duty	12	200	600 reduced load + aux tanks	All purpose O/S
Bell 412 SP	O/S transport	14	250	400	Ferry operations
EH101 Merlin	Military and civilian	30	275	500+	SAR and utility, ASW
Sea Hawk SH-60B	Military USN	10	200	400	SAR, ASW, Medivac
Sikorsky JayHawk (MRR)	USCG multi-mission with extended range	4 crew + 6	300	+45′ on scene	Multi-mission + SAR

There are, of course, many other types of helicopter in use. The ranges and passenger capacity given are only a guide. Factors influencing the range and number of persons carried will depend mainly on weather conditions, especially wind speed and operational characteristics of the individual aircraft.

Communications

Some of the larger helicopters are fitted to transmit and receive on 2182 kHz MF. The majority of search and rescue (SAR) aircraft are equipped with VHF/UHF RT, and cannot under normal circumstances work on the MF range. Should communications between ship and aircraft prove difficult, then a radio link via a coastal radio station may be established or Morse by Aldis lamp flashed direct to the helicopter.

Operational checklist

Safety checklist

For use with the *ICS Guide to Helicopter/Ship Operations*
    To be checked by the officer in charge. ☐
1 *General*
    (*a*) Have all loose objects within and adjacent to
        the operating area been secured or removed? ☐

(b) Have all aerials, standing or running gear above, and in the vicinity of, the operating area been lowered or secured?   ☐

(c) Has the Officer of the Watch been consulted about the ship's readiness?   ☐

(d) Are the fire pumps running and is there adequate water pressure on deck?   ☐

(e) Are fire hoses ready? (Hoses should be near to, but clear of, the operating area.)   ☐

(f) Are foam hoses, monitors and portable foam equipment ready?   ☐

(g) Are foam equipment operators, of whom at least two are wearing the prescribed firemen's outfits, standing by?   ☐

(h) Are the foam nozzles pointing away from the helicopter?   ☐

(i) Has a rescue party, of whom at least two are wearing firemen's outfits, been detailed?   ☐

(j) Is a man-overboard rescue boat ready for immediate lowering?   ☐

(k) Are the following items of equipment to hand?
   (i) portable fire extinguishers   ☐
   (ii) large axe   ☐
   (iii) crowbar   ☐
   (iv) wire cutters   ☐
   (v) red emergency signal/torch   ☐
   (vi) marshalling batons (at night)   ☐

(l) Has the correct lighting (including special navigation lights) been switched on prior to night operations?   ☐

(m) Is the deck party ready, and are all passengers clear of the operating area?   ☐

(n) Have hook handlers been equipped with strong rubber or suitable gloves and rubber-soled shoes to avoid the danger of static discharge?   ☐

2 *Landing On*

(a) Is the deck party aware that a landing is to be made?   ☐

(b) Is the operating area free of heavy spray or seas on deck?   ☐

(c) Have side rails and, where necessary, awnings, stanchions and derricks been lowered or removed?   ☐

(d) Where applicable, have portable pipes been removed and have the remaining open ends been blanked off?   ☐

(e) Are rope messengers to hand for securing the helicopter if necessary? (Note: only the helicopter pilot may decide whether or not to secure the helicopter.)   ☐

3 Tankers

Before carrying out the above checks the officer in charge should check that:

(a) *For tankers without an inert gas system*: tanks in, and adjacent to, the operating area have been vented to the atmosphere 30 minutes before the operation is due to start, thus releasing all gas pressure.

(b) *For tankers with an inert gas system*: the cargo tank internal pressure has been reduced to a level which will ensure that there is no discharge of gas during the helicopter operation.

(c) *For all tankers*: the tank openings have been resecured after venting.

Further information on marine helicopter operations can be obtained from the author's sister publication *Helicopter Operations at Sea* (2nd edition).

Sub-Sunk Procedure

British and many allied submarines are equipped with two indicator buoys for use in emergencies. The buoys are situated fore and aft, and can be released from inside the boat, should the submarine find herself in difficulties and unable to surface.

The sighting of an indicator buoy may be the first indication that a submarine is in difficulties. No time should be lost in warning the authorities of the situation, the possible rescue of survivors being dependent on time not being wasted.

It is suggested that surface craft adopt the following procedure on sighting an indicator buoy:

(a) Obtain own ship's position and advise by radio navy, coast-guard or police authorities, providing full details of the sighting.

(b) Do not stop engines, but remain in the area.

(c) Post lookouts to watch the indicator buoy and the surrounding area.

(d) Operate echo-sounding machine. Periodically bang on the lower hull to indicate the presence of surface craft to the submarine.

(e) Muster emergency boat's crew and have a boat made ready to recover possible survivors.

(f) Advise the medical officer to be prepared to treat possible survivors for shock and exposure.

The bottomed submarine may try to communicate with surface craft by use of pyrotechnic floats, which burn with flame and/or smoke on reaching the surface and serve as additional markers. The exact position of the submarine is essential if a rescue is to be effected.

Depending on conditions in the submarine, survivors may attempt to ascend to the surface at any time after the accident has occurred. Relevant factors to their survival will be the depth of

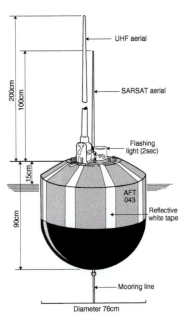

Figure 14.18 Type 0070: submarine indicator buoy.

NB. A programme is currently in place to change these buoys to an international orange colour and either of these buoys may be encountered.

water at the scene of the incident and the facilities aboard surface craft, i.e. recompression chamber availability, medical facilities, etc.

It is normal practice for survivors in this situation to wait before attempting to reach the surface, if conditions permit, until it is known that rescue craft are on the surface. The lack of air supply inside the submarine may, of course, make delays impossible.

Sub-Sunk: Indicator Buoys

The submarine indicator buoy (Type 0070) is made from expanded plastic, covered with a GRP skin to provide physical protection. It is semi-spherical in shape and floats with a freeboard of about 15 cm. The buoy is attached to the submarine by 1,000 m of braided line.

The buoy is recognized by longitudinal strips of red and white, retro-reflective tape and is allocated a three-digit serial number displayed on each side. Inscribed around the top part of the buoy are the words:

FINDER INFORM NAVY, COASTGUARD OR POLICE. DO NOT SECURE TO OR TOUCH

In addition to these marks the buoy will have two aerials; one for UHF and the second for SARSAT radio transmission. A flashing light, flashing approximately once every two seconds, is fixed on the top of the buoy and has a range of five nautical miles.

The UHF SARBE operates on 243 MHz. The emission will consist of three audio sweeps from 1600 Hz, down to not lower than 300 Hz, occupying a period of 1.2 seconds; the emission will then fall silent for 0.8 seconds. This transmission will continue for a minimum of 72 hours.

The 406 MHz UHF emission will consist of a SARSAT transmission.

15
Fire-Fighting

Introduction

With the introduction of ISM all anticipated shipboard emergencies can expect to be covered by procedural checklists. Fire on board is no exception, and as such will be supported by a procedural listing of potential activities to combat the emergency. In every case of fire at sea the following or similar actions are recommended:

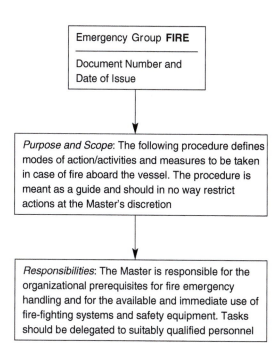

Emergency Group **FIRE**

Document Number and
Date of Issue

Purpose and Scope: The following procedure defines modes of action/activities and measures to be taken in case of fire aboard the vessel. The procedure is meant as a guide and should in no way restrict actions at the Master's discretion

Responsibilities: The Master is responsible for the organizational prerequisites for fire emergency handling and for the available and immediate use of fire-fighting systems and safety equipment. Tasks should be delegated to suitably qualified personnel

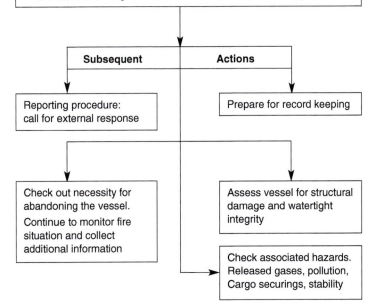

Initial Actions

Measures to be taken by the person who first observes the fire

Raise the alarm/inform bridge if possible/tackle fire immediately (if practical, i.e. small fire)

Measures by the Navigational Officer of the Watch

Raise the 'General Alarm', call the Master, engines placed on SBE. Bring the bridge to an alert status. Obtain a charted position. Proceed to fire station once relieved as OOW and report made to Master

Measures taken by the Ship's Master

Take the 'con' of the vessel following the report from the OOW. Introduce organized fire-fighting activities and retain mobile and fixed fire-fighting systems at ready. Order the activation of systems as and when necessary, following analysis of situation(s).

Communications (internal and external) following situation awareness.

Consider: ship's plans, cargo information, stability criteria, ballast/ tank transfer arrangements weather conditions, etc. as appropriate

Subsequent   **Actions**

Reporting procedure: call for external response

Prepare for record keeping

Check out necessity for abandoning the vessel.

Continue to monitor fire situation and collect additional information

Assess vessel for structural damage and watertight integrity

Check associated hazards. Released gases, pollution, Cargo securings, stability

NB. It should be borne in mind that many companies have their own interpretation regarding the content of emergency checklists, as described in Chapter 14, and above. Each checklist must be approved by the authority prior to being established.

Additional Requirements: Vessel in Port

- inform the Fire Service via the harbour control;
- remove all non-essential personnel from the vessel;
- have ship's plans and international shore connection ready to hand over to the Fire Service;
- inform the Fire Service of hazardous cargoes.

Equipment for On-Board Fire-Fighting

All 'Class 7' cargo vessels are required to be fitted with the following items:

- Fire main and water service pipes fitted with hydrants and with no other connection, other than for washing down. Such hydrants should always be accessible even when the vessel carries deck cargo.
- Sufficient appliances to ensure that two jets of water can be brought to bear to any part of the vessel.
- Fire pumps: cargo vessels of 1,000 grt and above must carry two independently driven fire pumps other than the 'emergency pump'. The capacity of each pump (except the emergency pump) shall not be less than 80 per cent of the total required capacity, divided by the minimum number of required pumps, and in any case not less than 25 m³/hour, and be capable of delivering the required two jets of water. (Passenger vessels over 4,000 grt must carry at least three fire pumps).
- The emergency fire pump shall have a total capacity of not less than 40 per cent of the total fire pump capacity, and in any case not less than 25 m³/hour and still retain a pressure on any other hydrant of 0.27 N/mm² (vessels 6,000+ grt) (0.25 N/mm² for vessels of 1,000–6,000 grt).
- Hydrants: the number and position of hydrants should be such as to provide the two jets of water required by the regulations, one of which by a single length of hose. (There are additional requirements for ro-ro vessels.) The pressure at any hydrant should not exceed that at which the hose can be correctly handled.
- Isolation valves must be fitted to separate the section of the fire main within the machinery space containing the main fire pump(s) from the rest of the fire main.
- Hoses: fire hoses made of non-perishable material should be provided for in the number of one hose for every 30 metres of ship's length with a minimum of five used only for the purpose of extinguishing fires. (Passenger vessels require one hose for every hydrant, as required by Reg. 5.) The requirement on hoses stated above does not include those required for engine/boiler rooms.
- Nozzles shall be of an approved dual-purpose type (i.e. spray/jet) which incorporates a shut-off.
- Fire extinguishers: vessels over 1,000 grt must have at least a minimum of five extinguishers for use in accommodation spaces. (There are additional requirements for engine/boiler rooms.)
- A fixed gas fire-extinguishing system (see pp. 541–545).
- A fire detection and fire alarm system must be fitted with manually operated call points capable of immediate operation. The system must be serviced by two independent sources of power and provided with a control panel located in the navigation bridge.
- A smoke detection system for vessels constructed after 1992.

NB. In cargo vessels below 1,000 grt, the number of hoses carried must be to the satisfaction of the administration.

One of the main pumps must be capable of remote starting from either the navigation bridge and the fire control station or have permanent pressurization of the fire main system by one of the main fire pumps.

NB. Ships below 1,600 grt may be waived this condition if the starting arrangement in the engine room is in an easily accessible position.

Passenger ships carrying more than 36 passengers must have hoses connected to hydrants at all times in interior locations.

- Fireman's outfits: all ships should carry at least two. Passenger vessel must carry at least two for every 80 metres of ship's length or part thereof, plus two sets of personal equipment. Vessels over 4,000 grt must carry four.
- International shore connection must be carried by all ships over 500 grt.
- A fire control plan must be carried by all ships. A duplicate set of fire control plans must be retained in a weathertight enclosure ready for use by shoreside fire-fighting personnel.

Additional Requirements for Passenger Vessels

- Hoses having a length of at least 10 metres.
- An automatic sprinkler system of an approved type to satisfy Regulation 12.
- Fire patrol system, plus manual alarms in passenger and crew spaces, together with a fire detection and alarm system to include non-accessible spaces. (Fire patrol members to be provided with two-way radios.)
- In each special category space three water fog applicators.
- Fixed fire-extinguishing system for cargo spaces intended for the carriage of vehicles.
- Escape route emergency lighting.
- A public address system with separate crew alarm.
- Fire main kept pressurized or readily accessible remote control of pumps.
- Extinguishers for control room spaces, e.g. navigation bridge, generator rooms, radio rooms, etc.
- In vessels with spaces for carrying motor vehicles, fire extinguishers suitable for oil fires, two for each 40 m of deck space.
- Galleys should be equipped with one extinguisher and one fire blanket (two of each if the galley area is more than 45 m^2).
- Mimic escape diagrams must be prominently displayed on the inside of cabin doors and in public spaces.
- Hoses located in internal spaces must be connected to the hydrant at all times.
- The hull, superstructures, bulkheads, decks and deck houses shall be constructed in steel or equivalent material. Special requirements exist for aluminium structures.

When empty, ro-ro vehicle deck spaces must have the capability to provide two jets of water from a single length of hose.

Roll On–Roll Off Vessels

- Cargo ships over 2,000 grt, with vehicle spaces, must be protected by a fixed CO_2 fire-extinguishing system. A water spray system is required for each cargo space having a deck above it; also for each space deemed to be closed but not capable of being sealed.

- A drainage and pumping system that will prevent the build-up of free surfaces is needed. (Ro-Pax vessels must have scupper discharge valves for operation from above the bulkhead deck. These must be kept open while the vessel is at sea.)
- Ro-ro spaces capable of being sealed must be fitted with a fixed gas extinguishing system.
- A separation must be provided between a closed ro-ro space and an adjacent open ro-ro space (unless the space is considered a closed ro-ro space over its entire length).
- A separation must be provided between a closed ro-ro space and the adjacent weather deck (unless the arrangement is as for the carriage of dangerous goods on the adjacent weather deck).
- Ro-ro cargo spaces must be fitted with a fixed fire detection and alarm system or a smoke detection system.
- Electrical equipment and wiring cannot be fitted in enclosed cargo spaces, closed vehicle deck spaces or open vehicle deck spaces unless it is essential for operational purposes.
- Special category spaces shall be treated as closed cargo spaces when dangerous goods are carried.
- Each ro-ro cargo space intended for the carriage of motor vehicles with fuel in their tanks for their own use must carry at least three water fog applicators; also, one portable foam applicator unit, provided that at least two such units are available in the ship for use in ro-ro spaces.
- In every ro-ro space portable fire extinguishers of a number to satisfy the administration should be located (at least one extinguisher being located at each access to such a space).
- Ro-ro spaces where crew are normally employed must be fitted with escape routes to satisfy the administration, but will be not less than two.
- Cargo spaces must have a ventilation system that is sufficient to provide ten air changes per hour when vehicles are inside the space. It must be fitted to have a rapid means of shutdown in the event of fire.

Tanker Vessels (Class 7T)

- For tankers of 20,000 tonnes deadweight or above, protection for cargo tank deck area and cargo tanks shall be achieved by a deck foam system and an inert gas system, respectively.
- Tankers carrying petroleum products having a flashpoint exceeding 60°C (closed cup test) must be fitted with a fixed deck foam system capable of supplying sufficient foam concentrate from monitors and applicators to ensure 20 minutes of foam generation in tankers fitted with an inert gas system, and 30 minutes of foam generation in vessels not fitted with inert gas systems.
- All tankers operating with a crude oil washing system must have an inert gas system.

- Cargo pump rooms must be provided with one of the following:
  - (*a*) a carbon dioxide system or halogenated hydrocarbon system;
  - (*b*) a high-expansion foam system suitable for the cargo carried;
  - (*c*) a fixed-pressure water-spraying system.

  The system must be capable of being operated from outside the pump room.
- All tankers must carry a portable instrument for the measuring of flammable vapour concentrations.

  There must be portable instruments additional to tank measuring devices for the measuring of oxygen and flammable vapour concentrations. Such instruments must be supported by calibration systems of both fixed and portable measuring instruments.
- Four fireman's outfits.
- Audible alarm system or automatic shut down of cargo pumps operable when predetermined limit of low pressure in the inert gas main is reached.
- Detailed instruction and maintenance manuals regarding the inert gas system and its application to the cargo tank system. To include fault guidance procedures in the event of failure of the inert gas system.

Small Fires

The containment and subsequent extinguishing of small fires will depend on three main factors:

1   Location and type of fire.
2   Number and availability of extinguishing agents (see Table 15.1).
3   Quick thinking and training of fire-fighters.

Most modern vessels have an adequate supply of portable extinguishers to tackle any small fire immediately and, if not able to extinguish the fire, at least contain it. On discovery of a fire, personnel should raise the alarm, no matter the size of the fire. This will allow back-up teams to equip themselves with more effective fire-fighting gear while containment of the fire is being attempted by the portable extinguishers.

These can be used to great effect when employed with common sense. The correct extinguisher for the job should be used, and probably the nearest extinguisher to the site of the fire will be the correct one. However, fire-fighters should check the colour and the labelled instructions on the outside of the extinguisher, since use of the wrong extinguisher could have fatal results for the operator. For example, a water extinguisher used on an electrical fire could cause severe burning and electric shock to the fire-fighter.

Think first and assess the situation, before taking action.

After raising the alarm, assess the type of fire and number of casualties, if any. Remove casualties if possible. Close down any ventilation. Obtain the nearest extinguisher considered correct for

Table 15.1 Extinguishing agents

	Extinguishing by cooling electrical conductors			Extinguishing by smothering		
				Non-electrical conductors		Non-conducting but toxic
Types of fire (according to the combustible material)	1 WATER Water/CO$_2$ Water/soda-acid	2 FOAM Chemical foam Mechanical foam	3 POWDER Dry sand Dry chem. powder CO$_2$	4 INERT GASES CO$_2$ Steam		5 VOLATILE LIQUIDS Carbon tetrachloride Methyl bromide, etc.
Type A Dry fires (wood, paper, textiles, etc.)	Yes	Yes (not particularly advised)	No (usable in special circumstances	No (usable in special circumstances)		No (for use in small fires)
Type B Fire in combustible liquids	No (only spray)	Yes	Yes	Yes		Yes – use in closed space: WARNING: Only advisable in small fires, in well-ventilated spaces or when the operator utilizes it in free air
Type C Fire in electrical equipment	No (only spray)	No	Yes	Yes		
Type D Fire in light metals	No (risk of combustion and projection of incandescent particles)	No (risk of combustion and projections of incandescent particles)	Yes (special dry chem. powder for fires of this type)	No		No (risk of combustion and projection of incandescent particles)
Type E Petroleum gas	No (only spray)	Yes	Ineffective due to pressure	Yes		Ineffective due to pressure
Type F Spontaneously combustible substances	Yes	Yes	Yes	As a temporary restraint		No

tackling that type of fire. Approach the seat of the fire close to the deck, allowing for the fact that heat rises. Have a standby man clear of the danger, ready to back up with further extinguishing agents. Emergency parties should be prepared to enter and relieve those first attempting to extinguish the blaze.

Subsequent Actions

Depending on the circumstances, cut off any power supply that may be live at the site of the fire. Rig hoses into the area to reduce the heat. Establish good communications as soon as possible. Prudent use of fire blanket, sand and scoop, wet towel or wet blanket can often prevent a major disaster occurring. Use the facilities available and make full use of improvised resources.

Accommodation Fires at Sea

Generally all major fires originate from either a smaller fire or an explosion. It is unlikely that fires within the accommodation will be caused by an explosion; consequently, speedy and efficient action to deal with the smaller fire will often prevent the larger, more crippling fire developing.

Regular training drills, well planned and using all the ship's equipment, are not only reassuring to passengers and crew members but an efficient method of making personnel familiar with the equipment available. Even if a fire is only suspected, the alarm should be raised immediately and crew members should be in no doubt that a false alarm would not result in punishment.

Accommodation fires generally occur in Class 'A' combustible material (see Table 15.1); bearing this in mind, the following course of action is recommended:

1   Raise the fire alarm (all types of fire).
2   Reduce speed (all types of fire).
3   Close down all mechanical ventilation.
4   Have fire-fighters, working in pairs, investigate and tackle the fire.
5   Isolate electrical 'live' circuits.
6   Surround the fire, attacking it from as many sides as possible with hoses.
7   Close all fire and watertight doors.
8   Approach the fire with the aid of breathing apparatus.
9   Have the Communications Officer standing by to transmit emergency or distress signals.
10. If traffic, weather and sea room will allow, bring the wind to a direction astern that will reduce the draught in the ship.

The order of events will, without doubt, vary with circumstances, and the actions of individuals will be dependent on the location of the fire and the facilities available in the vicinity. The above-mentioned procedure may be elaborated on.

Reduction of speed is necessary because the speed of the vessel through the water will provide continuous draught for the fire. This will provide oxygen for the fire, not the required starvation.

Closing down all mechanical ventilation will help to stop the passage of heat and smoke throughout the ship. Should heat or smoke be drawn in through passages, etc. it may become necessary to evacuate adjoining compartments. Again, starvation of the oxygen supply, effectively reducing the spread of heat, smoke and the fire itself, will be accomplished.

Fire-fighters need to operate in teams of not less than two, because the average person's courage in the face of danger is reduced considerably if he or she is alone. Two or more people may also be necessary to achieve a success, or at least better efficiency, than one.

Isolation of 'live circuits' is necessary because the dangers of water as an electrical conductor are well known, particularly when a strong jet of water is being brought into operation. Isolation of live circuits must be carried out before the fire is attacked with any water branch line, whether operating on jet or spray.

Surrounding the fire and attacking it, rather than operating from one side only, stops the fire being pushed from one region to another. It should not be forgotten that any fire has no less than six sides, and all six sides should be attacked whenever possible. This may only be in the form of boundary cooling of bulkheads, but the heat content and its effect are reduced.

Watertight and fire doors must be closed for any emergency when the hull is threatened. Not only is the passage of heat and smoke restricted, but subsequent casualties caused by the passage of fire or explosion can be greatly reduced. This is especially important on passenger vessels.

Breathing apparatus is essential for tackling accommodation fires, especially if internal fittings such as furniture containing poly-urethane foam are present. Toxic fumes from burning upholstery can be extremely hazardous for fire-fighters. Smoke helmets, for this reason, should not be worn, only the self-contained breathing apparatus.

Tackling the fire speedily is essential. Unless early location of the fire is made, fire-fighting may become extremely difficult. Rows of cabins and passageways tend to transmit heat and smoke quickly over a considerable area. Many of the cabins may form smoke traps, disguising the location of the fire to the fire-fighter or rescuer.

Initial actions are important, and these will depend on location and type of fire. If it is in a cabin, considerable build-up of heat may have already taken place, and entry could be disastrous if the interior has not been cooled off. This can usually be achieved by breaking open the bottom panel of the door and directing a jet to the deck head. This action will cause a deflection of the water jet and cool the interior of the cabin down prior to entry by fire-fighters behind a protective spray curtain. Indiscriminate smashing of ports and doors, however, should be avoided unless necessary to save life.

Galley Fires at Sea

The successful extinguishing of a galley fire will be more readily achieved if the location and method of using the available extinguishing agents are known beforehand. Freedom of access to these extinguishing agents is essential, and they should at no time be used for any other purpose than that for which they are designed.

A ship's galley will normally be equipped with several or all of the following extinguishing agents:

- foam extinguishers for oil-fired stoves (Figure 15.1)
- dry-powder extinguishers for electric stoves (Figure 15.2)
- CO_2 extinguishers and CO_2 gas to ductings

> All portable fluid fire extinguishers shall be of an approved type and shall have a capacity of not more than 13.5 litres and not less than 9 litres. Spare charges for extinguishers must be carried and no extinguisher must give off toxic gases in such a quantity as to endanger personnel.

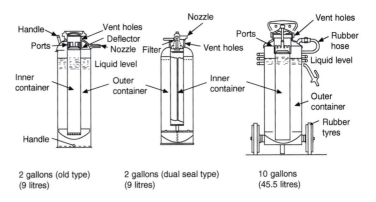

Figure 15.1 Sectional views of foam fire extinguishers.

2 gallons (old type) (9 litres)   2 gallons (dual seal type) (9 litres)   10 gallons (45.5 litres)

- inert gas compounds to deep fat fryers
- fire blanket
- sand and scoop in buckets or containers
- small hose reel and nozzle
- fire box, close to hand, containing hose, spray/jet nozzle, and fire axe.

Speedy and correct use of the above could reduce the risk of a major fire. Lack of thought in tackling the common chip-pan fire could result in the whole of the galley area becoming engulfed in flames, with the subsequent risk to catering personnel and to fire-fighters tackling the blaze. The majority of galley fires occur at the cooking stove, or from activities associated with the stove, e.g. lighting oil stoves, smoking when refilling the oil reservoir, over-heating pans of foodstuffs, especially fats, etc.

Human error is probably one of the main contributing factors – e.g. when pans of fat and such are left unattended; the escalation into a major blaze occurs when water is used as an extinguishing agent. A limited amount of forethought and training may prevent a serious outbreak by covering the open pan with a damp cloth, so cutting off oxygen from the blaze.

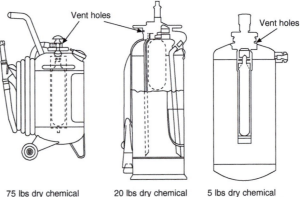

Figure 15.2 Chemical extinguishers.

75 lbs dry chemical extinguisher (34 kg)   20 lbs dry chemical extinguisher (9 kg)   5 lbs dry chemical extinguisher (2.5 kg)

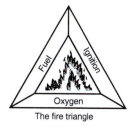

Figure 15.3 Making and breaking of fire triangle.

The fire triangle

All three elements of the fire triangle are needed to support combustion. The elimination of any one prevents fire.

The destruction of the so-called fire triangle (Figure 15.3) can be readily achieved in all small fires by the elimination of heat, fuel or oxygen. Major fires will respond in a similar manner when one of the three is nullified, but it may take considerably longer than with a minor outbreak. For instance, an oil drip tray turned into a blazing inferno by the direction of a jet of water onto it within the confines of the galley is almost impossible to control, whereas oxygen could easily have been excluded by deflecting foam onto its surface.

A clean, well-disciplined galley area will reduce the risk of fire. Regular drills and the training of crews, especially catering personnel, in correct fire-fighting procedure will reduce the risk still further.

Machinery Space Fires at Sea

The engine room of any vessel must be considered an extremely high-risk area, containing as it does certain items more susceptible to fire than any others on the ship. The majority of fires within the engine room are oil fires, Class 'B' or electrical fires, Class 'C' (see Table 15.1).

When an assessment of an outbreak of fire within the machinery space is made, detail as to the type and the extent, together with the location, must be thorough. A minor oil spillage will probably be a localized outbreak capable of being tackled by portable equipment, whereas a larger oil leak, say from a broken fuel pipe, may cause an extensive fire that can only be extinguished by use of a fixed smothering system such as CO_2 or foam.

Once an outbreak of fire inside the engine room has been discovered, a suggested course of action would be as follows:

1   Raise the fire alarm, or order somebody else to raise the alarm.
2   Inform the bridge at the earliest opportunity.

3   Investigate and tackle the fire immediately, if practicable.
4   Continue to fight the fire until the emergency party arrives at the scene.
5   Rescue injured persons as soon as practicable.
6   Establish supply of equipment – foam compound, etc.
7   Establish communication system, to include the bridge.
8   Attempt to contain the fire and extinguish by conventional means before use of fixed smothering apparatus.
9   Close down all ventilation, using non-essential personnel, once the alarm is raised and the location of the fire is established.
10   Close all watertight and fire doors as soon as possible after the alarm is raised.

The order of events will vary with circumstances, of course, and the recommendations given above must be used only as a guide. The actions taken by the Master, Chief Engineer, Engineering Officers, etc. will also be dependent on the type of machinery and the geography of the engine room; but any machinery space fire should be tackled after investigation has shown that the approach adopted will contain and possibly extinguish the outbreak. Conventional means of fighting the fire should be continued until supplies of foam compound are consumed or the available breathing air bottle supply runs out. Then, as a last resort, CO_2 or the equivalent should be injected in accordance with the fire-fighting plans of the vessel.

The actions needed to extinguish an engine-room fire should be taken, bearing in mind the limited supplies of conventional fire-fighting equipment on board. At the outset it should be assumed that a time will arrive when conventional fire-fighting methods can no longer be applied. To ensure the minimum amount of delay, therefore, any fixed fire-fighting installation should be made ready for operation at the earliest possible moment after the extent of the fire has been assessed.

Rescue of casualties should be a matter of priority, with due regard to the safety of rescuing personnel. Correct methods of gaining access to a fired area must be employed to prevent the fire spreading. Breathing apparatus should be employed to reduce the possibility of further casualties. First aid parties should be ready to treat any injury, especially burns. Regular drills will ensure that personnel, when attached to an emergency or stretcher party, know how to recognize burns and apply burn dressings.

Use of the breathing apparatus and stretchers within the confines of a compact engine room is not easy, and crew members should be exercised whenever drills are called to perform demanding tasks throughout the awkward parts of the vessel. The length of a contact line between any two fire-fighters should be tried and tested for adequacy when they are using engine-room ladders. Crews should be trained to use a messenger-location guide line when advancing into smoke-filled blind areas, bearing in mind that if the fire is deep-seated, say around the bottom plates, some breathing bottles will only last about 20 minutes.

Establishing efficient communications is one of the most essential requirements of tackling a fire at sea. There must be a link up from fire-fighters to the support personnel and to the bridge. In order for decisions to be taken, people in authority must be kept fully informed at all times of the situation. The decision to withdraw and inject, say CO_2, can only be made by someone who is aware of all the facts, especially those regarding supplies of equipment, condition of personnel, location of fire and danger of explosion.

Containment of the fire should first be attempted by use of conventional means, and the possibility of using a water spray from above the fire, as with a funnel fiddley construction, must be seriously considered. Not only will this produce a cooling effect before the injection, say, of CO_2, but also a steam cloud, causing a blanketing effect over the fire.

Watertight and fire doors should be closed as soon as possible for the safety of the vessel. Engine-room personnel should be well aware of emergency and tunnel escape systems.

Summary

On the discovery of the fire, the fire alarm must be raised, casualties removed from the scene and the fire investigated and tackled with primary equipment. Depending on weather conditions and the location of the fire, the oil supply should be cut off, emergency parties sent to the scene and boundary cooling should be started with the aid of emergency pumps and emergency generator.

Communications should be established to include the Master. The con of the vessel should be adjusted to minimize draught for as long as main engine power remains available. Any fixed extinguishing system should be made ready for immediate use, the Communications Officer told to stand by in the event of urgency or distress messages becoming necessary for transmission.

Ventilation, fire doors and watertight doors should be sealed, and overhead cooling of the fire scene should be carried out if possible. Personnel should be aware of particular hazards regarding smoke density in an already dark area, and the possibility of re-ignition from hot metal surfaces after they have assumed the fire to be out.

Fire Regulation Changes

In accordance with recent changes to SOLAS (1995 conference), IMO requires that passenger ships and ro-ro vessels of 500 gross registered tonnes (grt) and above, with category 'A' machinery spaces, be installed with an automatic water-based local extinguishing system in areas presenting high fire risk (Regulation II-2/10.5.6 and the MSC Cir.913).

The above is in addition to any fixed fire-fighting system required by the regulations. The time scale to comply with the recent changes to the regulations is as follows:

- New passenger vessels over 500 grt: July 2002
- New cargo vessels over 2,000 grt: July 2002
- Existing passenger vessels over 2,000 grt: October 2005

High-risk areas expected to be protected by spray nozzles include boiler fronts, above bilges and tank tops where oil fuel is likely to spread, oil fuel units like purifiers and clarifiers, hot fuel pipes near exhaust systems or similar heated surfaces.

Additionally, it should be noted that category 'A' machinery spaces of passenger vessels carrying more than 36 passengers must provide at least two suitable water fog applicators.

Mariners should be aware that 'water mist' fire protection systems use water as the fire-fighting medium in such a manner that it extinguishes fuel and other fires through cooling and oxygen depletion. The systems tend to cause minimal damage and prevent re-ignition by the cooling down of the space (see page 547).

Preventive Measures

The strategic siting of portable extinguishers, drip trays, hose boxes and hydrants can greatly reduce the risks of major fires developing. Regular testing drills of crew and equipment in unfamiliar surroundings and extensive precautions against accidental spillage will also go a long way towards preventing machinery space fires.

Cargo Space Fires at Sea and in Port

Fires in cargo spaces can generally be separated into two categories: (1) in dry cargo vessels with dry bulk cargoes or general cargo; and (2) in bulk oil, chemical or gas carriers. The methods of extinguishing fires will be dependent on the nature of the burning cargo and the other parcels of cargo around it, together with the location of the fire in relation to such features as engine rooms, pump room, paint lockers, etc. Often water directed onto the seat of the fire appears the most effective action, but this method of extinguishing should not be accepted without question, since the stability factors of the vessel have to be considered. Fixed gas systems can also be employed as an alternative option.

The following general approach is recommended:

1. Raise the fire alarm and inform the Master of the vessel.
2. Order the engine room to immediate 'standby' and reduce the vessel's speed.
3. Close down all ventilation to the fire area.

4 Assess the situation with regard to possible casualties and refer to the cargo plan to establish the type of cargo on fire.
5 Order the Communications Officer to stand by.
6 Commence boundary cooling and check adjacent compartments for additional fire risks.
7 Inject fixed CO_2 or other fire-fighting medium as per ship's fire plans.
8 Investigate port of refuge facility and inform shore authorities.

NB. The risk of opening up a battened down compartment when at sea must be considerable. To inspect the compartment would allow air to enter and feed the fire, with the possible flash-back situation occurring.

If the vessel were in a port, with extensive back-up facilities available, the method of approach would probably be different. Harbour authorities would have been informed and non-essential personnel could be disembarked to a place of safety.

A direct inspection and attack on the fire could be made by a full fire-fighting team with the view to possibly digging out the seat of the fire.

If the fire were found to be situated in, say, a 'tween deck rather than a lower hold, conventional methods of tackling the situation would be a prime consideration. However, hose branch lines should be pressurized and in plentiful supply before anyone enters the space. Breathing apparatus should be donned and ready for immediate use, with people wearing protective clothing. Emergency lighting should be on hand with reserve personnel. Bilge pumps should be in operation.

General advice for the conventional tackling of cargo space fires cannot be given in any direct form. Methods of approach will depend on the circumstances of the case. However, any seafarer who is trying to advance on a fire virtually blind should be aware of the following points:

- Never go into the compartment alone.
- Wear breathing apparatus when entering.
- Ensure that communications are established when entering the compartment.
- Shuffle only, do not walk. Keep body weight on the heel of one foot, using the other foot to detect forward objects.
- Keep the back of the hand in front of the face while advancing.
- Advance behind a spray curtain towards heated areas.
- Reduce damage by manually directing jets of water towards the fire area.
- Bear in mind that water is only effective when it is being turned to steam.
- When approaching the fire, work low down, reducing the smoke and rising heat effect.
- Remember that, at the fire point, the first jets striking the burning material will cause an expansion of surrounding gases. This will produce large volumes of steam and smoke at first, but this will soon clear and conditions improve.
- Keep regular checks on the fire containment by feeling the heat content and checking temperatures at different levels on

adjacent bulkheads. Spray branch lines can be most effective in boundary cooling to reduce temperatures, and their use lessens the risk of distortion of steel plate while covering a larger area than a jet.

Bulk Oil, Chemical or Gas Carriers

Fires on board these vessels are generally started by or associated with explosions. Conventional fire-fighting methods may be of little use other than providing a delaying action in order to carry out a successful abandoning of the vessel.

This is by no means a defeatist attitude. The main concern must be for the safety of life, and, to this end, if the use of hoses can provide valuable time to launch survival craft, then they have performed the most essential of functions. When used with spray nozzles, they can cool large areas of plate and perhaps stop the fire from spreading to accommodation areas. Foam appliances may also be used to prevent burning oil from reaching the accommodation decks.

With oil, chemical or liquefied gas cargoes, the development of toxic vapours is a distinct possibility. It is essential, therefore, that emergency action parties wear breathing apparatus and protective clothing. Where oil is known to be burning, the generation of dense black smoke will make fire-fighting extremely hazardous, and considerable thought must be given to the amount of oil, type of oil, surrounding cargoes, and access to the seat of the fire, before committing fire-fighters.

Efficient communications must be established, as with all fire situations, but walkie-talkie radios should be used with extreme caution in areas of explosive vapours unless they are of a safe type. Established communications between fire-fighters and the bridge could well assist the localizing and containment of the fire; the Master may turn the vessel stern to the wind, with the idea of reducing the draught within the vessel or even blowing the fire overside. The ship's head may need changing at a later stage in order to provide a lee for launching survival craft. These manoeuvres depend, of course, on the availability of main engine power and unimpaired steering gear.

Dust Explosions

With certain types of cargo where a heavy dust content is a prominent feature it is not uncommon to encounter large explosions and the creation of fireballs. These generally occur when attempts to extinguish a small fire by means of hose jets causes a massive disturbance of accumulated dust particles. The dust disturbed by the action of the hose is often thrown upwards and attempts to resettle above the fire area. The dust particles are so small that ignition can take place quickly while the dust cloud is still attempting to resettle. The whole dust area ignites, together with the oxygen content of the atmosphere, creating a fireball or explosion over an area, depending on the amount of dust disturbed and the heat generated.

Typical cargoes susceptible to dust explosion are grain-type cargoes, e.g. barley, wheat, etc. Should the size of the compartment be restricted, then a dust explosion could engulf the whole area and the hot gases could blow out the side of the vessel. The explosion would be violent and sudden, and if any fire-fighters were inside the compartment operating the hoses, they would suffer severe burns at the very least.

The cause of a dust explosion gives the answer to their prevention. Hose jets should not be used until the area has been thoroughly damped down by spray nozzle action. The dust will then congeal rather than remain in fine particle form.

Vessels carrying bulk cargoes such as grain, coal, etc. should take extra precautions in the event of fire to ensure that, if possible, the fire is contained at an early stage. However, fire-fighters should realize that at the bottom of empty or partially filled cargo holds dust tends to accumulate, and the risk of dust explosion at the bottom of a hold is far greater than at the top of any cargo.

Deep Tanks

General cargo is often carried in deep tanks. Should a fire develop either at sea or in port, deep tanks lend themselves to immediate flooding as a means of resolving the problem.

Counter-filling of opposing ballast tanks allows the fire to be extinguished without the stability of the vessel being in any way impaired. However, this method would produce a limited amount of free surface during the process of filling and pressing up the tank.

The tank could be flooded from the engine-room pipeline system by hose down the air pipe or by hose via the manhole into the tank, if accessible. Bilge pumps should have been checked and the cleanliness of hat boxes established before loading cargo, to ensure that the water could be pumped out after the fire was extinguished.

Summary

On the discovery of the fire, raise the alarm and attempt to identify the source and location of the burning material. Emergency parties should close down ventilation and check adjacent compartments. Hoses should be run out and boundary cooling started while stability factors are checked. Establish a communications link and supply lines after assessing the nature of the fire and method of extinguishing to be employed.

Check the cargo plan and identify potential hazards within the compartment. As a general rule, prepare fire-fighters for 'tween deck and/or lower-hold entry, depending on the fire's location. Prepare fixed smothering or extinguishing systems with the idea of inerting the space *except* for nitrate or sulphate cargoes. Steam smothering must never be used when explosives are present. Should nitrates,

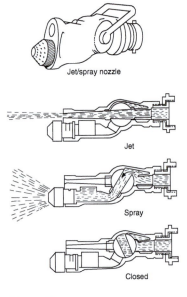

Jet/spray nozzle

Jet

Spray

Closed

Figure 15.4 Operation of combined jet/spray nozzle.

sulphates or explosives comprise the cargo, then flooding the compartment may be the only answer, together with the associated actions to maintain stability.

Preventive Measures

Prevent smoking during the loading or discharging of cargoes. Check on cargo ventilation requirements and liability to spontaneous combustion. Check out incompatible cargoes and their respective stowage areas.

Paint Room Fires

The problems concerning fires in paint rooms and other similar stores are usually related to one or more of the following:

- The location of the store, often well forward in the forecastle area.
- The size of the store, which is often quite small, limiting access and internal movement.
- The high risk of explosion, the possibility of excessive heat build-up and the risk of toxic overpowering fumes from the contents of the store.

Many fires in stores of this nature are caused by a combination of human error and spontaneous combustion. It is always so easy after a day's painting to clean off with a paraffin wad and throw it into a damp/wet corner. Spontaneous combustion action taking place several hours later may cause a minor flame that grows very quickly into a major fire.

Tackling the fire will be hampered by the plentiful supply of oxygen in the area, especially if the store is sited well forward. A fast heat build-up in a confined space can be expected, limiting the ability of fire-fighters to reach the seat of the fire. Breathing apparatus will be required to get in close, and the danger of chemical reaction with water as an extinguishing agent will be ever-present.

Boundary cooling of other bulkheads, deckheads and decks must be of prime consideration, together with the removal of stores adjacent to the fire area. Spray nozzles will probably be used to provide a heat shield for fire-fighters approaching the blaze. The best approach is at low-deck level, with the intention of directing a hose jet at the deckhead to effect a cooling reaction within the store.

Without doubt a carefully chosen course and the reduction of wind effect within the confines of the ship will greatly assist in fighting the fire. Good communications established as soon as possible between bridge, engine room and the site of the fire must be considered essential, preferably by walkie-talkie. Speedy action with regard to the gathering of equipment, water on deck, etc. is also essential to the effectual control of the fire. Use of a foam-based portable extinguisher in the early stages on material with an oil content could damp down a major outbreak until additional fire-fighting systems become available.

Casualties should be expected, probably people overcome by smoke, heat or toxic fumes, or a combination of all three. The main danger is the spread of the fire into the forward cargo holds. The method of tackling this should be one of direct attack.

Fixed Fire-Fighting Installations

- CO_2 gas system
- foam system
- pressurized water sprinkler system
- water mist fixed-pressure spray system.

When vessels in the past caught fire while at sea, they usually tried to fight the fire by conventional means. Obviously each case must be treated in the light of the circumstances prevailing at the time, with due consideration being given to the facilities available. In the author's view, attempts should be made in the case of engine-room fires to bring them under control before the injection of a fire-fighting gas medium, for the following reasons:

- Injection of, say, CO_2 gas would immobilize the machinery space and virtually leave the vessel without motive power and at the mercy of the weather.
- Once injection has taken place, it is unlikely that a second supply of gas could be made available. Therefore, as there is only one chance in most cases for the gas to take effect, this chance should not be wasted in the early stages. This is not to say that there should be any hesitation once it has been decided to use gas. Then speedy injection could be to the benefit of all.

Conventional fire-fighting methods in the way of hose/branch lines and foam installations within the machinery space may be the ideal fire-fighting medium. Breathing apparatus will be needed so that a plentiful supply of 'full air bottles' will be required; failing this, means of refilling (compressor) air bottles, located outside the machinery space, should be provided. In several cases valuable time has been bought by fighting a fire by conventional means until the air bottles for the self-contained breathing apparatus have run out. Time won in this way can be usefully employed in seeking out a safe anchorage or port having good fire-fighting facilities, or clearing away survival craft.

Cargo hold and tank space fires may, by their very nature, have to be treated as completely different sorts of fire. Relevant facts as to the available access to the fire area have to be considered, and flooding of a fire area may also be a worthwhile proposition, having due regard to the stability and free surface effects.

NB. Most paint rooms are now protected by a sprinkler water protection system.

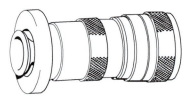

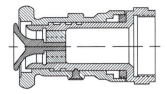

Figure 15.5 Section through jet/spray nozzle.

Steam Smothering Systems

In general the administration does not permit the use of steam as a fire-extinguishing medium in fixed fire extinguishing systems. However, where the use of steam is permitted, it should be used only in restricted areas as an addition to the required fire-fighting medium. The boilers supplying such steam will have an evaporation of at least 0.1 kg of steam per hour for each 0.75 m³ of the gross volume of the largest space to be protected.

In complying with these requirements the system shall be determined by and to the satisfaction of the administration.

Carbon Dioxide (CO₂)

This is probably the most popular of all the fixed fire-fighting systems employed at sea. The normal design incorporates a fixed bank of CO₂ container bottles whose contents can be directed, automatically or by direct manual operation, into any of the ship's protected spaces (Figures 15.6 and 15.7). Many systems are used in conjunction with a smoke-detector unit, the same sampling pipes guarding against smoke being used to inject the CO₂ gas via a three-way valve.

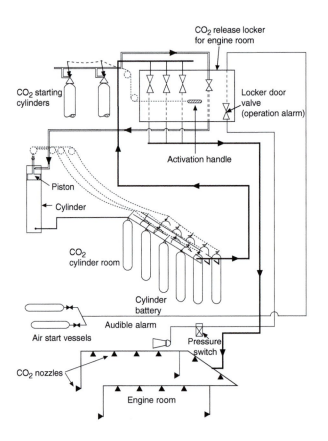

Figure 15.6 CO₂ total flooding system for engine room (bottle bank).

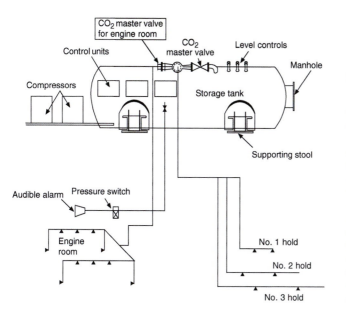

Figure 15.7 CO_2 total flooding system (bulk).

The storage tank, which is designed to hold in excess of ten tonnes of liquefied CO_2, is mounted on supporting stools, insulated by 150 mm of fire-resistant polyurethane foam, and finished in 1 mm galvanized steel plates. Refrigeration units may be attached to the tank or sited separately, giving the CO_2 a normal temperature of –20°C. Two independent units are the norm, working automatically, each being cooled by sea water and operated from external control panels. Alarm systems for each tank cover compressor failure, high pressure inside tank, low level in the tank and leakage through safety valves. The same sampling pipes guarding against smoke are used to inject the CO_2 gas via a three-way valve.

Fire-fighters should be aware that CO_2 is a smothering agent rather than an actual extinguishing one. The purpose of the gas is to deprive the fire of oxygen and by doing so break the fire triangle. Carbon dioxide is heavier than air and is usually injected into the upper levels of cargo holds and machinery spaces, all the oxygen being pushed to the upper levels as the CO_2 settles at the bottom of the space.

CO_2 gas has the following characteristics:

- It is a non-flammable gas.
- It is colourless.
- It is odourless.
- It is readily available in almost every port of the world.
- It is comparatively cheap.
- Systems may incorporate smoke detector units.
- It may be kept either in 45 kg cylinders or in bulk storage tanks.
- Normal temperature of liquefied CO_2 is –20°C.

Carbon Dioxide: Requirements

Cargo ships must carry available gas to give a minimum volume of free gas equal to 30 per cent of the gross volume of the largest cargo space protected on the ship.

In the case of machinery spaces, the quantity of carbon dioxide carried must be sufficient to give a minimum volume of free gas equal to the larger of the following volumes, either:

- 40 per cent of the gross volume of the largest machinery space protected.

NB. For machinery spaces the fixed piping shall be such as to permit 85 per cent of the gas to be discharged into the space within two minutes.

This volume is to exclude that part of the casing above the level at which the horizontal area of the casing is 40 per cent or less of the horizontal area, of the space concerned, taken midway between the tank top and the lowest part of the casing. Or

- 35 per cent of the gross volume of the largest machinery space protected, including the casing.

CO_2 Operation

Two separate controls must be provided for releasing the gas into a protected space and to ensure the activation of the alarm system. One control is used to discharge the gas, the other to operate the opening of a valve for the piping to convey the gas into the protected space.

The two controls shall be placed in a release box clearly identified for the particular space. Where the release box is kept locked, a key to the box shall be in a break-glass-type enclosure adjacent to the box position.

Fixed (Low-Expansion) Foam Systems (for Machinery Spaces)

When machinery spaces are fitted with a fixed, low-expansion foam system, it will be fitted in addition to the regular fire-fighting facilities stipulated by the regulations. The system must be capable of discharging through fixed outlets, in not more than five minutes, a quantity of foam sufficient to cover to a depth of 150 mm, the largest single area over which oil fuel is liable to spread. The foam distributed must be capable of extinguishing oil fires, and the system must be equipped with piping and valve operations for directing the foam to other main fire-risk areas within the protected space.

Fixed (High-Expansion) Foam Systems (for Machinery Spaces)

The high-expansion foam system must be capable of discharging through fixed outlets a quantity of foam sufficient to fill the greatest space to be protected, at a rate of at least 1 m in depth per minute. The quantity of available foam-forming liquid shall be sufficient to produce a volume of foam equal to five times the volume of the largest protected space. The expansion ratio of the foam shall not exceed 1,000 to 1.

The arrangement of the foam generator will be such that a fire in the protected space will not affect the foam generating equipment. The generator, its source of power, and the foam-forming liquid, together with the controls of the system, must be readily accessible and simple to operate.

Automatic Sprinkler, Fire Protection Systems

It is a requirement of the regulations that all passenger ships carrying more than 36 persons shall be equipped with an automatic sprinkler, fire detection and fire alarm system of an approved type, as specified in Regulation II-2.12.

Passenger vessels carrying less than 36 passengers will have either a fixed fire detection and alarm system or a similar automatic sprinkler, fire detection and alarm system, as previously stated.

Automatic sprinkler systems must be capable of immediate operation at all times without action by the crew. Each system must be capable of providing automatic visual and audible alarm signals at one or more indication units when any sprinkler comes into action. Any faults in the system must also be monitored by alarm sensors. When installed in passenger vessels the system must also indicate the location of any fire outbreak and monitoring equipment must be centralized in the navigation bridge or a main fire control station.

> NB. Fire control stations need to be adequately manned to ensure immediate response to alarm circuits.

The construction of the sprinkler system should be such that the 'sprinkler heads' should be grouped in sections of no more than 200 units. In passenger ship systems, sprinkler head sections should not serve more than two decks and should not be situated in more than one vertical zone. These conditions may be relaxed if the authority is satisfied that the protection of the ship against fire would not be reduced.

Each sprinkler section must be provided with a stop valve, protected against unauthorized use and a pressure sensor with indicator gauges at both the stop valve position and at the central control station.

Every sprinkler head must be placed in an overhead position within a pattern to maintain an average application rate of five litres of water per square metre per minute over the protected space. The operational temperature range in accommodation areas, with exceptions, will be from 68°C to 79°C.

Sprinkler heads, with a minimum of six to each section, must be placed not more than four metres apart and not more than two metres from a bulkhead in such a position as to prevent water projections being obstructed and allowing for all combustible material to be well sprayed. The overall arrangement of protection must be displayed on a plan which indicates the protected spaces, location by section and zone.

Systems will include a fresh water pressured tank and an independent powered pump. The pump, air compressor, alarm and detection operations will be supplied by not less than two separate power sources.

Pumps will be automatically activated by the pressure drop, before the standing charge of fresh water in the tank is exhausted. The pump, having a direct, independent suction to the sea, will be capable of maintaining the pressure level of the highest sprinkler to ensure continuous output for the simultaneous coverage of 280 square metres at the stated application rate.

Figure 15.8 Automatic sprinkler system.

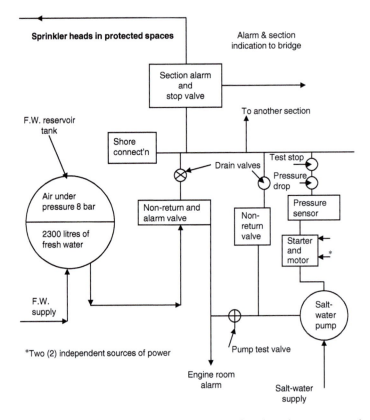

Additionally, every system must be fitted with an external connection from the ship's fire main and be provided with means of testing each section, the automatic operation of the pump and the monitoring devices.

Sprinkler Action

The sprinkler system is held in check under pressure (Figure 15.9). This pressure retains the water deluge by means of a glass bulb inside the sprinkler head within the protected space.

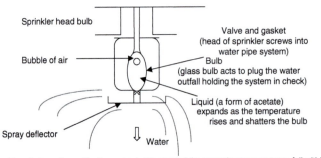

Figure 15.9 Sprinkler action.

In the event of a fire:

1   There would be a rise in temperature of the protected compartment.
2   The bubble of vapour inside the bulb of the sprinkler head is absorbed as the heat causes the acetate to expand.
3   Further increase in temperature causes the pressure to increase, breaking the glass bulb.
4   Air pressure in the storage tank expands and forces fresh water through the system. This flow of water activates the engine-room alarm by means of a flow-meter actuator. The water passes through the non-return valve to the 'section alarm valve'. Indication is then transmitted to the bridge via a flow-meter actuated alarm circuit.
5   As the fresh water level falls, so will the pressure and this pressure change is detected by the pressure sensor. Automatic transmission then activates the starter motor for the salt-water pump, which cuts in to relieve and take over the fresh water system.
6   Once the fire is out the section valve should be manually switched off.

Fixed Pressure Water-Spraying System

These pressurized water (fog) spraying systems are required to be installed on passenger vessels carrying more than 36 passengers with category 'A' machinery space. They are also fitted to vessels with special category spaces, generally ro-ro vessels.

The system must be capable of effecting an average distribution of water of at least five litres per square metre per minute in the protected space. Where increased application rates are required, these should be to the satisfaction of the administration.

The system shall operate in sections and the distribution valves should be in readily accessible positions outside the protected space. It should be retained in a charged position, such that any pressure drop would automatically bring the pump into action to supply the water requirement.

NB. Machinery spaces require two systems to comply with the regulations. Where a ro-pax vessel is involved with special category spaces, each of these spaces must additionally be protected by a water spray fog application.

The pump may be independently powered or supplied from the emergency generator. Where independent means are used to power the pump, this must not be situated in a position where it could be affected by the fire in the protected space.

These systems can extinguish fuel and other types of fire by means of cooling and oxygen depletion, causing minimal damage to plant. Re-ignition is prevented because the medium of water induces cooling from the onset. However, to be effective, distribution nozzles must be kept clear.

International Shore Connection

This is a fitment which is normally carried by all ships in order to provide a common link between shore hydrants and ships' fire mains (Figure 15.10).

Figure 15.10 International shore connection.

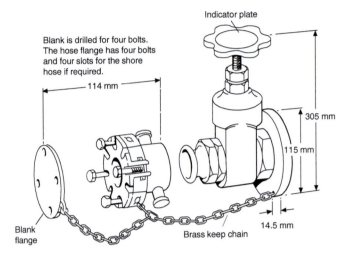

It is employed either aboard the vessel itself or ashore in conjunction with local Fire Service tenders, in the event of fire breaking out while the vessel is in port. The shore connection is usually situated in such a position as to be easily accessible to Fire Service officers, e.g. near the top of accommodation ladders, or the Mate's office.

Self-Contained Breathing Apparatus

The Siebe Gorman International Mk II, self-contained breathing apparatus (Figure 15.11) employs two cylinders of compressed air, which the wearer exhales direct to atmosphere. The cylinders are of a lightweight design so that, when fully charged, the apparatus complete with mask weighs only 38 lb (17 kg). The cylinder volume is four litres, providing enough air for 20 minutes when the wearer is engaged in hard work. Both cylinders have the same capacity.

The amount of work carried out by the wearer will obviously affect the consumption of air and consequently the time that person may continue working. The following are guidelines supplied by the manufacturer:

- Hard work rate: 40 minutes (twin cylinders).
- Moderate work rate: 62 minutes.
- At rest: 83 minutes.

Pre-operational Checks (Monthly)

1   Ensure that the bypass control is fully closed.
2   Open cylinder valves. The whistle will be momentarily heard as pressure rises in the set. Check cylinders are fully charged (pressure gauge inside green area).

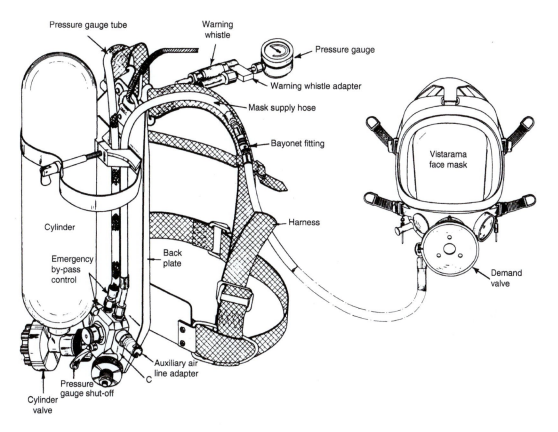

Figure 15.11 Siebe Gorman International Mark II compressed air breathing apparatus.

3 Any leaks in the apparatus will be audible and should be rectified by tightening the appropriate connections, but do not overtighten.

4 Close cylinder valves and observe pressure gauge. Provided it does not fall to zero in less than 30 seconds, the set is leak-tight.

5 Depress demand valve diaphragm to clear circuit of compressed air.

6 Close pressure gauge shut-off valve and reopen cylinder valves. The pressure should remain at zero. Reopen first valve.

7 Gently open emergency bypass control; air should then be heard to escape from the demand valve. Close control.

8 Close cylinder valves. Gently depress demand valve diaphragm and observe pressure gauge. When it falls to approximately 43 ats. (44.5 kg/cm²), the whistle should sound.

Preparation for Use

1 Demist mask visor with anti-dim solution.
2 Don the apparatus and adjust harness for a comfortable fit.

3   Open cylinder valves. Put on the mask and adjust to fit by pulling the two side straps before the lower ones.

4   Inhale deeply two or three times to ensure that the air is flowing freely from the demand valve and that the exhalation valve is functioning correctly. Hold breath and make certain that the demand valve is shutting off on exhalation or that leakage, if any, is slight.

5   Close cylinder valves and inhale until air in the apparatus is exhausted.

6   Inhale deeply. The mask should crush onto the face, indicating an airtight fit of both the mask and the exhale valve.

7   Reopen cylinder valves.

There are several manufacturers of breathing apparatus, and the sequence of operations may differ slightly from that described above. Caution in following correct procedures is advised in all cases, together with regular practice drills in the use of this type of emergency equipment.

Contents of Fireman's Outfits

Each fireman's outfit must contain the following:

- Waterproof protective clothing capable of protecting the skin from radiated heat from any fire and from burns and scalding by steam.
- Boots and gloves manufactured in rubber or other non-electrical conducting material.
- A rigid helmet providing impact protection.
- A safety lamp (electric) of an approved type with a minimum burning period of three hours.
- An axe of design and construction which satisfies the administration.

NB. In passenger vessels carrying more than 36 passengers the administration may require additional sets of personal equipment and breathing apparatus to be carried, having due regard to the type and size of the vessel (EEBDs).

Additionally, each breathing apparatus will have a harness arrangement and have a fireproof lifeline of sufficient length capable of being connected by a snap hook to the harness (see Figure 19.12).

An alternative to the self-contained breathing apparatus is permitted in the form of a smoke helmet or smoke mask provided with a suitable air pump and air hose length. Where the air hose needs to be greater than 36 m in length to reach any part of the holds or machinery space, this item may be substituted for a self-contained breathing apparatus.

Example: Cargo Fires

Liquid Natural Gas

Natural gas contains numerous component gases, but by far the greater percentage is methane (CH_4), which represents between

60 and 95 per cent of the total volume. This fact is important when considering the safety aspects for fire-fighters tackling a liquid natural gas (LNG) fire.

During the initial period of vaporization of the gas, ignition may be accompanied by a flash of varying proportions. However, because the velocity of propagation of a flame is lower in methane than in other hydrocarbon gases, it is unlikely that future ignition will have flash effect.

The fire-fighting plan should be well thought out in advance and a concentrated effort made rather than 'hit and run' tactics, as these will only consume the vessel's extinguishing facilities without extinguishing the fire. Before attempting to tackle a large fire, you should seriously consider allowing the fire to burn itself out.

Should an attempt to extinguish the fire be made, extensive use of 'dry powder' should be employed from as many dispensers as can be brought to bear. Fire-fighters should be well protected against heat radiation and possible flash burns, and approach the fire from an upwind direction. Powder dispensers should sweep the entire area of the fire, but direct pressure of powder jets onto the surface of the liquid should be avoided.

Should dry powder guns be used, fire-fighters should be well practised in their use and be prepared for some kick-back effect. They should also be made aware that there is no cooling effect from the use of dry powder, and that re-ignition after a fire has been extinguished is a distinct possibility.

In the initial stages it is always preferable to isolate the fire by shutting off the source of fuel. This may not, however, always be possible.

A final warning when tackling an LNG fire is that water should not be used directly, as this will accelerate vaporization of the liquid.

Plate 146 Two LNG carriers lie alongside each other outside the Dubai shipyard. Prominent features are the gas dome structures on the upper decks.

This is not to say that surrounding bulkheads and decks cannot be cooled down with water sprays, provided that water running off is not allowed to mix with burning LNG.

Cotton (Class 'F' Fire)

Cotton is a cargo liable to spontaneous combustion and one which is extremely difficult to bring under control. Cotton cargoes are such that they are shipped in bales of 500 or 700 lb (227 or 318 kg). A heavy cargo, cotton is often stowed in lower holds for stability reasons and to form a base for later cargo. It is a cargo in which the prevention of the fire initially is preferable to knowing how to tackle it, should it occur.

Cotton bales should be dry and free of oil marks, tightly bound and seen to be in good condition at the onset of loading. Stringent observation of 'no smoking' in and around cargo holds should be observed by stevedores and ship's personnel. Bare metalwork in holds should be covered to prevent moisture contact with cargo and the spar ceiling should be inspected to ensure that bales do not come into contact with the shell plate.

Should an outbreak of fire occur, the only sure way of extinguishing it is to dig out the affected area. This practice is not at all easy for crew members, who are inexperienced at handling heavy bales for any length of time. Deviation to a port for discharge may become the only alternative, depending on the size of the fire at the time of discovery and the ability to extinguish it.

If successful in digging out burning or smouldering bales of cotton, jettison them overboard. Re-ignition of cotton bales can occur, even after they have been totally immersed in water. Bales which appear to be extinguished will all too easily flare up after a thorough hosing down.

If breathing apparatus air supply is restricted and for other reasons it proves impossible to tackle the fire directly, containment should be the next consideration. This is probably best achieved by the battening down of the compartment and the injection of CO_2 while heading for a port of refuge with the necessary facilities. Boundary cooling should be carried out on as many of the six sides of the fire as are accessible. Any deviation of the vessel's course should be noted in the ship's log book.

Coal (Class 'F' Fire)

All coal cargoes give off an inflammable gas, and when this mixes with critical proportions of oxygen, then explosion and/or fire may be the end result. The gas given off by the coal is lighter than air and during the voyage it will work its way to the upper surface of the cargo. It is essential that coal is therefore provided with 'surface ventilation' to clear away any build-up of accumulated gas. Surface ventilation is achieved during the voyage by raising the outer corners of hatches or opening 'booby entrance hatches'. Steel hatch

NB. Coal fires when treated with water will generate considerable volumes of steam. This steam must be vented or the compartment may become pressurized.

covers should be raised on their wheels, provided at all times that weather permits such action. Ventilators should always be properly trimmed.

All types of coal, whether of the anthracite, lignite or brown coal varieties, are subject to spontaneous combustion. A close watch should be maintained on hold temperatures during passage and correct ventilation allowed to reduce temperatures in the event of over-heating. It is worth noting that coal increases its temperature by its absorption of oxygen. Correct ventilation for this cargo must therefore be considered to be surface ventilation only, for a limited period.

Should fire break out, early positive hose action will probably be the best way of containing it. However, personnel may not be able to spend much time on fire-fighting because of the excessive heat or the amount of smoke within the space. Breathing apparatus will be essential and the air supply in bottles may further restrict conventional means of fighting the fire.

The injection of CO_2 must be considered at an early stage, should conventional methods become impractical. It will be totally dependent on the size of the fire whether this agent will effectively extinguish it. At the very best it will contain the blaze to a degree and will certainly buy time for the Master to investigate safe port options. Alternatively, the final option would be to flood the space with water. Close investigation of the ship's 'damage stability notes' should be made before taking this action, with particular attention to the free surface effect of flooding such a large space, though in a compartment filled with coal there would be little free surface effect.

Hold preparation before loading coal will play a major part in averting a fire, and the following points are recommended:

1 Clean the hold space of residual debris.
2 Clean and test bilges.
3 Remove the spar ceiling.
4 Remove any dunnage clear of the space.
5 Make provision for obtaining temperatures at different levels of the cargo.
6 Trim the cargo throughout and on completion of loading.

Fish Meal (Class 'F' Fire)

Fish meal is a bagged cargo which is probably one of the most likely to catch fire while the ship is on passage from the loading port, due to spontaneous combustion. Experience has shown that vessels employed in the carriage of 'fish meal' must take stringent precautions when loading (Figure 15.12). Extensive ventilation channels must be allowed for at the onset of loading and these channels must not be allowed to become blocked by falling bags of cargo.

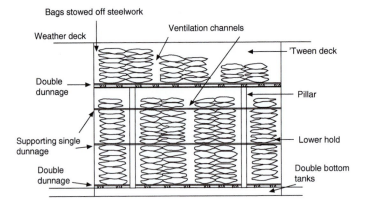

Figure 15.12 Fish meal stowage.

Deck Officers should be particularly aware that during loading bags should be sighted to ensure they are in good condition and dry. Telltale damp stains on the bags indicate that the cargo has been exposed to rain and that the contents are wet. These bags should be rejected at all costs.

Officers should be provided with injection thermometers for the purpose of testing bags during loading. Any batch with excessively high temperatures should also be rejected. Temperatures during the voyage should be taken at least twice per day and a watch maintained on the following temperatures for every space containing fish meal:

- hold temperatures
- ventilation inlet temperature
- ventilation outlet temperature
- ventilation channel temperature
- random bag selection temperature.

The hold should be thoroughly cleaned and steelwork covered with insulation paper before stowage begins. Bilge suctions and scuppers should also be inspected and tested before loading. Temperature and condition of bags should be checked at the onset of loading.

Large amounts of dunnage will be required for this cargo and where stowage is to commence on a steel deck, double dunnage must be laid. It is important that all dunnage is dry and free of oil marks. Single layers of dunnage should be placed at every height of seven bags. Ventilation channels of approximately 12 in. (30 cm) should separate double tiers of bags. Provision should be made for positioning thermometers at all levels of cargo, in all spaces containing fish meal.

If there is an outbreak of fire, close off all ventilation as soon as possible after the alarm has been raised. Make an immediate assessment of the fire area, and attempt to extinguish small fires, preferably by use of dry powder. If it is found that a major fire is already well established, ensure that all ventilation is cut off and inject CO_2. If this action fails to extinguish the blaze, then hose action may be the only alternative. Should hoses have to be brought to bear, then they should be as close to the fire as possible before being turned on. All bags of fish meal that are soaked by the hose action should be jettisoned as soon as the fire has been extinguished.

Hoses should not be used except as a last resort to save the bulk of cargo and the ship itself. Considerable spoilage of the bags will occur with hose action, and the likelihood of further outbreaks of fire becomes more probable rather than less. Correct stowage in the first instance, with continual checks on temperature conditions throughout the passage, will limit the chance of fire and give ample warning should it occur.

Roll On–Roll Off Vessels: Fire-Fighting Difficulties

When compared with the conventional vessel, ro-ro vessels are generally fitted with very large garage/vehicle decks. As such, the sheer size of the space requires large quantities of gas for total flood systems to be effective. Coupled with this, the necessity to seal the space and contain the gas volume is that much more demanding. Water drenching systems can be very effective but bring associated problems of free surface and slack water onto the vehicle deck. To cope with slack water it is essential that any incorporated drainage system has the capability to clear residual waters quickly and efficiently. Blocked drainage systems from burned residues are always a possibility following a fire in a protected space.

Conventional fire-fighting methods employing branch lines are extremely difficult for fire-fighters because of the limited access to deck areas. Cargo units are chained and retained as fixed obstructions to the potential hose party. Manoeuvring hose lengths around vehicle units to attain the fire scene would be heavy and awkward work. Dislodging chain lashings on associated units or fouling the hose and losing pressure would be an ever-present continuous problem.

Enclosed spaces must expect a degree of smoke build-up and any fire on a vehicle in close proximity to another vehicle must be expected to spread, depending on unit construction. Where fuel tanks are in the area, the risk of explosion is a real one to exposed fire-fighters. Therefore, fires in smaller private cars pose as great a threat as a box cargo unit. Fire-fighters with breathing apparatus and lifelines may find that the lifeline is an encumbrance with the close-stowed vehicles, and may restrict progress. The space may also acquire a harmful atmosphere from burning plastics or other similar synthetics or cargo combinations which in themselves may not have been classed as hazardous.

Once a fire has taken hold, the conditions could well slacken off lashings, especially so if tyres burn and deflate. If units come adrift the possibility of the 'domino effect' among other units in proximity is likely, especially in bad weather conditions. Some movement between units at sea is always visible and must be anticipated, but is usually controllable. Where a fire is established, conditions change overall and small movements can become accentuated and restrictive to personnel. As such, boundary cooling of a fire area may be restricted or even impossible to sustain.

Poor deck maintenance could also act against fire-fighters – where deck spaces are oily or greasy, they are liable to be inflamed more easily. Even when decks are well-maintained, water will generally make vehicle decks slippery and obstruct the progress of fire-fighters burdened with heavy hose equipment. Clearly, a case of good ship-keeping can pay dividends in the event of emergency incidents, as and when they may occur.

Where exposed or open decks are employed, the possibility of hazardous/toxic goods may be stowed. Leaking units, or units on fire, pose many of the problems stated above. Fire-fighters would probably be expected to tackle the incident from an upwind or side-on position to avoid toxics. This positioning may in itself prove difficult to attain from a tightly packed vehicle deck. However, the possibility of being able to jettison cargo is an alternative option, one which is generally denied to enclosed spaces. An exception to this would be if the vessel had side-loading facilities or a shell door arrangement. These options are included as extreme actions and would without doubt be difficult to achieve in the majority of cases. The jettison of cargo units in any event should be seen as an absolute last resort.

Fire Protection: Regulations

These are based on SOLAS Chapter II-2, Construction – Fire protection, fire detection and fire extinction.

Ro-ro cargo spaces

The regulations require that each ro-ro cargo space will be provided with a fixed fire detection and fire-alarm system. The system must be capable of detecting the onset of fire rapidly and the administration must be satisfied by the number, spacing and location of detection sensors.

Ro-ro cargo spaces capable of being sealed must be fitted with a 'fixed gas' fire-extinguishing system which will comply with the regulations. Ro-ro spaces which are not capable of being sealed must be fitted with a fixed-pressure water sprinkler system to protect all parts of any deck or vehicle platform in such a space. Such spaces and deck areas must have a drainage and pumping system which prevents the build-up of free surfaces. Additionally, all ro-ro cargo spaces must be provided with such numbers of portable fire extinguishers as the administration may deem sufficient.

At least one of these should be located at each access to such a cargo space. Where it is the intention to carry motor vehicles with fuel in their tanks, the space must be provided with extinguishers positioned at not more than 40 m intervals of deck length and:

- at least three water fog applicators;
- one portable foam applicator unit consisting of an air-foam nozzle of an inductor type capable of being connected to the fire main by a fire hose, together with a portable tank containing at least 20 litres of foam-making liquid and one spare tank (the nozzle must be capable of producing effective foam suitable to extinguish an oil fire at the rate of 1.5 m^3/min).

Closed ro-ro cargo spaces

Closed ro-ro spaces must be provided with an effective powered ventilation system which is capable of being rapidly shut down to effect closure in the event of fire. Ventilation ducts serving ro-ro spaces must be separated from other venting systems serving other cargo spaces.

Electrical wiring within such spaces must be intrinsically safe and be positioned at a height of not less than 450 mm above the deck. Other equipment which may present a source of ignition is not permitted.

Further detailed reference should be made to Regulation 53 of SOLAS.

16

Search and Rescue Operations

Introduction

With the introduction of the three manuals of IAMSAR, the civil aviation and the maritime industries were brought much closer together. The readership of this text will undoubtedly favour the maritime content, but should nevertheless take into account that the two industries can and do continue to work extensively together.

This mutual cooperation serves both the marine and aviation industries well. Although fixed-wing aircraft cannot, in general, recover from the surface, they can locate distressed parties early by coverage of a greater area. Surface craft can subsequently be brought in to recover when the situation is outside helicopter range.

The many search patterns and search and rescue (SAR) operations that have relied heavily on the involvement of aircraft are too numerous to mention. New technology has enhanced the range of rotary winged aircraft. Procedural changes have established all involved parties in uniform change across international borders. Such advances can only be in favour of those who will most need them, namely the potential survivor.

Action by Vessel in Distress

A ship in distress should transmit an appropriate distress alarm signal, followed by a distress message. This message should include the following main points:

- Identification of the vessel in distress.
- Position of the vessel in distress.
- Nature of the distress and the assistance required,
- Other relevant information to facilitate the rescue, e.g. number of persons leaving the ship, number remaining on board, Master's intentions, etc.

In addition to the main points mentioned above, further information regarding influencing factors should be passed on to assisting vessels. This may include:

- weather conditions in the immediate area of the ship in distress;
- details of casualties and state of injuries;
- navigational hazards, e.g. icebergs etc.;
- numbers of crew and passengers;
- details of survival craft aboard and of craft launched;
- emergency location aids available at the scene of distress and aboard survival craft.

A series of short messages is preferable to one or two long messages. Vessels in distress should use the time preceding a rescue attempt to minimize the risk of increased numbers of casualties. This could be done by reducing numbers aboard the stricken vessel by allowing non-essential personnel to disembark. Some companies now employ this technique as standard practice, but it should be used with extreme caution and must depend on weather conditions for the launching of survival craft and the degree of danger present aboard the parent vessel, bearing in mind that the mother ship provides the best form of protection while it remains sustainable.

Master's Obligations

In accordance with the International Convention for the Safety of Life at Sea, Masters have an obligation to render assistance to a person or persons in distress if it is within their power. Any Master of a vessel at sea, on receiving a signal for assistance from another ship, aircraft or survival craft, is bound to proceed with all possible speed to the scene of the signal. If possible, he should inform the distressed party that assistance is on its way. If the Master of a ship is unable, or under special circumstances considers it unreasonable or unnecessary, to proceed to the scene of distress, then he must enter that reason in the log book.

The Master of a vessel in distress which has made a request for assistance has the right to requisition one or more of those vessels which have answered his distress call. It will be the duty of the Masters of those vessels so requisitioned to comply with their call to assist and proceed with all speed to the distress scene.

The Master of an assisting vessel will be released from his obligations to assist when he learns that one or other vessels have been requisitioned and that, because they are complying, his own vessel is no longer required. He may also be released from further obligation to assist by an assisting vessel which has reached the distress scene and considers additional assistance is no longer required.

Obligations of Rescuing Craft

On receipt of a distress message any vessel in the immediate vicinity of the distressed vessel should acknowledge that the message has been received.

Should the craft in distress not be in the immediate area, then a short interval of time should be allowed to pass before acknowledgement of the distress signal is despatched so that other ships in close proximity may give prior acknowledgement.

The Master should immediately be informed that a distress message has been received, and whether acknowledgement has been sent by other vessels, together with the positions of the vessel in distress and would-be rescue craft. The Master will cause an entry to be made in the communications radio log book.

Bearing the latter statement in mind, the Master of any vessel in receipt of a distress message may repeat that message on any frequency or channel that he knows to be in common use in that area. (see Chapter 9).

When Assistance Is No Longer Required

Any casualty having despatched a distress message and finding that the assistance being provided is adequate may effectively reduce the level of communications to those prefixed by the urgency signal.

Any decision to reduce communications from a distress to an urgency level must be the responsibility of the Master in command of the distressed vessel, or his authorized representative. Receiving stations should bear in mind that a very urgent situation exists and the resumption of normal working conditions must be made with extreme caution. Table 16.1 illustrates types of signal.

Searching the Sea

Vessels may be employed in SAR activities alone or with other surface craft (Figure 16.1), or with aircraft. It can be expected that a specialized unit like a warship or military aircraft would assume the duties of the On Scene Co-ordinator (OSC) and coordinate the other search units in the area. Communications will be established on 2182 kHz or VHF channel 16, if possible. Failing this, a relay

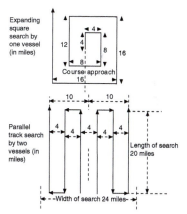

Figure 16.1 Sea search by one and two vessels.

Table 16.1 Emergency signals			
Type of message	**Prefix radiotelephone**	**Prefix radiotelegraph**	**Frequency/channel**
Distress	Mayday, Mayday, Mayday	SOS, SOS, SOS.	2182 kHz, channel 16 or any
Urgency	Pan Pan, Pan Pan, Pan Pan	XXX, XXX, XXX	other frequency at any time
Navigation warning	Securité, Securité, Securité	TTT, TTT, TTT	

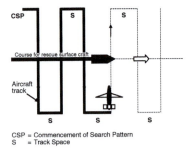

CSP = Commencement of Search Pattern
S   = Track Space

Figure 16.2 Coordinated surface search.

should be established between surface vessels and a coast radio station (CRS) to aircraft.

Surface vessels when engaged with aircraft in a coordinated search (Figure 16.2) could expect items of a specialist nature to be dropped into a search or rescue area. These items would probably be in the form of:

- parachute flares for illumination purposes;
- individual life rafts or pairs of life rafts joined by a buoyant rope;
- dye markers or flame floats;
- buoyant radio beacons and/or transceivers;
- salvage pumps and related equipment.

Should specialist units not be engaged in the search area then the Master of the vessel going to the assistance of the distressed vessel must assume the position of the OSC and communicate with the marine rescue coordination centre (MRCC).

Search Patterns: Choice and Aspects

Masters of ships called in to act as a search unit, or who find themselves designated as an OSC, may find the SAR mission coordinator (SMC) will provide a search action plan. However, this is not guaranteed, and the choice of the type of search pattern to employ may fall to the individual Master.

Clearly a choice of pattern will be influenced by many factors, not least the number of search units engaged and the size of the area to be searched. It will need to be pre-planned to ensure that all participants are aware of their respective duties during the ongoing operation. To this end the navigation officers of vessels can expect to play a key role within the bridge teams.

Establishing the search

Initially, the datum for the search area will need to be plotted. Where multiple search units are employed to search select areas, each area should be allocated geographic coordinates. This will reduce the possibility of overlap, time-wasting and assist reporting to eliminate specific sea areas.

Once the search area(s) has been established and an appropriate pattern confirmed, the 'track space' for the unit or units so engaged must be established. This must be selected to provide adequate safe separation between searching units while at the same time taking into account the following factors:

- the target size and definition;
- the state of visibility on scene;
- the sea state inside the designated search area;
- the quality of the radar target likely to be presented;

- height of eye of lookouts;
- speed of vessel engaged in search operation;
- number of search units (SUs) engaged;
- time remaining of available daylight;
- Master's experience;
- recommendations from MRCC;
- height above sea level (for aircraft).

Additional influencing factors are:

- night searches can be ongoing with effective searchlight coverage;
- the length of the search period may be restricted by the endurance of the vessels engaged;
- the target may be able to make itself more prominent if it retains self-help capability.

Pattern and respective track space should be selected with reference to the IAMSAR manuals, in particular Volume III.

The choice of Master and ship for OSC

When an SAR incident is instigated the MRCC is often faced with the choice of which ship and which Master is most suited to take the position of OSC. Ideally, a warship would be the first obvious choice because it has all the advantages and few of the disadvantages. Unfortunately, with isolated incidents especially, it is the commercial vessels that are more likely to be involved from the onset of the incident. With any SAR situation many variables could influence the selection of not only an OSC, but also the selection of individual SUs – the weather conditions and geography in relation to the position to name but two. Where a choice for the OSC has to be made, where no military unit is available, that choice will probably be made taking account of the following factors:

- Can the proposed vessel provide the necessary communication platform to carry out the task?
- Has the vessel the manpower to conduct its own ship routine and the additional emergency duties required of an OSC?
- What is the position of the vessel from the datum?
- Is the vessel equipped with the skills and equipment to carry through the duty?
- Has the vessel the necessary endurance?
- What is the experience of the Master?
- Is the nature of the cargo hazardous or perishable?
- What is the speed of the vessel?
- What is the draught of the vessel (relevant for shoal water operations)?
- Designation of OSC status requires the mutual agreement of the Master(s) where more than one ship is involved.

NB. The warship is ideal because it has adequate manpower, excellent plotting facilities, is often equipped with aircraft cover, no commercial pressure by way of perishable cargo, is self-sufficient with ample endurance, speed and manoeuvring capability together with medical and recovery methods if required. It has all these amenities, as well as probably the best communication links that any seagoing mobile is likely to possess.

Under the GMDSS legislation, vessels will be required to carry two search and rescue transponders (SARTs). These operate on 9 GHz for 3 cm radar. The effective range is approximately five nautical miles and their function is expected to enhance SAR operations. The radar signature from an SART would appear initially as a line of 12 dots. Turning to arcs on the observer's screen. This signature will change to a series of concentric circles as the range of the target is closed.

It should be borne in mind that the OSC is meant to coordinate SUs and may not be the first vessel on scene. Neither will it necessarily be involved in the actual recovery of persons from the sea. The function of the SUs is to recover persons from the water and, as such, they need recovery methods such as rescue boats.

Unless the OSC is jointly operating as an SU, this vessel would not necessarily require the immediate first aid equipment as desired by SUs. Neither does it have to be within the immediate area, although it is clearly an advantage to be as close as practical and certainly within RT-VHF range.

Aircraft in Distress

The distress message may vary with the time available from the onset of the emergency and the effective landing or ditching of the aircraft. However, when time permits, civil aircraft will transmit a distress call and subsequent distress message as follows.

Distress call by radiotelephony

1 The spoken words 'Mayday, Mayday, Mayday'.
2 The words 'This is …'.
3 The identity of the aircraft, spoken three times.
4 The radio frequency used in the transmission of the distress call.

Distress message

1 'Mayday'.
2 The call sign of the aircraft.
3 Information relating to the type of distress and the kind of assistance required.
4 The position of the aircraft and the time of that position.
5 The heading of the aircraft (true or magnetic).
6 The indicated air speed (in knots).
7 Any other relevant information which would aid and affect a recovery operation, e.g. intentions of the person in command, nature of any casualties, possibility of ditching, survival facilities available or not.

The term 'heading' when applied to an aircraft refers to the direction of the aircraft when in the air. Allowance must then be made for wind effect to ascertain the true direction over the sea. Indicated airspeed does not take into account the effect of the wind. This should be estimated to obtain a more realistic speed over the water. If the aircraft is to be ditched, the aircraft's radio transmitter may be left in the operative position, depending on circumstances.

Communication Between Surface Craft and Aircraft

Merchant vessels engaged in SAR operations with military aircraft should maintain a VHF watch on channel 16.

Surface vessels should use their normal call sign in communicating with an aircraft. Should the call sign of the aircraft be unknown, then the term 'Hawk', may be used in place of the aircraft call sign. When an aircraft is in the process of establishing communications with a surface craft without knowing the call sign of the vessel, the aircraft may use the inquiry call 'CQ' in place of the vessel's normal call sign.

Emergency Position-Indicating Radio Beacon

Every ship must be equipped with a satellite emergency position-indicating radio beacon (EPIRB) which can be activated from a site close to that position from which the ship is normally navigated, namely the ship's bridge, or otherwise, in an alternative location so that it can be operated remotely from that position.

The EPIRB must be capable of transmitting a distress alert through the polar orbiting satellite service operating in the 406 MHz band (although see exception, below).

The EPIRB must be stowed in an easily accessible position, capable of being manually released and placed into a survival craft. It must also have a float-free capability with an automatic activation facility.

Plate 147 The latest rotary wing aircraft assigned to British military forces: a triple-engine, five-blade rotor Merlin helicopter in operation with the Royal Navy.

NB. Every ship must have the capability to transmit distress alerts from ship to shore by using at least two separate methods. If the satellite EPIRB is used as the secondary means of distress alerting, and it is not remotely activated, it is acceptable to have an additional EPIRB installed inside the wheelhouse near to the conning position.

See GMDSS detail on p. 300.

Exception

Ships engaged in the sea area 'A' may carry, in lieu of the satellite EPIRB stated above, an EPIRB which shall be capable of transmitting a distress alert using DSC on VHF channel 70 and providing for location by means of a radar transponder operating on the 9 GHz band.

This EPIRB must similarly be capable of manual/automatic operation and have a float-free capability.

EPIRB Features

An EPIRB must:

- be of a highly visible colour, so designed that they can be used by an unskilled person. Their construction should be such that they may be easily tested and maintained and their batteries shall not require replacement at intervals of less than 12 months, taking into account testing arrangements;
- be watertight, and capable of floating and being dropped into the water without damage from a height of at least 20 m;
- be capable of manual activation and de-activation only;
- be portable, lightweight and compact;
- be provided with indication that signals are being emitted;
- derive their energy supply from a battery forming an integral part of the device and having sufficient capacity to operate the apparatus for a period of 48 hours. The transmission may be intermittent. Determination of the duty cycle should take into account the probability of homing being properly carried out, the need to avoid congestion on the frequencies and the need to comply with the requirements of the ICAO.
- be tested and, if necessary, have their source of energy replaced at intervals not exceeding 12 months.

Surface to Surface Rescue

Depending on circumstances, the options are the following:

1 Lower the ship's rescue boat and begin recovery.
2 Use a rocket line, messenger and hawser to draw survival craft off the distressed vessel.
3 Go alongside the distressed vessel.
4 Establish a tow if the stricken vessel will remain afloat.
5 Head to wind and part open stern door (ro-ro vessel) onto distressed vessel.
6 Use own life raft and drift survival craft towards distressed vessel on a towline.
7 Transfer personnel by breeches buoy.
8 Position rescue vessel's bow close to forecastle head of distressed vessel.

Use of Lifeboat/Rescue Boat

This is by far the most favoured method of taking people off a sinking vessel, though it is only practical in comparatively good weather. Attempting to put a lifeboat down at sea in anything over a force 6 would most surely endanger your own crew. This is not to say that it should not be attempted if no other method is available. Full use of the parent ship should be made to provide a lee for the boat when it is in the water. Transfer of personnel into a smaller craft, like a lifeboat or rescue boat, is extremely hazardous. Coxswains of rescue craft have found with experience that both vessels will probably ride easier with a following sea. To this end Masters are advised to conform to the heading and the speed dictated by the coxswain of the rescue craft. This is, of course, provided that the ship is able to manoeuvre.

Use of Rocket Line

Extreme caution should be used with this method after first establishing good communications. A rocket should not be propelled towards a tanker, but a tanker may propel one to the rescuing vessel. Further caution with use of the rocket line should be exercised if survivors are in the water or the surface contains floating oil patches. Do not attempt any transfer until a messenger line has established a strong towing hawser between the two vessels.

Securing the towing hawser to a survival craft like a life raft may prove difficult. It would be unwise to secure the hawser to the towing patch attached to the life raft as these towing patches have been known to pull adrift under excess weight. A possible method would be to punch a hole through the double floor of the raft and pass the towing hawser around the main buoyancy chamber. If this method is adopted, it would be wise to guard against rope burn by parcelling between the towline and the raft fabric with appropriate protective material. This method would mean the loss of watertight integrity inside the raft itself, but as it would not be expected to be in use for long, this would not be too serious, especially as the raft is being used for transportation and not for long-term survival.

Going Alongside

An appropriate method when the weather is so bad that the launching of a rescue craft would endanger your own crew members, this manoeuvre needs extreme care to avoid structural damage to either ship. Due consideration should be given before going alongside to the risks of fire, explosion or other similar effect arising from the distressed vessel. The possibility of escaping gas from some vessels must not be forgotten, and, to this end, the direction of the wind should be considered and the subsequent approach made with extreme caution.

Apart from the type of vessel in distress, which may vary, the structure, especially freeboard, will influence the decision to take the option of going alongside. The objective of removing personnel from a sinking vessel must be given priority, e.g. higher freeboard vessels like ro-ro moving onto a small fishing craft may well defeat the objective of saving life. Masters should consider whether any deliberate contact with the distressed vessel would be better made forward of the collision bulkhead rather than abaft the bulkhead.

Towing

This option may not always be available to a rescue vessel. The question of the distressed vessel's ability to remain afloat long enough to complete the operation will influence any Master's decision. In any event, where there is doubt, personnel would have to be removed.

Thought should be given to the prospect of beaching the distressed vessel if suitable ground is on hand and main engine power is still available to the stricken vessel. See Chapter 14 on beaching and Chapter 17 on towing.

Special Operations (Ro-Ro Vessel)

Today, with specialist trades engaged on the oceans of the world, certain vessels are specially equipped to tackle specific tasks. Bearing this in mind, a ro-ro vessel may find it possible to open her stern door partly to assist in a rescue operation. The construction of the stern door would be a determining factor, namely, the freeboard to the level of the 'hinge' must be adequate to allow such action.

It should be borne in mind that special circumstances could call for bold but not foolhardy action. Once the stern door is opened, even by the smallest amount, watertight integrity of the vehicle deck is lost. Should a main engine failure occur or hydraulics fail to operate the locking of the door when required, the watertight integrity of the ship would be lost for an indefinite period.

The recovery of physically fit survivors by means of scrambling nets over a part-opened stern door/ramp cannot be ruled out as being a viable method of rescue. Use of bow thrusters to maintain the ship's head into wind would greatly assist the operation. This method would obviously be dependent on the circumstances at the time, especially the weather conditions, but may prove more acceptable than launching own boats or causing a swamping situation by going alongside a smaller vessel with an incompatible height of freeboard.

Use of Own Life Rafts

This method could be used in circumstances where the distressed craft had no life rafts of her own or when a connection with a rocket line cannot be established. A similar method of securing the towline to the raft as that already described on p. 567 is recommended.

The disadvantage of this particular method is that control of the raft, drifting towards the distressed vessel, will be difficult, especially when compared with transfer by use of the established messenger, as described on p. 568. The use of oil should be considered if sea conditions warrant such action, but caution should be exercised, especially if there are survivors in the water or about to enter the water.

Use of Breeches Buoy

This is a very doubtful proposition and would be extremely difficult to carry through successfully. The operation is complicated and requires crews to be well practised and experienced in the ways and methods of transfer and replenishment at sea. Exceptional ship-handling would be required by the rescuing vessel, and it would be unlikely for the average merchant vessel to have the required expertise and equipment to complete such an operation. This is not to say that it could not be achieved, but even a naval vessel, well practised in transfer by jack stay, would expect to encounter some difficulty with such a rescue operation. The weather conditions would undoubtedly decide the matter. In bad weather it would be impossible, and in fine weather the use of lifeboats or rafts would be a better proposition (see also p. 567).

Vessels in Contact

This option may be compared with going alongside, but the advantage is that your own crew are removed from the dangers associated with putting a survival craft into the water. It may be an appropriate option when the freeboards of both vessels are different, so that the height of the forecastle head deck is above that of the distressed vessel. By the added use of scrambling nets over the bow onto the distressed vessel survivors may be recovered.

The structure of the majority of ships might make this method possible because of the increased scantlings and additional strength in the fore end. Superficial damage may occur and this should be considered before attempting the operation. Skilled ship-handling will be required to bring about a successful conclusion.

Use of Oil

In special operations such as those described above the prudent use of oil on the water surface can be dramatically effective. The type of oil recommended is a light vegetable or animal oil, or even light diesel oil if that is all that is available. Fuel oil should not be used. After oil has been used, a statement should be entered into the oil record book and ship's log book.

Pyrotechnics

Smoke signals, rockets and distress flares may all attract the attention of rescuers to those in distress (see Figures 16.3 to 16.7).

Rocket Parachute Flare

These shall be contained in a water-resistant casing having brief instructions or diagrams printed on the outside regarding their operation.

The rockets when fired vertically reach an altitude of not less than 300 m and, at or near the top of its trajectory, shall eject a parachute flare.

The flare will burn bright red in colour, and will burn with an average luminous intensity of not less than 30,000 cd. The burning period should be not less than 40 s, and have a descent rate of not more than 5 m/s. The parachute should not be damaged while burning.

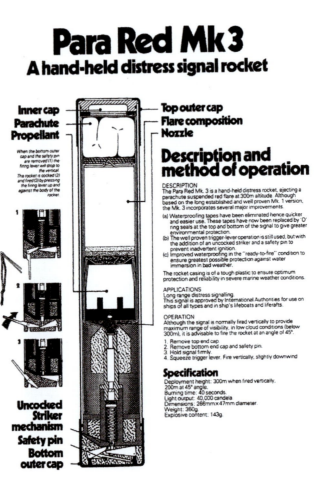

Figure 16.3 Distress signal rocket.

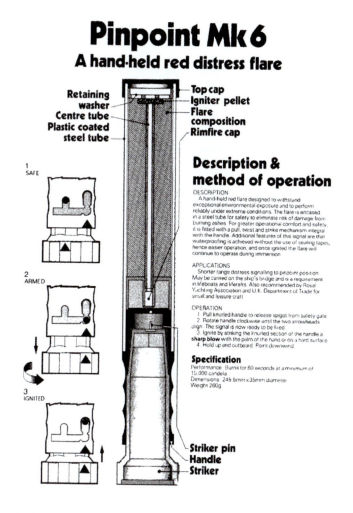

Figure 16.4 Distress flare.

Hand Flares

These shall be contained in a water-resistant casing having brief instructions or diagrams illustrating their operation printed on the outside.

The hand flare shall burn with a bright red colour with an average luminous intensity of not less than 15,000 cd. The burning period shall be not less than one minute and should continue to burn after being immersed for a ten-second period under 100 mm of water.

Buoyant Smoke Floats

These shall be contained in a water-resistant casing having brief instructions or diagrams regarding their operation printed on the outside of the case.

A buoyant smoke float should not ignite in an explosive manner but when activated in accordance with the manufacturer's

Figure 16.5 Buoyant smoke signal.

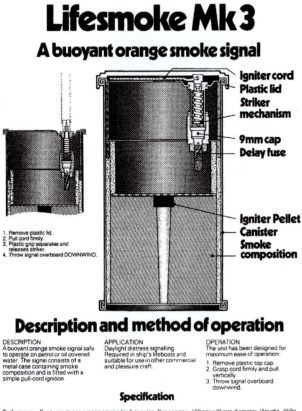

Lifesmoke Mk 3

A buoyant orange smoke signal

Igniter cord
Plastic lid
Striker mechanism

9mm cap
Delay fuse

Igniter Pellet
Canister
Smoke composition

1. Remove plastic lid.
2. Pull cord firmly.
3. Plastic grip separates and releases striker.
4. Throw signal overboard DOWNWIND.

Description and method of operation

DESCRIPTION
A buoyant orange smoke signal safe to operate on petrol or oil covered water. The signal consists of a metal case containing smoke composition and is fitted with a simple pull-cord ignition.

APPLICATION
Daylight distress signalling. Required in ship's lifeboats and suitable for use in other commercial and pleasure craft.

OPERATION
The unit has been designed for maximum ease of operation:
1. Remove plastic top cap.
2. Grasp cord firmly and pull vertically.
3. Throw signal overboard downwind.

Specification

Performance: Produces dense orange smoke for 3 minutes Dimensions: 160mm × 85mm diameter Weight: 459g
Note: A 4-minute smoke duration version is available to special order

instructions it should emit smoke of a highly visible colour at a uniform rate for a period of not less than three minutes when floating in calm water It should not emit any flame during the time of the smoke emission, neither should the signal be swamped in a seaway. It must be constructed in a manner so as to emit the smoke when submerged in water for a period of ten seconds when under 100 mm of water.

Breeches Buoy

Provided that the vessel is within striking distance of the shoreline, rescue may be attempted by means of the breeches buoy. However, the use of a single buoy to transport a single survivor in this day and age is extremely questionable.

The extensive use of helicopters has all but replaced the very risky breeches buoy method, although a life raft was successfully employed in a breeches buoy system in Canadian waters in the 1990s to transport 20 persons at a time from a wreck to the shoreline.

Initial contact by a ship's rocket line-throwing apparatus limits the activity to a distance of about 250 metres unless the

> NB. The use of helicopters has generally superseded the use of breeches buoy operations.

coastguard/shore party is equipped with a more powerful rocket launching system with a longer line.

Rocket Line-Throwing Apparatus

The rocket line-throwing apparatus (Figure 16.6), once fired, will be affected by the force of the wind acting on the rocket line. The rocket, however, should be aimed directly at the target or if anything a little downwind of the target, but never into the wind. The manufacturers must build into the rocket a limit of deflection. When the deflection is taken into account it equates to approximately 10 per cent of the range of the apparatus (23 m in normal range of 230 m, either side of target). The weight of line acts as a drag on the flight of the rocket, providing essential weight to the directional flight.

Efficient communications between target and operator should first be established to ensure that it is perfectly safe to fire a rocket

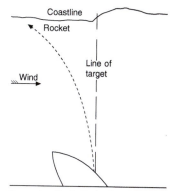

Incorrect method – Rocket fired into wind

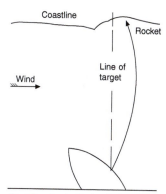

Correct method – Rocket fired at target

Figure 16.6 Firing rocket and line.

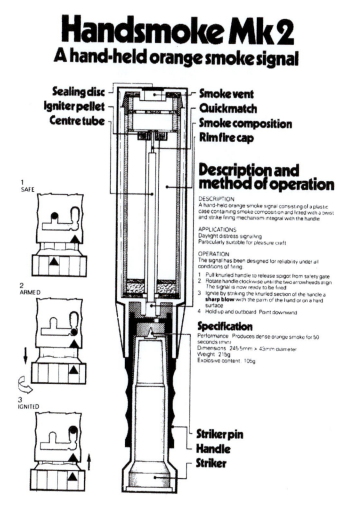

Handsmoke Mk 2
A hand-held orange smoke signal

Sealing disc
Igniter pellet
Centre tube

Smoke vent
Quickmatch
Smoke composition
Rimfire cap

Description and method of operation

DESCRIPTION
A hand-held orange smoke signal consisting of a plastic case containing smoke composition and fitted with a twist and strike firing mechanism integral with the handle.

APPLICATIONS
Daylight distress signalling
Particularly suitable for pleasure craft

OPERATION
The signal has been designed for reliability under all conditions of firing.

1 Pull knurled handle to release spigot from safety gate
2 Rotate handle clockwise until the two arrowheads align. The signal is now ready to be fired
3 Ignite by striking the knurled section of the handle a **sharp blow** with the palm of the hand or on a hard surface
4 Hold up and outboard. Point downwind

Specification
Performance Produces dense orange smoke for 50 seconds (min)
Dimensions 245.5mm × 43mm diameter
Weight 215g
Explosive content 105g

1 SAFE

2 ARMED

3 IGNITED

Striker pin
Handle
Striker

Figure 16.7 Smoke signal.

towards the target. Should a tanker or gas carrier be the target, the firing of a rocket may prove hazardous. The signal 'GU' may be exhibited to mean 'It is not safe to fire a rocket'.

Regulations Affecting the Line-Throwing Appliance

Every line-throwing appliance shall be capable of throwing a line with reasonable accuracy and carrying the line at least 230 m in calm weather conditions. It should comprise not less than four projectiles and four lines each with a breaking strength of not less than 2 kN. The rocket in the case of a pistol-fired rocket or the assembly, in the case of integral rocket and line, should be contained in water-resistant casing. All the equipment is then contained in a weather-proof container.

Every cargo and passenger ship must be provided with a line-throwing appliance as stated, in compliance with Regulation 18

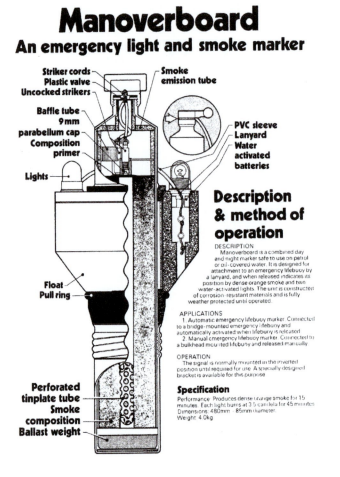

Figure 16.8 Light and smoke marker for 'man overboard'

SOLAS, and section 7.1 of the LSA code. Instructions in the use of this equipment must also be provided for the use of personnel.

Communications

To carry out any rescue operation, efficient communications between all parties are essential. They may be established in many forms, but radiotelephone and VHF are the best. Should these be unavailable owing to damage, loss of power or weather interference, alternative methods must be adopted. These may be one or a combination of the following:

- Shouting, word of mouth, distance and weather permitting
- Morse code by flash light or sound
- International flag hoists
- Morse by flags.

The Work of HM Coastguard (HMCG)

The HM Coastguard operates around the coastline of the United Kingdom and is the authority responsible for coordinating all civil and maritime SAR activity. The organization operates with 21 MRCC constantly manned watch/communication stations and sub-stations They cover 16 areas and an appropriate scale of rescue equipment is maintained at the various stations. The organization is further supported by the Royal National Lifeboat Institute (RNLI), some civil helicopter support including military assistance and available shipping in range.

A continuous listening watch is broadcast on 156.8 MHz, the VHF distress, urgency, safety and calling frequency (VHF channel 16) for a range of up to 30 miles offshore of the UK coastline. Additional communication links are maintained on 2182 kHz, the MF distress frequency from selected centres, which provides 150 miles ground wave coverage from the UK coastline, while an electronic watch is maintained on VHF, DSC channel 70. All MRCC stations are also equipped with telephone, fax and telex services. The UK has been declared fully operational for sea area 1 (GMDSS) since 1998.

The Coastguard operates SAR helicopters based at Sunburgh in the Shetland Isles, Lee-on-the-Solent and daylight flight capabilities from Portland. It also has four emergency towing vessels (ETVs) prudently stationed to provide emergency towing facilities where a risk to the environment is likely from shipping involved in grounding or similar incident.

When an SAR incident is instigated the Coastguard has several emergency agencies that can be contacted to assist and advise on the prevailing circumstances. These include:

- all shipping in the immediate and affected area;
- ARCC Kinloss, which can call in military units of the Royal Navy and Royal Air Force;

New proposals for 2015 intend to provide SAR helicopter coverage and emergency towing vessels by private companies.

- Coastguard helicopters as appropriate;
- the RNLI lifeboat stations and inshore rescue craft;
- the AMVER organization based in New York;
- Auxiliary Coastguard units and rescue teams;
- emergency towing vessels of the Coastguard organization;
- MRCCs to coordinate operations.

Operational Aspects

- The Coastguards can order ships to divert for the purpose of SAR.
- Communications between the RNLI and the Coastguard are very good.
- Helicopter use is restricted as bad weather conditions (winds over 50 knots) and fog sometimes prevent helicopters becoming airborne.

Procedure on Receipt of Distress Signal

1   Signal received by telephone, police, radio, VHF or by visual sighting.
2   Acknowledge receipt of the distress message by orange smoke, four white star shells, maroons or radio.
3   Raise the alarm and advise RNLI of the situation. Alert potential rescue forces. Ask a helicopter to stand by. Alert the local Coastguard station.
4   Nominate the OSC.
5   Obtain assessment report from the OSC.
6   Despatch rescue equipment if required, e.g. rafts, pumps, etc.
7   Start systematic plot of the distressed ship's position. Obtain tidal information and weather information to ascertain probable rate of drift.
8   Enter events as they occur in Coastguard log book.
9   Update report from OSC.
10 Order additional rescue forces to the scene or specialist units if required, e.g. lifeboat, helicopter, Nimrod, etc. Should an operation be set in motion, the MRCC would order up specialist units as soon as a full assessment of the situation had been made.

AMVER Organization

The automated mutual-assistance vessel rescue (AMVER) system is a ship position-reporting system operated by the US Coastguard covering the whole of the Atlantic and Pacific Oceans. Other systems are in operation, e.g. AUSREP for the Australian coast, but the AMVER system is more familiar to mariners in the northern hemisphere.

The AMVER organization exists to provide information regarding the positions and intended future movements of merchant

HM COASTGUARD RESCUE COORDINATION CENTRES

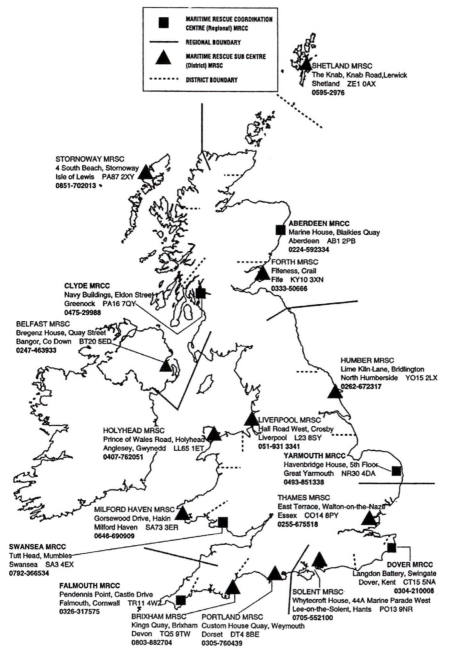

MARITIME RESCUE COORDINATION CENTRE (Regional) MRCC

REGIONAL BOUNDARY

MARITIME RESCUE SUB CENTRE (District) MRSC

DISTRICT BOUNDARY

SHETLAND MRSC
The Knab, Knab Road, Lerwick
Shetland   ZE1 0AX
0595-2976

STORNOWAY MRSC
4 South Beach, Stornoway
Isle of Lewis   PA87 2XY
0851-702013

ABERDEEN MRCC
Marine House, Blaikies Quay
Aberdeen   AB1 2PB
0224-592334

FORTH MRSC
Fifeness, Crail
Fife   KY10 3XN
0333-50666

CLYDE MRCC
Navy Buildings, Eldon Street
Greenock   PA16 7QY
0475-29988

BELFAST MRSC
Bregenz House, Quay Street
Bangor, Co Down   BT20 5ED
0247-463933

HUMBER MRSC
Lime Kiln Lane, Bridlington
North Humberside   YO15 2LX
0262-672317

HOLYHEAD MRSC
Prince of Wales Road, Holyhead
Anglesey, Gwynedd   LL65 1ET
0407-762051

LIVERPOOL MRSC
Hall Road West, Crosby
Liverpool   L23 8SY
051-931 3341

YARMOUTH MRCC
Havenbridge House, 5th Floor
Great Yarmouth   NR30 4DA
0493-851338

THAMES MRSC
East Terrace, Walton-on-the-Naze
Essex   CO14 8PY
0255-675518

MILFORD HAVEN MRSC
Gorsewood Drive, Hakin
Milford Haven   SA73 3ER
0646-690909

SWANSEA MRCC
Tutt Head, Mumbles
Swansea   SA3 4EX
0792-366534

DOVER MRCC
Langdon Battery, Swingate
Dover, Kent   CT15 5NA
0304-210008

FALMOUTH MRCC
Pendennis Point, Castle Drive
Falmouth, Cornwall   TR11 4WZ
0326-317575

SOLENT MRSC
Whytecroft House, 44A Marine Parade West
Lee-on-the-Solent, Hants   PO13 9NR
0705-552100

BRIXHAM MRSC
Kings Quay, Brixham
Devon   TQ5 9TW
0803-882704

PORTLAND MRSC
Custom House Quay, Weymouth
Dorset   DT4 8BE
0305-760439

Figure 16.9 HM Coastguard stations. The Coastguard has seven rescue helicopters based around the UK coastline at Stornoway and Sunburgh airports, Lee-on-the-Solent and Portland. The Coastguard also has use of air sea rescue helicopters provided by the Royal Navy, Royal Air Force, United States Air Force and the Irish Coast Guard.

vessels making offshore voyages. This information can be coordin-ated in the event of a maritime emergency and an effective rescue attempt.

Voluntary information from participating vessels is transmitted to the Coastguard Centre, New York, via selected radio stations in the form of one of four standard messages (Figure 16.10). The information is fed to a computer which constantly updates the par-ticulars on every vessel to its next port of call. Position-fixing of vessels is done by the 'dead reckoning' method, based on the ves-sel's last known speed and course. Masters are requested to update their positions every 48 hours.

HM Coastguard operations out of bases or MRCCs are divided into three regions:

- Scotland and Northern Ireland
  - Aberdeen
  - Belfast
  - Clyde
  - Forth
  - Shetland
  - Stornoway
- Wales and west of England
  - Brixham
  - Falmouth
  - Holyhead
  - Liverpool
  - Milford Haven
  - Swansea
- East of England
  - Dover
  - Humber
  - London
  - Portland
  - Solent
  - Thames

Standard Types of AMVER Message

1   The complete Type 1 report consists of 11 parts, together with any pertinent remarks, with enough information to initiate a plot. It may be considered a sailing plan sent before, at, or immediately after departure, as soon as communications have been established.
2   This is considered a position report and includes the date, time and the position of the vessel. Occasional position reports are necessary during long passages to ensure that the computer will predict future positions with acceptable accuracy.
3   This is the arrival report, despatched on arrival at the harbour entrance of the port of destination. Should communications have ceased, then the computer will automatically terminate

AMVER

| | MESSAGE | Form Approved OMB No. 04-R3073 |

Automated Mutual-assistance VEssel Rescue (AMVER) System
"that no call for help shall go unanswered"

1 Name	2 Call sign	3 Type

4 Position	5 Date-Time GMT

6 Sailing Route

6 Sailing Route

6 Sailing Route

7 Speed	8 Destination	9 ETA

10 Call sign of commercial radio station guarded this voyage (please list twice)

11 Medical personnel onboard this voyage (Doctor, Paramedic, no medic)

To ensure that no charge is applied, all AMVER messages should be passed through specified AMVER radio stations

Dept. of Trans., USCG, CG-4796 (Rev. 6-78)
Previous editions are obsolete

Department of Transportation
United States Coast Guard

Message Types & Format

TYPE 1 – The complete Type 1 report consists of eleven parts and any pertinent remarks and contains the information necessary to initiate a plot. It is called an initial AMVER message and may be considered a movement report or sailing plan. Type 1 reports may be sent immediately prior to departure, immediately after departure, or as soon as adequate communications can be established.

TYPE D – The Type D report is a deviation report and need include only information which differs from that previously reported. It is sent when the actual position will vary more than 25 miles from the position which would be predicted based upon data contained in previous reports. It may indicate a change of route, course, speed, or destination and can include any pertinent remarks.

TYPE 2 – The Type 2 report is considered a position report and includes the date and time of the position. It may contain additional entries and remarks. During long passages, it is suggested that Type 2 reports be submitted at 36-hour intervals to insure accuracy of the computer plot, Parts 6, 7, 8, and 9 may be omitted from the message if desired. Positions are also extracted from weather reports from ships participating in the international weather observation program.

TYPE 3 – The Type 3 report is an arrival report and is sent upon reaching the harbor entrance at port of destination. Parts 6, 7, 8, and 9 may be omitted. If communications cannot be established, the computer will automatically terminate the plot at the predicted time of arrival. However, the report is desired to increase the accuracy of the plot. Type 3 reports are especially desired upon arrival at the harbor entrance of United States ports.

Name 1	Call sign 2	Report type 3	Position 4	Date-time 5	Sailing route 6	Speed 7	Destination 8	ETA 9	Commercial station 10	Medical personnel 11
Name of ship	Radio call	1, D, 2 or 3	Latitude & Longitude to nearest 10th degree (name of point may be used if convenient, i.e. Ambrose)	Date-Time GMT of position, (Use 6 digit, i.e. 041800. first 2 are month last 4 are GMT hours and minutes)	Latitude & Longitude to nearest 0.1 degree of each turn point along intended track, Use "RL" for rhumb line or "GC" for Great Circle before each point to show method of sailing. When track is to be coastal, state "Coastal".	To Nearest 0.1 kt	Next Port of Call	Esti-mated Time of Arrival at destina-tion GMT	Call sign of commercial station to be worked on voyage. (List twice)	Doctor, Paramedic, or No Medic.

ANY VESSEL OF ANY NATION DEPARTING ON AN OFFSHORE PASSAGE OF 24 HOURS DURATION OR GREATER IS ENCOURAGED TO BECOME A PARTICIPANT IN THE AMVER SYSTEM BY SENDING APPROPRIATE AMVER MESSAGES.

Figure 16.10 AMVER message.

the plot. However, the report is desired to check the accuracy of the plotting system.

4   This is the deviation report, and need only include information which differs from the previous report. It is transmitted when the actual position will vary more than 25 miles from the predicted position as based on previous reports. It may include a change of course, speed or destination, together with any pertinent remarks.

Passenger Vessels: Decision Support System (For Ships' Masters) ref, 1996 Amendments to SOLAS, Effective July 1998 (Chpt III, Regulation 29)

This recent amendment to the regulations is now effective for all passenger vessels and makes it a mandatory requirement that a 'decision support system' for emergency management will be provided on the navigation bridge. This system shall, as a minimum, consist of a printed emergency plan or plans to cover all foreseeable emergencies.

These emergency situations should include, but not be limited to:

*   fire situations
*   damage to the vessel
*   pollution
*   unlawful acts which threaten the safety of the ship and the security of passengers and crew
*   personnel accidents, e.g. man overboard
*   cargo-related accidents
*   emergency assistance to other vessels.

The emergency procedures established within the plans are expected to provide support and guidance to ships' Masters when making decisions to handle any emergency or combination of emergencies.

The system should be user-friendly and may also employ the use of computer-based systems for provision of relevant data in an emergency. Masters may also find that the following references to the IAMSAR manuals, and Chapter 8 of the International Safety Management (ISM) Code, where applicable, may be useful when called upon to resolve such situations.

Search and Rescue Plans for All UK Passenger Ships (Additional ref., MSN 1761 (M))

Under SOLAS amendments, it is now a requirement for all passenger vessels sailing in UK waters to carry a plan for co-operation with the SAR services. The requirement is one affecting both inland waters and tidal waters and the guidelines for developing this plan may be found by reference to MSN 1721.

The plan could expect to cover the essential parties that would be engaged in an emergency operation, namely:

- the shipping company
- the ship(s)
- the MRCC
- the SAR facilities
- the media and training/exercising elements.

The plan would be expected to contain a list of contents, an introduction and a description of the expected cooperation between relevant parties and the following elements.

The Company (Element of Plan)

The name and address with emergency contact numbers would be designated, together with a 24-hour initial and alternative communication/contact arrangements. There would also be additional communications by way of direct telephone or fax links providing positive contact with relevant company personnel.

Included will be route chartlets providing details of the ship's passage plan and the available services along such routes, together with the boundaries of respective search and rescue regions (SRRs).

Liaison arrangements between the company and relevant RCCs will allow the provision of incident information to be passed to field operators. Also, a checklist of specific information, such as:

- numbers of persons on board
- cargo carried
- bunkers available, etc.

The company would also make available liaison officers with access to any supporting documentation, such as ship's plans, fire-fighting arrangements, etc., that may be required by the flag state authorities.

The Ship(s) (Element of Plan)

This would contain the general particulars of the vessel and include:

- maritime mobile service identity (MMSI), call sign, country of registration, type of vessel, gross tonnage, overall length, maximum draught, service speed, maximum number of persons allowed on board and the number of crew carried;
- communication equipment carried;
- profile plans and deck plans which would highlight life-saving facilities and fire-fighting equipment;
- helicopter landing/winching areas with any approach sectors and the types of aircraft for which the deck is constructed
- rescue methods intended for use to recover persons from the sea or from other vessels;
- a descriptive colour scheme of the vessel.

The MRCC (Element of Plan)

- Details of SAR regions along the route.
- Chartlets which illustrate the SRRs in relation to the ship's area of operation.
- SMC and summary of function.
- Designation of the OSC by definition and selection criteria.
- A summary of functions.

SAR Facilities (Element of Plan)

- MRCC/SCs along the route together with the addresses/contacts of relevant stations.
- Communication equipment available and frequencies of operation (watched or unwatched), together with a contact list inclusive of MMSIs, call signs, telephone, fax and telex numbers.
- A general description of SAR facilities along the route of both air and surface units, with any additional support units by way of fast rescue craft, heavy/light or long-/short-range helicopters, fire-fighting units or other relevant vessels.
- A communication plan.
- A search plan.
- Medical advice/assistance contacts. Advisory contacts for toxics, chemicals, fire-fighting, and shoreside reception facilities.
- Means of informing next of kin.

Final Plan Elements

- Media relations and periodic exercises.

Plate 148 A Sea King helicopter working with the Royal Navy in an anti-submarine role off the English coastline.

Abbreviations for Use in SAR Operations

A/C	aircraft
ACC	area control centre
ACO	aircraft coordinator
AES	Aeronautical Earth Station
AFN	aeronautical fixed network
AFTN	aeronautical fixed telecommunications network
AIP	aeronautical information publication
AIS (i)	automatic identification system
AIS (ii)	aeronautical information services
AM	amplitude modulation
AMS	aeronautical mobile service
AMS(R)S	aeronautical mobile satellite (route) service
AMSS	aeronautical mobile satellite service
AMVER	automated mutual assistance vessel rescue
ANC	air navigation commission
ARCC	aeronautical rescue coordination centre
ARSC	aeronautical rescue sub-centre
ATC	air traffic control
ATN	aeronautical telecommunications network
ATS	air traffic services
CES	Coast Earth Station
COSPAS	space system for search of vessels in distress
CRS	coast radio station
C/S	call sign
C/S	creeping line search
CSC	creeping line search coordinated
CSP	commence search point
CW	continuous wave
DF	direction finding
DMB	datum marker buoy
DME	distance measuring equipment
DR	dead reckoning
DSC	digital selective calling
ELT	emergency locator transmitter
EPIRB	emergency position indicating radio beacon
ETA	estimated time of arrival
ETD	estimated time of departure
ETV	emergency towing vessel
FIC	flight information centre
FIR	flight information region
FM	frequency modulation
FRC	fast rescue craft
F/V	fishing vessel
GES	Ground Earth Station
GHz	gigahertz
GLONASS	global orbiting navigation satellite system
GMDSS	global maritime distress and safety system

GNSS	global navigation satellite system
GPS	global positioning system
GS	ground speed
gt	gross tonnage
HF	high frequency
IAMSAR	International Aeronautical and Maritime Search and Rescue Manual
ICAO	International Civil Aviation Organization
IFR	instrument flight rules (instrument flying rating)
ILS	instrument landing system
IMARSAT	International Mobile Satellite Organization
IMC	instrument meteorological conditions
IMO	International Maritime Organization
INS	inertia navigation system
INTERCO	International Code of Signals
ITU	International Telecommunications Union
JRCC	Joint (aeronautical and maritime) Rescue Coordination Centre
kHz	kilohertz
LES	Land Earth Station
LKP	last known position
LUT	local user terminal
m	metres
MCA	Maritime Coastguard Agency
MCC	mission control centre
MF	medium frequency
MHz	megahertz
MMSI	maritime mobile service identity
MRCC	maritime rescue coordination centre
MRSC	maritime rescue sub-centre
MSI	maritime safety information
M/V	merchant vessel
NBDP	narrow-band direct printing
NM	nautical mile
OSC	on scene coordinator
OSV	offshore supply vessel
PIW	person in water
PLB	personal locator beacon
POB	persons on board
R	search radius
RCC	rescue coordination centre
RF	radio frequency
RSC	rescue sub-centre
R/T	radiotelephone
RTG	radio telegraphy
SAR	search and rescue
SARSAT	search and rescue satellite-aided tracking
SART	search and rescue transponder
SC	search coordinator
SCC	SAR coordinating committee

SDP	search data provider
SES	Ship Earth Station
SITREP	situation report
SMC	SAR mission coordinator
SOLAS	International Convention for the Safety of Life at Sea
SPOC	SAR point of contact
SRR	search and rescue region
SRU	search and rescue unit
SS	expanding square search
SU	search unit
T (i)	true course
T (ii)	search time available
TAS	true air speed
TLX	teletype
TS	track line search
UHF	ultra-high frequency
UIR	upper flight information region
USAR	urban search and rescue
USCG	United States Coastguard
UTC	coordinated universal time
V	SAR facility ground speed
VFR	visual flight rules
VHF	very high frequency
VMC	visual meteorological conditions
VOR	VHF omni directional radio range
VS	sector search
WMO	World Meteorological Organization

17

Ship-Handling: Equipment

Introduction

How can anybody possibly provide details of the innovations that have so dramatically affected the business of ship-handling? There have been so many they are too numerous to count. Rudders alone have passed through an age of renovation, giving ships the ability to virtually turn in their own length, while steerable pods and directional thrusters have dispensed with the need for some rudders completely.

Jet propulsion has entered the ferry market and active fins dominate ship stabilization. Controllable pitch propellers (CPPs) are no longer the novelty they once were and one-man bridge (OMB) control is the norm. All these sit alongside unmanned machinery spaces (UMSs) that continue to drive our ships ever forward.

For the ship-handler, the hardware has been provided in the form of better facilities and more effective equipment. The fact is that there is much more of it, by way of bow thrusters (stern thrusters), single screw and twin screw, single rudders and twin rudders; the list goes on. Dare we forget, however, that workhorse of the sea with its single right-hand fixed pitch propeller. It is still active and remains operationally sound on many trades. The industry has been forced to change and I would like to think for the better. It is now the task of our mariners to handle the modern ship in these still early days of technological advance.

Terms and General Definitions

Cavitation

This is the empty cavity caused by the ship's displacement which is re-filled by water flowing down the ship's sides as the vessel moves ahead.

Con

'To take the con' is an old-fashioned term meaning to take control of the navigation of the vessel. It is still occasionally encountered, especially in the United

States (conn). Submarines take the term 'conning' tower from the same word. It also implies conducting the navigation of the vessel by giving orders to the helmsman.

Drag

This is the frictional resistance caused by the ship's hull; in some parts of the world drag is also used to describe the difference between the forward and after draughts.

Headreach

This is the distance a vessel will continue to travel forwards after the main engines have been stopped.

Helmsman

The helmsman is a person designated to steer the vessel manually. Helm orders used to apply to the tiller, but after the First World War began to be accepted and related to the rudder.

List

A vessel which is not on an even keel in the upright position, but heeled over to one side or another, is listing.

Pitch

This is the axial distance moved forward by the propeller in one revolution through a solid medium. Measurement may be achieved by use of a pitchometer or by practical calculation in drydock.

Pivot point

A position aboard the vessel about which the ship rotates when turning is called the pivot point. In ships of conventional design (midships accommodation) the pivot point was approximately one-third the length of the vessel measured from forward when the ship was moving ahead. The pivot point changed when the vessel was going astern.

Quartermaster

In the modern merchant service a senior helmsman is called a quartermaster, especially in large passenger vessels. The term was previously applied to a petty officer who assisted the Master and officers.

Set

This is a term used to describe the movement of a vessel which is being influenced by tide or current so that the course being steered is not truly representative of the track the vessel is making. A vessel under such influence is often described as 'setting down'.

Sheer

(a) The angle that a ship will lie to her cable when at anchor; (b) used to describe the upward curvature of the uppermost continuous deck – hence the term 'sheer strake'; (c) a vessel may also sheer away from her intended heading, so making a sharp alteration of course.

Squat

A term to describe the bodily sinkage of a vessel when underway and making way, squat is most noticeable in shallow water. Its value will vary proportionally to the square of the speed of the vessel. A vessel affected by squat may experience an increase in the forward or aft draughts and a subsequent change of trim when making way through the water, with possibly critical consequences to the handling and steerage of the vessel.

Stopping distance

This is the minimum distance that a vessel needs to come to rest over the ground. Speed trials for new tonnage normally include test runs to provide information to watch officers showing the time and distance a vessel will take to stop (a) from full ahead after ordering main engines to stop, and (b) from crash full astern (emergency stop) (see Figure 17.1).

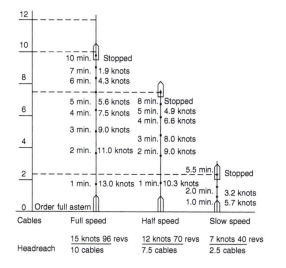

Figure 17.1 Stopping distance of a vessel with the particulars given.

18,000	dwt
Draught forward	28'3" (8.6 m)
Draught aft	30'3" (9.2 m)
Crash stop	from full ahead to full astern

Way

When a vessel starts her main engines and begins to increase speed, she gathers way. When she is moving forward over the ground, irrespective of speed, she is making way. When the speed of the vessel will still affect and obtain a correct response from the rudder, she has steerage way. When the vessel is moving over the ground in the direction of the stern, under reverse propulsion, she has stern-way. When a vessel is underway and making way and proceeding at a speed considered excessive for the prevailing conditions and situation, she has too much way. Finally, underway, as defined by the Regulations for the Prevention of Collision at Sea, means that a ship is not at anchor, or made fast to the shore, or aground.

Weather side

The side of the vessel exposed to the wind. Seas may be seen to break over the weather side as opposed to the lee side, which is the sheltered, favoured side of the vessel.

Yaw

Wind and sea astern of the vessel make 'yawing' about a real possibility, but this is not to say that the vessel would not behave in a similar manner if the weather was affecting the vessel from another direction. A vessel is said to yaw when, either by accident or design, her head falls off the course she is steering. Yawing is often a consequence of a following wind and sea.

Factors in Ship-Handling

The construction of the modern vessel will soon be such that CPP, bow thrust unit, stabilizers, etc. become the norm, rather than novelties. However, the initial cost of installing specialized equipment of this nature is very high. Consequently, it will be some time before every owner accepts the new developments in equipment as standard.

Controllable

- Main engine power
- propeller or propellers – fixed or controllable pitch
- anchors
- mooring ropes
- rudder movement
- bow thrust (if fitted)
- bow rudder (if fitted)
- tugs (may be classed as controllable only as long as they respond as requested; ship-handlers may find that tugs should be included in the following list of uncontrollable factors.

Plate 149 Central bridge control console set into the integrated navigation bridge of a modern high-speed craft.

Uncontrollable

- The weather
- tide and/or current
- geographical features such as shallow water, floating obstructions, bridges and ice accretion
- traffic density.

High-Speed Craft Categories

The IMO high-speed craft code was introduced in 1994 and had mandatory implementation in 1996. Under the auspices of the code, high-speed craft were placed into one of three categories:

Category 'A' Craft

Category 'A' is defined as any high-speed passenger craft carrying not more than 450 passengers and operating on a route where it has been demonstrated to the satisfaction of the flag or port state that there is a high probability that in the event of an evacuation at any point of the route, all passengers and crew can be rescued safely with the least of:

1   the time to prevent persons in survival craft from suffering exposure causing hypothermia in the worst intended conditions;
2   the time appropriate with respect to environmental conditions and geographical features of the route; or
3   four hours.

Category 'B' Class

Category 'B' is defined as any high-speed passenger craft other than a Category 'A' craft, with machinery and safety systems arranged such that, in the event of damage, disabling any essential machinery and safety systems, in one compartment, the craft retains the capability to navigate safely.

A Cargo Craft Class

This is defined as any high-speed craft other than a passenger craft and which is capable of maintaining the main functions and safety systems of unaffected spaces, after damage in any one compartment on board.

Maximum Speed Formula

Speed must be equal to, or exceed 3.7 × the displacement corresponding to the design waterline in metres cubed, raised to the power of 0.1667 (metres per sec).

Applicable to most types of craft, this corresponds to a volumetric Froude number greater than 0.45.

High-Speed Craft

Chapter II of the high-speed craft code draws attention to the potential hazards that may affect high-speed design craft, when manoeuvring at speed:

- Directional instability is often coupled to roll and pitch instability.
- Broaching and diving in following seas, at speeds near to wave speed, is applicable to most types of craft.
- Bow diving and planing, both in mono-hulls and catamarans, is due to dynamic loss of longitudinal stability in relatively calm seas.
- Reduced transverse stability with increased speeds in mono-hulls.
- Porpoising of planing mono-hulls being coupled with pitch and heave oscillations can become violent.
- Chine tripping is a phenomenon of planing mono-hulls occurring when the immersion of a chine generates a strong capsize moment.

- Plough-in of air cushion vehicles either longitudinally or transversely as a result of bow or side skirt tuck under or sudden collapse of skirt geometry, which in extreme cases could cause capsize.
- Pitch instability of SWATH (small water-plane area twin hull) craft due to the hydrodynamic moment developed as a result of the water flow over the submerged lower hulls.
- Reduction in the effective metacentric height (roll stiffness) of surface effect ship (SES) in high-speed turns compared to that of a straight course, which can result in sudden increases of heel angle and/or coupled roll and pitch oscillations.
- Resonant rolling of SES in beam seas, which in extreme cases could cause capsize.

Specific design features incorporated at building can go some way to overcome the above effects and enhance safer stability conditions and manoeuvring aspects.

Rudders

Unbalanced Single Plate Rudder

The rudder stock and all pivot points (pintles and gudgeons), including the bearing pintle, lie on a straight line. It is no longer used for large constructions because of alignment problems, but is occasionally seen on smaller vessels – coastal barges and the like. The rudder is defined as being 'unbalanced' because the whole of the surface area is aft of the turning axis.

Semi-Balanced 'Mariner'-Type Rudder

This is a very popular rudder for modern tonnage, especially for the container-type vessel and twin-screw vessels. The term 'semi-balanced' refers to the amount of surface area forward of the turning axis. If the proportion of surface area is less than 20 per cent forward of the axis, then the rudder is said to be semi-balanced.

Balanced Bolt Axle Rudder

The surface area of the rudder is seen to be proportioned either side of the 'bolt axle'. In fact, the amount of surface area will vary, but generally does not exceed 25–30 per cent forward of the axle. The advantage of a balanced rudder is that a smaller force is required to turn it so that smaller steering gear may be installed at lower running cost. Ideally a reduction of torque is achieved because the rudder is turning about a more centralized position, which would not be experienced with, say, the unbalanced plate rudder already mentioned. Balanced rudders are of streamlined construction, which reduces drag.

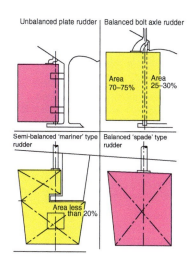

Figure 17.2 Types of rudder.

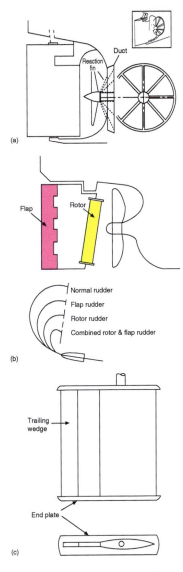

Figure 17.3 *(a)* Reaction fin with duct arrangement; *(b)* combined rotor cylinder with improved flap; *(c)* Schilling rudder.

Balanced Spade-Type Rudder

A very widely used rudder, especially in vessels engaged on short voyages such as ferries and roll on–roll off (ro-ro) ships. It is, however, not common in other types of vessel. The main disadvantage is that the total weight of the rudder is borne by the rudder bearing inside the hull of the vessel.

Modern Advancements: Rudder Construction

The duct arrangement forward of the propeller has been tried and tested and shown to improve hull resistance as well as homogenize the flow onto the propeller. Today's concept is to build in a reaction fin which will generate a reverse flow onto the propeller, and so improve the efficiency of the propeller by reducing transverse effects. This is a new development which is said to counter the effects of cavitation, reduce vibration and improve manoeuvrability.

Figure 17.3(a) (inset) shows a perspective view of the assembly.

Combined Rotor Cylinder with Improved Flap

Developments with rotors and flaps have greatly affected the turning circles and the manoeuvrability of vessels so equipped. The main disadvantage of these comparatively recent advances within the marine industry is the additional maintenance problems because of the added moving parts. Comparable turning arcs would seem beneficial to specialist craft, which require more in the way of demanding manoeuvres. The cost of installation is an obvious disadvantage for the ship's owners when compared with a more conventional rudder (Figure 17.4).

The Schilling Rudder

The 'schilling rudder' may be a fully hung spade or lower pintle simplex type. The balance is usually about 40 per cent. The rudder angle, at the full helm position, is 70–75°, providing the vessel with the manoeuvrability to turn on its own axis.

The hydrodynamic shape of the rudder helps to extract the slipstream from the propeller at right angles when at the maximum helm position. This capability, which employs the main engine power, virtually acts as a stern thruster, providing an effective sideways berthing facility. The build of the rudder is quite robust and with no moving parts it is relatively maintenance free, if compared to the rotor or flap types. Other features include the end plates, which help to reduce pitching and absorb impact, so giving limited protection to the main body of the rudder. The 'trailing wedge' reduces the yaw of the vessel, providing course stability with the minimum of helm movements necessary.

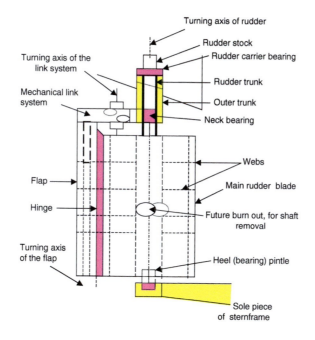

Figure 17.4 Parts of example flap rudder construction. General arrangement of 'BECKER' standard type flap rudder.

Further development has led to the 'VecTwin system' which employs twin rudders with the single propeller. A conventional wheel is used to operate rudders in tandem when engaged on passage and a single joystick control for more involved manoeuvres. The system can provide 70 per cent of the ahead thrust in a sideways direction and up to 40 per cent astern thrust. The large helm angles are capable of reducing the stopping distance by 50 per cent of a conventional ship when the rudders are in the 'reflectance', 'clamshell' position.

Propellers

Right-Hand Fixed Propeller

It is not within the scope of this text to discuss every type of propeller operational in present-day vessels. Some knowledge of the theory of the right-hand fixed propeller, however, should be sufficient for the reader to understand the properties of other propellers.

The marine propeller could be considered as something like a paddle wheel. Figure 17.5 shows the reaction and basic forces which affect the surface area of the propeller blades.

The result of the side components, caused by the rotation, may be resolved with the ahead motion created by the pitch angle of the blade. If the two forces are themselves resolved, then it would be seen that the stern of the vessel is moved to starboard. Consequently, if the stern is moved to starboard, then the bow may be seen to move to port. The opposite will happen when the vessel is moving astern, the only difference being that the side component will be

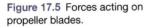
Figure 17.5 Forces acting on propeller blades.

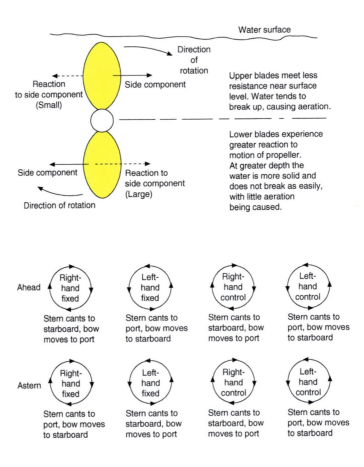

Water surface

Direction of rotation

Reaction to side component (Small)

Side component

Upper blades meet less resistance near surface level. Water tends to break up, causing aeration.

Side component

Reaction to side component (Large)

Lower blades experience greater reaction to motion of propeller. At greater depth the water is more solid and does not break as easily, with little aeration being caused.

Direction of rotation

Ahead	Right-hand fixed	Left-hand fixed	Right-hand control	Left-hand control
	Stern cants to starboard, bow moves to port	Stern cants to port, bow moves to starboard	Stern cants to starboard, bow moves to port	Stern cants to port, bow moves to starboard

Astern	Right-hand fixed	Left-hand fixed	Right-hand control	Left-hand control
	Stern cants to port, bow moves to starboard	Stern cants to starboard, bow moves to port	Stern cants to starboard, bow moves to port	Stern cants to port, bow moves to starboard

Figure 17.6 Propeller action when viewed from astern (transverse thrust effects.)

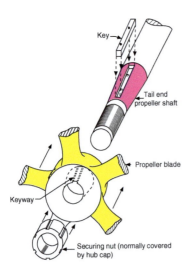

Key

Tail end propeller shaft

Propeller blade

Keyway

Securing nut (normally covered by hub cap)

Figure 17.7 Method of securing conventional propeller.

resolved with a sternway motion and the result will be that the stern cants to port, with the bow going to starboard (see Figure 17.6).

Securing the propeller is shown in Figure 17.7.

Controllable Pitch Propeller

This is probably one of the most practical, and certainly most valuable, advances in the marine industry over the last 20 years (Figures 17.8 and 17.9). The advantages of the CPP over and above the conventional fixed propeller are as follows:

- A reduction in fuel costs and consumption is achieved by the regular fixed turning speed of the shaft. The main machinery operates under optimum conditions per tonne of fuel burned.
- Expensive diesel fuel is saved by the use of 'shaft alternators' linked to the constant-speed rotating shaft. Auxiliary generators, though still carried, are not required for the normal loads that would be expected aboard a conventional vessel with a fixed propeller.

Plate 150 Controllable pitch propeller of a twin screw ship seen exposed in drydock. The picture shows the blades bolted to the hub by four bolts per blade. This allows each blade to be individually removed/replaced for maintenance, if required.

- Should the propeller be damaged, spare propeller blades are carried and can be relatively easily fitted. Should only one blade be damaged, then the pitch of the propeller can be increased in order to return to port under the vessel's own power, though at a reduced speed.
- A distinct ship-handling advantage is obtained by being able to stop in the water without having to stop main engines.
- The need for compressors and for compressed air for use in starting 'air bottles' is greatly reduced.
- The Watch Officer gains more direct control over the vessel's speed for anti-collision purposes.

The advantages of the CPP system are now accepted, and one may wonder why more ships are not so fitted. The main reason is the very high cost of installation at the building stage, and the even higher cost of installing a CPP as a structural alteration to an existing vessel. A minimum shaft horsepower must also be present from the main machinery in order to obtain fuel economy and efficient running.

Manufacturers have accepted the needs of the industry by incorporating several features to improve the system.

Automatic disengaging of the propeller blades can now be achieved in some designs. This is especially desirable, for example, in small harbours, when the eddies from a constant rotating shaft may cause concern to small-boat owners or moored yachts near the stern.

Emergency override controls may be included, and usually are, to enable the automation system to be bypassed and bring direct load on to the pitch actuator of the propeller.

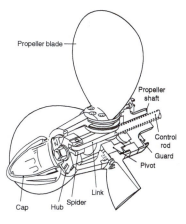

Figure 17.8 Controllable pitch propeller.

Twin-Screw Single Rudder Vessels

Twin-screw vessels are normally designed with their propellers equidistant from the fore and aft line. Usually both are outward

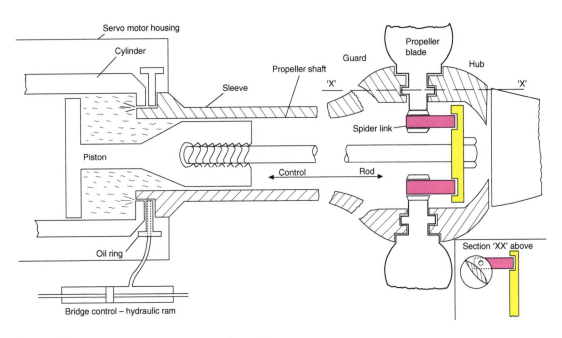

Figure 17.9 Controllable pitch propeller: working parts.

turning, the starboard propeller being right-hand fixed and the port propeller left-hand fixed. Many modern vessels are now constructed with twin CPPs, especially fast ferries and the like.

The twin-screw vessel is usually easier and simpler to handle than the conventional single-screw vessel. The transverse thrust on a single screw vessel strongly affects the steering capability, but with twin screws the forces tend to counteract each other, preventing the steering problems experienced by the single screw vessel (see Figure 17.10).

A distinct advantage of twin screws, apart from the increased speed created, is that if the steering gear breaks down, the vessel can still be steered by adjusting the engine revolutions on one or other of the propellers. When turning the vessel, for instance, one propeller can go ahead while the other is going astern (Figure 17.11).

Machinery 'Pod Technology' (Pod Propulsion Units)

Several recent cruise ships have moved towards 'pod propulsion units' as a means of main power, and the more recent buildings in the ferry sector reflect the potential use of 'pod technology' for many different sizes of vessels of the future. The compactness of the 'pod' and the associated benefits to passenger/ferry operations would seem to offer distinct advantages to ship-handlers, operators and passengers alike.

Plate 151 Structural support for twin screw/twin rudder vessel seen in drydock where each tail end shaft is supported by struts extending from each quarter of the hull.

Some of the possible advantages from this system would be in the form of:

- Low noise levels and low vibration within the vessel.
- Fuel efficiency with reduced emissions.
- Good manoeuvring characteristics and tighter turning circles as when compared with a similar ship operating with standard shaft lines and rudders.
- Reduced space occupied by bulky machinery, making increased availability for additional freight or passenger accommodation.

Vessel seen from astern, engines moving AHEAD   Vessel seen from astern, engines moving ASTERN

Figure 17.10 Twin-screw propulsion.

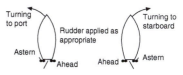

Figure 17.11 Steering by engines.

Plate 152 Modern ferry with twin screw (CPP) and twin rudder arrangement seen exposed in drydock.

Plate 153 Main engine control room for modern twin screw passenger ferry.

- Simpler maintenance operations for service or malfunction (pods are easy to remove/and replace).

Machinery pods are usually fitted to the hull form via an installation block. Each vessel has customized units to satisfy the hydrodynamics and the propulsion parameters. Propeller size and the revolutions per minute (rpm) would also need to reflect the propulsion requirements to the generator size with electric 'azipod units' (Figure 17.12).

Azipod propulsion systems provide the action of pulling rather than pushing the vessel through the water. A typical twin propeller azipod configuration would consist of three main diesel generators driving an electric motor to each propeller, with full

Plate 154 Twin propeller and twin rudder arrangement seen exposed on a vessel in drydock.

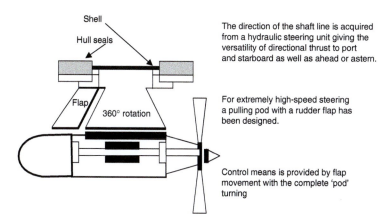

Shell

Hull seals

Flap

360° rotation

The direction of the shaft line is acquired from a hydraulic steering unit giving the versatility of directional thrust to port and starboard as well as ahead or astern.

For extremely high-speed steering a pulling pod with a rudder flap has been designed.

Control means is provided by flap movement with the complete 'pod' turning

Figure 17.12 Pod propulsion unit.

bridge control transmission. Power ranges start from about 5 MW up to 38 MW, dependent on selected rpm. Adequate built-in redundancy is accounted for by providing three generators for only two propellers.

It is reported that ship-handling is easier, turning circles are comparatively tighter than vessels fitted with conventional rudders and example speeds of 25 knots ahead, 17 knots astern and 5 knots sideways provides excellent harbour manoeuvring.

Varieties of pod designs are rapidly entering the commercial market supported by associated new ideas to improve fuel efficiency and provide better performance. Many are water-cooled, eliminating the need for complex air-cooled systems, while the Siemens–Schottel Propulsion (SSP) system has propellers at each end of the pod rotating in the same direction.

Propeller Slip

When the vessel is moving ahead, the propeller exerts pressure on the water to create the forward motion. Propeller slip occurs because water is not a solid medium and there is some 'slip' related to it.

Slip may be considered as the difference between the speed of the vessel and the speed of the engine. It is always expressed as a percentage:

$$\text{Propeller slip} = \frac{\text{Actual forward speed}}{\text{Theoretical forward speed}}\,(\text{per cent})$$

The calculated value of slip will be increased when the wind and sea are ahead, and if the vessel has a fouled bottom. The differing values of slip are especially noticeable after a vessel has been cleaned in drydock.

Theoretically, a vessel should never have a negative slip, but this may occur in one or more of the following conditions: a strong following sea, a following current or a strong following wind.

Example 1

During a 24-hour period of a voyage a ship's propeller shaft was observed to turn at 87 rpm. The pitch of the propeller was 3.8 m.

The observed ship's speed over the ground was ten knots for the same 24-hour period. Calculate the value of the propeller slip during this period. (A nautical mile equals 1852 m.)

$$\text{Slip (per cent)} = \frac{\text{Engine distance} - \text{Ship's distance}}{\text{Engine distance}} \times 100$$

$$\text{Engine distance} = \frac{\text{Pitch} \times \text{rpm} \times 60 \times 24}{1852}$$

$$= \frac{3.8 \times 87 \times 60 \times 24}{1852}$$

$$= 257.054$$

$$\text{Ship's distance} = 24 \times 10$$

$$= 240$$

$$\text{Slip} = \frac{257.054 - 240}{257.054} \times 100$$

$$= +6.6 \text{ per cent}$$

Example 2

A propeller has a pitch of 4.5 m. The ship steams for a period of 18 hours at 115 rpm and then steams for a further six hours at the reduced speed of 100 rpm. After the full 24-hour period the logged distance indicates 330 miles, but the log is known to have a 2 per cent negative slip. Calculate the propeller slip. (A nautical mile equals 1852 m.)

$$\text{Propeller slip (per cent)} = \frac{\text{Engine distance} - \text{Ship's distance}}{\text{Engine distance}} \times 100$$

From the log:

$$\text{Ship's distance} = \left(\frac{330}{100} \times 2\right) + 330$$

$$= 336.6 \text{ nautical miles.}$$

$$\text{Engine distance} = \frac{4.5 \times 115 \times 18 \times 60}{1852} + \frac{4.5 \times 100 \times 6 \times 60}{1852}$$

$$= 301.8 + 85.33$$

$$= 387.13 \text{ nautical miles}$$

$$\text{Propeller slip} = \frac{387.1 - 336.6}{387.1} \times 100$$

$$= \frac{50.5}{387.1} \times 100$$

$$= +13.05 \text{ per cent}$$

Turning Circles

General Definitions

Advance

The forward motion of the ship from the moment that she starts the turn, i.e. the distance travelled by the vessel in the direction of the original course from starting the turn to completing the turn.

Diameters

The greatest diameter scribed by the vessel from starting the turn to completing the turn (ship's head through 180°) is the tactical diameter. The internal diameter of the turning circle where no allowance has been made for the decreasing curvature as experienced with the tactical diameter is the final diameter.

Transfer

The distance which the vessel will move, perpendicular to the fore and aft line at the commencement of the turn. The total transverse movement lasts from the start of the turn to its completion, the defining limits being known as the transfer of the vessel when turning (Figure 17.13).

Advice for Helmsman and Officer of the Watch

1   A deeply laden vessel will experience little effect from wind or sea when turning, but a vessel in a light or ballasted condition will make considerable leeway, especially with strong winds.
2   When turning, the pivot point of the vessel is often situated well forward of the bridge and may produce the effect of the vessel turning at a faster rate than she actually is.

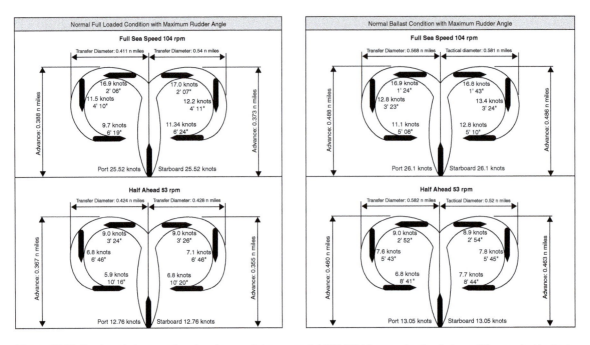

Figure 17.13 Turning circle examples: 'modern container vessel' 4,318 TEU (comparing loaded condition against ballast condition with a maximum rudder angle or hard over at 35°).

Details of ship and operation:

Length OA	292 m	Deadweight at design draught	61,787 MT	Calm weather.
Breadth	21.7 m	Lightship tonnage	20,560 MT	No current.
Draught	13.5 m	Service speed	24.3 knots	Clean hull.

(Water depth for turn example twice that of the draught.)

Plate 155 New hull design has caused major changes to many marine aspects, not least ship-handling and manoeuvring, but also to rescue and survival recovery operations.

3   A vessel trimmed by the stern will steer more easily, but the tactical diameter of the turn is increased.
4   A vessel trimmed by the head will decrease the diameter of the turning circle but will become difficult to steer.
5   If a vessel is carrying a list, the time taken to complete the turn will be subject to delay. A larger turn will be experienced when turning into the list.

Factors Affecting Turn

If the vessel is fitted with a right-hand fixed propeller, she would benefit from the transverse thrust effect, and her turning circle, in general, will be quicker and tighter when turning to port than to starboard. The following factors will affect the rate of turn and the size of turning circle:

- structural design and length of the vessel;
- draught and trim of vessel;
- size and motive power of main machinery;
- distribution and stowage of cargo;
- even keel or carrying a list;
- position of turning in relation to the available depth of water;
- amount of rudder angle required to complete the turn;
- external forces affecting the drift angle.

Structural design and length. The longer the ship, generally the greater the turning circle. The type of rudder and the resulting steering effect will decide the final diameter, with the clearance between rudder and hull having a major influence. The smaller the clearance between rudder and hull the more effective the turning action.

Draught and trim. The deeper a vessel lies in the water, the more sluggish will be her response to the helm. On the other hand, the superstructure of a vessel in a light condition and shallow in draught is considerably influenced by the wind.

The trim of a vessel will influence the size of the turning circle in such a way that it will decrease if the vessel is trimmed by the head. However, vessels normally trim by the stern for better steerage and improved headway and it would be unusual for a vessel to be trimmed in normal circumstances by the head.

Motive power. The relation between power and displacement will affect the turning circle performance of any vessel in the same way that a light speedboat has greater acceleration than a heavily laden ore carrier. It should be remembered that the rudder is only effective when there is a flow of water past it. The turning circle will therefore not increase by any considerable margin with an increase in speed, because the steering effect is increased over the same period. (The rudder steering effect will increase with the square of the flow of water past the rudder.)

Distribution and stowage of cargo. Generally this will not affect the turning circle in any way, but the vessel will respond more readily

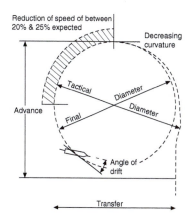

Figure 17.14 Turning circle.

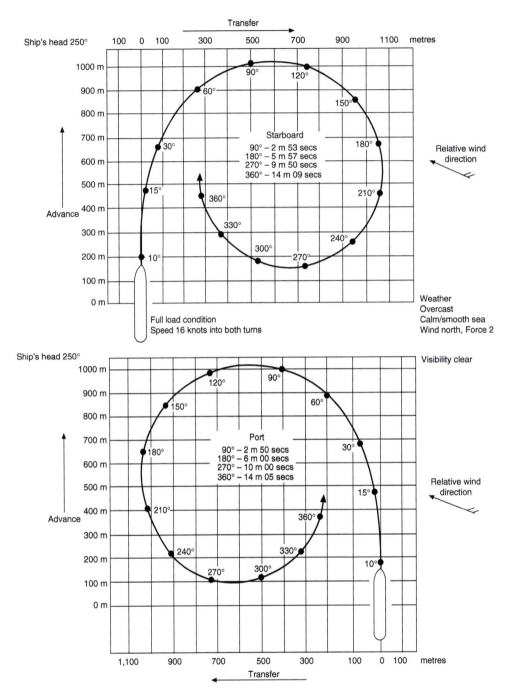

Figure 17.15 Turning circle of VLCC. Particulars of vessel: gross tonnage 133,035; net 108,853; dead weight 270,665; length OA 338.6 m; breadth 53.6 m; mould depth 26.4 m; load draught 20.6 m; main engine power 22,380 kW.

Plate 156 The new anchor handling/supply vessel *Pacific Retriever* seen testing systems while on trials.

if loads are stowed amidships instead of at the extremities. Merchant ship design tends to distribute weight throughout the vessel's length. The reader may be able to imagine a vessel loaded heavily fore and aft responding slowly and sluggishly to the helm.

Even keel or listed over. A new vessel when engaged on trials will be on an even keel when carrying out turning circles for recording the ship's data. This condition of even keel cannot, however, always be guaranteed once the vessel is commissioned and loaded. If a vessel is carrying a list, it can be expected to make a larger turning circle when turning towards the list, and vice versa.

Available depth of water. The majority of vessels, depending on hull form, will experience greater resistance when navigating in shallow water. A form of interaction takes place between the hull and the sea bed which may result in the vessel yawing and becoming difficult to steer. She may take longer to respond to helm movement, probably increasing the advance of the turning circle, as well as increasing over the transfer. The corresponding final diameter will be increased proportionately.

Rudder angle. Probably the most significant factor affecting the turning circle is the rudder angle. The optimum is one which will cause maximum turning effect without causing excessive drag.

If a small rudder angle is employed, a large turning circle will result, with little loss of speed. However, when a large rudder angle is employed, then, although a tighter turning circle may be experienced, this will be accompanied by a loss of speed.

Drift angle and influencing forces. When a vessel responds to helm movement, it is normal for the stern of the vessel to traverse in opposing motion. Although the bow movement is what is desired, the resultant motion of the vessel is one of crabbing in a sideways direction, at an angle of drift.

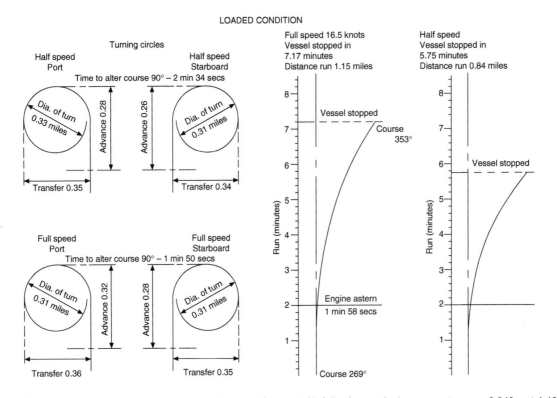

Figure 17.16 Turning circle and stopping distance of vessel with following particulars: gross tonnage 8,243; net 4,461; dead weight 13,548; five-hatch general cargo; length OA 161 m; breadth 22.97 m; loaded draught 13.1 m; main engine power 8,952 kW.

When completing a turning circle, because of this angle of drift, the stern quarters are outside the turning circle area while the bow area is inside the turning circle. Studies have shown that the 'pivot point' of the vessel in most cases describes the circumference of the turning circle.

Steering Gear Operations

Steering methods have changed considerably over the years, but the basic function has remained and that has been to move the rudder to change the movement of the ship's head. Even this basic function has had effective competition from the rotable thrusters, controllable water jets and the more recent steerable pod technology. However, new concepts have not yet completely dominated the steerage of ships and many vessels are fitted with conventional rudder movement in order to control the vessel's heading.

Telemotor Transmission

One of the early methods of steering which employed two single rams must be considered obsolete within today's developments. It has been replaced by four ram systems operated by dual steering motors in the form of an electro-hydraulic system (Figure 17.17).

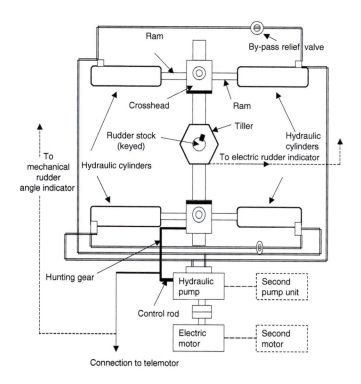

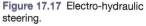

Figure 17.17 Electro-hydraulic steering.

Steering Applications

Most steering systems aboard commercial shipping are incorporated with an automatic steering unit, which includes follow-up, non-follow-up, remote and automatic modes of steering (in addition to manual control) (Figures 17.18 and 17.19).

Follow-Up (FU) Mode

When the position of the wheel is changed the rudder will begin to move and will keep moving until it reaches the ordered position as indicated by the helm. If the helm is returned to midships or any other position, the rudder will immediately 'follow up' and respond to the new command, taking up a new position.

Non-Follow-Up (NFU) Mode

Where the non-follow-up tiller steering is employed, the rudder moves in the desired direction from activation of the tiller, the position of the rudder being verified by means of the rudder indicator. If the tiller is returned to the midships position the rudder remains in the present position, and does not follow the tiller movement back to the midships position. The rudder will only be caused to move from the present position with further activation when a new port or starboard command is received.

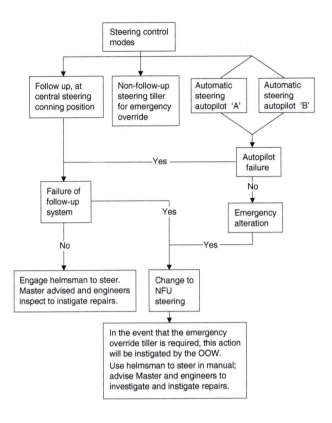

Figure 17.18 Steering procedures.

Remote Mode

Bridge wing stations very often contain pilotage controls. To effect one of these 'remote stations' the autopilot remains in 'auto' until one of the remote stations is selected by an ACCEPT control. Once this is activated, the autopilot goes into a 'standby' mode and will be so indicated on the main display. Should it be necessary to switch between remote stations, activating the ACCEPT control switch activates the required station. Most systems incorporate a tiller NFU controller within each remote station.

Automatic Mode

This mode provides automatic heading control, taking account of dynamic parameters such as speed and the head movement. This is achieved by adapting the steering control output to provide a course with minimal rudder motion. Individual functions for weather and sea conditions are selected inputs from the autopilot.

Steering Test Applications

The regulations require that a ship's steering gear will be tested by the ship's crew within 12 hours before departure. Such a test will include the operation of the following:

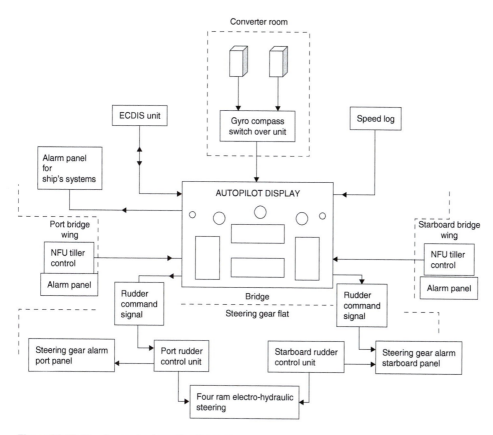

Figure 17.19 Steering and automatic pilot stations.

- the main steering gear system;
- the auxiliary steering gear system;
- remote steering gear control systems;
- the steering positions located on the navigation bridge;
- the emergency power supply;
- the remote steering gear control systems, power failure alarms;
- the steering gear unit, failure alarms;
- the automatic isolating arrangements and other associated automatic equipment.

The above checks must include the full movement of the rudder to its maximum capability. The communication system between the navigation bridge and the steering gear compartment must also be inspected and found to be satisfactory.

It is expected, as well as being a requirement, that operating instructions regarding the changeover procedures from one system to another (manual to automatic, and vice versa), should be displayed in block diagram format on the bridge. All ship's officers are expected to be familiar with these operations.

Emergency Steering Drills

The regulations stipulate that an emergency steering gear drill must take place at intervals of three months in order to practise emergency procedures. Such drills must be recorded in the official log book.

Steering Applications

If and when the steering gear develops a defect, the options open to ship's personnel would generally be defined by the company checklist for just such an occurrence (Figure 17.20). However, the reader should be aware of what is available on board his or her vessel and in general each ship will have the following steering elements to consider:

- *Main steering gear* (operating on single steering motor), open water.
- *Main steering gear* (engages second steering motor to retain control). In the event of defect on a single motor it must be considered as an emergency to lose steerage control. Therefore, the second motor may well be considered the emergency steering alternative.

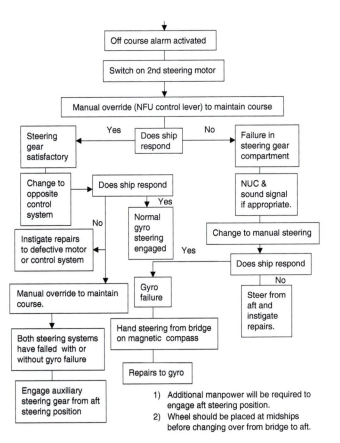

Figure 17.20 Steering gear failure: fault finder.

- *Auxiliary steering system* based in the steering flat is operated directly over the rudder stock to effect turning of the rudder (for the purpose of the Regulations, this is deemed the emergency steering system and must permit the vessel to be conned from a position remote from the bridge).
- *Jury steering* is an improvised steering system; when all other options have failed, the ship employs extraordinary measures to steer the vessel, e.g. steering by engines for twin-engine vessels; use of anchors below the surface to act as drag weights; drogue weight deployed from one quarter to another.

Auxiliary/Emergency Steering

The alternative positions of steering the ship in the event that the main systems have been rendered inoperable vary in type. The most popular method would seem to be secondary hydraulic oil tanks and pumps, situated in the steering flat itself. These can be connected quickly to replace the bridge transmission system and can be operated by manual button controls via the communication system.

Older ships may still be fitted with mechanical means of moving the rudder via use of the aft mooring/docking winches (as illustrated in Figure 17.23).

Loss of Rudder

The loss-of-rudder situation could affect any mariner at any time and all the steering systems on the ship will be to no avail if there is

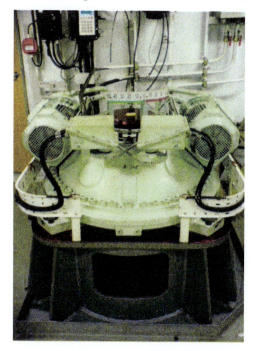

Plate 157 Head casting of rotary vein steering unit.

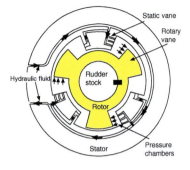

Figure 17.21 Rotary vane hydraulic steering.

nothing aft to turn. Provided power is still available, jury steering of one kind or another may be possible. However, with the size of vessels today, the practical answer to the problem is to call in the assistance of tugs. This is especially so if the ship is large, as with a VLCC or large ore carrier.

Rotary Vane Steering

Rotary vane steering is a compact steering unit which is situated on top of the rudder stock. A rotor is 'keyed' onto the stock and the whole is encased by steel jacket known as a 'stator'. The concept allows FU and NFU modes to operate with either electric or hydraulic transmission systems.

Model variations allow rudder angles of 2 × 35°, or 2 × 60°, with options of up to 90°. The system tends to act as a self-lubricating rudder carrier as well as generating the turning movement to the rudder. This is achieved by oil being delivered under pressure to one side of the blades of the rotor. With the rotor being keyed to the rudder stock, when the rotor is caused to turn, so does the stock.

Clearly the direction of turn will be affected by the direction of the pressurized oil affecting the rotor blades. Therefore, in theory the rotor and the stock can turn only one way, namely in the direction of the pressurized oil. However, if the directional flow of the oil is reversed, by reversing the rotation of the oil pump, then the rotor will also be caused to turn in the opposite direction. This pump reversal from one direction to another provides the necessary directional flow to cause movement to port and starboard.

The oil under pressure is kept contained within the unit by the stator. The stator is dynamically sealed and is leak-free, generally providing an effective, alternative steering mechanism within the created pressure chamber.

Bow/Stern Thruster Units

Elliott White Gill 360° Thrust and Propulsion Units

Over the last 20 years thrust units have proved themselves in all aspects of ship-handling. Advances in design, power and control have all led to the development of bigger thrusters and better performance.

The Elliott White Gill 360° thruster unit (Figure 17.24) has some distinct advantages over the conventional 'tunnel thruster'. Not only can the force of the thrust be directed as the operator desires, but with its location totally submerged all the time there is little chance of damage from surface obstructions.

The position of installation is so far beneath the water surface that the performance is not impaired by heavy weather. Pitching or heavy rolling have little or no effect as the intakes rarely break the surface, if at all. Limited maintenance is required, with the unit being readily accessible from within the vessel. No part of the unit projects beyond the lines of the hull.

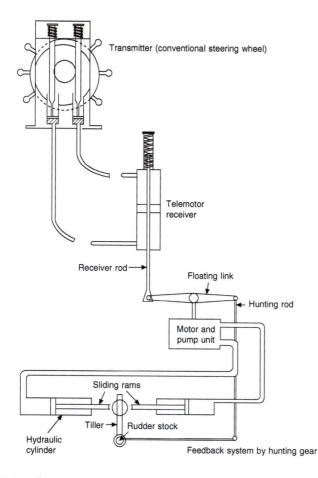

Figure 17.22 Telemotor transmission (two ram).

Elliott White Gill 360° Trainable Thrust Units

The main ship-handling features of the 360° trainable thruster (Figure 17.25) are:

- the thruster may be used as an auxiliary means of power or propulsion, being employed for both propulsion and steering of the vessel;
- it is capable of turning a vessel in its own length and turning 'broadside' on without resorting to the use of main engines and/ or rudder;
- remote control of the thruster unit is achieved from a main control bridge panel; additional bridge wing control panels may be fitted as required;
- the thrust capacity of up to 17 tonnes can hold the vessel on station even in bad weather or heavy sea conditions.

Steerable 360° Thrusters

Azimuth thrusters have become extremely popular within certain sectors of the marine industry, especially so in offshore operations

Figure 17.23 Alternative steering methods in event of breakdown.

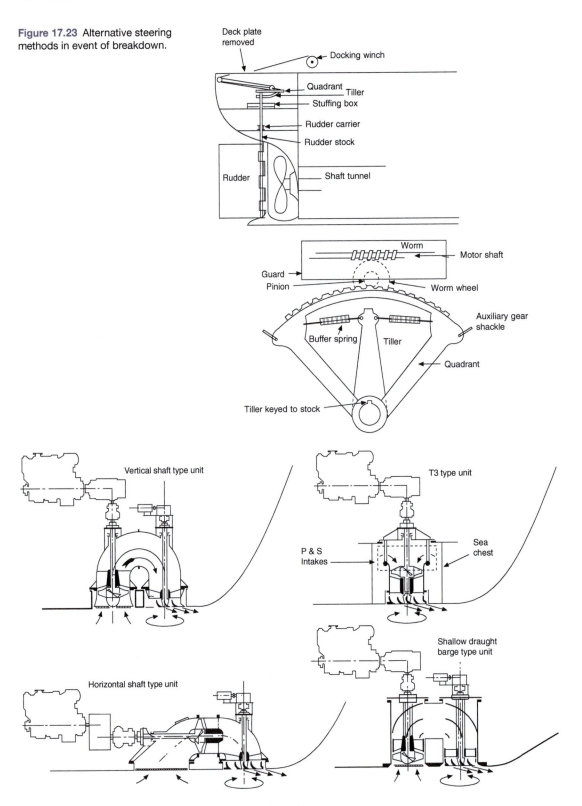

Figure 17.24 Elliott White Gill 360° thrust and propulsion units.

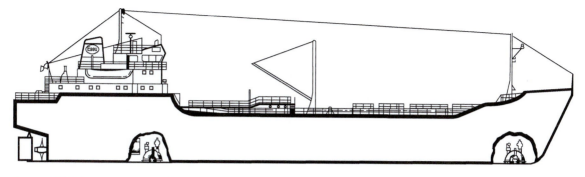

Figure 17.25 Elliott White Gill 360° trainable thrust units.

with dynamic positioning methods coming to the fore. They tend to provide precise manoeuvring and are often fitted to tugs at the time of building or retrofitted to meet specific needs of the area of operation (Figure 17.26).

Some smaller vessels are now fitted with the more powerful thrusters as main-line propulsion systems, as on the lines of the 'Kort Nozzle' design. However, the larger vessels, like the Class 1 passenger liners, tend to incorporate thruster variants in conjunction with 'pod propulsion units' (see Pod Propulsion on pp. 598–601).

The ferry sector of the industry has for some time relied on thruster units. Although initially considered expensive units to fit, they have more than shown their worth by allowing Masters to conduct their own pilotage, without the costly assistance of tugs at

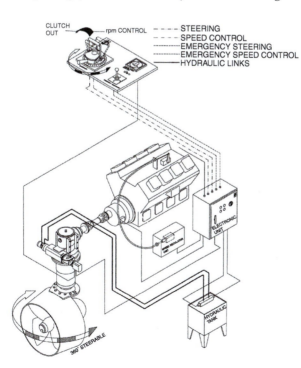

Figure 17.26 Example of the operation of a 360°, steerable thruster, manufactured by Holland Roer-Propeller.

Plate 158 Twin tunnel bow thrust units exposed for maintenance on a vessel on the blocks in drydock. Tunnel thruster units usually operate on 50 or 100 per cent thrust and are brought into operation from the central bridge console or from remote bridge wing operating stations. The units are usually submerged below the waterline at the bow or stern regions and protective grates (not shown) tend to keep surface debris clear of impellers.

every instance of entering and leaving port. One or two drawbacks should be noted by the mariner, in that they are externally fitted and as such are presented as appendages to the hull. The positions should therefore be found on the 'drydocking plan', and care must be taken to position the docking blocks clear of the thruster units to ensure a damage-free docking. The other problem is when navigating in close proximity to a vessel fitted with a rotable thruster unit, the possibility of mooring ropes being sucked into the rotors becomes real.

Waterjet Propulsion

Many ferries are now fitted with two, three or four waterjets in addition to main-engine propulsion (Figure 17.27). The additional power and manoeuvrability achieved with these units has become widely accepted throughout the industry, especially in the high-speed craft sector.

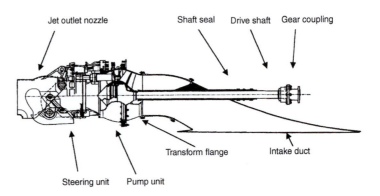

Figure 17.27 Waterjet propulsion.

Different manufacturers tend to specific designs, but the advantages and features of the units are generally common. The MJP waterjets, for example, are manufactured in stainless steel castings with intake diameters of 450–1350 mm, capable of delivering an engine brake power from 500 to 10,000 kW.

A high thrust pump unit operates on comparatively lower fuel consumption, which also produces a high reverse thrust capability. They are fitted with a full electro/hydraulic control package which provides alternative steering, autopilot, wing stations and various rpm interfaces.

Special features include a flexible drive shaft allowing ±0.25° to take account of movement within the ship's structure when underway. Hydraulic rams provide a steering angle of ±30° in a rapid response time, while the reverse bucket mechanism is said to achieve full astern from full ahead (or vice versa) in five seconds (Model MJP650).

Fin Stabilizers

There are three principal methods of reducing roll by means of stabilization available to the shipowner:

1   active fin – folding (Figures 17.28 and 17.29) or retractable type;
2   fixed (non-retractable) fin;
3   free surface tanks.

All systems have their merits, but the fin types would appear to be unrivalled when fitted to vessels engaged at speeds in excess of 15 knots. Should the vessel be operating at low speeds or at anchor in an exposed position, then a free surface tank system may be better suited for the nature of the work.

Plate 159 Two passenger vessels, the *Seawing* and the *Salamis*, moored alongside the quay with head lines and spring lines.

Plate 160 Fixed (non-retractable) stabilizer fin. Seen exposed on a vessel in drydock.

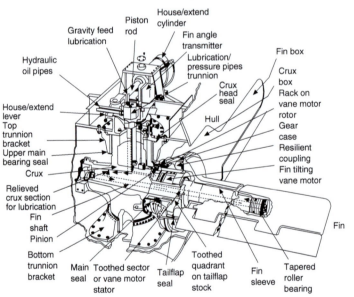

Figure 17.28 Fin stabilizers.

Manoeuvring with Mooring Lines

The main function of mooring lines, be they wire or fibre ropes, is to retain the vessel in position. However, there are times when they may be used in the turning or manoeuvring of the vessel, as when entering a dock or coming off quays (see Figures 17.30 and 17.31).

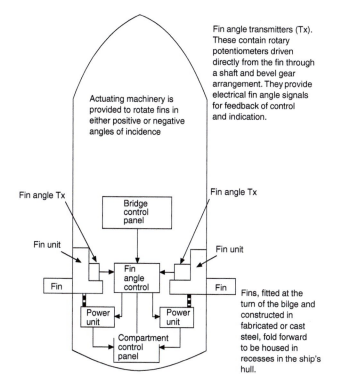

Fin angle transmitters (Tx). These contain rotary potentiometers driven directly from the fin through a shaft and bevel gear arrangement. They provide electrical fin angle signals for feedback of control and indication.

Actuating machinery is provided to rotate fins in either positive or negative angles of incidence

Fins, fitted at the turn of the bilge and constructed in fabricated or cast steel, fold forward to be housed in recesses in the ship's hull.

Figure 17.29 Folding fin stabilizer unit.

Fairleads

The roller fairlead is often encountered as a double or even treble lead, but is also found as a single lead on a stand or pedestal (Figure 17.32). It is in common use aboard a great many modern vessels, where it is generally referred to as an 'old man' or a 'dead man' because of its static pose. It has proved its usefulness in mooring operations for altering the lead of a rope or wire through very sharp angles.

Maintenance should be on a regular basis with regard to greasing and oiling about the axis. The pedestal should be painted at regular intervals to prevent corrosion. Should a lead of this type become seized, it is normal to soak the moving parts in release oil and then attempt to free the roller lead by use of a mooring rope to the warping drum end, so creating a friction drive.

Universal Multi-Angled Fairlead

This fairlead (Figure 17.33) consists of two pairs of axial bearing rollers, one pair in the vertical plane and the other pair in the horizontal. The main advantage of this type of lead is that it provides a very wide angular range, not only in the horizontal and vertical planes but also in any oblique plane. The main disadvantage of the lead is that it requires regular maintenance in the way of periodic greasing through grease nipples at each end of the rollers. When compared with the panama lead, the rollers respond when mooring lines are under tension, so friction is reduced, whereas the panama lead has no moving parts and friction may cause limited damage.

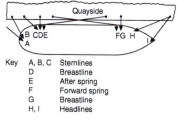

Key		
A, B, C	Sternlines	
D	Breastline	
E	After spring	
F	Forward spring	
G	Breastline	
H, I	Headlines	

Figure 17.30 Example of moorings used to secure vessel.

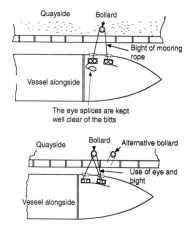

Figure 17.31 Mooring rope used as a bight (*above*) and as an eye and a bight (*below*).

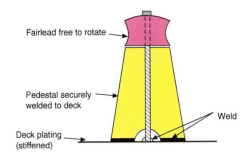

Figure 17.32 Roller fairlead.

Plate 161 The passenger cruise ship *Seawing* moored starboard side to, alongside.

Universal leads are regularly found on the quarter and shoulder areas of the vessel for the multiple use of spring or head and stern lines.

Panama Lead

This type of lead is very common aboard modern vessels. It may be a free-standing lead, as shown in Figure 17.34, in which case the underdeck area is strengthened, or it may be set into bulwarks and strengthened by a doubling plate. The lead is one favoured by seafarers because the rope or wire cannot jump accidentally when under weight.

Bollards (Bitts)

The term 'bollard' is usually applied to a mooring post found on the quayside and 'bitts' to the twin posts found on ships (Figure 17.35).

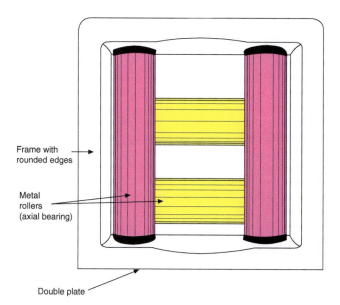

Frame with
rounded edges

Metal
rollers
(axial bearing)

Double plate

Figure 17.33 Universal multi-angled fairlead.

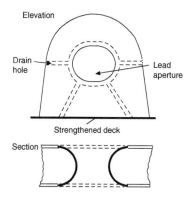

Elevation

Drain
hole

Lead
aperture

Strengthened deck

Section

Figure 17.34 Panama lead.

Plate 162 Single (old man) roller lead. Such examples can be motorized to work as a capstan.

Plate 163 Triple roller fairlead on the fore mooring deck of a passenger vessel.

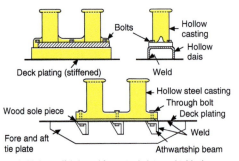

Figure 17.35 Bollards and bitts.

18
Ship-Handling: Manoeuvring and Mooring Operations

Introduction

In the previous chapter numerous examples of ship-handling hardware were highlighted. However, it should be realized that the equipment shown has been limited to one or two examples common to the commercial market. There are many manufacturers of propellers, rudders, machinery, etc., but in the end the performance is judged on how the handler employs the manoeuvring aids at his or her disposal.

The elements under direct control of the ship-handler remain the same as ever, just as those elements without control remain the same. However, it would be nice to think that with modern technology and increased power ratios, the task of manoeuvring has been made that much easier in today's world.

These days many more ship's Masters are carrying out their own pilotage while working under a pilot exemption certificate for specific ports. They need the bow thrust units and the controllable pitch propellers (CPPs) of the modern generation to dock and undock the modern ships. The river and island ports around the world have become more accessible because of up-to-date manoeuvring aids, bringing trade and prosperity to areas that were previously denied.

The larger vessel and the faster vessel are now totally integrated within the industry and although basic principles still apply, new equipment now tends to dictate the methods of berthing and mooring. It must only be expected that the men on the bridge will keep abreast of new developments.

Berthing

Let us assume that no tugs are available and that the ship has a right-hand fixed propeller (see Figures 18.1 to 18.14).

Plate 164 The *Calypso* passenger cruise vessel lies starboard side to, moored by head lines and rope springs while the ro-ro vehicle ferry *Mykono∑* turns off the berth to navigate stern first to the link span ramp.

Plate 165 Example of the aft mooring deck aboard a ro-pax ferry. Tension winches with warping drums to service through international roller fairleads to each quarter. Mooring ropes coiled on deck with hatchway access to the rope store in the foreground.

Berthing your Own Vessel, Starboard Side to, Between Two Ships Already Secured Alongside, with an Onshore Wind

Assume your own vessel is in position (1) stemming the ebb tidal stream (Figure 18.3). A mooring boat is available, and the vessel is fitted with a right-hand fixed bladed propeller.

1 The vessel must be manoeuvred to stem the tidal current as in position (1).
2 Manoeuvre the ship to position (2) parallel to the moored vessel 'A'.
3 Run the ship's best mooring rope from the starboard quarter to the quayside with the aid of the mooring boat (2) and keep this quarter rope tight, above the water surface.
4 Slow astern on main engines – then stop. This movement should bring the vessel's stern in towards the quayside by means of 'transverse thrust'. The bow should therefore move to starboard, outward to bring the current onto the ship's port bow, position (3).
5 The vessel should turn with current effective on the port side and no slack given on the quarter rope to complete an 'ebb swing', position (4). The wind should affect the port bow, blowing the ship rapidly towards the quayside. To check this movement towards the quay, let go the offshore anchor.
6 Run the forward head line (position 5) and draw the vessel alongside from the fore and aft mooring positions, easing the weight on the cable as the vessel closes the quay.

If an 'onshore wind' is present the first bow line would probably need to be carried aft to allow it to be passed ashore or, alternatively, use the same mooring boat which was initially employed.

If no mooring boat is available for berthing, the quarter line could be passed, first to the moored vessel 'A', and then onto the quayside.

Clearing a Berth

Let us assume that no tugs are available and that the ship has a right-hand fixed propeller (see Figures 18.4 to 18.7).

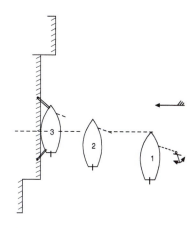

Figure 18.1 Berthing, wind onshore, tidal conditions calm.

1 Stop the vessel over the ground in a position with the ship's bow approximately level with the middle of the berth. Let go offshore anchor.
2 Control the rate of approach of the vessel towards the berth by ahead movements on main engines, checking and easing out anchor cable as required. Try and keep the vessel parallel to the berth.
3 Check cable within heaving line distance of the berth. Make fast fore and aft. Slack down cable when alongside.

NB. Where the manoeuvre is required when an 'offshore wind' is present, use of the offshore anchor would reduce the rate of approach towards the berth (as shown).

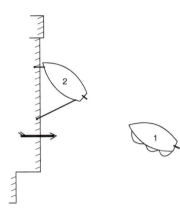

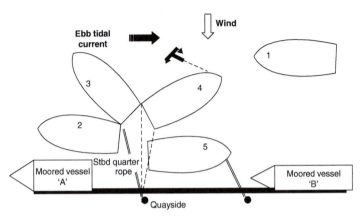

Figure 18.2 Berthing, wind offshore, tidal conditions calm.

Figure 18.3 Berthing a vessel starboard side to, between other ships.

1 Approach berth at a wide angle to reduce wind effect and prevent the bow from paying off.
2 Slowly approach berth and maintain position over ground.
3 Pass head line and stern line together from the bow area.
4 Stay dead slow astern on main engines, ease head line and at the same time take up the weight and any slack on the stern line. Draw the vessel alongside and secure. Depending on the strength of the wind, it would be advisable to secure a breast line forward as well as additional lines fore and aft as soon as practicable.

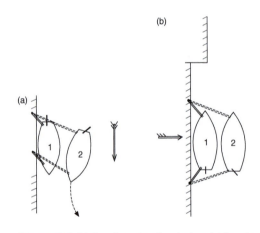

Figure 18.4 *(a)* Clearing a berth, wind and tide astern.

1 Single up to stern line and forward spring.
2 Main engine astern, ease out on stern line until stern is well clear of quay.
3 Let go and take in stern line. Let go forward.
4 When well clear of quay, stop main engine. Put rudder to port, and go ahead on main engine.

(b) Clearing a berth, no tugs available, right-hand fixed propeller.

1 Single up to a head line and stern line.
2 Let vessel blow off the quay: keep the vessel parallel to the quay by checking and controlling lines forward and aft.
3 When clear of the quay, let go fore and aft lines. Half ahead followed by full ahead on main engines if circumstances permit. Rudder applied as appropriate.

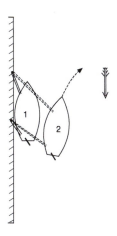

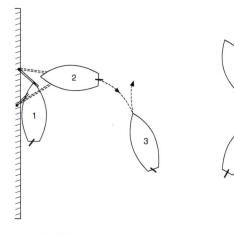

Figure 18.5 Clearing a berth, wind and tide ahead.

Figure 18.6 Clearing a berth, port side to, no wind or tide.

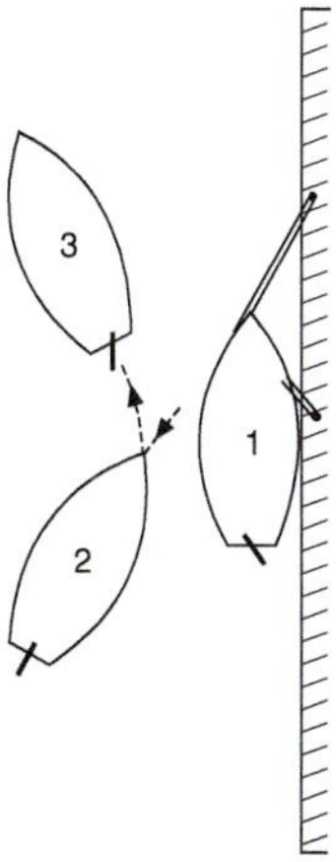

Figure 18.7 Clearing a berth, starboard side to, no wind or tide.

1 Single up to a head line and aft spring.
2. Ease away head line, rudder to starboard. With the tidal effect between the bow and the quayside the ship's bow should pay off.
3. Ease out on head line, slow ahead on main engines, take in head line and pick up slack on aft spring. Let go and take in aft spring. Use engine and rudder as appropriate.

1 Single up forward to an offshore head line and forward spring.
2 Keeping the weight on the forward spring, heave on the head line in order to cant the stern away from the quay wall. The stern will make a more acute angle with the quay if the main engine is ordered 'dead' slow ahead and the rudder put hard to port. Care should be taken to avoid putting the stem against the quay wall, especially if the vessel is of a 'soft nose' construction. Let go in the forepart.
3 Put main engines astern and allow the vessel to gather sternway to clear berth.

1 Single up forward to an offshore head line and forward spring.
2 Heave on the head line to bring the stern away from the quay wall. It may be necessary to double up the forward spring with the intention of using an ahead engine movement, allowing the spring to take the full weight, and effectively throwing the stern out from the quay. Let go smartly forward, main engines astern. When vessel gathers sternway, stop.
3 When clear forward, put rudder hard a-port, and main engine full ahead.

Securing to Buoys

No tugs are available and the ship has a right-hand fixed propeller (see Figures 18.9 to 18.12).

Mooring

The term mooring is used in conjunction with the securing of the vessel, either by two anchors or to a mooring buoy. The term is often used when vessels are moored to a jetty or quay by means of mooring ropes (Plate 166). The term may be considered, therefore, to be rather a loose one, applying to several methods of securing a ship. Most seafarers consider it to mean 'mooring with two anchors', in the form of a running moor, standing moor or open moor.

Entering A Dock

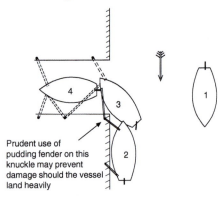

Prudent use of pudding fender on this knuckle may prevent damage should the vessel land heavily

Figure 18.8 Entering dock, wind and tide astern.

1 The vessel should turn 'short round' (Figure 18.22) or snub round with use of starboard anchor. The ship will then be in a position of stemming the wind and tide and should manoeuvre to land 'port side to' alongside the berth below the dock (2).
2 Secure the vessel by head lines and aft spring to counter tide effect and keep her alongside.
3 Put main engines slow ahead to bring the 'knuckle' of the dock entrance midships on the vessel's port side. Pass a second head line from the starboard bow across the dock entrance to the far side. Take the weight on this head line. Let go aft spring. As the vessel comes up to the knuckle, ease the port head line until the ship's head is in the dock, then heave on the port head line to bring ship parallel to sides of dock.
4 Carry up head lines alternately from each bow. Send out stern line and forward spring once the vessel is inside the dock. Stop main engines and check ahead motion as appropriate.

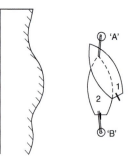

Figure 18.9 Securing to buoys, no wind or tide.

1 Approach the buoy 'A' slowly, with the buoy at a fine angle on the starboard bow, to allow for transverse thrust when going astern.
2 Stop the vessel over the ground and pass head and then stern lines. Align vessel between buoys 'A' and 'B' by use of moorings, and secure fore and aft.

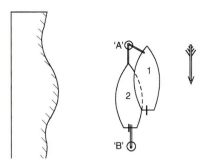

Figure 18.10 Securing to buoys, wind and tide ahead.

1 The vessel should stem the tide and manoeuvre to a position with buoy 'A' just off the port bow. It may be necessary for the vessel to turn short round or snub round on an anchor before stemming the tide. Adjust main engine speed so that the vessel stops over the ground. Pass head line.
2 Although an astern movement of main engines would cause the bow to move to port, if required, holding onto the head line would achieve the same objective by allowing the tide/current to effect the desired movement from position '1' to position '2'. Pass stern line once vessel is aligned between the two buoys 'A' and 'B'.

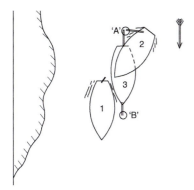

Figure 18.11 Securing to buoys, wind and tide astern.

1 Vessel under sternway, stern of the vessel seeking the eye of the wind. Use of rudder may assist to bring buoy 'A' onto the starboard quarter.
2 Run stern line from starboard quarter and make fast.
3 The vessel could expect to be moved by wind and tide to a position between the two buoys. The vessel may then be secured forward by head lines to buoy 'B'.
4 The success of this manoeuvre will, of course, depend on the strength of wind and tide. It might be necessary to turn the ship around to stem wind and tide, or, if the ship is to lie in the direction shown, it might be necessary to turn the ship and secure the bow to the other buoy shown and allow her to swing with the change of tide. Care should be taken that any stern lines are kept clear of the propeller when the vessel is navigating stern first.

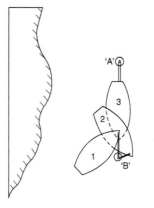

Figure 18.12 Securing to buoys, no wind or tide.

1 Approach buoy 'B' at a fine angle on the starboard bow. Pass head line and overrun the buoy about a third of the vessel's length from the bow. Hold onto the head line to check the vessel's headway. Allow the head line to act as a spring.
2 Rudder hard a-starboard, main engines ahead to turn the vessel about buoy 'B'.
3 Astern movement on engines will cause the port quarter to close towards buoy 'A'. This motion will further be assisted by the transverse thrust effect of the propeller. When the vessel is aligned between buoys, secure fore and aft.

Plate 166 Passenger vessel moored portside by headlines and breastline (source: Shutterstock).

Letting Go from Buoys

No tugs are available and the ship has a right-hand fixed propeller (see Figures 18.13 and 18.14).

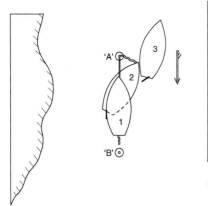

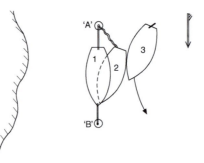

Figure 18.14 Letting go from buoys, wind and tide astern.

Figure 18.13 Letting go from buoys wind and tide ahead.

1 Let go stern line from buoy 'B'. When clear aft, apply starboard helm and go dead slow ahead on main engines.
2 As the vessel's bow moves to starboard, ease the head line. When clear of buoy 'A', let the head line go forward.
3 Main engines ahead, port rudder.

1 Slack stern line to see if the vessel will 'cant' away from buoy 'A'.
2 If the vessel cants, let go head line, with main engines half astern, allow vessel to gather sternway.
3 When the vessel clears buoy 'A', let go stern line. Main engines ahead once stern line is clear of propeller, helm hard a-port.
   If the vessel will not 'cant', let go the head line and heave the vessel close up to buoy 'A'; put rudder hard a-port, let go aft, with main engines full ahead. Once headway is gathered, make sharp helm movement to hard a-starboard to throw the stern clear of the buoy.

Rigging Slip Wires

The purpose of the slip wire is to enable the vessel to let herself go, at any time, without being dependent on the port's linesmen to clear lines from bollards. It is generally always the last line to let go. In some circumstances a slip rope may be used (see Figure 18.15).

Slip wires tend to run easily when letting go and heaving taut, but the wire is heavy and often difficult to handle. A strong messenger must be employed to heave the eye back aboard when rigging, because the wire will not float as a rope may, and there may be a long drift between the bow or stern and the bollard buoy.

Slip ropes are easier to handle and manipulate through the ring of a mooring buoy, but they are bulky and slow in running because of surface friction between the rope and buoy ring. They generally float on the surface when going out to the buoy and when being heaved back aboard; this fact considerably reduces the weight on the messenger.

Whether a wire or rope is to be used, a prudent seaman will always seize the eyes of the slip to allow clear passage through the ring of a mooring buoy. As the mooring boat travels towards the

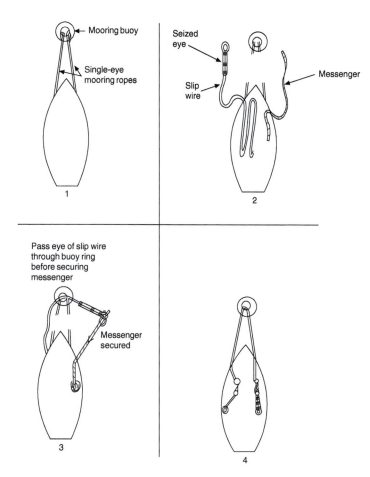

1 — Mooring buoy

Single-eye mooring ropes

1

Seized eye

Slip wire

Messenger

2

Pass eye of slip wire through buoy ring before securing messenger

Messenger secured

3

4

Figure 18.15 Rigging of slip wires.

A port mooring boat will be required for this operation, together with a lifejacket for the man engaged in buoy jumping and dipping the lines through the ring of the buoy.

1 Secure the forward or after end to the buoy in order to steady the vessel before passing the slip wire.
2 Prepare the slip wire beforehand by seizing the eye of the wire to enable it to pass through the ring of the buoy. Flake a messenger to the mooring boat with the slip.
3 Dip the slip wire through the ring of the buoy and secure the messenger to it. Once the small boat is clear, signal the vessel to heave the slip wire aboard, via the messenger.
4 Once the slip wire is aboard, release the messenger and turn up the slip wire on the bollards. Do not place eyes over bitts, as this may restrict letting go when weight is on the wire.

buoy, pay out both slip wire and messenger. A man wearing a life-jacket should then 'jump the buoy', pass the seized end of the slip through the ring, then secure the messenger to the small part of the eye of the slip wire. The messenger should never be passed through the ring of the buoy first, for this may cause the hitch to jam in the ring of the buoy when heaving back aboard. Signal to the officer in charge aboard the vessel to heave away on the messenger and bring the slip wire back aboard. Detach the messenger and turn up both parts of the slip wire in 'figure eights' on the bitts. Do not put the eyes of the slip on the bitts, as this would make letting go difficult if weight is on the wire.

Operation

Arrange the slip wire in long flakes down the deck length, then pass the eye down into the mooring boat. Additional slack on the wire should be given to the boat and coiled down on the boat's bottom boards. This provides the boat handler with slack to ease the weight, should the slip become snagged aboard. Pass a messenger into the mooring boat with the slip wire, but do not make the messenger fast to the slip wire at this stage.

Mooring Operations and Deployment of Anchors

Introduction

The anchor plan

An anchor plan should be established between the interested parties, namely: the ship's Master/Captain or offshore installation manager (OIM), the officer in charge of the anchor party, or the Master of Anchor Handling Vessel (AHV). It would be expected that these key personnel would inform relevant crew members through an established chain of command regarding relevant criteria.

In the construction of any anchor plan the following items must be worthy of consideration:

- The intended position of anchoring of the vessel.
- The available swinging room at the intended position.
- The depth of water at the position, at both high and low water times.
- That the defined position is clear of through traffic.
- That a reasonable degree of shelter is provided at the intended position.
- The holding ground for the anchor is good and will not lend to 'dragging'.
- The position as charted is free of any underwater obstructions.
- The greatest rate of current in the intended area of the anchorage.
- The arrival draught of the vessel in comparison with the lowest depth to ensure adequate under-keel clearance.
- The choice of anchor(s) to be used.
- Whether to go to 'single anchor' or an alternative mooring.
- The position of the anchor at point of release.
- The amount of cable to pay out (scope based on several variables).
- The ship's course of approach towards the anchorage position.
- The ship's speed of approach towards the anchorage position.
- Defined positions of stopping engines and operating astern propulsion (single-anchor operation).
- Position monitoring systems confirmed.
- State of tide ebb/flood determined for the time of anchoring.
- Weather forecast obtained prior to closing the anchorage.
- Time to engage manual steering established.

NB. Many large vessels with 15–20-tonne anchors would not generally 'let go' the anchor but 'walk it back' all the way in a fully controlled manner.

When anchoring the vessel it would be standard practice to have communications by way of anchor signals prepared for day and/or night scenarios. Port and harbour authorities may also have to be kept informed if the anchorage is inside harbour limits or inside national waters.

Single anchor: procedure

The Master, or pilot, should manoeuvre the vessel to the desired position, and take all way off, so that the vessel is stopped over the ground. She should be head to the wind and/or tide, and have her anchor walked back out of the hawse pipe, on the brake ready for letting go. The bridge should be informed that the anchor is on the brake of the windlass, or cable holder, and is ready for the order to 'let go'.

The engines should be operated to give stern way to the vessel. The Master should check overside and see the stern wake, about halfway up the vessel's length, and know that stern way is being made through the water, before giving the order to 'let go'. The officer in charge of the anchor party should order the brake to be taken off and allow the cable to run out with the weight of the anchor. The idea is to lay the cable out in length along the sea bottom, and not cause it to pile up on itself.

The officer in charge should start to apply the brake once enough cable has run out to prevent it falling on top of the anchor. The procedure is to check on the cable periodically, by applying the brake, while the vessel drops astern, either under engine power or through the action of the tide, and lays the required length of cable.

Communication from the forecastle head to the bridge should be maintained by walkie-talkie, loud hailer/phone, or by the ringing of the ship's forward bell.

Deep Water Anchoring

Should it become necessary to anchor the vessel in an emergency, or in anchorages such as the Norwegian fjords, the Master may be forced to a deep water anchorage. Should this occur, the anchor should not be let go in the normal manner, but walked back all the way to the sea bed.

The anchor party should have an idea of the depth of the water and be able to estimate when the anchor is on the bottom. As the vessel drops astern once the anchor begins to hold, the cable should be seen to grow slightly. However, because of the considerable weight of cable in the vertical up and down position, it will be increasingly difficult to see the cable snatching or growing to indicate that the vessel is brought up.

The officer in charge of the anchor party should be aware of the total length of cable that the vessel is equipped with to each anchor. Bearing this in mind, a close watch should be kept on the amount

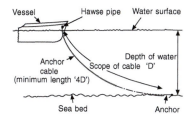

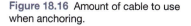

Figure 18.16 Amount of cable to use when anchoring.

NB. Masters or officers in charge should consider that taking the vessel into an anchorage must be considered a bridge team operation.

of cable being paid out. Many vessels have the joining shackle of the last length (shackle) of cable painted a distinctive colour for this reason.

When lowering away, under no circumstances should the anchor cable be allowed to run out from the brake. Control could be lost with the excessive weight of cable in use, resulting in possible loss of anchor and cable, together with serious damage and possible injury.

Laying/Carrying Out Anchors

Only exceptional circumstances would generally dictate this operation, such as stranding or beaching. Success will be dependent on several or possibly all the following points:

- size and weight of anchor;
- size, type and weight of cable or warp;
- size and capabilities of ship's own boats;
- depth of water;
- type of holding ground and geographic features;
- experience of crew members;
- prevailing weather at time of operation;
- availability of specialized equipment, including tugs.

With the ever-increasing number of large vessels in operation, the exercise of carrying out an anchor would in many cases become irrelevant, unless ships are specifically equipped to do the job. An example of this makes this apparent if, say, a 20-tonne anchor of a VLCC was to be carried out by means of the ship's boats. Even with large boats the gunwales would be nearly awash, and even if the anchor was allowed to submerge, in most cases (depending on the type of anchor) little benefit would be gained from buoyancy effects.

Vessels carrying very large anchors in the 15–25-tonne range would have little alternative to obtaining the assistance of a tug or other similar craft capable of accepting and releasing such a load. Masters engaged on such operations should bear in mind the additional weight factor of cable and/or warps, together with shackles, etc.

With a smaller vessel equipped with the corresponding weights of anchors, i.e. under 15 tonnes, the possibility of employing one as a stream or kedge anchor is not beyond the bounds of feasibility. The ship will be required to have lifting gear capable of hoisting the anchor, together with boats capable of transporting it.

Special operations like this will, of course, depend on the circumstances at the time, but as a general rule two boats will be required to support an anchor of any size. Slipping arrangements must be adequate to release the anchor from between the boats in a controlled manner.

When letting go a stream anchor, a long enough lead must be used, and allowance must also be made for the anchor to drag before becoming 'hung up'. A mistake in the position of letting go will, without doubt, double the work load, for what is, in any event, a lengthy operation.

Safety factors to be taken into consideration should include making a risk assessment, prior to the following:

1   Commence operations at a time to coincide with a maximum period of daylight.
2   Limit the number of men in the boats when carrying out to an essential number.
3   Use experienced seamen who have been briefed in full about the operation.
4   Lifejackets on relevant personnel, especially in the boat.
5   Buoy the anchor, prior to letting go (an additional boat may be useful here).
6   Use a stopper arrangement over the warp or cable to prevent whiplash on release.
7   Any strongbacks used should be braced and of adequate strength.
8   Prior to release, check that the warp will fall clear of boats.

Clearing a Foul Hawse

The object of this operation is to remove the foul turns in the two anchor cables caused by the vessel turning with the tide change continually in the same direction (Figure 18.17). It is a lengthy operation and should be started as soon as the vessel has swung and is riding at her new position; this will provide a six-hour interval before the tide turns again and the vessel assumes another position. To this end all preparatory work should be carried out before the vessel swings.

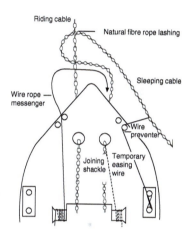

Figure 18.17 Foul hawse.

Position of the Lashing

Although in the following procedure it is said that the cables should be hove up to bring the foul turns above the water, and the lashing should be placed about the two cables below the turns, the mariner should be aware that there is a case to be made for securing the lashing above the turns. Should there be any doubt that the tidal stream may catch the vessel and cause her to swing during the oper-ation, then her sleeping cable could well become the riding cable. If this occurred and the lashing was secured below the turns, then the full weight of the vessel at anchor would come on to the fibre lash-ing, a highly undesirable condition.

The lashing itself should be a natural fibre lashing, as it will cut easily, rather than a synthetic rope, which may cause the blade to slip on its surface. If a knife and manhelper are not to be used to cut the cables adrift, then soak the lashing in petrol or other flammable liquid and burn the lashing adrift. This procedure may be assisted by heaving on both cables to break the lashing as it becomes weaker.

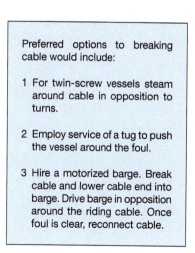

Preferred options to breaking cable would include:

1   For twin-screw vessels steam around cable in opposition to turns.

2   Employ service of a tug to push the vessel around the foul.

3   Hire a motorized barge. Break cable and lower cable end into barge. Drive barge in opposition around the riding cable. Once foul is clear, reconnect cable.

A manhelper is a long wood pole, usually of bamboo. Paint brushes, metal wood scrapers, knives, etc., are often secured to the end of the pole for the purpose of extending the handle of the implement. Securing is achieved by fitting the tool/implement to the pole by using two jubilee clips. This lashing, depending on its position, may be secured from either a boat or by a man in a bosun's chair. Should the latter be used, extra care should be taken to see that the man is removed from the chair before the cables splay apart.

Alternative Method

A method mainly employed by the Royal Navy was to use two wires, one from each bow, as messengers about the fouled cables. This method is effectively the same as the one described, but the use of two wires tends to expedite the work.

Clearing a Foul Hawse: Procedure (Figure 18.17)

1 Heave up on both cables to bring the foul turns above the water and lash both cables together below the turns with a natural fibre lashing. This lashing will prevent the turns from working themselves further down the cable.
2 Pass a wire preventer around the sleeping cable, down from the turns. This will reduce the weight on the turns, and serve to secure the sleeping cable should the end be lost. This preventer should be passed in such a manner that it may be slipped from the deck when the foul is cleared.
3 Walk back on the sleeping cable to bring the next joining shackle conveniently forward of the windlass. Rig a temporary easing wire to a point forward of this shackle so that it can take the weight of the sleeping cable when the shackle is broken.
4 Reeve a wire rope messenger from the windlass drum overside; the wire should be a cargo runner or other similar wire.
5 Take a half turn about the riding cable with this cargo runner wire. This turn should be made in the opposite direction to the foul turns in the cables.
6 Pass the end of the wire messenger up through the hawse pipe of the sleeping cable and secure it to the end of the sleeping cable.
7 Heave away on the wire messenger, and at the same time ease out on the easing wire, heaving the end of the sleeping cable up towards the fairleads, thus removing a half turn from the fouled cables.
8 This procedure should be repeated, removing half a turn at any one time, until the fouled cables are clear.
9 When all the turns are clear, haul in the sleeping end of the cable and rejoin with the joining shackle on deck.

10 The preventer wire should be slipped and cleared; the lashing should be cut by using a sharp knife with a manhelper. Heave away on both cables, picking up any slack.

To Weigh Anchor by Deck Tackle

This operation would be necessary if the windlass for some reason could not be used.

A heavy-duty tackle, of safe working load (SWL) minimum 15 tonnes, should be secured to bitts about the break of the forecastle head. This tackle should be sited as near as possible to give a clear lead, lying as near parallel to the line of cable as is practical. A heavy-duty wire pendant should be shackled to the moving block of the deck tackle (see Figure 18.18). A joggle shackle, if carried, should secure this pendant to the cable, while the downhaul of the deck tackle is led to a cargo winch drum.

The length of the operation will depend on the amount of cable that is paid out, but in any event the task will be made easier by an overhauling purchase rigged on the opposite side. The purpose of this is to reduce the work of manually overhauling the deck tackle as each length of cable is hove in.

There will no doubt be variations of this method, and the position of deck tackle, overhauling purchase, snatch blocks, etc. will be dependent on deck fitments. The idea of the heavy-duty pendant is that this will not foul any deck fitments. If the tackle was secured directly to the cable, then its passage in heaving on the cable might become restricted by the sheer size of the blocks.

Alternative

If the vessel is fitted with a windlass, a friction drive may be set up between one of the forward cargo winches and the drum end of the windlass. However, should cable holders be fitted, this may not be practical. It is possible that if power is lost on the windlass, it may also be lost on cargo winches. If this is the case, then manual power, together with an improvised capstan, might be considered. Failing all else, buoy the anchor cable and slip.

Anchor Recovery: Loss of Windlass Power

Should a vessel lose the use of cable holder or windlass in a situation with the anchor down, recovery of the anchor may prove difficult but not impossible. The method indicated in Figure 18.19 shows an alternative to rigging recovery tackle as previously described.

This method eliminates the use of overhauling gear and reduces the time to effect recovery. However, the system could only be employed on certain conventional vessels equipped with lifting

Prior to breaking and dipping cable end to clear a foul hawse, Masters should attempt to steam around cables or employ a tug to push the vessel around in opposition to the foul turns in the cables.

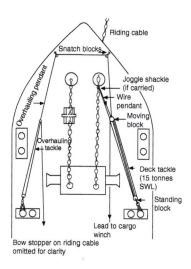

Figure 18.18 Weighing anchor by deck tackle.

Figure 18.19 Anchor recovery after loss of windlass power.

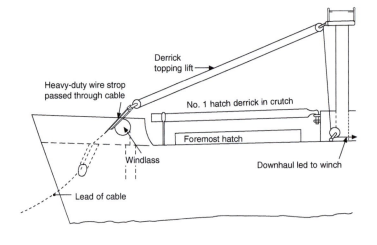

gear forward. Figure 18.19 shows that the operation is rigged for the topping lift of the derrick to be secured to the cable above the hawse pipe, the downhaul of the topping lift then being led via lead blocks to a cargo winch.

The topping lift is secured to the cable by means of a heavy-duty steel-wire strop or snotter passed through the cable links and shackled to the moving block. The windlass should be out of gear, with the cable being held on the brake. Start heaving on the topping lift and take the weight off the windlass, then release the brake. Continue to heave on the topping lift, allowing the gypsy of the windlass to freewheel, walking the cable back into the spurling pipe.

Place the brake hard on, detach the wire strop and resecure to repeat the process for as many times as it takes to recover the full scope of cable. Prudent use of engines 'ahead' could ease the weight factor and speed the recovery operation.

Hanging Off an Anchor

There may be occasion in the ship's life to detach the anchor from the cable, either for mooring to a buoy or for towing operations, where the bare end of cable is required. The method of obtaining the bare end of cable was associated with 'catting the anchor', where a vessel was equipped with a 'clump cathead'. A modern vessel will either be able to detach the cable while leaving the anchor secured in the hawse pipe or it will become necessary to 'hang the anchor off' (Figures 18.20 and 18.21).

The operation of hanging off an anchor is generally carried out to clear the hawse pipe and allow the bare end of the remaining cable a more suitable lead. The object of the exercise is to hang the anchor off aft of the hawse pipe, preferably about the break of the forecastle head, so as not to foul the cable forward.

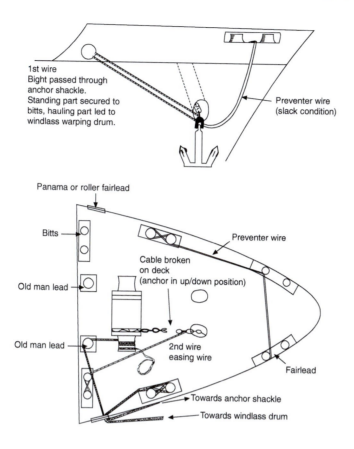

Figure 18.20 Hanging off an anchor from shoulder position.

1st wire
Bight passed through anchor shackle.
Standing part secured to bitts, hauling part led to windlass warping drum.

Preventer wire (slack condition)

Panama or roller fairlead

Bitts

Preventer wire

Cable broken on deck (anchor in up/down position)

Old man lead

Old man lead

2nd wire easing wire

Fairlead

Towards anchor shackle

Towards windlass drum

2nd easing wire

1st wire

Preventer wire (taut)

Ganger length   1st shackle length of cable

Figure 18.21 Hanging off an anchor. Once the full weight of the anchor is taken up on the first wire, the preventer should be hove taut. The easing wire should be secured forward of the next joining shackle, and the weight subsequently taken up on this. The joining shackle should now be 'broken' on deck and the bare end of the cable walked back to clear the hawse pipe. The bare end of cable is retrieved back aboard the vessel by hawser and bosun's chair operation from the fairlead, the hawse pipe being left clear to allow the use of the remaining cable inside the locker.

Sequence of Operation

1   The anchor should be walked back clear of the hawse pipe.
2   With the aid of a man in a bosun's chair, a heavy wire should be passed through the anchor crown 'D' shackle, this wire being led from the shoulder at a point from which it is intended to suspend the anchor. (The wire should be of sufficient SWL to

accept the full weight of the anchor and a limited amount of cable.)

3 This wire should be secured aft of the forecastle head, one end being turned up on bitts while the other is turned onto the windlass drum (with heavy anchors, both parts should be turned up on bitts).

4 Rig a preventer wire in a slack condition well forward of operations in case the first wire should part once the cable is broken.

5 The anchor cable should be walked back to allow the first wire to accept full weight of the anchor. The first wire is now in the up/down position.

6 Continue to walk back on the cable to bring the next joining shackle on deck, securing this length in short bights. Engage bow stopper or other cable-securing arrangements.

7 Rig a second easing wire forward of the joining shackle, and take the weight of the amount of cable between the anchor and the joining shackle on deck.

8 Break the joining shackle.

9 Clear away cable securings, and walk back on the easing wire to bring the bare end clear of the hawse pipe.

10 Rig a hawser, with the aid of a bosun's chair, to recover the bare end inboard via the fairlead, thus leaving the hawse pipe clear.

> With any specialist anchor operations a risk assessment should always be conducted prior to commencing operations.

The cable joining shackle should not be broken until the first wire has been secured (both parts of the bight), because if control of the first wire is lost, and the cable has already been broken, then the possibility of losing anchor and a length of cable becomes more than probable. This probability is increased with very heavy anchors, e.g. 20 tonnes.

Lost Anchor and Resecuring of Spare Anchor

Should a vessel be in the unfortunate position of having lost her port or starboard bow anchor, then she is considered to be unseaworthy for the purpose of marine insurance. It therefore becomes essential to resecure the spare anchor as soon as is practicable.

Methods of securing the spare anchor to an end of cable will, in most cases, need considerable thought, depending on several facts:

- size and weight of the spare anchor;
- position of stowage of the spare anchor in relation to the forecastle head;
- type of structure of the vessel, and obstructions between the spare anchor and forecastle head;
- the size and capability of lifting gear aboard the vessel, i.e. SWL of derricks/cranes, etc.
- whether tugs are readily available.

- capacity of the ship's own boats to transport the spare anchor from a possible aft position to a forward position;
- whether the ship can make a port of refuge with ease and exercise a 'carry up' the quay operation, employing shore transport, e.g. a flat-top truck.
- the prevailing weather and experience of crew members if the resecuring is to take place at sea.

Alternative Methods

1. When in port, or if the vessel can be expected to make a port, employ shoreside cranes to offload the spare anchor on to a flat-top truck or wagon. Transport the anchor along the quayside the length of the vessel and resecure the anchor on the quay after paying out some slack on the affected cable.
2. A large vessel carries very heavy anchors. Transport for such heavy weights must be available, e.g. 20-tonne anchor. It must be achieved in a safe manner to begin with.
3. Should a tug or barge be readily available for hire, considerable time and labour could be saved by engaging such a craft for the purpose of completing the operation.
4. Emergency rigs of heavy-duty 'sheer legs' straddling the spare anchor could be considered, but this would depend on the weight of the anchor, not to mention the fact that materials for rigging heavy 'sheers' may not be available.
5. Another possibility would be to employ the lifting facilities of another vessel. This is a particularly attractive alternative when another vessel from the same company is close by and is in a position to moor alongside; the lifting gear, as well as the mode of transport for the spare anchor, are provided by the second vessel.
6. The ideal craft for the operation is, of course, a floating crane, especially if the spare anchor is stowed in an awkward position, say on the afterdeck.
7. Should the vessel be near a drydock, the facilities of the drydock could be employed. However, it would be extremely costly to enter drydock for the one specific purpose of resecuring a spare anchor. Full consideration would have to be given to other needs, such as hull inspections and surveys, etc.

Turning Vessel Short Round

The ship has a right-hand fixed propeller (see Figure 18.22).

Running Moor

In all ship-handling situations the vessel should stem the tide if control is to be maintained. The running moor operation (Figure 18.23) is no exception to this rule, and should the tidal stream be astern

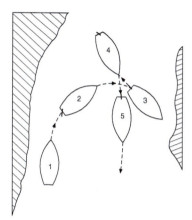

Figure 18.22 Turning a vessel short round.

The vessel is equipped with a right-hand fixed propeller, and, when turning 'short round', she would turn more easily to starboard than to port.

1. Start the manoeuvre from the port side of the channel to provide the maximum distance for the headreach movement of the vessel.
2. Rudder hard a-starboard, main engines full ahead. Stop engines. Do not allow the vessel to gather too much headway.
3. Rudder midships, main engines full astern.
4. As sternway is gathered, the bow of the vessel will cant to starboard while the port quarter will move in opposition, owing to the effects of the transverse thrust. Stop engines.
5. Rudder to starboard, engines ahead.

of the vessel, then she should be manoeuvred to stem the tide, either by turning short round or snubbing round on an anchor. A running moor procedure is as follows:

1   Speed over the ground should be 4–5 knots, preferred depth of water being dependent on draught, and good holding ground chosen if possible. Let go the weather anchor so that the vessel will be blown down from the anchor cable before she reaches the desired position.

2   Continue to make headway, paying out the cable of the anchor which has been let go. Continue to pay out the cable up to eight or nine shackles, depending on the amount of cable carried aboard and the depth of water. The vessel will overrun the desired mooring position.

3   The vessel should start to drop astern as the engines are stopped. Let go the lee anchor and pay out the cable. Start heaving away on the weather anchor cable to bring the vessel up between the two anchors. The vessel may require an astern movement on the engines to begin drawing astern.

In comparison with the standing moor the ship's machinery is running and operational throughout the manoeuvre. In the standing moor the vessel's machinery could well be out of action, standing still, while the vessel drops astern with the tidal stream.

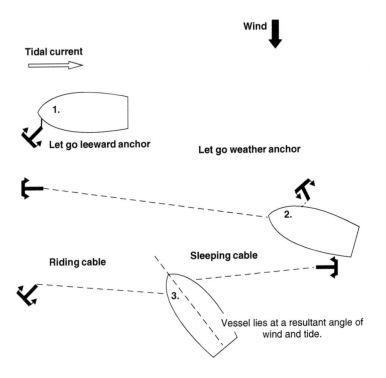

Figure 18.23 The running moor (ship must first stem the tide).

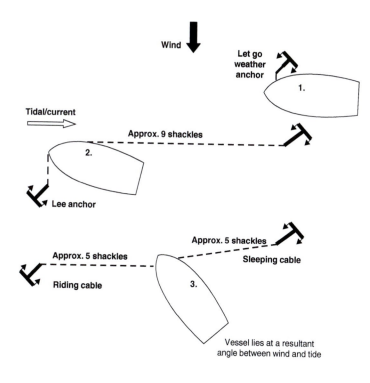

Figure 18.24 The standing moor (vessel must first stem the direction of tide).

Standing Moor

The vessel must stem the tide in order to retain control of the operation (Figure 18.24), which proceeds as follows:

1 The vessel should be head to tide, stopped over the ground. Sternway should be gathered either by the tidal stream or operating astern propulsion. Let go the lee anchor (riding cable) and allow the vessel to drop astern. Pay out the anchor cable as sternway is gathered, up to 8–9 shackles, depending on the amount of cable carried aboard and the depth of water.

2 Take the sternway off the vessel by use of engines ahead and checking on the weight of the cable. Order maximum helm away from the released anchor, and engines ahead to cant the vessel before letting go the weather anchor (sleeping cable). The mariner should continue to use engines ahead or astern as necessary to ease the weight on the windlass as the vessel heaves on the riding cable.

3 Continue to heave on the riding cable and pay out the sleeping cable until the vessel is brought up between the two anchors.

A standing moor is sometimes preferred to a running moor when the tidal stream is very strong. The standing moor, in theory, could also be carried out by just allowing the tidal stream and the windlass to do the work, if the vessel was without engine power.

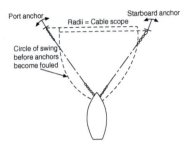

Figure 18.25 Open moor.

The main danger of mooring with two anchors is the possibility of causing a foul hawse when the vessel swings with the turn of the tide. To reduce this most undesirable condition the Royal Navy tends to use a mooring swivel, joining the two cables. Merchant vessels would not generally carry such a swivel unless it is intended to secure the vessel to a semi-permanent mooring over an indefinite period of time.

Open Moor

The open moor (Figure 18.25) is used extensively when additional holding power is required. It would be employed when a single anchor would not provide enough weight to hold the vessel and prevent the ship from dragging in a non-tidal water.

Possibly the best method of approach is to stem the current and position the vessel to let go the windward anchor. Once this first anchor has been 'let go', pay out on the cable with simultaneous 'ahead movements on engines' to manoeuvre the vessel towards a position of letting go the second anchor. Extensive use of rudder and engines may be required to achieve this second desired position.

Once the second position is attained, let go the second anchor, order astern movement of the engines, and pay out on the second anchor cable. The first anchor cable will act as a check until both cables have an even scope; once this situation is achieved then cables can be paid out together as required to obtain the final position of mooring.

Masters should bear in mind that with this method the first anchor may be turned out of the holding ground when the vessel gathers sternway after the second anchor has been released. To this end it may become prudent to check both cables prior to coming to rest, so ensuring that both the second and the first anchors are bedded in and holding.

Baltic Moor

The vessel should approach the berth with the wind on the beam or slightly abaft the beam. The stern mooring wire should be secured in bights by light seizings in the forward direction to join the ganger length of the anchor cable before the approach is begun. Then proceed as follows:

1   Manoeuvre the vessel to a distance off the berth of two or three shackles of cable. This distance will vary with the wind force and expected weather conditions.
2   Let go the offshore (starboard) anchor. The weight of the anchor and cable will cause the sail twine securing on the mooring wire to part, and as the cable pays out so will the stern mooring wire.
3   Let the wind push the vessel alongside while you pay out the cable and the stern wire evenly together.
4   Use the ship's fenders along the inshore side between the vessel and the quay, then pass head and stern lines as soon as practical.

5   Secure head and stern lines on the bitts before taking the weight on the anchor cable and the stern mooring wire. This tends to harden up the inshore (port) moorings.

One reason behind the Baltic moor is that many ports experience strong onshore winds.

When the vessel comes to let go and depart the port, unless she is fitted with bow thrust units, the Master may encounter difficulties in clearing the berth. However, heaving on the anchor cable and on the stern mooring will allow the vessel to be bodily drawn off the quay. Once clear of the berth, full use can be made of engines and helm to get underway.

The main disadvantage of this moor is that time is required to let the stern mooring go from anchor/cable. To this end the size of shackle used and the possibility of allowing it to pass up the hawse pipe are critical factors. Alternatives are to find a lee for the vessel for the purpose of disengaging the stern mooring.

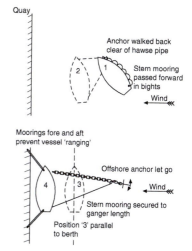

Figure 18.26 Baltic moor.

Mediterranean Moor

This moor is carried out usually for one of two reasons – either quay space is restricted and several vessels are required to secure, or a stern loading/discharge is required (as for a tanker). The object of the manoeuvre is to position the vessel stern to the quay with both anchors out in the form of an open moor. The stern of the vessel is secured by hawsers from the ship's quarters to the quay.

This type of mooring (Figure 18.27) is not unusual for tankers using a stern load or discharge system. However, a disadvantage to the dry cargo vessel lies in the fact that cargo must be discharged

Plate 167 Passenger/vehicle ferries lie stern to quay in Mediterranean moors in the Greek island of Rhodes.

Figure 18.27 Mediterranean moor.

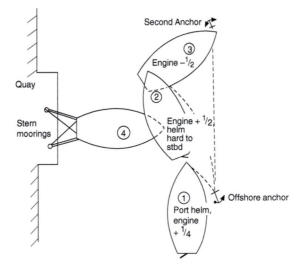

into barges. It is not a favourable position in bad weather and there is a distinct possibility of fouling anchor cables, especially when other vessels are moored in a similar manner close by. The procedure is as follows:

1 Approach the berth as near parallel as possible to the quay. Let go the offshore anchor. Main engines should be ahead and dead slow.
2 Rudder should be positioned hard over to turn the vessel away from the quay. Continue to let the cable run, and pay out as the vessel moves ahead. A check on the cable as the vessel starts to turn would accentuate the turn and produce astern-to orientation for the vessel. Stop main engines.
3 Let go the second anchor and come astern on main engines, paying out the cable on the second anchor. As the vessel gathers sternway, recover any slack cable on the offshore anchor. Stop engines and check the sternway on the vessel, as required, by braking on the cables (astern movement from position 3 will generate transverse thrust effect to turn the aft part into the quay).
4 Manoeuvre the vessel to within heaving line distance of the quay by use of engines and cable operations. Pass stern moorings to the quay. Tension on the moorings is achieved by putting weight onto the cables once the moorings have been secured on bitts.

Mediterranean Moor: Modern Vessel Equipped with Twin CPP and Bow Thrust

The more up-to-date vessel equipped with twin propellers and bow thrust unit(s) would expect to make the Mediterranean moor look quite simple. Neither is it necessary to always approach with the

Plate 168 The vessel *Nefeli* seen in a Mediterranean moor, each anchor deployed as close to the fore and aft line as practicality permits.

quay on the port side of the vessel (as for right-hand fixed propeller vessels).

With enhanced manoeuvring equipment like CPPs together with 'bow thrust', the approach to the mooring can be made with either port side or starboard side to the quay at a distance of about two ships' lengths (Figure 18.28).

Both anchors should be walked back clear of the hawse pipes and held on the windlass brakes. Once the centre position of the intended berth is reached the way should be taken off the ship prior to commencing a turn in the offshore direction.

The inshore engine should be placed ahead, with the offshore engine placed astern. Maximum bow thrust of 100 per cent should be given in order to turn the vessel about the midships point.

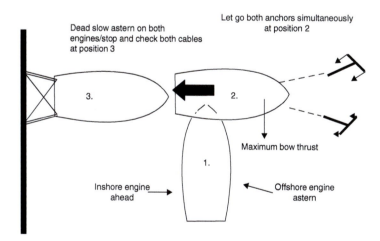

Figure 18.28 Manoeuvring a vessel into position.

At position '3' the vessel should be within heaving line range of the quayside and moorings can be passed to secure the stern. Once all fast aft, tension cables at the fore end to render stern moorings taut.

Dredging Down

A vessel is said to be 'dredging down' when she is head to the wind and/or tide (stemming the tide), with an anchor just on the bottom. The amount of cable out is limited to the minimum to put the anchor on the bottom.

Dredging down occurs when the vessel is not moving as fast as the current, which makes the rudder effective and allows the ship to manoeuvre. It is normal to expect a crabwise motion of the vessel over the ground, which is often employed for berthing operations. Used in conjunction with bold helm, the direction of the ship's head can be appreciably changed.

Snubbing Round

A vessel can turn head to tide without too much difficulty, provided that there is sufficient sea room to do so. Should the sea room not be available then a tighter turn will be required. This can be achieved by means of one of the ship's anchors, in the operation of snubbing round on the weight of the cable.

It is most frequently practised when the vessel has the tidal stream astern or in berthing operations. The vessel's speed should be reduced so that she can just maintain steerage way. Let go either the port or starboard anchors, at short stay, and allow the cable to lead aft, dragging the anchor along the bottom. The cable will act as a spring, reducing headway, and canting the bow round towards the side from which the anchor was let go. The Master or pilot of the vessel should supplement this anchor/cable action by use of maximum helm and increase in engine power to bring the vessel through 180°. The anchor party should be briefed on the operation beforehand, and know when to apply the brake to the cable, so giving the check on the vessel's forward motion that is necessary to complete the turn.

If the manoeuvre is attempted with too much headway on the vessel, excessive weight will be brought onto the cable as the vessel turns, which could result in the cable parting. In general practice, the anchor is let go to about a shackle, depending on the depth of water. The brake is then applied to start the turning motion on the vessel.

Anchoring in an Emergency

A vessel is approaching a channel in reduced visibility, speed five knots. The Officer of the Watch receives a VHF communication that the channel has become blocked by a collision at the main entrance (Figure 18.29). What would be a recommended course of action when the vessel was one mile from the obstructed channel, with a flood tide of approximately four knots running astern?

1   Assuming the vessel to have a right-hand fixed propeller, put the rudder hard a-starboard and stop main engines. The vessel would respond by turning to starboard. The anchor party should stand by forward to let go starboard anchor.
2   Let go starboard anchor. Full astern on main engines to reduce headreach. Letting go the anchor would check the headway of the vessel and act to snub the vessel round. Stop main engines.
3   Full ahead on main engines, with rudder hard a-starboard. Ease and check the cable as weight comes on the anchor. Once the vessel has stopped over the ground, go half ahead on main engines, allowing the vessel to come up towards the anchor and so relieve the strain on the cable. Heave away on the cable and bring the anchor home. Clear the area and investigate a safe anchorage or alternative port until channel obstruction is cleared.

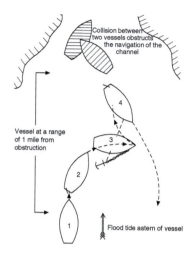

Figure 18.29 Emergency anchor to avoid obstruction.

Interaction

Most vessels will at one time or another experience some form of interaction with another vessel, perhaps through navigating in shallow water or passing too close to an obstruction. In this age of the big ship, Masters and pilots should know exactly what interaction is and what the results of its occurrence may be.

Interaction is the reaction of the ship's hull to pressure exerted on its underwater volume. This pressure may take several forms (Figures 18.30 to 18.33).

Interaction in Narrow Channels

Vessels navigating in narrow channels (Figures 18.34–18.36) may also see telltale signs of interaction, e.g. when passing another vessel which is moored fore and aft. The interaction between the vessels will often cause the moored vessel to 'range on her moorings'. A prudent watchkeeper on that vessel would ensure that all moorings were tended regularly and kept taut. The experienced ship-handler would reduce speed when passing the moored vessel to eliminate the possibility of parting her mooring lines.

Another telltale sign, again in a narrow channel such as a canal, may be noticed when a vessel is navigating close to the bank. As the vessel proceeds, a volume of water equal to the ship's displacement is pushed ahead and to the sides of the vessel. The water reaches the bank and rides up it. Once the vessel has passed, the water falls back into the cavity in the ship's wake. The interaction in this case is between the hull of the ship and sides of the bank. An increase in squat may be experienced because of the loss of water under the vessel's keel. This may even bring about the vessel grounding.

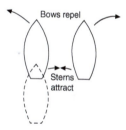

Figure 18.30 Overtaking when two vessels are passing too close to each other on parallel courses. Interaction may occur when the vessels are abeam, resulting in deflection of the bows and attraction of stern quarters, with dangerous consequences.

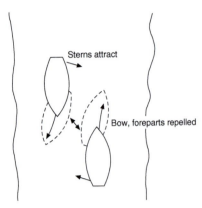

Sterns attract

Bow, foreparts repelled

Figure 18.31 Interaction between two vessels on reciprocal courses.

The period of time in which interaction is allowed to affect both vessels is limited because the pressures and water cushions created only last during the period of passing. When vessels are on reciprocal courses, the length of time that the vessels are actually abeam of each other is short (as opposed to one vessel overtaking another).

No problems arise when both vessels have ample sea room. However, in narrow channels there is the danger of grounding or collision as bows are repelled and sterns pulled towards each other.

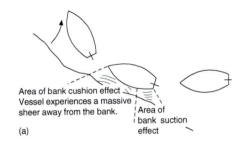

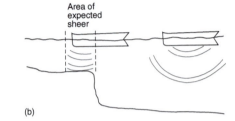

Area of bank cushion effect.
Vessel experiences a massive sheer away from the bank.

Area of bank suction effect

(a)

Area of expected sheer

(b)

Figure 18.32 Situations involving interaction.

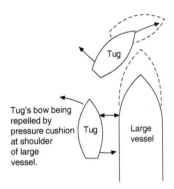

Tug

Tug's bow being repelled by pressure cushion at shoulder of large vessel.

Tug

Large vessel

Figure 18.33 Interaction between large vessel and tug.

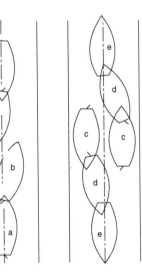

Figure 18.34 Recommended passing positions for two vessels in opposition in narrow channel.

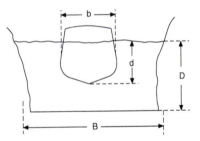

Figure 18.35 Blockage factor.

Blockage factor $= \dfrac{b}{B} \times \dfrac{d}{D}$

Example
Let b = 45′; B = 100′; d = 26′; D = 78′;

$\therefore$ Blockage factor $= \dfrac{45}{100} \times \dfrac{26}{78} = \underline{0.15}$

The effects may be reduced by a reduction in speed, provided steering is not impaired by such action.

Attention is drawn to MGN 18 regarding *Interaction between Ships*.

Shallow Water Effects and Squat

When a vessel enters shallow water, she experiences a restricted flow of water under the keel, which causes an apparent increase in the velocity of water around the vessel relative to the ship's speed. Consequently, an increase in the frictional resistance from the ship's hull will result.

If the increase in the velocity of water is considered in relation to the pressure under the hull form, a reduction of pressure will be experienced, causing the ship to settle deeper in the water. The increase in the frictional resistance of the vessel, together with the reduction of pressure, may result in the ship 'smelling the bottom'. A cushion effect may be experienced, causing an initial attraction towards shallow water, followed by a more distinct 'sheer' away to deeper water.

Where shallow water is encountered in confined waters, e.g. channels and canals, a 'blockage factor' (see Figure 18.35) must be taken into account. Ships may sink lower in the water when the blockage factor lies between 0.1 and 0.3; this, combined with a change of trim from the shallow water effect, is generally expressed as 'squat'. The result of a vessel squatting will be a loss of clearance under the keel, making steering and handling difficult.

Vessels navigating with a blockage factor between 0.1 and 0.3 push a volume of water ahead. This water, carried back along the sides of the channel to fill the void left astern of the ship, is often referred to as the 'return current'. The rate of the returning water has an effect on the ship's speed, and the maximum speed that the vessel can reach becomes a limited factor known as 'canal speed'.

Figure 18.32(a) shows interaction occurring between a vessel and a bank, sometimes referred to as a bank cushioning effect. A vessel with helm amidships may create an area of increased pressure between her hull and the bank. The result is that the vessel appears to be repelled from the bank while her stern is apparently sucked into the bank, with obvious dangers to rudder and propellers.

Figure 18.32(b) shows interaction occurring between the vessel's hull and the sea bed when in shallow water (shallow water effect). When approaching a shallow water area, a vessel may initially be attracted to the shelving or the obstruction. However, as pressure builds up between the hull and the sea bottom, the vessel may experience a sudden and decisive sheer to one side or the other. Rudder effect may also be reduced by turbulence caused by a reaction from the sea bottom.

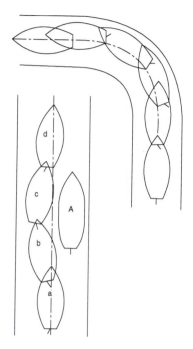

Figure 18.36 Recommended positions when rounding a bend (*above*) or overtaking another vessel (*below*) in narrow channels.

Interaction – Engaging with Tug's (Figure 18.33)

1 As the tug approaches the larger vessel to collect the towline, its bow is repelled by the shoulder of the larger vessel.
2 Counter helm is applied to correct the outward motion of the tug.
3 As the tug moves ahead under the bow of the larger vessel, it experiences an attraction to the larger vessel accentuated by the tug carrying the counter helm.
4 Unless prompt action is taken by the helmsman on the tug, the two vessels could collide, with the tug passing in front and under the larger vessel's bow, into an area of less pressure.

Influencing Factors on Squat

- The speed of the vessel.
- The rpm in relation to the 'canal speed'.
- The type of bow construction, which will affect the bow wave and distribution of pressure.
- The position of the longitudinal centre of buoyancy (LCB), near or through which the downward force of squat will probably act.

Squat may occur by the head or by the stern. If the LCB is aft of the centre of flotation, a squat by the stern would be expected; and if the LCB is forward of the centre of flotation, the vessel would be expected to settle by the head.

The strongest influence on the amount of squat will be the speed of the vessel. As a general guide, squat is proportional to the square of the speed. A reduction in speed will lead to a corresponding reduction in squat.

The limits for vessels passing when navigating in narrow channels can often be extremely fine (Figure 18.34). Both vessels are recommended to reduce speed in ample time in order to minimize the interaction between ship and ship and ship and bank. Provided a sensible speed is adopted, it should prove unnecessary to alter the engine speed while passing, thus keeping disturbance and changing pressures to the minimum as the vessels draw abeam.

In normal circumstances each vessel would keep to her own starboard side of the channel (ab and cde in Figure 18.34), and good communications should be established before the approach to ascertain exactly when the manoeuvre will start. Efficient port/harbour control can very often ease situations like this simply by applying forward planning to shipping movements.

High-Speed Craft and Safe Speed (Ref., Regulation 6, ColRegs)

It should be realized from the onset that the Collision Regulations are applicable to all vessels, inclusive of high-speed craft. This application also includes Regulation 6, 'Safe Speed', which in turn must also be construed in conjunction with the other relevant remaining regulations.

The question of what constitutes a safe speed for a high-speed craft must therefore be considered in the same light as a conventional vessel would consider safe speed. The fact remains that a high-speed vessel must still retain the ability to move out of trouble just as a conventional vessel needs to avoid close-quarter situations. The letter of the law within the ColRegs is designed to avoid close-quarter situations and many of these can be avoided by not only a reduction of speed but also an increase of speed.

Such a statement is not meant to be controversial, but is meant to highlight that an increase of speed can be just as effective in avoiding close quarters as a decrease in speed. Such action, however, should not be taken without long-range radar scanning beforehand, and should not be sustained for an indefinite period. Neither should a decision of this nature be made without full information regarding the immediate environment.

Bearing in mind that a high-speed craft, like 'hydrofoils' on the 'plane', moving at over 40 knots, which then encounters poor visibility may reduce to, say, 15 knots. In so doing her mode changes to that of full displacement, and she can no longer assume the same manoeuvrability as when she is operating at increased speed.

The use of high speed, in good visibility, can and is well used to take early action to avoid close-quarter situations. However, once poor visibility is encountered, Watch Officers must be aware of the need to be able to stop their vessel within half of the visible range. Again this option is not being advocated by this author; on the contrary, to bring the vessel to a dead stop can in certain circumstances be more hazardous than maintaining ship manoeuvrability. What is being highlighted is that stopping, or increasing speed, are alternative actions to decreasing speed and should not be dismissed out of hand. They are and remain, options, and the circumstances of each scenario will dictate what is considered prudent at the time.

Watch Officers are reminded, however, that Regulation 6 is not a stand-alone regulation, and the ColRegs also stipulate that:

> Assumptions should not be made on the basis of scanty information, especially scanty radar information.

Further reading on the operation of high-speed craft may be obtained from the high-speed craft code and from the author's sister publication, *Marine Ferry Transports: An Operator's Guide*.

Night vision equipment

The need to detect targets, day and night and in all weathers, has been the classic challenge to virtually all mariners. With infrared imaging devices these problems have to some extent been relieved and these have been developed as an asset to the Watch Officer at sea. Litton Marine Systems have developed a typical example known as AMIRIS, the Advanced Maritime Infrared Imaging System, which allows day- or night-time detection, in all visibility conditions.

The camera unit is built as an all-weather unit and comes equipped with the following facilities: two axes joystick control, 360° auto or manual pan viewing, window wipe/wash control, manual focus, video interface, recording ability, colour monitor and radar interface capability.

Additional uses of the equipment can be seen as with ship security alongside in port, navigation in ice regions, a high-speed aid to navigation, enhanced radar target information and underway security in poor visibility.

Working with Tugs

The function of the tug is to assist the pilotage of a vessel. This function has brought many types of tug into being, the most common being the oceangoing tug and the smaller dock tug (Figure 18.37 and Plate 174). Extensive use of supply vessels in the dual-purpose role of supply and towing have caused design and construction firms to add towing facilities to many supply vessels.

The very nature of the employment of tugs underlines the fact that tremendous weight and stresses have come into play, with consequent risk to operators. Many accidents have occurred in the past on mooring operations, and a considerable number of these have been during the use of tugs and their towlines.

Safe Handling of Towlines

- Seamen should never stand in the close vicinity of a towline when stress is seen to be in the line.

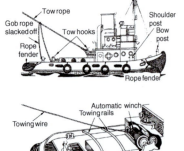

Figure 18.37 Dock tug.

Plate 169 A harbour tug assisting the manoeuvre of a vessel in confined waters.

- Towlines should always be let go in a controlled manner (by use of rope tail from wire eye) to ensure that the tug's crew are not endangered.
- Sharp-angled leads should be avoided.
- Chafe on towlines should be avoided, especially over long periods, by parcelling the towline and lubricating any leads employed. Means of adjusting the length of the towline to avoid continual wear and tear, or in the event of bad weather, should be provided.
- It is not considered good seaman-like practice to secure the eye of a tug's wire over the vessel's bitts. The control of the station is then passed to the tug, and the ship becomes dependent on the tug's Master to come astern. Effectively this eases the weight on the towline and allows the ship's personnel to slip the tow. However, in an emergency, if the eye had been secured over the bitts, the ship's personnel would not have been able to release the towline.
- When a ship's tow rope is released from a stern tug, in the majority of operations main engines should be turning ahead. The screw race will tend to push the towline well astern and clear of the propeller. This also occurs with a towing wire when fitted with a nylon pennant. The majority of man-made fibre ropes float as they are stretched astern, providing the officer on station with more handling time to bring the towline aboard without fouling the propeller.
- After any towline has been secured by turns aboard the vessel, the weight should be taken to test the securing before the start of actual towing operations.
- Efficient communications should be established between the bridge, the tug, and the officer on station, before starting the tow.

Girting or Girding

This is a term used to describe a tug being towed sideways by the vessel she is supposed to be towing. The danger arises when the towing hook is close to midships. The height of the towing hook is an important factor, as are the speed and rate of swing of the towed vessel.

This situation could be extremely dangerous if the tug's gunwales are dragged under by the force of the vessel under tow acting on the towline, especially if the weather deck of the tug has open hatchways. If in an emergency the tug's stern cannot be brought under the towline very quickly, the tow should be slipped (see Figure 18.38).

Long-Distance Towing

Should a vessel have to be towed, owing to engine failure or some other reason, then she will require secure towing arrangements

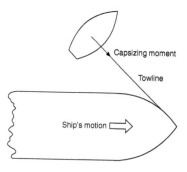

Figure 18.38 Girting or girding.

Plate 170 Three hydrofoils of the Virtu Ferry Company lie moored alongside each other in Valetta harbour, Malta. When in confined waters such craft revert to full displacement mode and their manoeuvring speeds are reduced accordingly.

Plate 171 Seamen handle the ship's stern towline on the after deck of the tug.

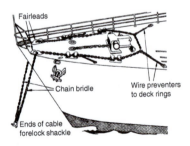

Figure 18.39 Method of securing chain bridle.

aboard. Experience has shown that if an efficient method of securing is established at the beginning of the towing operation, considerable time and effort will be saved at a later date in the event of the towline parting.

One suggested method of forward securing is by means of a chain cable bridle, constructed from small chain cable if available (Figure 18.39).

Preventer wires or relieving tackles, with the weight taken up, should be secured to the bight of the bridle before the towline is secured to it by a heavy-duty shackle. Ample grease or other lubricant should be applied to the fairleads and bollards which are expected to take the full weight of the bridle once it is connected to the towline.

The bearing surface of the chain bridle could be adjusted if relieving tackles are used instead of preventer wires, and that would prevent continuous chafe at any one point on the bridle. Lubrication and stress on the bridle should regularly be checked, but personnel should in general avoid the vicinity of the towline and bridle when weight is being taken up by the towing vessel.

The preparation of the chain cable bridle is a lengthy one and mariners should take account of the manpower required and the time to complete the operation before expecting to get underway. Securing the bridle is a lengthy process even in ideal weather conditions, but should the towline part, say in heavy weather, the mariner may find the task of resecuring the tow even more difficult.

Examples 1 and 2 in Figure 18.40 show use of composite towlines employing tug's towing spring and the towed vessel's anchor cable. Example 3 is more appropriate for the smaller type of vessel.

NB. It is unlikely that commercial vessels would carry spare chain (anchor cable would be too heavy to manhandle) to rig a chain bridle. However, it could be delivered by the oceangoing tug on arrival.

Plate 172 Ro-ro ferry manoeuvring inside harbour limits in Klaipeda, Lithuania (source: Shutterstock).

Plate 173 Gob rope in use with a ship's towline on the afterdeck of a docking tug. Tension is achieved in the gob rope by means of a centreline capstan. The use of the 'gob rope' generates a shewing motion rather than a capsize motion on the tug.

Plate 174 Large passenger vessel
engaged in berthing portside to with tug
assistance (source: Shutterstock).

Alternative Towing Methods

See Figure 18.40.

1   The towing vessel's insurance wire towing spring can be
     combined with the anchor cable of the vessel under tow.
     The wire from the towing vessel can be secured about the

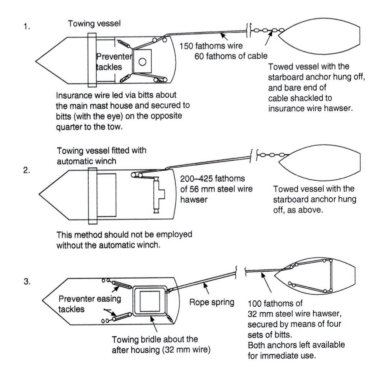

1.

Towing vessel

150 fathoms wire
60 fathoms of cable

Preventer
tackles

Insurance wire led via bitts about
the main mast house and secured to
bitts (with the eye) on the opposite
quarter to the tow.

Towed vessel with the
starboard anchor hung off,
and bare end of
cable shackled to
insurance wire hawser.

2.

Towing vessel fitted with
automatic winch

200–425 fathoms
of 56 mm steel wire
hawser

Towed vessel with the
starboard anchor hung
off, as above.

This method should not be employed
without the automatic winch.

3.

Preventer easing
tackles

Rope spring

100 fathoms of
32 mm steel wire hawser,
secured by means of four
sets of bitts.
Both anchors left available
for immediate use.

Towing bridle about the
after housing (32 mm wire)

Figure 18.40 Towing methods.

aft mast housing, the deck house or the poop itself. Sharp leads will need to be well parcelled and protected by wood to prevent chafe and the tow parting. The main disadvantage of using the anchor cable of the towed vessel is that the anchor is usually hung off at the shoulder, and the vessel under tow cannot use this anchor in an emergency. This fact may not seem important at the onset of the tow, but the anchor could play an important role in reducing the ship's momentum once the destination has been reached. The obvious advantage of employing the anchor cable is that the length of towline can be adjusted by direct use of the windlass. The anchor may remain in the hawse pipe, with the cable passing through the centre lead (bullring if fitted).

2 An alternative method of towing is possible when the tug is fitted with an automatic winch. The handling of the towline is made relatively easy once the cable or chain bridle of the vessel under tow has been secured. The lengthening and shortening of the towline is carried out by manual operation of the winch, while the tension in the towline is controlled automatically under normal towing conditions. This method should not be attempted by vessels using a conventional docking winch, as the additional strain brought to bear on the axis of the winch could render it inoperative.

3 A wire towing bridle can be used. In this method the towing bridle is secured to the vessel doing the towing operation, not the vessel being towed. This bridle is rigidly secured in position by preventer tackles and set around the after housing (poop area). Sharp corners should be well parcelled to prevent chafe and lubricants applied to bearing surfaces of the towline whenever necessary.

4 A combination of 'rope spring' and steel wire hawser is employed, with the wire hawser being secured around four sets of bitts. The main advantage of this method is that both anchors are left ready for use, but adjusting the length of the towline can prove a lengthy and sometimes dangerous task.

Composite Towline

The 'composite towline' is established by use of the ship's anchor cable being secured to a towing spring from the towing vessel. In order to connect the two, it may be necessary to hang the anchor off and bare the end of cable. The anchor can be 'hung off' in the hawse pipe by its own securing devices if the vessel is fitted with a 'centre lead' in the bows. Alternatively, with no centre lead the anchor could be 'hung off' at the shoulder, from the break of the forecastle head.

The Master may decide not to hang the anchor off at all but leave it *in situ*. This would then provide a steep catenary to the towline and could provide a dampening effect and reduce yawing movement of the towed vessel.

Figure 18.41 Composite towline.

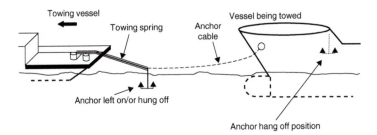

Use of Two Tugs

This method (Figure 18.42) has the obvious advantage of giving more power on the towlines and increasing the speed of the tow. However, the expense of employing two tugs instead of one is considerable, especially if one tug can manage the job, though taking a little longer. Certain heavy ULCC and VLCC vessels would, of course, need two or more tugs.

The use of two tugs, one off each bow, has the effect of reducing the yaw of the vessel under tow. Towlines secured on each side will vary in length and construction but should be such as to lead approximately 30° away from the fore and aft line of the parent vessel. This method is often used for towing floating drydocks and the like as it achieves greater manoeuvrability.

Emergency Towing Arrangements for Large Tankers

In November 1983 the IMO adopted resolution A535(13) regarding emergency towing arrangements applicable to new tankers of

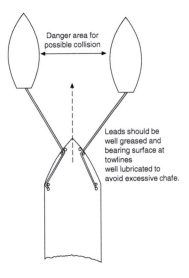

Figure 18.42 Towing by two tugs.

Plate 175 Anchor bitts and forward centre lead arrangement to accommodate emergency towing arrangement.

20,000 grt or over. Amendments now make emergency towing arrangements compulsory for all tankers of 20,000 grt or over.

The new amendment for all ships over 500 grt and passenger ships capable of carrying more than 12 passengers enters into force on 1 January 2010. Ref. MSC 1/Circ 1255. SOLAS CH II Reg 1/3-4 Emergency Towing Procedures.

Major Towing Components

For the tug or towing vessel. The towline, a pennant, a chafing chain, a fairlead and a towing gear connection or strongpoint.

For the vessel being towed. A system which facilitates ease of connection which is capable of connecting and releasing aboard the towed vessel in the absence of main power. A standardized point of connection between the towline and the chafe chain should be used.

The Applied Requirements

The requirements to fit emergency towing arrangements has changed effective from 1 January 2010. Legislation now requires that all ship types (not just tankers) over 500 grt and all passenger vessels carrying 12 or more passengers must be fitted with emergency towing arrangements. These arrangements must ensure rapid deployment both in the fore and aft positions.

The aft arrangement

This should be pre-rigged and be capable of being deployed in a controlled manner, inside harbour conditions in not more than 15 minutes. The pick-up gear for the aft towing pennant should be designed for at least manual deployment, by one man, assuming the absence of any power in adverse weather conditions which might prevail at the time.

The forward arrangement

This should be capable of being deployed in harbour conditions in not more than a period of one hour. The design should be such as to be at least capable of securing the towline to the chafing chain using a suitable pedestal roller lead.

Author's note: The main difference between the forward arrangement and the aft arrangement is that the aft fitting has a towing pennant and pick-up gear. The forward element only requires the chafing chain from the strongpoint to the pear-link on the end of the chafing chain. Many ships are in fact fitting the complete arrangement with the towing pennant forward and aft inclusive and relieving the tug from engaging with his towline at the forward end.

NB. Forward emergency towing arrangements which comply with the aft arrangement conditions may be acceptable. In either case, the chafing chain should be stowed in such a manner as to permit rapid connection to the strongpoint.

Purpose of provision

The purpose of the emergency towing arrangements fitted to ships is to facilitate a quick and easy way to take a vessel under tow in the event that she starts to founder.

The popular anchor point employed is commonly known as a 'Smit Bracket', which allows the crew to deploy the chafing chain and/or towing facility for potential pick up by tug(s) when the vessel can be held in a position away from a lee shore danger. This provision was initially a requirement for tankers (1983). However, the guidelines were accelerated with loss of the *Braer* off the Shetland Islands, in 1993.

Location and Geometry

The position of the strongpoint and fairlead should be such as to allow towing from either side of the bow or stern. The axis of the towing gear should be, as far as practical, parallel to and not more than 1.5 m either side of the centreline. The fairlead should be positioned in relation to the strongpoint so that distance between them is not less than 2.7 m and not more than 5 m so sited that the chafing chain lies approximately parallel to the deck when under strain.

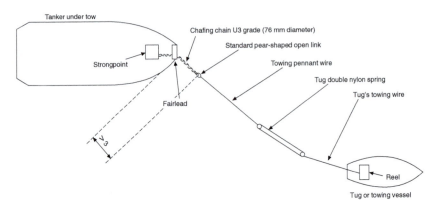

Figure 18.43 Example of emergency towing arrangements for large tankers. Additional reference: Volume III, *Command Companion of Seamanship Techniques* by D.J. House.

Plate 176 Bow section showing centre panama lead (bullring) which can be employed for towing and could eliminate the need to hang an anchor off at the shoulder (source: Shutterstock).

19
Tanker Operations

Introduction

A large percentage of merchant fleets have considerable tonnage in the way of tanker-type vessels, the majority of which are carriers of crude oil and the associated oil products. Oil cargoes are usually carried in very large crude carriers (VLCCs) up to 350,000 deadweight tonnage (dwt). Other refined products from the oil industry, e.g. diesel oils, kerosene and gas oils, are normally carried in the smaller product carrier vessels. It is accepted practice that the majority of vessels carry the same type of cargo year after year. This avoids contamination of different categories of oil, and the need to carry a variety of gear for different cargoes, and more than one pumping arrangement.

The obvious hazards in the carriage of these cargoes are fire with or without explosion, the emission of toxic vapours and oil pollution. Tanker personnel need to understand the following points about the dangers of their calling.

Tanker Vessels

General Petroleum Tankers

Crude oil carriers, mainly the larger type of tanker, comprise:

- supertankers of 50,000–160,000 tonnes;
- VLCCs of 160,000–300,000 tonnes;
- ultra large crude carriers (ULCCs) of 300,000 tonnes and over.

Refined Product Carriers

Mainly smaller tankers up to 50,000 tonnes, these are divided into carriers of *clean oils* such as motor spirit, naphtha, kerosene and gas oil; and carriers of *black oils* such as fuel oil, diesel oils and furnace oils. Carriers of one category never move into the other.

Plate 177 The largest tanker in the world is the *Jahre Viking*, seen manoeuvring with tugs outside the Dubai drydock. The *Jahre Viking* has now been converted to a floating storage and production unit.

Specialized Carriers

Vessels falling into this category generally require specialist construction and operating procedures. Examples include gas carriers, chemical carriers, liquid (molten) sulphur carriers, bulk wine carriers and bitumen carriers.

Tanker Hazards and Precautions

Fire and Explosion

The most dangerous condition of an oil tank is when the cargo has been discharged, and before any tank cleaning and gas freeing has been carried out. When a cargo tank is full, the possibility of fire is present but explosion risk is quite small since the air/hydrocarbon vapour atmosphere above the oil is small. When the tank is empty, however, the air/hydrocarbon vapour atmosphere is at its maximum. If the atmosphere is within the flammable limits for air/hydrocarbon mixtures, the tank is extremely susceptible to explosion.

Oxygen analyser and explosimeter

Oxygen analysers come in several forms. Some are electrically operated and can give continuous readings of oxygen content of the atmosphere being sampled. Others measure oxygen content of the sample by chemical reaction and will only last for a limited number of tests before renewal of the chemical is required. These oxygen analysers are used to check whether there is sufficient oxygen in an enclosed space to support life, as well as in other tanker operations where oxygen content is required.

An explosimeter is designed to measure the flammability of a gas sample. The instrument is calibrated to read from 0 to 100 per cent of the lower flammable limit of the gas sample. For most air/hydrocarbon atmospheres in normal tankers the lower flammable limit (LFL) is about 1 per cent hydrocarbon gas by volume, while its upper flammable limit (UFL) is about 10 per cent hydrocarbon gas by volume.

Fixed fire-fighting systems

These systems are permanently installed systems with a specific coverage and function. They take many forms, of which the following are some of the more commonly used.

Fixed foam systems utilize a centralized foam tank to supply a fixed system covering a specific area that may be susceptible to oil fires. Typical areas covered by fixed foam systems are engine-room and boiler-room bilge areas and pump-room bilge areas. A tanker's main cargo deck is often covered by fixed foam monitors supplied from a centralized foam tank. The most commonly used foam is of a protein-based type which uses water as its drive and combining agent. Specialized foams such as Hi-Ex (high-expansion) foam may be found but are not, as yet, the most common type.

Foam is the best agent for extinguishing oil fires as it floats across the surface of burning oil and thus reaches wherever the oil itself reaches. Once the surface of the oil is covered, the fire is effectively smothered. The foam blanket must be maintained until the oil temperature has fallen sufficiently to reduce the risk of re-ignition.

Fixed CO_2 (carbon dioxide) systems utilize a centralized CO_2 room containing the required number of CO_2 gas cylinders to provide coverage for the area being protected. The CO_2 bottles are pressurized and thus provide their own drive when released. The CO_2 is used to displace air from a fire, thus smothering it. CO_2 is a non-conductive agent and can thus be used in the area of electrical switchboards, and is also used for pump-room smothering, though electrostatic charging has to be guarded against.

Fixed BCF (bromochlorodifluoromethane) is a vaporizing liquid which is an excellent smothering agent, but is costly and, like CO_2, is only usable once and is not fitted to many vessels on the large scale employed with most fixed systems. It is sometimes found as a fixed system in emergency generator rooms of the diesel-operated type.

Fixed dry powder systems are frequently found in emergency diesel generator rooms. The dry powder smothers the surface of the fire; it is non-conductive and may be used on electrical equipment, though it may cause damage due to its abrasive properties.

Water wall systems are often fitted at the front of tanker deckhouses to stop radiant heat affecting the accommodation owing to a cargo tank fire. Water is pumped through spray nozzles high on the accommodation front and falls as a water curtain.

Inert gas is not specifically a fire-fighting system but may be utilized in cargo tanks. It is often possible to fit a portable bend to allow inert gas into a pump-room space.

Portable fire-fighting equipment

Small portable fire-fighting equipment is designed to act as first aid in the event of a fire. If a fire is tackled at an early stage, it may be put out by small, portable equipment. If a fire is larger than can be handled by small, portable equipment, then larger equipment requiring greater manpower must be used or one of the fixed systems may be resorted to, if available. Small, portable equipment is supplied in many types similar to the fixed systems except in their size.

Common sources of ignition

- *Smoking*: particularly dangerous are striking matches or lighters.
- *Galleys*: oil or electric, i.e. burners, toasters, etc.
- *Accommodation electrical equipment*: this is not usually designed to be gastight, so flammable gas must be kept out.
- *Metals*: ferrous and non-ferrous metal tools should be used with care. Non-ferrous metal tools are only marginally less likely to cause incendive sparks and have other drawbacks such as the danger of 'thermite' spark or flash which may be incendive in the presence of flammable gas. *Spontaneous combustion* can occur with most combustible organic materials, some being more susceptible than others. Auto-ignition was previously discussed, and is a serious hazard where oil may come into contact with hot surfaces (see pp. 338–339, 536–537 and 541).
- *Ship/shore electrolytic*: may occur in sufficient intensity to cause incendive sparks at a ship's manifold and insulating pieces are frequently used to reduce this risk.
- *Static electricity*: can be generated by fluid flow.
- *Lightning*: thunderstorms or electrical storms should cause a cessation of operations while they are in the vicinity of a tanker.

Intrinsically safe

A circuit, or a part of a circuit, is intrinsically safe when any spark or thermal effect, produced normally or accidentally, is incapable, under prescribed test conditions, of causing ignition of a prescribed gas or vapour.

Tanker Safety Guides

The *Tanker Safety Guide* issued by the International Chamber of Shipping should be available for reference on all tankers.

Plate 178 The BP tanker *British Reliance* being escorted and manoeuvred by four escort tugs.

Tanker owners also issue their own rules and regulations and these should be read and adhered to.

Emergency plans and procedures

Emergency plans and procedures to cover all foreseeable situations should be drawn up and then practised at regular intervals. On first joining a ship, the prudent mariner will familiarize himself with the plans and procedures for the ship and his position in those plans. Careful attention should be paid during practice sessions so that the need to think in an emergency is kept to a minimum; everyone should automatically do the correct thing even under the stress of a possible panic.

Ship/shore checklist

Table 19.1 Ship/shore (tanker) officer's example checklist

VESSEL BERTH No.	DATE TIME			
1 Are SMOKING regulations being observed?				
2 Are GALLEY requirements being observed?				
3 Are NAKED LIGHT requirements being observed?				
4 Are electric cables to portable equipment disconnected from power?				
5 Are the ship's main transmitting aerials switched off?				
6 Are hand torches of an approved type?				
7 Are portable R/T sets of approved design?				
8 Are all external doors and ports in the amidships accommodation closed?				
9 Are all doors and ports in the after accommodation that are required to be closed in fact closed?				
10 Are ventilators suitably trimmed with regard to prevailing wind conditions?				
11 Are unsafe air conditioning intakes closed?				
12 Are window-type air conditioning units disconnected?				
13 Is the ship securely moored and agreement reached on use of tension winches?				
14 Are cargo/tanker hoses in good condition?				
15 Are cargo/bunker hoses properly rigged?				
16 Are unused cargo/bunker connections blanked?				
17 Is stern discharge line (if fitted) blanked?				
18 Are sea and overboard discharge valves (when not in use) closed and lashed?				
19 Are scuppers effectively plugged?				
20 Is the agreed ship/shore communication system working?				
21 Are all cargo/bunker tank lids closed?				
22 Are cargo tanks being loaded or discharged open to atmosphere via the agreed venting system?				
23 Are fire hoses and equipment ready for use?				
24 Are emergency towing wires correctly positioned?				
25 Is the ship ready to move under its own power?				
REMARKS:				

We have checked with each other the items on the above checklist and have satisfied ourselves that the entries we have made are correct to the best of our knowledge.

CHECKED BY:

            (for ship)   (for terminal)

Dangers of Petroleum Spirit

Should petroleum ignite, it is the vapour given off by the liquid that burns, not the liquid itself. However, temperatures are such that the liquid vaporizes quickly. As with all fires, the vapour will only burn if the air/oxygen supply has access to the fire. The volume mixture of vapour to air defines the upper and lower explosive limits, which normally lie between 1 and 10 per cent.

Once a liquid is burning, volumes of gas will be given off. The consequences of the build-up of this gas, especially in an enclosed space, could be disastrous and lead to explosion. Risk of explosion is more likely with the expansion of the gas within the space.

Petroleum vapour can have a variety of effects on the human body, depending on the quantity. Some types of vapour are toxic, and if inhaled in a large enough quantity could prove fatal. If a lesser quantity is inhaled the person exposed may develop the symptoms of a drunken state. A person's sense of smell may also be affected by petroleum vapour, so reducing the body's warning systems for detecting the presence of gas.

General Definitions

Clean Ballast

This is water carried in a cargo tank that has previously been thoroughly washed, as have the pipelines serving it.

Dirty Ballast

This is water carried in an unwashed tank or ballast that has been contaminated with oil from another source, e.g. loading through dirty pipelines will contaminate ballast.

Flashpoint (of an Oil)

The lowest temperature at which the oil will give off vapour in quantities that when mixed with air in certain proportions are sufficient to create an explosive gas.

Gas Lines (Gas Freeing)

When loading, the air or gas originally inside the tank must be allowed to escape or pressurization will occur. The gas lines connect at the top of the tank and allow this air or gas to escape.

Ignition Point (of an Oil)

This is the temperature to which an oil must be raised before its surface layers will ignite and continue to burn.

Inert Gas

This is gas that is low in oxygen content obtained from the funnel uptake or an inert gas generator. It is pumped into tanks while cargo is being discharged to reduce the risk of fire or explosion.

Manifold

This is the point on either side of the vessel where the ship's pipelines are connected to shore pipelines. It is usually about the midships point. The valves on the ship's pipelines near the intersection with the manifold are known as the manifold valves.

Permanent Ballast

This is water carried in tanks specifically designed for its carriage, known as permanent ballast. There must be no possibility of the tanks being connected to the oil cargo system. This ballast is not to be confused with the permanent ballast in dry cargo vessels, e.g. concrete blocks.

Pipelines

These are the pipes used to move cargo around, both when loading and discharging and for any other transfer operation undertaken.

Sounding

The distance from the bottom of the tank to the surface of the liquid in that tank is the sounding. Ullage plus sounding equals the depth of the tank.

Ullage

The distance from the measuring point at the top of the tank to the surface of the liquid in that tank.

Volatile liquid

This is liquid that has the tendency to evaporate quickly, and has a flashpoint of less than 60°C.

Cargo-Handling Equipment

Automatic Tank Ullaging Gauge

An automatic gauge sited at the cargo tank and giving continuous readings of ullage, this may be modified to give a remote reading in a control room. It may be operated by spring-tensioned tape and float, by pressure tube, by ultrasonic transducer or any other approved device.

IMO adopted resolution A.774(18) based on 1991 guidelines for ship's ballast water

All participating ships must implement a ballast water and sediment management plan and carry a ballast water record book. Many countries already comply and the UK is ratifying the agreement in 2012. New ships and existing vessels are now being fitted with an approved ballast water management system.

Flameproof Gauze

Portable brass gauzes are used to cover open sighting ports to stop any flame or spark from flashing back into the tank. Permanent types are fitted in gas lines to prevent flashbacks.

Linen Tape and Brass Weight or Wood Block

Hand-operated, this tape may be used for ullaging or sounding when fitted with a brass weight, but only for ullaging when fitted with a wood block. This type is prone to slight errors, owing to stretch, particularly when the tape is old.

Sighting Port Spanner

The sighting port on the top of a tank hatch may have various types of closing arrangement which usually require a special type of spanner (see Figure 19.1).

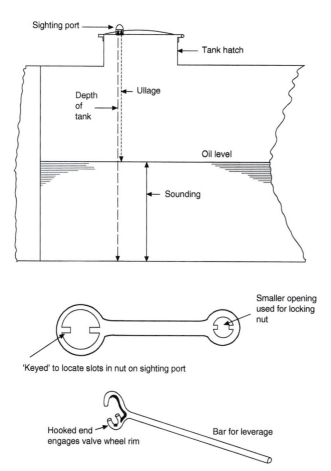

Figure 19.1 Tank measurements (*above*) and port spanner and wheel key *(below)*.

Sounding Rod

This rod is usually about 1 m in length and made of brass. It may comprise several short lengths linked together for use in sounding tubes that have bends in them, or it may be solid for other tanks and straight tubes. The rod is attached to a suitable length of line, and may only be used for soundings. Solid rods are frequently used for final 'dips' of oil cargo tanks to find the amount of residues remaining.

Specific Gravity (SG) Glass and Hydrometer

This is a tall sample glass into which is poured a sample of oil from the tank. The hydrometer is then used to measure the specific gravity of the oil, which is used, in conjunction with the oil's temperature and volume, in obtaining the weight of the cargo.

Steel Tape and Brass Weight

Hand-operated, this combination may be used for ullaging or sounding as required. It must not be used where electrostatic charges are present in tank atmospheres, as lightning-like arcing can be a source of ignition.

Temperature Can and Thermometer

The temperature can is made of brass with a narrow neck and weighted base so that it will sink on entering the oil and thus fill with oil. The can is left in the oil for sufficient time to reach the same temperature as the oil, it is then hove up and a thermometer is used to find the temperature. In some cases the thermometer is left in the can when it is in the oil. The can may be adapted for use in obtaining samples of the oil, though a similar but larger sampling can is generally used for this purpose. The temperature of the oil cargo is used in calculating the weight of the cargo.

Ullage Stick

This simple graduated stick about 2 m in length has a cross-piece at the zero graduation mark. The cross-piece rests on the coaming of the ullage opening (usually the sighting port). The stick is used when open loading to finish off a tank, as it only measures the last 2 m of ullage.

Wheel Key

A metal bar with a hook type of end, the wheel key engages with the rim of a valve wheel so that increased leverage is obtained for opening or closing the valve (see Figure 19.1).

Tank measurements are shown in Figure 19.1.

Whessoe Tank Gauge

The function of the gauge (Figures 19.2 and 19.3) is to register the ullage of the tank at any given time, in particular when the liquid level in the tank is changing during loading or discharging. The gauge is designed to record not only at the tank top, but also in a central control room, a transmitter being fitted to the gauge head for this purpose. A particularly useful addition to oil tankers with numerous tanks, it allows the reading of all tanks to be carried out at one central control room.

The unit is totally enclosed and of rugged construction in a non-ferrous metal. Inside the housing is a calibrated ullage tape, perforated to pass over a sprocket wheel and guided onto a spring-loaded tape-drum. The tape extends into the tank and is secured to a float of critical weight. As the liquid level rises or falls, the tape is drawn into or extracted from the drum at the gauge head.

The tape-drum is internally spring-loaded and provides a constant tension in the tape, at the float connection, regardless of the amount of tape paid out. A 'counter window' for displaying the tape and fitted at the gauge head allows ullage readouts at any time. (The counter chamber is oil-filled.) Located inside each tank are a pair of guide wires, each secured to an 'anchor bar' welded to the tank bottom. The upper end of the wires is secured to cushion springs beneath the gauge. When the level indicator is not in use, the float is stowed in a locked position under the gauge head. An automatic lock-up arrangement of the float is achieved once the float is raised to its full extremity. Should this fail, the float can be secured manually.

General Operations and Procedures

Procedure when Loading

Clean ballast is discharged overside, dirty ballast is pumped ashore. Tanks should be stripped as dry as possible; they are usually inspected by shore representatives before loading begins. Some tanks may remain empty, and these will usually be about midships. During discharge of clean ballast, flexible shore hoses are connected to the ship's manifold. After tank inspection, ship's lines are set, and loading is started slowly into one tank. When cargo is seen to be coming on board, more tanks are opened and the loading rate increased. A few minutes before completion, 'stand by' is given to the jetty man, and the rate is slowed down. The shore personnel are told to stop operations prior to required cargo levels being reached.

Procedure when Discharging

Ullages and such are taken and the quantity of cargo on board is calculated. Ship's lines are set for cargo discharge. Simultaneously with these operations, flexible shore hoses are connected to the ship's manifold. When shore indicates that it is ready to receive, the

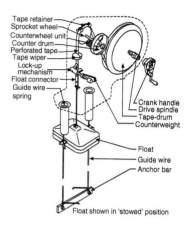

Figure 19.2 Whessoe tank gauge.

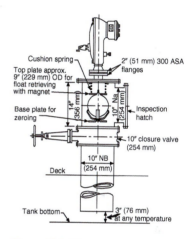

Figure 19.3 Whessoe tank gauge (Model No. 3303) for use with liquefied gas cargoes.

appropriate tank suction valve and the manifold gate valve are opened. Pumping is then started slowly, a careful watch being kept on the back pressure gauge (indication of valve shut ashore). When it has been established that cargo is being discharged, further pumps may be started and more tanks opened (subject to limitations made by shore).

Ballasting

When a tanker has no cargo to carry on a voyage, or there is insufficient cargo to provide enough sinkage for safety and to ensure the propeller is underwater, then the tanker can take ballast on board from the water in which it is floating.

Tank Cleaning

Where, for any reason, a cargo tank requires cleaning, then modern practice is to use fixed or portable tank cleaning machines which automatically direct high-pressure jets about the tank in order to wash off oil and oil residues. The washings fall to the bottom of the tank and are then pumped to a special tank called a slop tank, where they are retained until it is possible to discharge them to shore facilities (see Figures 19.4 to 19.6).

Safe Entry of Enclosed Spaces

Before one can enter a tank, it must be washed to remove oil so that no further gas can be generated. The tank must then be ventilated, which is usually accomplished by the use of water-driven portable fans or by a large fan in the engine room whose air can be directed to the tank required by the ship's gas lines. Even where a tank has had no oil in it and is clean, the atmosphere must be changed for fresh air, as even rust in a tank will extract vital oxygen from the air in it. Where inert gas has been used, this must be blown out, as it contains far too little oxygen to support life. See Guidelines on p. 685.

Action on Operating Failure

Where *any* operational failure occurs and no one of appropriate authority is present, then the safest course of action is to *stop* operations immediately. Any inconvenience caused by this action is negligible when compared to the possible consequences of continuing in a dangerous state, e.g. fire, explosion, overflow and/or pollution.

Duties of Officer of the Deck During Loading or Discharging

1 Tend moorings.
2 Tend the gangway and supervise the watchman's duties.
3 Calculate the rate of loading.

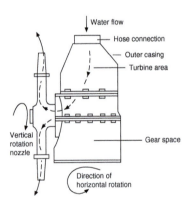

Figure 19.4 Simplified Butterworth 'K' machine. For the cleaning of small and medium tanks. General particulars are as follows: small and compact, weighing less than 50 lb (23 kg); cleaning range of 10 m; fluid capacity of 31.5 m³ per hour; wash cycle times range from 50–23 minutes for inlet pressure of 3.5–12.3 kg/cm².

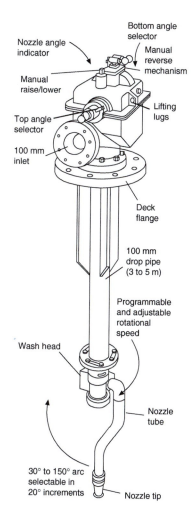

Nozzle angle indicator

Bottom angle selector

Manual reverse mechanism

Manual raise/lower

Top angle selector

Lifting lugs

100 mm inlet

Deck flange

100 mm drop pipe (3 to 5 m)

Programmable and adjustable rotational speed

Wash head

Nozzle tube

30° to 150° arc selectable in 20° increments

Nozzle tip

Figure 19.5 Butterworth 'Lavomatic SA' machine.

A deck-mounted system for crude oil washing (COW) on large, medium and small tanks, the Butterworth machine has been in use since 1973. It is a fast, economical system of tank washing which meets the standards of MARPOL, IMO and the UK marine authority.

Tank cleaning may begin as soon as the cargo level falls below the wash head. A selection of washing arcs are available from –30° to 150° at 20° increments.

The machine requires 133 wash-head revolutions for a maximum cycle from –30° to 150° and back to –30°. Programme speed can be varied automatically and the system incorporates a manual override for nozzle direction and position.

Washing times can be considerably reduced by prudent adjustment of the revolving wash head. Typical settings would be, say, 1 rpm in the harder-to-clean lower regions of the tank, and 2.5 rpm in the upper, easier-to-clean regions. The discharge rate of the machine would depend on the nozzle size, but the following examples are issued by the manufacturer:

- Inlet pressure 10 kg/cm² (150 psi).
- 38 mm nozzle discharges about 150 m³ per hour.
- 29 mm nozzle discharges about 100 m³ per hour.

4 Check fire wires are correctly rigged and ready for use.
5 Check regularly for oil pollution.
6 Ensure that no hot work, naked lights or unsafe lights are left exposed.
7 See ship to shore checklist is observed.
8 See all tank lids are closed and safe venting is being carried out.
9 Ensure oil loading into correct tanks.
10 Maintain state of heel and trim as required.
11 See that all oil hoses are supported and correctly secured.
12 Take care that electrically operated apparatus is used only if intrinsically safe.
13 Permit use of safe torches only.
14 See that means of access to accommodation are closed.
15 Ensure all scuppers are plugged and drained.
16 Have fire-fighting equipment ready for immediate use.
17 Allow no unauthorized personnel aboard the vessel.
18 See that the loading plan is observed and followed.

Figure 19.6 Washing cycle selection.

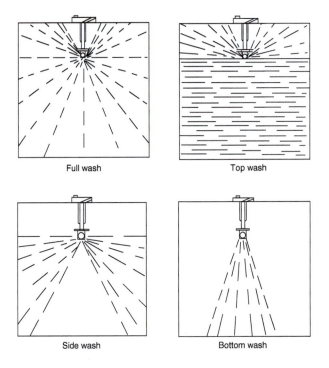

Full wash

Top wash

Side wash

Bottom wash

19 Have sufficient personnel available for topping-off purposes.
20 Ensure red light or 'B' flag is displayed.
21 Have unused lines blanked off.
22 Make sure all company and port regulations are observed.
23 Establish efficient communications with shoreside personnel.
24 Keep deck log book up to date with current entries.
25 Maintain general deck watch, especially for changes in weather.
26 Check that the inert gas system is isolated.
27 Avoid unnecessary pump-room entry. Have a standby man available when entering.
28 See that drip trays are positioned.
29 Enforce no-smoking regulations.
30 Forbid overboard discharges.

Tanker Layout and Ventilation

Figure 19.7 gives a bird's eye view of tanker layout, and Figures 19.8 to 19.10 cover ventilation.

Gas Freeing (Tanker Vessels)

Gas freeing the ship is a requirement in preparation for entry into drydock or where tank entry is a need. It is usually achieved by one of two methods:

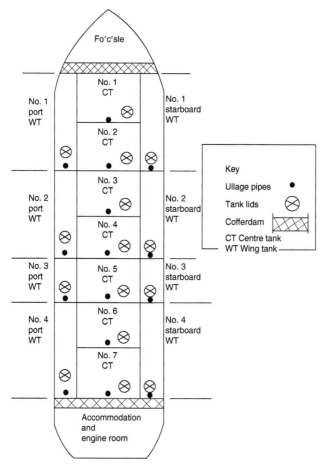

Figure 19.7 Ullage pipe and tank lid plan.

1 by means of portable water-driven fans; or
2 by an eductor driven by air or steam.

It is, however, pointed out that it can also be achieved by using the inert gas fans in conjunction with the fresh air intake, although the use of the inert gas fans is not as fast as the two stated methods. In any event, prior to gas freeing the tanks should be thoroughly purged with inert gas to bring the tank atmosphere below the critical dilution line. Previously, gas freeing used to be carried out by means of canvas 'windsails', but these have generally been superseded.

Two accepted methods are employed: either the displacement or the dilution method. The displacement method makes use of available pipe work which is open at the bottom of the tank, e.g. purge pipe or cargo line. This operation is achieved because petroleum vapours are heavier than air and tend to accumulate at the bottom of the tanks. The dilution method relies on fans to blow air into the bottom of the tank in order to 'dilute' the vapours.

Figure 19.8 Gas venting system.

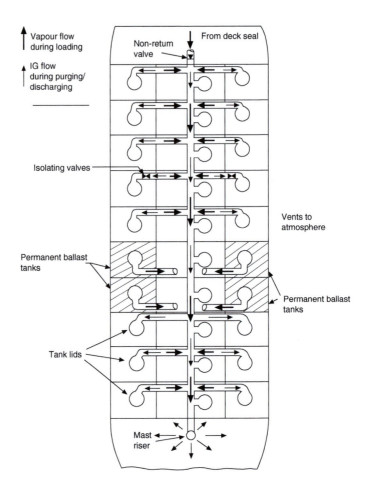

NB. The gas freeing operation should not be attempted until the tank is purged with inert gas below the 'critical dilution line', in order that the tank's atmosphere does not pass through the flammable zone with dilution by air.

Tanks should be vented during gas freeing in such a way as to produce an exit velocity which will adequately take any gaseous vapours clear of the deck area. Accommodation access points during this time should be securely closed off to prevent toxic vapours entering the accommodation areas.

Once tested, by means of 'gas detectors', the ship will be issued with a 'gas free' certificate.

Health and Safety

Accidents Due to Tanker Operations and Ship Design

Openings in the decks need to be covered and secured, which requires studs and nuts which project above deck level as much as 50 mm. These projections are usually obvious, but during night-time operations, especially at sea, they can provide a serious danger,

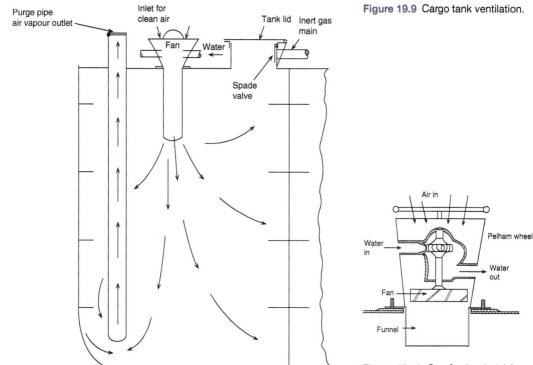

Figure 19.9 Cargo tank ventilation.

Figure 19.10 Gas-freeing (axia) fan.

Plate 179 Mimic diagram for tanker pipeline systems seen in a cargo control room of a tanker vessel.

not only of stubbed toes, but of broken toes or even broken ankles and legs. During working hours safety shoes should be worn, never loose fitting shoes or soft or open types of footwear.

Pipelines are ever-present on the decks of tankers; many are small and low and may be stepped over, but many are large and their tops may be over 1 m above deck level. The larger pipelines

Plate 180 The *Stolt Petrel*, a small coastal tanker fitted with free-fall lifeboat at the stern, navigating in restricted waters.

will have walkways constructed across them at regular intervals and these should be used. To jump on top of pipelines to cross from one side of the deck to the other is extremely dangerous, as they are usually painted with gloss paint and falls between or from pipelines can cause all types of fractures and even death. Overhead pipes and associated steelwork pose a very real threat and great care should be taken when passing under these. Safety helmets should be worn in the working environment.

Rarely does oil cargo find its way onto the tanker's decks, but much of the equipment used on the tanker may require lubricating oil and the vessel may also use hydraulic oil for operating certain machinery. Oil leaks and seepages can form an almost invisible sheen on decks, which can be like ice. If water is present also, i.e. rain or sea water, then the situation becomes even more dangerous.

During operations on deck there is often the temptation to run from one area to another, especially if urged on by other persons. Running *must* be avoided as all the various dangers are accentuated for a running person. If you are being relied upon to complete an operation and you run to do it, and fall, there is a very good chance you will be at least unconscious; everyone is then in danger since the operation is out of immediate control. If you were walking and fell, it is far less likely that you would be totally disabled and, apart from a few seconds of delay, the operation could be completed.

Inert Gas

Because inert gas is low in oxygen content, generally 5 per cent or less, it not only reduces fire hazards but also forms an asphyxiation risk. The human body is used to air containing 21 per cent oxygen and the average exhaled air is still in the region of 17 per cent oxygen; below 17 per cent content the air is no longer adequate for active life, and as the percentage falls the danger of death by asphyxiation rises. Where the presence of inert gas is suspected, the

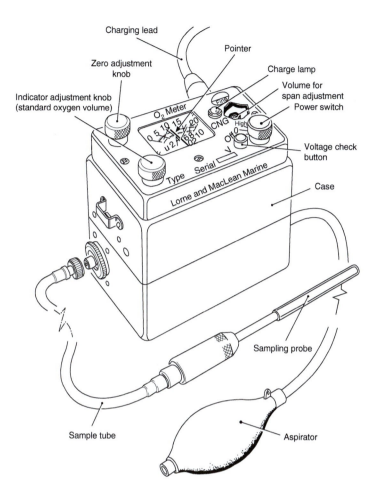

Figure 19.11 Oxygen test meter.

atmosphere should be tested for oxygen content before entry; if in doubt as to whether inert gas could or has been present, test anyway (see Figure 19.11). The compartment in question should also be continuously ventilated with fresh air.

Skin Contact with Hydrocarbons

Hydrocarbons are dangerous in many ways and in varying degrees, depending on their type. Certain hydrocarbons are thought to cause skin cancer if prolonged skin contact is maintained. Protective clothing such as gloves and boiler suit should be worn and kept clean. A dirty boiler suit increases skin contact time for any hydrocarbon with which it may be impregnated.

Inhalation and/or Swallowing of Hydrocarbons

Hydrocarbons and hydrocarbon vapours are, in varying degrees, toxic (poisonous). Hydrocarbon vapours are heavier than air and will displace air in a compartment from the bottom upwards, so that anyone entering a compartment will be at risk from asphyxiation through reduced oxygen levels and also poisoning from hydrocarbon toxicity. While a re-entry into fresh air will remove the threat of asphyxiation, the toxic effect may remain. Certain hydrocarbons, e.g. of the aromatic family, can have a permanent cumulative effect and are particularly dangerous.

Where hydrocarbon gases are present, care must be taken to reduce the inhalation of these gases. The dangers due to swallowing hydrocarbons are also severe. Not only is long-term toxicity of the body's cells a danger, but immediate permanent damage to the throat, stomach and internal organs can also result.

Where a person is working in an atmosphere thought to be gas-free but starts to show symptoms similar to a drunken state, i.e. giggling, singing, lack of coordination, general fooling around, etc. he is showing the first signs of hydrocarbon poisoning. A rescue procedure should be adopted immediately so that the person can be removed safely from the poisonous atmosphere.

Protective Equipment

Compressed air breathing apparatus (CABA) comprises a face mask supplied with air from an air bottle carried by the user. In some cases the air may be supplied via a filter from a compressed air deck line (ALBA), but the user should also have a fully charged air bottle with him which will automatically continue to supply him with air should the deck air line supply fail. This provision is necessary to allow the person time to evacuate the space he is working in.

Automatic oxygen resuscitating equipment (Rescuepac) comprises oxygen bottles with automatic metering valves that will automatically supply a collapsed person with oxygen at the correct rate. This equipment is a powerful item of rescue equipment and should always be readily available in case of mishap when work is carried out in enclosed spaces.

Escape Sets and Other Rescue Equipment

Escape sets are small emergency escape breathing device (EEBD) sets kept in positions where hydrocarbons may be released owing to operational failure, and they allow a person in that position sufficient air to effect an escape from the compartment. A typical position for an escape set would be at the bottom of a tanker's pump room.

Smoke helmets are mainly used for rescue and fire-fighting, but may also be used for working purposes. The user wears a mask connected via a pipe to a bellows that must be situated in fresh air. The bellows may be mechanically operated but is more usually foot-operated. The pipe should be no longer than a length through which the user can draw air even if the bellows fails.

Lifelines and safety harnesses, the former steel cored and the latter made of terylene webbing, should be used where necessary.

Collapsed Person in Enclosed Space

Where any person or persons are working in an enclosed space, they must have a person outside the space whose sole responsibility is to watch them working to ensure their safety. If a person is seen to collapse in an enclosed space, the alarm must be raised immediately so that a rescue team with protective equipment (Figure 19.12), resuscitating equipment, lifelines and agreed communication systems can enter the space and carry the person to the nearest fresh air source (Figure 19.13). It is essential that the observer does not enter the compartment; he must raise the alarm and entry into the space must be made only by the rescue team with the correct equipment.

A gas detector operation is shown in Figure 19.14.

Adequate steps by a responsible officer prior to allowing entry into an enclosed space must be taken to include:

- a risk assessment being conducted; and
- a permit to work being completed and issued.

Guidelines for the Use of Marine Safety Card No. 1

Responsibility for safety, both at the time of entry of any tank or other enclosed space and during the entire operation, rests with the Master or responsible officer. This responsibility covers conditions of work for shore-based employees as well as for members of the ship's crew. The Master or officer makes sure that adequate steps have been taken to eliminate or control the hazards. He must also make sure that all personnel understand the nature of such hazards which remain, and the precautions to be followed.

Enclosed spaces include any tank, cargo space or compartment in which toxic, inert, asphyxiating, flammable or other dangerous gases may accumulate, or oxygen may be deficient, such as:

1 any space containing or having last contained combustible or flammable cargo or gases in bulk;
2 any space containing or having last contained cargoes of a poisonous, corrosive or irritant nature;
3 spaces in tankers immediately adjacent to the spaces referred to in (1) and (2) above;
4 cargo spaces or other spaces that have been closed and/or unventilated for some time;
5 storerooms or spaces containing noxious or harmful materials;
6 spaces that have been fumigated.

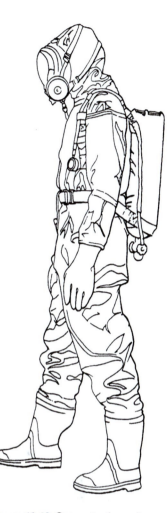

Figure 19.12 Gas protection suit. The gas protection suit shown here is designed for work in highly toxic atmospheres, e.g. in cargo tanks, etc. During such operations, the suit gives protection to the complete body. The suit is manufactured from extensible, abrasion-proof material, which is a highly durable polyester fabric, neoprene coated on one side. It is a one-piece suit, enabling the wearer to don it quickly without assistance. Entry is effected through a diagonal aperture which is sealed with a gas-tight waterproof zip fastener. The sleeves are equipped with gas-tight cuffs or may be provided with gloves. The full-vision facemask, with the universal, pneumatic seal and speech diaphragm, is bonded to the suit, allowing easy fitting for self-contained and airline breathing sets.

Figure 19.13 Pump-room rescue operation.

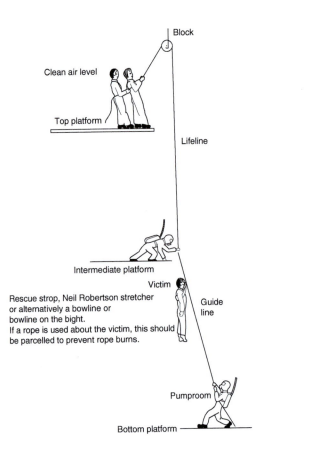

Figure 19.14 Combustion gas detector (catalytic filament type).

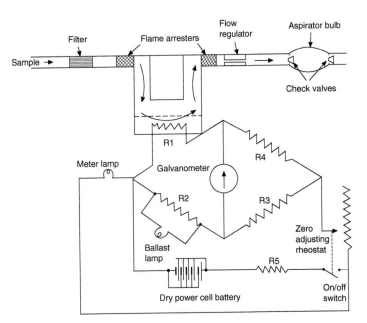

The hazards inherent when working in an enclosed space can be avoided or overcome if the following rules are applied properly each and every time a space is entered:

1   Establish a definite system of pre-planning for enclosed space entry and a crew instruction programme.
2   Prepare the space for entry by physically isolating it, cleaning it to remove contaminants and testing to ensure absence of such contaminants.
3   Use a checklist, backed up by a permit system. The checklist should only be issued to another crew member after the Master or responsible officer is satisfied personally with the precautions taken, personal protective equipment to be used and procedures to be followed.

The Marine Safety Card No. 1 (Figure 19.15) serves as a method of reminding all concerned of recommended procedures and also as a checklist to ensure that all existing hazards are considered and evaluated, and, where necessary, the correct protective measures taken. It has been designed so that it may be used on board all types of ship, from the largest tanker to a small coastal vessel. Instructions and advice listed on the card are not intended in any way to take the place of other rules and recommendations on board the ship; it is intended to reinforce these. It may also be used in conjunction with a permit-to-work system.

It is recommended that the cards be issued only when the need for their use arises. When in use, the cards should be completed properly as instructed. Any relaxation is likely to result in diminishing respect for their use, with a resulting decline in the standards of safety achieved. In order to operate successfully, the Marine Safety Card must receive support from senior ship's personnel; the response of other crew members will obviously be influenced by this.

It may sometimes be necessary for a person to enter an enclosed space that is known to contain an unsafe atmosphere. This practice should only be allowed when it is essential or in an emergency. On no account should routine work be carried out under such conditions. Section 1 of the checklist should be completed by the Master or a responsible officer and the card should then be handed to the person who is to enter the space for completion of Section 2. Section 3 should be checked jointly by the responsible officer and the person who is to enter the space on every occasion that breathing apparatus is used.

It should be remembered that rescue and resuscitation equipment should be tested at the time of inspection and check.

The card is finished with a matt surface on the checklist side. It is recommended that a soft pencil is used to make the checks. After use the card should be cleaned with a rubber, tissue or damp cloth.

The card is issued by the General Council of British Shipping.

MARINE SAFETY CARD No. 1

Entering Cargo Tanks, Pump Rooms, Fuel Tanks, Coffer-dams, Duct Keels, Ballast Tanks or similar enclosed compartments.

General Precautions

Do not enter any enclosed space unless authorised by the Master or a responsible officer and only after all the appropriate safety checks listed on the reverse of this card have been carried out.

The atmosphere in any enclosed space may be incapable of supporting human life. It may be lacking in oxygen content and/or contain flammable or toxic gases. This also applies to tanks which have been inerted.

The master or a responsible officer MUST ensure that it is safe to enter the enclosed space by:

(a) ensuring that the space has been thoroughly ventilated by natural or mechanical means; and
(b) where suitable instruments are available, by testing the atmosphere of the space at different levels for oxygen deficiency and/or harmful vapour; and
(c) where there is any doubt as to the adequacy of ventilation/testing before entry, by requiring breathing apparatus to be worn by all persons entering the space.

WARNING

Where it is known that the atmosphere in an enclosed space is unsafe it should only be entered when it is essential or in an emergency. All the safety checks on the reverse side of this card should then be carried out before entry and breathing apparatus must be worn.

Protective Equipment and Clothing

It is important that all those entering enclosed spaces wear suitable clothing and, that they make use of protective equipment that may be provided on board for their safety. Access ladders and surfaces within the space may be slippery and suitable footwear should be worn. Safety helmets protect against falling objects and, in a confined space, against bumps. Loose clothing, which is likely to catch against obstructions, should be avoided. Additional precautions are necessary where there is a risk of contact with harmful chemicals. Safety harnesses/belts and lifelines should be worn and used where there is any danger of falling from a height.

There may be additional safety instructions on board your ship, make sure that you know them.

Further information on safe entry into enclosed spaces is contained in the Code of Safe Working Practices for the Safety of Merchant Seamen and the ICS Tanker Safety Guides.

Issued by the General Council of British Shipping,
30–32 St Mary Axe, London, England EC3A 8ET.

© 1975

Figure 19.15 Marine Safety Card No. 1.

SAFETY CHECK LIST

Before entering any enclosed space all the appropriate safety checks listed on this card must be carried out by the master or responsible officer and by the person who is to enter the space.

N.B. For routine entrance of cargo pump rooms only those items shown in red are required to be checked.

SECTION 1

To be checked ☑ by the master or responsible officer

1.1   Has the space been thoroughly ventilated and, where testing equipment is available, has the space been tested and found safe for entry ?   ☐

1.2   Have arrangements been made to continue ventilation during occupancy of the space and at intervals during breaks ?   ☐

1.3   Are rescue and resuscitation equipment available for immediate use beside the compartment entrance ?   ☐

1.4   Have arrangements been made for a responsible person to be in constant attendance at the entrance to the space ?   ☐

1.5   Has a system of communication between the person at the entrance and those in the space been agreed ?   ☐

1.6   Is access and illumination adequate ?   ☐

1.7   Are portable lights or other equipment to be used of an approved type?   ☐

When the necessary safety precautions in SECTION 1 have been taken, this card should be handed to the person who is to enter the space for completion.

SECTION 2

To be checked ☑ by the person who is to enter the space

2.1   Have instructions or permission been given by the master or a responsible officer to enter the enclosed tank or compartment ?   ☐

2.2   Has SECTION 1 been completed as necessary ?   ☐

2.3   Are you aware you should leave the space immediately in the event of failure of the ventilation system ?   ☐

2.4   Do you understand the arrangements made for communication between yourself and the responsible person in attendance at the entrance to the space ?   ☐

SECTION 3

Where breathing apparatus is to be used this section must be checked jointly by the responsible officer and the person who is to enter the space.

3.1   Are you familiar with the apparatus to be used ?   ☐

3.2   Has the apparatus been tested as follows ?
    (i)   Gauge and capacity of air supply
    (ii) Low pressure audible alarm
    (iii) Face mask – air supply and tightness   ☐

3.3   Has the means of communication been tested and emergency signals agreed ?   ☐

Where instructions have been given that a responsible person be in attendance at the entrance to the compartment, the person entering the space should show their completed card to that person before entering. Entry should then only be permitted provided all the appropriate questions have been correctly checked ☑.

Figure 19.15 Continued

Inert Gas System

The purpose of an inert gas system (IGS) (Figure 19.16) is to blanket the surface of the cargo (or ballast) and prevent a mixture of air and hydrocarbons causing fire or explosion within the tank space. The gas is supplied by means of an 'inert gas generator' or extracted from 'boiler flue gases' taken from the main boiler uptakes. Remotely controlled 'butterfly valves' allow the extraction of the gas from port and starboard boiler uptakes before its entry via the scrubbing tower, demister unit then water seal (Figure 19.17) before entering the space.

The cooled, clean inert gas is drawn off from the scrubbing tower by conventional centrifugal fan units capable of delivering sufficient gas to replace cargo during discharge at the maximum pumping rate plus 25 per cent and to maintain a positive pressure at all times. The gas will enter the tank after passing through a 'deck-mounted water seal', which is specifically incorporated into the system to prevent hydrocarbon gases flowing back up the line. The deck water seal unit is fitted with a steam heater for operations in cold weather to prevent freezing.

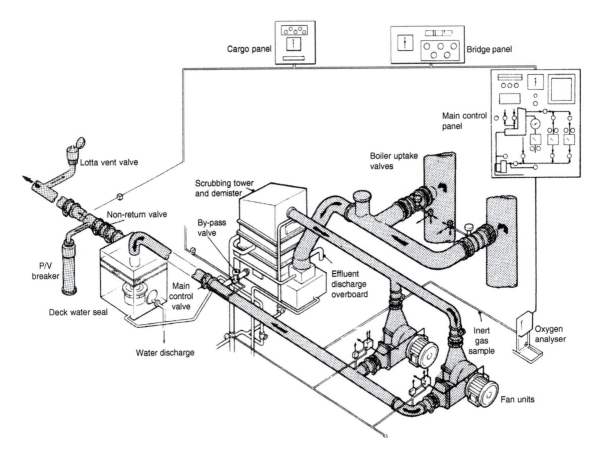

Figure 19.16 Inert gas system.

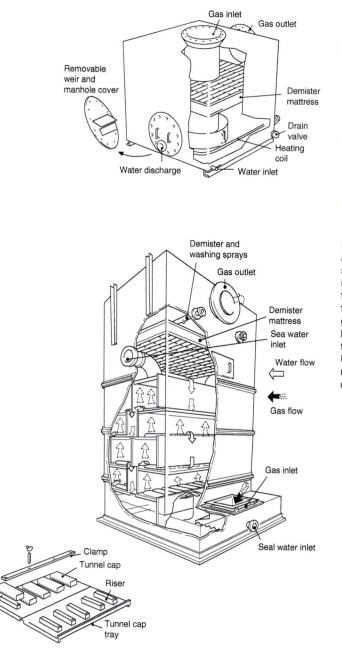

Gas inlet

Gas outlet

Removable
weir and
manhole cover

Demister
mattress

Drain
valve

Heating
coil

Water discharge

Water inlet

Figure 19.17 Deck-mounted water seal. This deck-mounted water seal prevents hydrocarbon gases flowing back to the uptakes. This safety feature is additional to the seal in the bottom of the scrubber and the gas non-return valve in the inert deck main. Sea water is used for the water seal and it is continuously pumped into the base at four tons per hour, the correct level being maintained by a weir. A coarse polypropylene demister mattress is fitted above the water seal to remove any water carryover. All internal surfaces are ebonite rubber lined, cured *in situ*.

Demister and
washing sprays

Gas outlet

Demister
mattress

Sea water
inlet

Water flow

Gas flow

Gas inlet

Seal water inlet

Clamp

Tunnel cap

Riser

Tunnel cap
tray

Figure 19.18 Scrubbing tower and demister. The purpose of the scrubber is to cool and remove unwanted elements from the boiler flue gas. Water is introduced from the top of the scrubber, while the gas enters via a water seal at the bottom. This water seal also serves to cool the gas as it enters the tower. Up to 98 per cent of acids (SO_2) are removed under normal operational conditions.

Additional safety features included in the system are a mechanically operated 'non-return valve' and a pressure/vacuum breaker fitted to prevent over or under pressurization of the cargo tanks. Alternatively the pressure/vacuum breakers may be fitted individually to each tank.

Venting of tanks during loading or when purging hydrocarbons is achieved by vent valves or masthead risers. As cargo levels rise during the process of loading the inert gas is vented into the atmosphere.

Composition of Boiler Flue Gases

The average composition of flue gases employed as inert gases and obtained from main or auxiliary boilers is as follows: CO_2, 13 per cent; O_2, 4 per cent; SO_4, 0.3 per cent; the remaining percentage is made up of nitrogen and water vapour. Such a mixture will not support combustion, and is therefore suitable for use as an inert gas once it is cooled and cleaned.

Advantages

- Safety: risk of fire and/or explosion reduced.
- Reduced corrosion: tank corrosion is inhibited by the low oxygen content of the gas.
- Faster cargo discharge: the increased tank pressure created during the period of discharge by the introduction of the inert gas into the tank speeds up the discharge operation.
- Tank washing time is reduced because it is possible to wash with high-capacity fixed guns in an inert atmosphere. Crude oil washing (COW) is also possible under these conditions.
- Fresh air purging of the tanks: the system can provide large volumes of fresh air to cargo tanks very quickly, which is beneficial for maintenance and tank inspections.
- Cheap and readily supplied (funnel exhaust gases).
- Compatible for use with certain chemicals which react with oxygen.

Disadvantages

- Installation cost is high initially, with additional expense incurred for general maintenance.
- Danger to personnel due to the lack of oxygen within the tank.
- Reduced visibility inside the tanks.
- Additional cost of an inert gas generator required for use when main engines are not in use, e.g. when in port.
- Danger of flammable gases returning towards the boiler if water seal and non-return valve are not properly maintained.
- Improved purity required in inert gases for use with chemicals, i.e. need for nitrogen, with the additional expense that this purchase incurs.

Requirements for Inert Gas Systems

Additional reference should be made to the Revised Guidelines for Inert Gas Systems adopted by the Maritime Safety Committee, June 1983 (MSC/Circ. 353). In the case of Chemical Tankers, reference Resolution A. 567 (14) and A. 473(XIII).

1 Tankers of 20,000 dwt and above, engaged in carrying crude oil, must be fitted with an IGS.

2 Venting systems in cargo tanks must be designed to operate to ensure that neither pressure nor vacuum inside the tanks will exceed design parameters for volumes of vapour, air or inert gas mixtures.

3 Venting of small volumes of vapour, air or inert gas mixtures, caused by thermal variations affecting the cargo tank, must pass through 'pressure vacuum valves'. Large volumes caused by cargo loading, ballasting or during discharge must not be allowed to exceed design parameters. A secondary means of allowing full flow relief of vapour, air or inert gas mixtures to avoid excess pressure build-up must be incorporated, with a pressure sensing and monitoring arrangement. This equipment must also provide an alarm facility activated by excess pressure.

4 Tankers with double hull spaces and double bottom spaces shall be fitted with connections for air and suitable connections for the supply of inert gas. Where hull spaces are fitted to the inert gas permanent distribution system, means must be provided to prevent hydrocarbon gases from cargo tanks entering double hull spaces. (Where spaces are not permanently connected to the IGS appropriate means must be provided to allow connection to the inert gas main.)

5 Suitable portable instruments and/or gas sampling pipes for measuring flammable vapour concentrations and oxygen must be provided to assess double hull spaces.

6 All tankers operating with a COW system must be fitted with an IGS.

7 All tankers fitted with an IGS shall be provided with a closed ullage system.

8 The IGS must be capable of inerting empty cargo tanks by reducing the oxygen content to a level which will not support combustion. It must also maintain the atmosphere inside the tank with an oxygen content of less than 8 per cent by volume and at a positive pressure at all times in port or at sea, except when necessary to gas free.

9 The system must be capable of delivering gas to the cargo tanks at a rate of 125 per cent of the maximum rate of discharge capacity of the ship, expressed as a volume.

10 The system should be capable of delivering inert gas with an oxygen content of not more than 5 per cent by volume in the inert gas supply main to cargo tanks.

11 Flue gas isolating valves must be fitted to the inert gas mains, between the boiler uptakes and the flue gas scrubber. Soot blowers will be arranged so as to be denied operation when the corresponding flue gas valve is open.

12 The 'scrubber' and 'blowers' must be arranged and located aft of all cargo tanks, cargo pump rooms and cofferdams separating these spaces from machinery spaces of category 'A'.

13 Two fuel pumps or one with sufficient spares shall be fitted to the inert gas generator.

14 Suitable shut-offs must be provided to each suction and discharge connection of the blowers. If blowers are to be used for gas freeing they must have blanking arrangements.

15 An additional water seal or other effective means of preventing gas leakage shall be fitted between the flue gas isolating valves and scrubber, or incorporated in the gas entry to the scrubber, for the purpose of permitting safe maintenance procedures.

16 A gas regulating valve must be fitted in the inert gas supply main, which is automatically controlled to close at predetermined limits. (This valve must be located at the forward bulkhead of the foremost gas-safe space.)

17 At least two non-return devices, one of which will be a water seal, must be fitted to the inert gas supply main. These devices should be located in the cargo area, on deck.

18 The water seal must be protected from freezing and prevent backflow of hydrocarbon vapours.

19 The second device must be fitted forward of the deck water seal and be of a non-return valve type or equivalent, fitted with positive means of closing.

20 Branch piping of the system to supply inert gas to respective tanks must be fitted with stop valves or equivalent means of control for isolating a tank.

21 Arrangements must be provided to connect the system to an external supply of inert gas.

22 Meters must be fitted in the navigation bridge of combination carriers which indicate the pressure in slop tanks when isolated from the inert gas main supply. Meters must also be situated in machinery control rooms for the pressure and oxygen content

Plate 181 A VLCC tanker seen in transit through the Dardanelles.

of inert gas supplied (where a cargo control room is a feature these meters would be fitted in such rooms).

23 Automatic shutdown of inert gas blowers and the gas regulating valve shall be arranged on predetermined limits.

24 Alarms shall be fitted to the system and indicated in the machinery space and the cargo control room. These alarms monitor the following:

- low water pressure or low water flow rate to the flue gas scrubber;
- high water level in the flue gas scrubber;
- high gas temperature;
- failure of the inert gas blowers;
- oxygen content in excess of 8 per cent by volume;
- failure of the power supply to the automatic control system, regulating valve and sensing/monitoring devices;
- low water level in the deck water seal;
- gas pressure less than 100 mm water gauge level;
- high gas pressure;
- insufficient fuel oil supply to the inert gas generator;
- power failure to the inert gas generator;
- power failure to the automatic control of the inert gas generator.

Mooring Large Tankers

Anchoring

The requirements of large tankers in the way of deeper water and heavier anchor equipment than other vessels have become obvious areas of consideration since the arrival of the first 100,000 dwt vessel. Now much larger tonnage, such as the VLCC- and the ULCC-type vessels, have appeared, further consideration is warranted.

The depths of tanker anchorages throughout the world usually range from 20 to 30 fathoms (36.6–55 m). The minimum amount of cable that a large vessel may expect to use must be considered about six times the depth of water, i.e. 120 to 180 fathoms (220–330 m), provided all other conditions are favourable. Since the length of chain cable required by the Classification Society is 351 m for the largest ships, it can be seen why Masters are reluctant to use anchors.

In all fairness to the shipowners, the majority have equipped their vessels with adequate reserves of cable, and it is not uncommon for vessels to carry 20 shackles (550 m) of chain cable on each anchor. If conditions were such that ten times the depth of water would be an appropriate amount of cable to use, this would limit the vessel to anchoring inside depths of 55 m.

Having considered the amount of cable to be used in anchoring, mariners should look at any weak links in the system. They do exist, and are encountered usually at the windlass with the braking system, or at the anchor itself with respect to its holding power.

Most of the information regarding the anchor arrangements for large vessels has come from experience gained on smaller vessels. In many respects the experience has been transferable, but in other areas new concepts of safe handling have had to be developed. Controlling the speed of a running cable by use of a band brake on the windlass is no longer acceptable. The momentum achieved, say, by a 15-tonne anchor with added weight of cable, free running, is too great to handle.

In order to control the great weights of anchor and chain, the chain velocity and the consequent friction, hydraulically operated braking systems have now been devised. The modern designs are such that the faster the cable runs, the greater the pressure created on the braking system. Other commercially available systems employ disc brakes and limit switches governing the speed of the windlass. The more common practice now is to walk heavy anchors back, in gear, all the way to the sea bed.

Types of Anchor

There are many types of anchor presently in use aboard most kinds of larger vessel, not just large tankers. Various weights of anchors with different sizes of cable have been tried and tested in all conditions over the last few years. The AC 14 anchor, popular not only with warships but also with large passenger liners, would appear to be the most suitable to combat the kinetic energy of, say, a ULCC moving slowly over the ground.

Seafarers engaged in the mooring of large vessels will no doubt be aware of the many variables which could affect the operation before the 'brought up' position is reached. The holding ground, weight of chain and the weight of the anchor itself will influence the time that the anchor is dragging before it starts to hold, assuming that the anchor does not become snagged or hung up on a rocky bottom.

The old idea that the amount of cable paid out is what holds the vessel is still true for VLCCs and ULCCs, but vessels fitted with an anchor of high holding power will have a distinct advantage. Masters and berthing pilots should be wary once the anchor has held, especially one of good holding power, of the possibility of parting the chain cable by excessive ship-handling movements. The problem is accentuated when the external uncontrollable forces of current, wave motion and wind are present in a manoeuvring operation.

If the berthing situation is one where anchors may be used, full consideration of their use should be made before the operation is executed. Prudent use of tugs' mooring lines, bow thrust units, main engine propulsion and an efficient mooring launch will undoubtedly help in ship-handling operations with this type of vessel.

Mooring Systems

Offshore terminals where tankers of all sizes are required to load and discharge via single point moorings (SPMs) are now an accepted fact of the oil tanker trade. Complete rope assemblies for securing an SPM are commercially available, and they are made to provide not only maximum strength but also a high energy absorption capacity to counteract heavy and repeated loadings. The general design may vary to take chafe into account either at the buoy end or at the vessel's 'pick up' end. Vessels are very often fitted with purpose-built bow stoppers for accepting the fairlead chains.

Failing this, tankers are secured by nylon braidline strops or flexible (6 × 36) galvanized steel wires turned directly onto bitts (see Figure 19.19).

Plate 182 Tanker moorings and floating pipeline seen connected to a floating storage unit (FSU).

Plate 183 Chafe chain moorings secured with AKD stoppers aboard a tanker vessel engaged in oil transfer from floating storage unit (FSU).

Plate 184 Single point mooring buoy as seen through chain mooring pipe lead. Floating oil pipe is also seen at surface which would be led aft to connect with manifold.

Plate 185 Upper deck of a large tanker seen in the Setubal Drydock, Lisbon, Portugal.

Plate 186 The floating production, storage and offloading system (FPSO) *Seaway Falcon*.

Plate 187 Gas tanker seen at sea Upper structure of tank spaces, prominent feature of the LNG carriage trade.

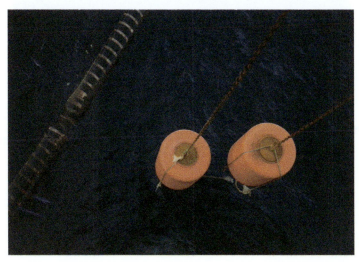

Plate 188 Example pick-up buoys in use next to a floating pipeline from a floating storage unit (FSU).

Oil Pollution

This subject is presented under the following headings:

- loading/discharging of bulk oil;
- compulsory insurance for vessels carrying persistent oil in bulk;
- reception facilities for oily waste;
- reporting of pollution incidents;
- penalties and offences with regard to oil pollution incidents;
- prevention of oil pollution;
- the prohibition of oil discharges into the sea from ships;
- résumé of existing oil pollution regulations and what can be expected for the future.

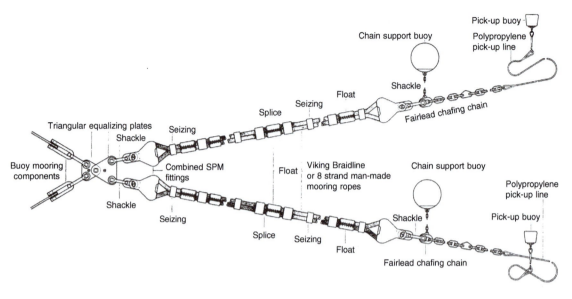

Figure 19.19 Typical mooring system used by buoy operators in a conventional single point mooring.

Loading and Discharge of Bulk Oil

The Master of any vessel has final responsibility for the correct loading and safe carriage of his cargo. However, it is accepted practice that his responsibilities are delegated to ship's officers, the Chief Officer being generally given that of Cargo Operations Officer for the vessel.

Loading of cargoes is expected to comply with all provisions of the stability booklet. The slack tanks should be noted and kept below the limiting number. Free surface build-up in slack tanks in the past has sometimes made the vessel 'unstable' while loading or discharging. Should any abnormal list develop during loading, oil cargo may overflow. This undesirable situation could be exacerbated by an imbalance in the quantity of water in ballast tanks, the combined effect of free surface in too many tanks and the added free surface effect from partly filled fresh water and ballast tanks.

Cargo Officers will require the following information when about to load bulk oil:

- cargo specifications and special characteristics, e.g. lead content;
- loading temperatures, together with flashpoints and specific gravity of oil;
- nominated quantities and tank order of loading;
- maximum shore loading rate and maximum back pressures at the manifold;
- communication system and emergency stop arrangements;
- number of hoses to be employed, with their respective size for each grade of oil;

Plate 189 Pipeline/manifold connection aboard a large tanker vessel.

- loadline figures for bunkers, boiler feed, stores, etc. to ensure that the vessel's draught conforms to loadline regulations when passing through the various 'zones' (these calculations will determine quantity of cargo loaded).

The terminal will require the following information:

- types of previous cargoes carried and the method of tank cleaning employed;
- maximum loading rate and topping off rate that the vessel can handle;
- maximum back pressures at the manifold;
- cargo loading plan, tank disposition and cargo quantities;
- order of loading or discharging;
- quantity of ballast for discharge and quantity of slops (oily waste), together with their disposition;
- method of venting;
- precautions against static;
- cargo specifications and ballast time for the vessel.

The Cargo Officer should take the following precautions against accidental oil spillage or leakage:

1 See moorings are tended throughout operations, and hose lengths sufficient to allow for ranging. Close off all valves not in use.
2 Carry out regular checks on cargo tanks, especially during the topping off period.
3 Plug scuppers before starting, draining off any excess water.
4 Provide drip trays at the manifold.
5 Blank off all lines and connections not in use.
6 Draw up contingency plans in the event of spillage.

SOPEP – Ship's Oil Pollution Emergency Plan, Ref. MGN 110.

In the event of spillage, e.g. a burst hose length, proceed as follows:

1 Stop all cargo operations. Sound the alarm.
2 Prevent oil or vapour entering the engine room.
3 Inform the harbour authority, terminal manager and adjacent shipping.
4 Enter details of the incident into the oil record book.
5 Close access doors and shut down ventilation systems.
6 Consult spillage contingency plans (SOPEP).

The following pump-room precautions should be taken:

1 Avoid loading through the pump room.
2 Ensure that all drain plugs and strainer covers, etc. are secure before loading.
3 Inspect pump glands regularly for leakage and the overheating of bearings.
4 Test level alarms before they are employed.

Transfer of oil from a vessel while in port cannot be undertaken before the following procedure is carried out:

1 Written permission must be obtained from the harbourmaster. In some ports the vessel may have to be moved to a special berth before permission will be granted.
2 Port by-laws must be observed at all times.
3 All overboard discharges must be secured when connected to oil transfer pumps before transferring takes place.

Compulsory Insurance

Insurance regulations are laid down by the International Convention on Civil Liability for Oil Pollution Damage, 1969, which came into force from 19 June 1975, and the Merchant Shipping Act (Oil Pollution), 1971, as amended by section 9 of the Merchant Shipping Act, 1974. The regulations state that a vessel of whatever registry, when carrying more than 2,000 tons of persistent oil, shall not be allowed to enter or leave a port in the United Kingdom without a certificate of insurance (or other financial security). This also applies to UK ships entering any other country's territorial waters with more than 2,000 tons of persistent oil, in bulk, as cargo.

Non-persistent oils include motor spirit, kerosene and the lightest fractions of the refining process. If they were to be deposited close to a coastline they might contribute to pollution of the beach areas, but if deposited at a reasonable distance out at sea they evaporate or otherwise disappear.

Animal and vegetable oils are assimilated by the sea water or physically by animal life within the sea water, whereas a *persistent oil* will not break down with sea water, and remains for indefinite periods floating on the surface. The mineral oil derivatives most likely to cause contamination are fuel oils and waxy crude oil waste, together with diesel. These particular grades when discharged at sea do not dissipate completely, but leave a 'film' over the surface which gradually coalesces to form thick, rubbery lumps of lower specific gravity than that of sea water. The lumps float with tide, winds and currents.

Persistent and non-persistent oils are graded by the authorities in relation to the nuisance value of the type of oil when mixed with sea water.

Certificates are issued by the government authority of the country whose flag the vessel sails under. In the United Kingdom the certifying authority is the MCA, Insurance Division. Satisfactory evidence must be produced to the certifying authority for the issue of the certificate. Non-compliance with the regulations for obtaining a certificate of insurance (or not being covered by financial security) may cause the vessel to be detained or fined, or both when the fine would not exceed £50,000.

Reception Facilities for Slops

Dirty ballast and oily waste are the main constituents of slops and the main problem in pollution control. From the early days of pollution control it has been the responsibility of the oil companies or the tankers themselves to solve the problem of dealing with waste products.

Many ports have now established reception facilities for slops, but there are as many without such means. To offset the immediate problem, tankers allocate one or more of their cargo tanks for the storage of waste products. This temporary storage lasts only until the vessel is able to pump the contents of the 'slop tank' ashore into purpose-built receptacles.

During tank-washing procedures, the oily waste rises to the surface, leaving clean (relatively oil-free) water underneath. The pumping of this water via an oily water separator certainly eases and reduces the problems of volume in the slop tanks. The problems of wax and sludge remain and have to be handled by shoreside facilities.

Waste may be classified as:

- dirty ballast water;
- tank-washing residues;
- sludge and scale (from tank cleaning operations);
- oily mixtures contaminated by chemical cleansing agents;
- contaminated bilge water;
- sludge from purification of fuel or lubricating oils.

Signatories to the Convention for the Prevention of Oil Pollution have established a reporting scheme whereby Masters of vessels may enter a report on port facilities. Reports on reception facilities for oily waste products should be submitted to the shipowner and then forwarded to the national administration (in the UK the MCA). MGN 82 (M + F) gives further details.

Reporting of Pollution Incidents

Oil spillage reporting arrangements have been practised by UK-registered vessels for some considerable time, but the Marine Environment Protection Committee of IMO has recently applied the reporting scheme to cover spills of substances other than oils.

Masters and other observers should report any of the following:

- An accident in which actual spillage of oil or other harmful substance occurs, or may occur.
- Any spillage of oil or other harmful substance observed at sea.
- Any vessel seen discharging oil in contravention of the International Convention for the Prevention of Pollution Regulations 1983–85.

Such incidents or slicks which may affect coastlines or neighbouring states should be reported to the nearest coast radio station. In the United Kingdom reports should be directed to the Coastguards via the coast radio station. Pollution reports should be made as quickly as possible and in plain language to the Marine Pollution Control Unit (MPCU) of the MCA.

They should contain the following information:

- name of the reporting ship;
- name of the ship, if known, causing the pollution (whether or not this is the reporting vessel);
- time and date of the incident or observation;
- position of the incident or observation;
- identity of the substance, if known;
- quantity of spill (known or estimated);
- wind and sea conditions.

Penalties and Offences

Under the United Kingdom Merchant Shipping Act, 1997, and the Merchant Shipping (Oil Pollution) Act 1996, as amended, the owner or Master of a ship from which oil has been illegally discharged into the sea is liable, on summary conviction, to a fine not exceeding £250,000, or on conviction on indictment to a fine.

The shipowner can limit and escape liability if he is not at fault and can prove that the discharge was:

- through an act of war or natural phenomenon beyond his control;
- any other person, causing damage or intending to cause damage, who is not employed by the company or an agent of the company, was responsible. This covers the shipowner against terrorism and such activity;
- due to any authority not maintaining navigational equipment to the proper specifications.

Standard Scale of Fines for Summary Offences

These fines were established for 1998, and are updated periodically by the Home Secretary using Statutory Instruments.

Level on Scale	Amount of Fine
1	£ 200
2	£ 500
3	£ 1000
4	£ 2500
5	£ 5000

For serious maritime offences, on summary conviction a maximum fine of £50,000 could be imposed. Such offences include:

- Concealment of British nationality.
- Causing the ship to appear British.
- Failing to render assistance following collision.
- Where a ship is considered dangerously unsafe.
- Failure to comply with government directives.
- Entering or leaving a UK port without a valid Oil Pollution Insurance Certificate.
- Proceeding to sea in contravention of a Detention Order.
- Proceeding against the direction of flow of a TSS.
- Carrying an excess number of passengers over that stipulated by the Passenger Certificate.

Under the Merchant Shipping Act (Prevention of Oil Pollution) Amended 1997, a £250,000 offence exists for illegal discharge from ships, on summary conviction.

The owner can limit liability to approximately £56 per ton of the vessel's tonnage or approximately £5,800,000, whichever is the lesser. Should the shipowner be at fault, then he or she cannot limit their liability.

For many offences under the Merchant Shipping Act the fines incurred range from £200 to £5,000 on summary conviction of the offence, together with an unlimited fine and imprisonment on conviction on indictment.

As regards insurance, the carriage of more than 2,000 tons of persistent oil in bulk as cargo without valid insurance or other valid financial security is an offence. The penalty on summary conviction is a fine not exceeding £50,000 and possible detention of the ship.

It is also an offence if the Master fails to produce a certificate of insurance. The Master is liable on summary conviction to a fine not exceeding £2,500. Should a vessel fail to carry a certificate of insurance, then the Master of the vessel is liable on summary conviction to a fine not exceeding £2500. If a person directed by the regulations fails to deliver a certificate of insurance (surrender the certificate to the correct authority), then that person is liable on summary conviction to a fine not exceeding £1,000. Regarding the movement of oil, vessels are required to be fitted with items of equipment that prevent the discharge of oil into the sea. Such equipment must comply with the standards specified in the Oil in Navigable Waters Act. Should these provisions be contravened, the owner or the Master of that vessel is guilty of an offence. The penalty on summary conviction is a fine not exceeding £2,500, or on indictment, a fine.

Transferring oil at night may be an offence. No oil should be transferred at night – between sunset and sunrise – to or from a vessel in any harbour in the United Kingdom unless the requisite notice has been given in accordance with the Oil Pollution Act, or the transfer is for the purposes of the Fire Service. On summary conviction the offender is liable to a fine not exceeding £500.

Failure to report a discharge of oil is an offence. It is the duty of the owner, Master or occupier of the land about which a discharge of oil occurs to report such discharge. Any person so concerned who fails to make such a report is guilty of an offence, and on summary conviction to a fine not exceeding level 5 on the standard scale (£5,000).

Failure to comply with instructions from the Secretary of State, or his designated agent, to avoid pollution from the result of a shipping casualty is an offence. Should any obstruction occur, the person causing that obstruction, on summary conviction, may be subjected to a fine not exceeding £50,000, or on conviction or indictment to a fine.

Failing to carry an oil record book, as required by the regulations, is an offence, and the owner or Master shall be liable to a fine not exceeding £500 on summary conviction.

Failure to keep proper records is an offence, subject on summary conviction to a fine not exceeding £500 for the person who is responsible.

Deliberately making a false or misleading entry in the oil record book or in any other similar records is an offence. The penalty on summary conviction is a fine not exceeding £500 or imprisonment for a term not exceeding six months, or both, or on conviction on indictment a fine or imprisonment for a term not exceeding two years or both.

Failure to produce the oil record book is an offence, subject on summary conviction to a fine not exceeding £200.

Any person who obstructs the duty of an inspector who is acting with the power of inspection concerning oil records is guilty of an offence. He or she is liable on summary conviction to a fine not exceeding £500.

It is a requirement for records to be retained for a minimum period of two years by the authority designated by the regulations. If those responsible for the custody of records fail in this duty, they may be liable on summary conviction to a fine not exceeding £500.

> NB. Since the *Exxon Valdez* pollution incident in Alaska (1989), shipowners have not been allowed to limit liability in the United States in the event of oil pollution accidents.

> NB. The level of fines is continually being updated and figures quoted can expect to change on the recommendations of the UK Home Secretary.

Prevention of Oil Pollution

Regulations from IMO now specify the installation of oily water separators aboard all non-tanker-type vessels over 400 grt. There are many types of oily water separators available, each providing clean water discharge well below the 15 parts per million of oil in water requirement.

Depending on size, capacity will vary with the model being used, from 0.5 m³ per hour up to 60 m³ per hour. The primary purpose of oily water separators is to prevent pollution, but the value of the recovered oil should not be overlooked.

The *Torrey Canyon* disaster in March 1967 demonstrated the need for pollution control and increased research into prevention methods. It also highlighted the need for new ideas and methods of containment in pollution incidents.

The enclosure of any spillage by use of some form of barrier was widely investigated and subsequently tried. Some degree of success was achieved when small spillages were encountered and good weather prevailed at the time. However, over large areas the time required to establish the barrier was found to be excessive, and barrier equipment needed to encircle a large area would not always be readily available. The controlling factor in the containment of oil spillage by a floating barrier is undoubtedly the weather.

Strong detergents have been tried on many occasions in 'clean-up operations' after spillage has occurred. The main disadvantage of this method is that the detergent used must be effective in breaking up the oily substance quickly, but very few achieve this result. Large quantities of detergent are required and the cost of using this method is high. Difficulties also arise with dispensing detergent over a wide area and achieving full coverage.

One would think, after the many lessons that have been given, it would be found cheaper and more practical to train personnel and equip modern ships to prevent pollution occurring in the first place. However, the consequences of collision or accident will always need to be dealt with by external agencies.

Prohibition of Oil Discharge into the Sea from Ships

With certain exceptions, no discharge of oil into the sea may take place within the territorial waters of the United Kingdom. This applies to ships of any flag. It is also forbidden for ships registered in the United Kingdom to discharge oil into the sea anywhere else in the world.

Notable exceptions are as follows:

- The ship is proceeding on a voyage.
- The ship is not within a special area.
- The oil content of the effluent does not exceed 15 ppm.
- The ship has in operation the filtering equipment and the oil discharge and monitoring and control system as required by the regulations.
- The ship is more than 12 miles from the nearest land.

In the event of proceedings being brought against the owner or Master of a vessel, special defences may apply in the following circumstances:

- If it can be proved that any discharge was made for the purpose of securing the safety of any vessel, of preventing damage to any vessel or cargo, or of saving life.

- If it can be proved that the discharge occurred in consequence of damage to the vessel, and that as soon as practicable after the damage occurred all reasonable steps were taken for preventing or (if it could not be prevented) for stopping or reducing the escape.
- If it can be proved that the escaped oil or mixture was caused by reason of leakage, that neither the leakage nor any delay in discovering it was due to any want of reasonable care, and that as soon as practicable after the escape was discovered all reasonable steps were taken for stopping or reducing the leak.

Management of Ship's Waste/ Garbage

(Ref., The Prevention of Pollution, Garbage Regulations 1988. S.I. No. 2292 applicable to all UK ships) (MARPOL Annex V) (Merchant Shipping Port Waste Reception Facilities, Regulations 1997, S.I. No. 3018)

The problem of waste disposal and management of the same is one of high consideration by every responsible person, and especially the shipping companies operating the large passenger vessels and the ferry companies. Waste cannot, at this moment in time, be eliminated totally, but by prudent planning and use of correct equipment and shoreside facilities, together with staff training, the problem can most certainly be reduced considerably.

Suggested On-Board Waste Controls

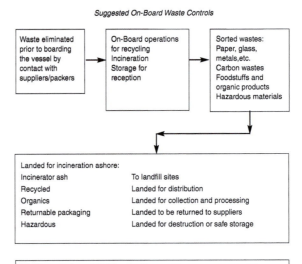

Suggested On-Board Waste Controls

| Waste eliminated prior to boarding the vessel by contact with suppliers/packers | On-Board operations for recycling Incineration Storage for reception | Sorted wastes: Paper, glass, metals,etc. Carbon wastes Foodstuffs and organic products Hazardous materials |

Landed for incineration ashore:

Incinerator ash	To landfill sites
Recycled	Landed for distribution
Organics	Landed for collection and processing
Returnable packaging	Landed to be returned to suppliers
Hazardous	Landed for destruction or safe storage

NB. Every vessel over 400 GT certified to carry 15 persons or more must now carry a Garbage Management Plan (GMP) and maintain a Garbage Record Book recording the treatment of ship's garbage (Additional Ref., MGN 1720 M & F).

Prevention of Pollution by Garbage

Ref. Annex 5, MARPOL.

Penalty

Mariners should be aware that under the 1997 amendment order to the M.S. (Prevention of Pollution) (S.I. 1997/No. 2569), any person guilty of an offence of causing pollution through garbage shall be liable on summary conviction to a fine not exceeding £250,000.

Additional reference

Further reference with regard to the prevention of pollution by garbage can be found in the MARPOL résumé in the following chapter. Classes of garbage are stipulated, and the import of records regarding the disposal of garbage to avoid causing an offence, is included.

Ballast Water Treatment Systems

All seagoing vessels take on ballast water to provide better stability and improved ride conditions. The larger cruise ships, bulk carriers and cargo ships are not an exception to this concept and could expect to be equipped with a ballast water treatment system to satisfy the environmental conditions of the future. It is now realized that taking on board and changing ballast water in different locations poses one of the top threats to the world's oceans. It is estimated that up to 4,500 different species of micro-organisms can be transported in a ship's ballast. A lot of bacteria, larvae and microbes die during transit, but some survive to invade the local environment where discharge takes place.

The IMO is taking the problem seriously and has introduced new legislation for member countries since 2002. A recent development has been the 'micro kill' separator which can remove sediment from water during ballasting. It addresses the problem of not introducing non-indigenous species into ballast water by ensuring that micro-organisms are not allowed to enter ballast tanks. The system also treats ballast in transit by ultraviolet (UV) light, which destroys or renders inactive various biological organisms prior to de-ballasting. The system has already been installed on two cruise ships, thought to be the first examples of water ballast treatment systems placed aboard operational vessels.

It is suggested that all new tonnage will in future be equipped with a ballast water management plan to reduce the threat of marine pollution caused by ballast water being introduced to a non-native environment. Short-haul ferries are generally not seen as being major culprits compared to, say, the large bulk cargo carriers. However, cargo vessels, tankers and cruise ferries operating over longer range can be expected to draw all shipping sectors into

NB. Examples of this can be seen in the United States where the European 'Zebra Mussel' (*Dressiena polymorpha*) has infested many internal waterways and has generated control costs of $1 billion since 1989. In southern Australia 'Asian Kelp' (*Undaria pinnatifida*) is invading new areas of the sea bed and displacing the natural bottom growth.

NB. The effectiveness of UV systems is not yet fully known and continues to be monitored closely. It is highly probable that results from UV irradiation could be greatly improved if coupled with a hydro-cyclone pretreatment operation.

any new regulations affecting ballast controls. Ballast water management plans would address treatment procedures approved by IMO and acceptable to marine administrations. Where procedures affect trim, stability or bending/shear forces on a hull they would also require approval by Classification Societies. Various treatment options are currently being studied.

These tend to fall into two categories:

1 Biocidal chemical: ultraviolet irradiation, electrical discharge, oxygen deprivation, ultrasound, chlorination, ozone and organic acids. Biocidal products are defined as 'active substances and preparations containing one or more active substances, put up in form in which they are supplied to the user, intended to destroy, deter, render harmless, prevent the action of, or otherwise exert a controlling effect on any harmful organism by chemical or biological means'.
2 Physical: ocean water exchange (currently the only system which has international acceptance by countries which have controls in place), exchange/salinity increase, heating and filtration/hydro-cyclone systems.

The effectiveness and operating costs based on cubic metres of water vary considerably between the options listed above. Clearly, the costs against bulk carriers which are frequently engaging in ballast operations will be considerably higher than the occasional and lesser-used methods employed by operational ferries. However, capital costs for plant can still expect to add to voyage costs in the future.

Record keeping of ballast movement

Masters are now obliged to maintain a 'ballast management system' aboard their vessels to provide a record of the position of taking ballast on board and in what quantity. Such records are to be retained in a ballast management log book similar to the garbage record, and operations should be controlled by a water ballast management plan.

20

The Application of MARPOL and the Prevention of Pollution

Introduction

The International Convention for the Prevention of Pollution from Ships (1973) was adopted by the International Conference on Marine Pollution as convened by IMO, inclusive of Protocol I, and reports on incidents involving harmful substances and Protocol II, on arbitration. This convention was subsequently modified in 1978 by the Protocol Relating to Tanker Safety and Pollution Prevention. The convention so modified in 1978 is known as MARPOL. This has since been modified by the Protocol of 1997.

Abbreviations and Acronyms within MARPOL

BCH	bulk chemical code
CAS	condition assessment scheme
CBT	clean ballast tank
COW	crude oil washing
ECA	emission control area
EDI	electronic data interchange
EEBD	emergency escape breathing device
EEDI	energy efficiency design index
GHGs	greenhouse gases
HCFC	hydro chlorofluorocarbons
IACS	International Association of Classification Societies
IAPP	International Air Pollution Prevention (certificate)
IBC	International Bulk Chemical (code)
ICS	International Chamber of Shipping
IGS	inert gas system
IMDG	International Maritime Dangerous Goods code
IMO	International Maritime Organization
IOPP	International Oil Pollution Prevention (Certificate)
MARPOL	The Convention for the Prevention of Pollution from Ships

MEPC	Marine Environment Protection Committee
MPCU	Marine Pollution Control Unit (of MCA)
NLS	noxious liquid substances
ORB	oil record book
PL	protective location
SBT	segregated ballast tank
SEEMP	ship's energy efficiency management plan
SOPEP	ship's oil pollution emergency plan
VOCs	volatile organic compounds

Definitions for Use (within the Understanding of MARPOL)

Administration

The government of the state under whose authority the ship is operating.

Associated Piping

The pipeline from the suction point in a cargo tank to the shore connection used for unloading the cargo; this includes all the ship's piping, pumps and filters which are in open connection with the cargo unloading line.

Bulk Chemical Code

The code for the construction and equipment of ships carrying dangerous chemicals in bulk.

Centre Tank

Any tank inboard of a longitudinal bulkhead.

Chemical Tanker

A ship constructed or adapted primarily to carry a cargo of NLSs in bulk; includes an oil tanker as defined by Annex I of MARPOL, when carrying a cargo or part cargo of NLSs in bulk (*see also* Tanker).

Clean Ballast

Ballast carried in a tank which since it was last used to carry cargo containing a substance in Category A, B, C or D, has been thoroughly cleaned and the residues resulting therefrom have been discharged and the tank emptied in accordance with Annex II of MARPOL.

Combination Carrier

A ship designed to carry either oil or solid cargoes in bulk.

Continuous Feeding

Defined as the process whereby waste is fed into a combustion chamber without human assistance while the incinerator is in normal operating condition with the combustion chamber operative temperature between 850°C and 1,200°C.

Critical Structural Areas

Locations which have been identified from calculations to require monitoring, or from service history of the subject ship, or from similar or sister ships to be sensitive to cracking, buckling or corrosion, which would impair the structural integrity of the ship.

Crude Oil

Any liquid hydrocarbon mixture occurring naturally in the earth whether or not treated to render it suitable for transportation. This includes:

- crude oil from which certain distillate fractions may have been removed; and
- crude oil to which certain distillate fractions may have been added.

Dedicated Ship

A ship built or converted and specifically fitted and certified for the carriage of:

- one named product;
- a restricted number of products each in a tank or group of tanks such that each tank or group of tanks is certified for one named product only or compatible products not requiring cargo tank washing for change of cargo.

Domestic Trade

A trade solely between ports or terminals within the flag state of which the ship is entitled to fly the national flag, without entering into the territorial waters of other states.

Discharge

In relation to harmful substances or effluent containing such substances, discharge means any release howsoever caused from a ship and includes any escape, disposal, spilling, leaking, pumping, emitting or emptying.

Emission

Any release of substance subject to control by Annex VI, from ships, into the atmosphere or sea.

Garbage

All kinds of victual, domestic and operational waste, excluding fresh fish and parts thereof, generated during the normal operation of the ship and liable to be disposed of continuously or periodically except those substances which are defined or listed in other Annexes to the present convention.

Good Condition

A coating condition with only minor spot rusting.

Harmful Substance

Any substance which, if introduced into the sea, is liable to create hazards to human health, to harm living resources and marine life, to damage amenities or to interfere with legitimate use of the sea, and includes any substance subject to control by the present convention.

Holding Tank

A tank used for the collection and storage of sewage.

Incident

Any event involving the actual or probable discharge into the sea of a harmful substance, or effluents containing such a substance.

Instantaneous Rate of Discharge of Oil Content

The rate of discharge of oil in litres per hour at any instant divided by the speed of the ship in knots at the same instant.

International Trade

A trade which is not a domestic trade as defined above.

Liquid Substances

Substances having a vapour pressure not exceeding 2.8 kp/cm^2 when at a temperature of 37.8°C.

Miscible

This means soluble with water in all proportions at wash-water temperatures.

Noxious Liquid Substance

Any substance referred to in Appendix II of Annex II, of MARPOL, or provisionally assessed under the provisions of Regulation 3(4) as falling into Category A, B, C or D.

NO_x Technical Code

The Technical Code on control of Emission of Nitrogen Oxides from Marine Diesel Engines, adopted by the Conference, Resolution 2 as may be amended by the Organization.

Oil

Petroleum in any form, including crude oil, fuel oil, sludge oil refuse and refined products (other than petrochemicals, which are subject to the provisions of Annex II).

Oily Mixture

A mixture with any oil content.

Oil Tanker

A ship constructed or adapted primarily to carry oil in bulk in its cargo spaces. This includes combination carriers and any 'chemical tanker' as defined by Annex II, when it is carrying a cargo or part cargo of oil in bulk.

Organization

The Inter-Governmental Maritime Consultative Organization.

Product Carrier

An oil tanker engaged in the trade of carrying oil other than crude oil.

Residue

Any NLS which remains for disposal.

Residue/Water Mixture

Residue in which water has been added for any purpose (e.g. tank cleaning, ballasting, bilge slops).

Segregated Ballast

That ballast water introduced into a tank which is completely separated from the cargo oil and fuel oil system and which is permanently allocated to the carriage of ballast or to the carriage of ballast or cargoes other than oil or noxious substances.

Sewage

This means:

- drainage and other wastes from any form of toilet, urinals and WC scuppers;
- drainage from medical premises (dispensary, sick bay, etc.) via wash basins, wash tubs and scuppers located in such premises;
- drainage from spaces containing living animals;
- other waste waters when mixed with drainage as listed above.

Ship

A vessel of any type whatsoever operating in the marine environment; this includes hydrofoil boats, air cushion vehicles, submersibles, floating craft and fixed or floating platforms.

Shipboard Incinerator

A shipboard facility designed for the primary purpose of incineration.

Slop Tank

A tank specifically designated for the collection of tank drainings, tank washings and other oily mixtures.

Sludge Oil

Sludge from the fuel or lubricating oil separators, waste lubricating oil from main or auxiliary machinery, or waste oil from bilge water separators, oil-filtering equipment or drip trays. The latest MARPOL amendments relating to on-board management of oil residues (sludge) have been adopted, providing definitions for holding tanks for oil residue, sludge tanks and oily bilge water tanks.

SO_x Emission Control Area

An area where the adoption of special mandatory measures for SO_x emissions from ships is required to prevent, reduce and control air pollution from SO_x and its attendant adverse impacts on land and sea areas. SO_x emission control areas shall include those listed in Regulation 14 of Annex VI.

Special Area

A sea area, where for recognized technical reasons in relation to its oceanographical and ecological condition and to the particular character of its traffic, the adoption of special mandatory methods for the prevention of sea pollution by oil is required. Special areas include: the Mediterranean Sea, the Baltic Sea, the Black Sea, the Red Sea, the Gulf Area, the Gulf of Aden, the North Sea, the English Channel and its approaches, the Wider Caribbean Region and Antarctica.

Substantial Corrosion

An extent of corrosion such that the assessment of the corrosion pattern indicates wastage in excess of 75 per cent of the allowable margins, but within acceptable limits.

Suspect Areas

Locations showing substantial corrosion and/or locations considered by the attending surveyor to be prone to rapid wastage.

Tank

An enclosed space which is formed by the permanent structure of the ship and which is designed for the carriage of liquid in bulk.

Tanker

An oil tanker as defined by Regulation 1(4) of Annex I, or a chemical tanker as defined in Regulation 1(1) of Annex II of the present convention.

Wing Tank

Any tank which is adjacent to the side shell plating.

Application of MARPOL

The present MARPOL convention applies to ships entitled to fly the flag of a party to the convention. It also applies to ships which are not entitled to fly the flag of a party but which operate under the authority of a party.

The convention does not apply to 'warships'. However, each party shall ensure that such ships owned or operated by that party act in a manner so far as is reasonable and practicable in accordance with the present convention.

1 Duty to Report

The Master or person in charge of a ship involved in an 'incident' shall report the details of the incident without delay to the nearest

NB. In the event that the ship so involved is abandoned, or the report from such a ship is incomplete or unobtainable, the owner, charterer, manager or operator of the ship, or their agents, assume the obligation of reporting the incident.

coastal state. Such a report should contain details to the fullest extent possible and include the following:

- The identity of the ship involved.
- The time, type and location of the incident.
- The quantity and type of harmful substance involved.
- Assistance and salvage measures required.

Such reports should contain any additional information and any details on developments affecting the incident.

2 Surveys for Tankers

Every oil tanker of 150 tons gross tonnage and above and every other ship of 400 grt and above will be subject to the following surveys:

- an initial survey
- a renewal survey
- an intermediate survey
- an annual survey
- an additional survey following any repair.

3 Issue or Endorsement of International Oil Pollution Prevention Certificate (IOPP)

The IOPP will be issued after an initial or renewal survey in accordance with the provisions of Regulation 4 of Annex I, MARPOL, to any oil tanker of 150 gross tons or over and to any other ships of 400 grt and above, which are engaged in voyages to ports or offshore terminals under the jurisdiction of other parties to the convention.

4 Duration and Validity of the IOPP Certificate

The IOPP Certificate shall be issued for a period specified by the administration, which shall not exceed five years.

If a ship, at the time of expiry of the Certificate, is not in a port in which it is to be surveyed, the administration may extend the period of validity of the Certificate. However, this extension would only be granted to permit the ship to complete her voyage to the port in which she is to be surveyed.

Vessels engaged on short voyages, which have not been extended under the previous provision, may have the IOPP extended for a period of up to one month from the date of expiry.

The IOPP Certificate will cease to remain valid if:

- the relevant surveys are not completed in accordance with the regulations;

The parties to the convention have obligations under a series of Articles, of various Protocols and Amendments. The following provides details on a working selection of these. Further reference should be made to the MARPOL 73/78, Consolidated Edition of 2003 (IMO publication).

- the Certificate is not endorsed in accordance with the regulations;
- the ship is transferred to the flag of another state, in which case a new Certificate must be issued by that state.

5 Control of the Discharge of Oil

Subject to the regulations affecting special areas and discharge of oil to save life and preserve the safety of the ship, any discharge of oil or oily mixture into the sea from ships is prohibited, except when the following conditions are satisfied:

(a) for an oil tanker, except as provided for in part (b) of this paragraph:
  (i) the tanker is not within a special area;
  (ii) the tanker is more than 50 nautical miles from the nearest land;
  (iii) the tanker is proceeding on route;
  (iv) the instantaneous rate of discharge of oil content does not exceed 30 litres per nautical mile;
  (v) the total quantity of oil discharged into the sea does not exceed 1/30,000 of the total quantity of the particular cargo of which the residue formed a part;
  (vi) the tanker has in operation an oil discharge monitoring and control system and a slop tank arrangement as required by the regulations.
(b) from a ship of 400 grt and above, other than a tanker and from machinery space bilges excluding cargo pump-room bilges of an oil tanker unless mixed with oil cargo residue:
  (i) the ship is not within a special area;
  (ii) the ship is proceeding on route;
  (iii) the oil content of the effluent without dilution does not exceed 15 parts per million; and
  (iv) the ship has in operation oil discharge, monitoring and control equipment (if the ship is not so equipped, these conditions for discharge will not be allowed to apply).

The above conditions do not apply to segregated or clean ballast or mixtures with less than 15 parts per million and do not originate from cargo pump-room bilges and are not mixed with oil cargo residues. Neither must discharges into the sea contain chemicals or other substances introduced for the purpose of circumventing the conditions of discharge.

Oil residues that cannot be discharged in accordance with the regulations must be retained on board or discharged to designated reception facilities.

Any visible sightings of oil on or beneath the surface of the water in the vicinity of the ship will be investigated, together with

NB. In the case of a ship of less than 400 grt, other than an oil tanker, the administration should ensure that oil residues are stored on board for discharge into reception facilities or will meet the requirements stipulated above.

relevant oil discharge records to ensure that a breach in the regulations has not taken place.

6 Oil Discharge and Monitoring Equipment

Ships of 400 grt must be fitted with oil-filtering equipment which does not allow the discharge into the sea of any oily mixture which has an oil content exceeding 15 parts per million.

Ships of 10,000 grt and above must be provided with oil-filtering equipment and with alarm arrangements that will automatically stop any discharge of oily mixture when the content of the effluent contains in excess of 15 parts per million of oil content.

These requirements may be waived if the vessel is trading exclusively in a 'special area', and the ship is fitted with a holding tank capable of total retention on board of all oily bilge waters; the contents of this tank being held to discharge into reception facilities and that the administration is satisfied that adequate ports with reception facilities are available on the vessel's route.

The IOPP would be endorsed to the effect that monitoring and filtering equipment is not carried in lieu of the holding tank and that correct records are maintained of discharges in the oil record book.

7 Reception Facilities

The government of each party to the convention is obligated to provide reception facilities for ships with oily residues and oily mixtures. These provisions must be situated at oil-loading terminals, repair ports and in other ports where the needs of shipping require reception for oil residues and mixtures.

Examples where reception facilities are required include the following:

- Ports where crude oil is loaded.
- Ports and terminals where oil, other than crude oil, is loaded at an average quantity of more than 1,000 metric tons per day.
- All ports with ship repair yards or tank cleaning facilities.
- All ports and terminals which handle ships provided with sludge tanks, as per the regulations.
- All loading ports for bulk cargoes which handle combination carriers.

All the above ports with designated reception facilities must ensure that the facility provided is adequate for the needs of shipping using the port. Every oil tanker will be equipped on the open deck with a discharge manifold connection to reception facilities for the discharge of dirty ballast water or oil-contaminated water.

8 Crude Oil Washing

Every new crude oil tanker over 20,000 tons deadweight shall be fitted with a cargo tank cleaning system using crude oil washing (COW).

Every oil tanker so equipped with a COW system must also be provided with an operation and equipment manual for the system. This manual must be revised if changes are made to the system.

Where a ship operates with a COW system it must have an inert gas system (IGS) for every cargo tank and slop tank on board the vessel (see Chapter 19).

New tankers fitted with segregated ballast tanks or COW must have oil piping designed and installed so that oil retention in the line is minimized. Also, means must be provided to drain all cargo pumps and oil lines at the completion of cargo discharge; where necessary, by connection to a stripping device. Line and pump drainings must be capable of being discharged both ashore and to a cargo tank or slop tank.

The occasion of COW must be entered in the oil record book, and include the port or ship's position where the operation was carried out. The tanks being washed should be identified and the number of machines engaged should be stated, along with the time of start and completion of washing, together with the wash line pressure and pattern used.

9 The Oil Record Book

Every oil tanker of 150 grt and above and every ship of 400 grt and above shall be provided with an oil record book: Part I (machinery space operations).

Every oil tanker of 150 grt or above will also be provided with an oil record book: Part II (cargo/ballast operations).

Entries into the oil record book should reflect any movement of oil in or out of the vessel's tanks and be recorded on a tank-to-tank basis if appropriate. All entries will be signed by the officer or officers in charge of the operation and each page will be signed by the Master.

The oil record book should be kept readily available for inspection at any reasonable time by an officer of the government of a party to the convention. Such an officer may take a copy of an entry made into the oil record book and such a copy certified as a true copy by the Master may be used in judicial proceedings as evidence.

10 The Ship's Oil Pollution Emergency Plan (SOPEP)

Every oil tanker of 150 grt or above and every ship other than a tanker over 400 grt or above shall carry on board a shipboard oil pollution emergency plan, approved by the administration.

The plan must be written in the working language of the Master and the officers. The plan should meet the guidelines of the Organization and include the following:

- The procedure by the Master or officer in charge of the vessel to report an oil incident as required by the articles.
- A list of the authorities or persons to be contacted in the event of an oil pollution incident (namely: designated persons ashore).
- A detailed description of the action to be taken immediately by persons on board to reduce or control the discharge of oil following the incident.
- The procedure and point of contact on the ship for coordinating shipboard action with national and local authorities in combating the pollution.

Where applicable, for ships carrying NLSs, this plan may be combined with the shipboard marine pollution emergency plan.

11 Loading Tankers and the Ship's Intact Stability

The vessel should be loaded with all cargo tanks filled to a level corresponding to the maximum combined total of the vertical moment of volume, plus the free surface inertia moments at 0° heel, for each individual tank. Cargo density should correspond to the available cargo deadweight at the displacement at which transverse 'KM' reaches a minimum value, assuming full departure consumables and 1 per cent of the total water ballast capacity. The maximum free surface moment should be assumed in all ballast tanks.

For the purpose of calculating GM_0, liquid free surface correction should be based on the appropriate upright free surface inertia moment. The righting lever curve may be corrected on the basis of liquid transfer moments.

12 Double Hull Tanker Design Example

Figure 20.1 outlines an example of a Double Hull Tanker diagram.

13 Categories of Noxious Substances

For the purpose of the MARPOL Regulations, NLSs are divided into four categories:

Category 'A'. NLSs which, if discharged into the sea from tank cleaning or de-ballasting operations, would present a major hazard to either marine resources or human health or cause serious harm to amenities or other legitimate users of the sea and therefore justify the application of stringent anti-pollution measures.

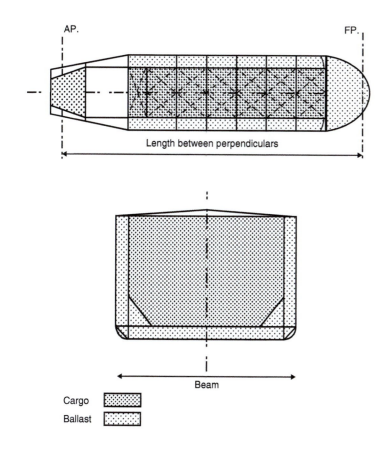

AP.

FP.

Length between perpendiculars

Beam

Cargo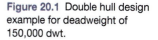

Ballast

Figure 20.1 Double hull design example for deadweight of 150,000 dwt.

Category 'B'. NLSs which, if discharged into the sea from tank cleaning, de-ballasting, etc., would justify the application of special anti-pollution measures.

Category 'C'. NLSs which, if discharged into the sea from tank cleaning, de-ballasting, etc., would cause minor harm to amenities or other legitimate uses of the sea and therefore require special operational conditions.

Category 'D'. NLSs which, if discharged, would present a recognizable hazard to either marine resources or human health or cause minimal harm to amenities or other legitimate users of the sea and therefore require some attention in operational conditions.

> NB. The discharge of any of the above categories of NLSs is prohibited except where the special conditions are complied with (Ref., Regulation 5, of Annex II, MARPOL).

14 Discharge Exceptions (Ref., Regulation 11, Annex I, and Regulation 6, Annex II, MARPOL)

The discharge into the sea of either oil/oily mixtures or NLSs for the purpose of securing the safety of the ship or saving life at sea or discharge resulting from damage to the ship or equipment is classed as an exception, provided that all reasonable precautions

Plate 190 The *Anna Bella* gas carrier seen alongside a river berth discharging the cargo via the flexible shore link.

have been taken after the occurrence of the damage or discovery of the discharge for the purpose of preventing or minimizing the discharge.

An exception is where the Master or owner acted with intent to cause damage or recklessly and with knowledge that damage would be the probable result.

Where oil or substances are discharged, when being used for the purpose of combating specific pollution incidents, this will be subject to the approval of the government in whose jurisdiction the incident lies.

15 Cargo Record Book

All ships designated to carry NLSs must be provided with a cargo record book. The record book must be completed on a tank-to-tank basis and include:

1 the loading of cargo;
2 internal transfer of cargo;
3 unloading of cargo;
4 cleaning of cargo tanks;
5 ballasting of cargo tanks;
6 discharge of ballast from cargo tanks;
7 disposal of residues to reception facilities;
8 discharge into the sea or removal by ventilation of residues.

16 Certification for Ships Carrying Noxious Liquid Substances

An International Pollution Prevention Certificate must be issued by the administration following an initial, or a renewal, survey and will be valid for a period not to exceed five years.

The certificate may be extended by the administration if the ship is not in the port in which it is to be surveyed. However, the extension will only be granted for the purpose of the ship completing the voyage to the port of survey and then only where it appears proper and right to do so.

17 The Carriage of Harmful Substances at Sea: in Package Form

The carriage by sea of harmful substances in package form is prohibited, except in accord with Annex III of MARPOL.

Each party to the convention must issue detailed requirements for the marking, labelling, documentation, stowage, quantity limitations and exceptions for preventing or minimizing pollution of the marine environment by harmful substances.

The labelling and documentation of the packaged harmful substances must bear the correct technical name. The marking of the package should also include the words 'MARINE POLLUTANT'.

The labelling must be durable so that the markings would remain identifiable after at least three months' immersion in sea water.

The package(s) supplied for shipping shall be accompanied by a signed certificate or declaration that the shipment offered for carriage is properly packaged and labelled and in a condition as not to present a hazard to the marine environment.

> NB. Empty packages which have been used for the carriage of harmful substances shall themselves be treated as harmful substances unless they have been treated to ensure they contain no residues.

18 Packaged Harmful Substances

Packages of harmful substances must be stowed in a manner so as not to present a hazard to the marine environment without impairing the safety of the vessel or the persons on board.

It should be noted that certain substances may be prohibited for carriage or be limited in quantity. Due consideration to the size, construction and equipment of the ship could directly influence the quantity shipped.

19 Jettisoning of Harmful Packages

It is prohibited to jettison harmful packaged substances except where necessary for the purpose of securing the safety of the ship or for saving life at sea.

20 Regulations to Prevent Pollution by Sewage from Ships

Reference to Annex IV of MARPOL is applicable to all ships over 200 grt and ships less than 200 grt which are certified to carry more than ten persons. Such ships are required to be surveyed and be issued with an International Sewage Pollution Prevention Certificate valid for a period not exceeding five years. The Certificate may be extended by five months but only for the purpose of completing the voyage to the state in which it is to resurveyed.

The vessel must be equipped with a sewage treatment plant or holding tank of adequate capacity approved by the administration. Pipelines to permit discharge to reception facilities must be fitted with a standard shore connection as stipulated by the regulations.

Discharge of sewage into the sea is prohibited except where the ship is discharging comminuted and disinfected sewage using an approved system, at a distance of more than four nautical miles from the nearest land, or sewage which is not comminuted or disinfected at a distance of more than 12 nautical miles from the nearest land. This is provided that, in any case where the sewage has been stored in holding tanks, it shall not be discharged instantaneously, but at a moderate rate when the ship is on route and proceeding at not less than four knots.

The ship must be in the waters under the jurisdiction of a state and be discharging sewage in accordance with less stringent requirements as may be imposed by such state.

Ships may discharge to secure the safety of the ship or also to save life.

> NB. The sewage treatment plant should be so approved as not to result in visible floating solids, nor cause discolouration of the surrounding water.

21 The Prevention of Pollution by Garbage from Ships

Reference to Annex V of MARPOL, inclusive of Resolution MEPC 201(62) and the 2012 guidelines for the implementation as set out by the Marine Environment Protective Committee to come into force 1st January 2013, applicable to all ships.

Additional products are now covered by the annex and include all of the following:

> Plastics, cooking oils, metals, glassware, wood, rubber and aluminium cans. All are capable of being recycled.
> Other products include: food waste, medical waste, out-of-date pyrotechnics, animal waste, animal carcasses, cargo residues from bulk cargoes, cleaning agents, packaging materials, incinerator ash and paper products.

Where disposal of designated garbage is made through a Port reception facility, it must be segregated prior to landing. The amounts/tonnages must be disclosed and recorded with the outcome as to whether it is being recycled, used for landfill, compacted and stored, or to be incinerated. Receipts for landed garbage should be obtained from the officer in charge of the reception facility and these should be retained within the ship's garbage records.

Disposal into the sea of garbage outside special areas

Disposal at sea is now generally prohibited, with some exceptions to commuted food waste after it has been passed through a commuter or grinder and made as far as practical away from the nearest land. Commuted garbage must be capable of passing through a screen with openings no greater size than 25mm mesh.

Special areas include: the Mediterranean Sea, the Baltic Sea, the Black Sea, the Red Sea, the North Sea, the Arabian Gulf area, the wider Caribbean area and Antarctica.

22 Disposal of Garbage in Special Areas

This is prohibited except: when necessary to secure the ship and the safety of life; any escape of garbage is due to damage to the vessel provided all reasonable precautions have been taken before and after the occurrence.

> NB. Reception facilities for receiving ships' garbage must be established within special areas.

23 Garbage Management Plan and Garbage Record Book

Every ship of 400 grt and above and every ship certified to carry 15 persons or more shall carry a 'garbage management plan'. Such a plan will contain written procedures for the crew regarding the collection, storage and processing of garbage on board the vessel.

All ships similarly qualified must also carry a 'garbage record book'. This record book should contain entries concerning each discharge of garbage, each time and place incineration takes place and the amount of garbage disposed of. Where shoreside reception facilities are employed, receipts for the amount of garbage disposed of should be obtained from the port operator of the facility. Such receipts should be retained in the record book for a period of at least two years.

24 Regulations for the Prevention of Air Pollution from Ships

Reference Annex VI of MARPOL. All ships of 400 grt and above will be subject to survey and be issued with an International Air Pollution Prevention Certificate which shall be valid for a period not exceeding five years. An extension of five months to this period may be granted for the purpose of allowing the ship to complete its voyage to the port of survey.

Any deliberate emissions of ozone-depleting substances is prohibited. Emissions of nitrogen oxides, applicable to diesel engines with an output of more than 130 kW, other than to save the security of the ship or to save life, are prohibited (applicable to ships constructed or converted after 1 January 2000). (This does not apply to emergency diesel engines, lifeboat engines and any device or equipment intended to be used solely in case of emergency.)

Sulphur oxides: the sulphur content of any fuel oil used on board shall not exceed 4.5 per cent mass by mass

Incinerators: on-board incineration will only be allowed with an approved incinerator. Incineration of specified substances as per Regulation 16, Annex VI, is prohibited.

Incinerators must be provided with a manufacturer's operating manual and personnel responsible for the operation of the incinerator shall be trained and be capable of implementing the guidelines provided by the manual.

Author's Note

It is not the intention to reproduce line by line the MARPOL text. However, marine students may find the salient points, as listed in this chapter, useful to gain an overview that will lend to a more detailed study of particular, relevant points.

Additional reference on pollution prevention measures of the marine environment might also be obtained from any of the following.

Statutory Instruments

1996   No. 282 The Merchant Shipping (Prevention of Pollution) (Law of the Sea Convention) Order 1996.

1996   No. 2154 The Merchant Shipping (Prevention of Pollution) Regulations 1996.

1997   No. 2367 Merchant Shipping Safety (Dangerous Goods and Marine Pollutants) Regulations 1997.

1997   No. 2569 The Merchant Shipping (Prevention of Pollution) (Amendment) Order 1997.

1999   No. 2567 The Merchant Shipping (Accident Reporting and Investigation) Regulations 1999.

Marine Guidance Notices

MGN 110   Oil Pollution Emergency Plans.
MGN 82   Oil Pollution Regulations.
MGN 143   MARPOL 73/78, Standards related to Incineration of Waste.

Merchant Shipping Notices

700   Oil pollution prevention on tankers, separation of cargo oil piping system, from the sea.
1577   Liability of shipowners for oil pollution damage.
1196   *Marine Pollution: Manual on Oil Pollution.*
1374   Oil pollution: compulsory insurance.
1197   Oil record books.
1643   Overboard discharges.
1678   Special waste regulations.

21

Appendix I: Mariner's Self Examiner (for the Training of Deck Officers, MN)

The headings to each section indicate to which examinations the following questions and answers are applicable. Candidates preparing for marine examinations are advised that nautical examiners are permitted to ask questions from lower grade qualifications if necessary to assess an understanding of any particular topic.

Efficient Deck Hand

1 What natural fibre ropes are known to you?

Ans. Manila, hemp, sisal, coir and cotton.

2 What type of stopper would be applied to a multi-plait, synthetic mooring rope?

Ans. A 'West Country Stopper', sometimes called a 'Chinese Stopper' is suitable for this type of rope.

3 How would you identify the safe working load (SWL) of a five-ton metal block?

Ans. The SWL is found stamped on the binding (side straps) of the block, together with the 'certificate number'. The SWL will also be indicated on the block certificate retained by the ship's Chief Officer.

4 What would you use seizing wire for, on the deck of the vessel?

Ans. Seizing wire is a thin multi-purpose wire employed regularly around the deck of the ship for mousing shackles, marking shackle lengths on the anchor cable, or securing of bottle screws to prevent accidental loosening.

5 How would you break and separate a 'Kenter lugless' joining shackle?

Ans. In order to break the Kenter joining shackle, it would be necessary to 'punch and drift' the 'spile pin'. The movement of the spile pin would clear the 'lead pellet' retainer. It would then be possible to knock out the centre stud between the two halves of the shackle. Once the stud is removed knock the two sides of the shackle at the side to cause them to separate.

6   What safety precautions would you adopt when rigging stages to work overside?

Ans.   It would be expected practice to consult the Code of Safe Working Practice for working aloft or overside and in any event it would be a requirement to complete a permit to work for any overside operations.

Additionally the following safety checks would be made:

1   Inspect the condition of the stage to ensure that it is free of cracks, grease or other faults.
2   Load test the stage itself to four times the amount of the intended load.
3   Check the gantlines are in good condition, without powdering or mildew and ensure they are of adequate length to provide slack and overhaul for the task.
4   Rig a side ladder alongside the stage position.
5   Always wear a safety harness and safety line to a separate anchor point.
6   Have a standby man waiting on the stages at deck level to raise the alarm in the event of incident.
7   Always rig the stage over water, never over a dock or quay.
8   Have the downhauls of the gantline run on opposite sides of the stage and secure the bowlines high to provide the stage with stability.
9   Have a lifebuoy close at hand for emergency use by the standby man.

NB. If the gantline is in any way suspect cut new lengths.

7   What marks would you find on the 'hand lead line'?

Ans.   The hand lead line is marked by:

- 2 and 3 fathoms: 2 strips and 3 strips of leather, respectively
- 5 and 15 fathoms: white duck
- 7 and 17 fathoms: red bunting
- 10 fathoms: leather with a hole in it
- 13 fathoms: blue serge
- 20 fathoms: 2 knots.

8   How would you secure the rope tails of the pilot ladder to the deck ring bolts?

Ans.   Where ring bolts are provided at the pilot station, the tails would be secured by round turns and two half-hitches.

9   What type of construction would you expect the cargo runner wire to be on a 10-tonne SWL derrick?

Ans.   Cargo runner wires are usually 24 mm diameter flexible steel wire rope (FSWR). The construction would be six strands by 24 wires per strand (6 × 24 wps) laid up around a jute or hemp core/heart.

10   After heaving the anchor home into the hawse pipe, describe how the anchor would be secured before proceeding to sea?

Ans.   Once the anchor is tight into the hawse pipe it would be secured by:

1   Turning on the brake, hard up.
2   Securing the devil's claw and tightening the bottle screw.

3 Securing the guillotine bar or the compressor, bow stopper in position.
4 Sealing the hawse pipe covers shut.
5 Sealing the spurling pipe covers shut.
6 Passing additional chain lashings through the cable and securing.

Old tonnage may not have spurling pipe covers and these would be subsequently sealed by a fabric pudding and covered in cement.

11 How would you break out a new coil of mooring rope or a new mooring wire?

Ans. Use a turntable, suspended on a swivel, which will permit the coil to be rotated freely. This will allow the wire or rope to be flaked from the coil without causing kinks.

12 What is the difference between the head block and a heel block of a conventional derrick?

Ans. The head block will be fitted with an oval eye, swivel becket, at the crown, while the heel block will be fitted with a 'duck bill' fitment to accommodate the gooseneck.

13 When is a steel wire rope considered unfit for service and would be subsequently condemned?

Ans. A steel wire rope is condemned when 10 per cent of the wires, in any eight diameter lengths, is broken.

14 When placing a chain stopper, what hitch would be used to start the securing and why is it used?

Ans. A chain stopper is started by use of a 'cow hitch'. The reason this hitch is employed is because it will not 'jam' when it is time to release the stopper.

15 What precautions would you take when securing a 'bosun's chair' for use in a vertical lift operation?

Ans. The use of chairs invariably will involve a height either above decks, overside or into tanks. As such, use of a chair would require the completion of a 'permit to work', issued by the ship's Chief Officer. The chair itself would be secured by a 'double sheetbend' to the gantline. The gantline and the chair would be inspected for overall condition and if found satisfactory would be used in conjunction with a lizard.

Hoisting and lowering of chairs should not be carried out by the use of winches. A lowering hitch is the normal practice of operation.

All tools and gear should be secured on lanyards or in a safety tool belt.

In every case the person in the chair should be fitted with a safety harness and safety line secured to a separate anchor point.

16 Where would you expect to find a 'monkey's fist' and what is its purpose?

Ans. A 'monkey's fist' is found in the end of a heaving line to provide weight to the line, so that when thrown it is given directional force.

NB. Company policy may dictate that the anchor is left in gear on the windlass.

NB. If a clove hitch is used instead, this may not be easily freed on release.

NB. Additional reference should be made to the Code of Safe Working Practice.

17   When rigging and tensioning a wire rope with a 'bottle screw' (USA – turnbuckle), how would you lock the bottle screw to prevent adverse movement?

Ans.   The bottle screw can be secured by either:
- using locking nuts on the screw threads;
- use of a locking bar stretched between the eyes of the screw; or
- by interwoven seizing wire from the ends through the mid part.

18   When doubling up a derrick, a floating, moving block is set into the cargo runner. What 'tackle' is made by this rig?

Ans.   Doubling up the derrick would turn the hoist from a single whip, into a 'gun tackle' having a power gain of 2, rove to disadvantage.

19   When rigging a 'pilot ladder', what safety equipment would be rigged alongside to assist the ascent of the pilot?

Ans.   Pilot ladders would be rigged with stanchions and manropes on either side. A lifebuoy and heaving line would also be kept readily available and pilot operations would be conducted with communications available to the ship's bridge.

20   What is the purpose of a back splice in a rope end and is there an alternative procedure?

Ans.   Unless the rope end is secured in some manner the strands of the rope would become unravelled and the rope would become useless. An alternative to a back splice would be to use a 'whipping' to prevent the end from fraying, the most secure being a 'sail maker's whipping'.

Proficiency in Survival Craft and Rescue Boats

1   What is the construction of lifeboat falls and when are they serviced?

Ans.   Lifeboat falls are constructed in extra flexible steel wire rope (of 6 × 36 wps) or wirex. They are renewed every five years or whenever necessary.

2   What are the two functions of the ship's rescue boat?

Ans.   The rescue boat should be capable of recovering persons from the water and marshalling other survival craft.

3   What is the standard length of the painter in an inflatable life raft?

Ans.   SOLAS require a minimum of 15 metre painter length; however, manufacturers of life rafts fit a standard of 25 metres minimum.

4   What type of gas is used to inflate the life rafts?

Ans.   Life rafts are inflated from a gas bottle containing CO_2 with a small amount of nitrogen included.

5   How would you beach a life raft?

Ans.   Taking the beach in any type of craft should preferably be carried out during the hours of daylight. In the case of beaching a life raft, it would be prudent to ensure the following:

1 Order personnel to don and secure lifejackets.
2 Inflate the double floor of the raft.
3 Open the entrances and man the paddles.
4 Stream the drogue and check the rate of drift onto the beach.

6   At what depth would you expect the hydrostatic release unit of a life raft to activate?

Ans.   Release units activate at a depth of four metres below the surface.

7   The regulations require that certain ships must carry a six-man life raft either forward or aft, in addition to the normal survival craft carriage requirements. What two conditions must exist for the carriage of this raft?

Ans.   The ship must have all aft, or all forward accommodation and have a 100 m length away from the stowage of remaining survival craft.

8   What type of pyrotechnics are carried in survival craft and what is their period of validity?

Ans.   Survival craft are expected to carry:
- six hand held flares
- four rocket parachute flares
- two orange smoke canisters

All pyrotechnics are valid for a three-year period.

9   Can a ship's lifeboat be a designated rescue boat and if so, what additional equipment would it be expected to carry?

Ans.   Yes, a ship's lifeboat can be a designated rescue boat and as such must be supplied with the following additional equipment:
- sufficient paddles to make headway in calm seas;
- two buoyant rescue lines and coits (length 30 m);
- a searchlight capable of illuminating a light-coloured object at night at a distance of 180 metres;
- a buoyant line of 50 metres in length and of sufficient strength to tow a life raft as per the regulations;
- sufficient TPAs for 10 per cent of the number of persons the boat is permitted to carry, or two, which ever is the greater.
- first aid outfit in a waterproof container;
- means of recovering persons from the sea.

10   What does the abbreviation TEMPSC stand for?

Ans.   TEMPSC represents: totally enclosed motor propelled survival craft.

11   Enclosed lifeboats are fitted with a compressed air bottle. What is the function of this compressed air?

Ans.   The quantity of air inside the bottle is sufficient for the full complement of the lifeboat to be able to breath normally during a period of launching and evacuation from a vessel while the survival craft is in a battened-down condition.

It contains sufficient air for about ten minutes, during which time it is expected that the boat would clear to an upwind position and permit normal ventilation to take over, away from any toxic or harmful substances/gases.

12   What checks would you make prior to starting the diesel engine of a motorized lifeboat?

Ans.   Before starting the engine of any boat the coxswain should check that the propeller is clear and not obstructed. There should be adequate fuel in the tank and sufficient lubricating oil in the sump. Different manufacturers of lifeboat engines operate different systems of starting. Some designs using a hydraulic start require accumulators to be pumped and charged, while other systems employ a battery/key start operation.

13   What emergency muster signals would you expect to be made to indicate the following scenarios: (*a*) boat stations, (*b*) fire on board, (*c*) abandon ship?

Ans.   Boat stations would be indicated by seven or more blasts on the ship's whistle, siren or bell system followed by a continuous sounding.

Fire on board the vessel would be indicated by a continuous ringing of the ship's fire alarm bell.

'Abandon ship' would be indicated by word of mouth by the Master, the officer in charge or a designated deputy. This may be transmitted by a public address system.

14   How and when are life rafts serviced?

Ans.   Inflatable life rafts are serviced at intervals of 12 months by an approved MCA servicing agent.

The service period may be extended by five months if the position of the vessel does not permit this schedule.

15   How would you expel water from inside a partially flooded lifeboat?

Ans.   The boat must be equipped with a manual pump, two buckets and a bailer. These could all be employed to empty water from the boat.

16   How would you launch a davit-launched life raft?

Ans.   Davit-launched systems are designed to launch several inflatable life rafts which are generally stowed in a rack, close to the launch station. The procedure is as follows:

1   Manhandle the life raft to a position under the davit fall (assuming it is the first raft).
2   Pull out the head shackle clear of the raft canister and secure the locking hook of the fall to the shackle.
3   Pull off the container-retaining lines and secure these either side of the launch station.
4   Pull out the bowsing in lines and secure them to the deck cleats provided.
5   Pull out the 'short painter' and take a temporary turn.
6   Check that the overside area is clear.
7   Hoist the raft in its canister on the davit.

8 Turn the davit outboard to a position approximately 70° away from the fore and aft line.

9 Pull on the painter to cause the life raft to inflate, overside.

10 Once inflation is complete, tension up on the bowsing in lines.

11 The person in charge would check that the access doors are open and that the raft is free of gas. The buoyancy chambers would be sighted to be in good condition and passengers would be advised against sharp objects as they board in a stable manner.

12 Once fully loaded, the bowsing lines would be cast off into the raft along with the short painter. Overside would be finally checked to be sighted and clear.

13 The raft would be lowered towards the surface.

14 Once the life raft has been lowered to approximately 2 m above the surface, the release 'red lanyard' is pulled sharply to remove the locking tongue of the hook arrangement.

15 As the raft continues being lowered to the surface, the weight comes off the hook as the buoyancy takes the weight of the raft at the surface, and immediately releases the life raft from the fall wire.

NB. On lowering, men should be stationed on paddles near the access points, while the coxswain would be on standby to activate the release from the fall.

NB. Once the first life raft has been released the fall and hook arrangement are returned to deck level. The hook would be recovered from overside by means of the 'tricing line' to allow the process to start again with a second or subsequent life raft.

17 Why does a davit-launched life raft have two painters, a short one and a standard long one?

Ans. The purpose of the short painter is for use in the davit-launched mode. The long painter is installed for using the raft in the general way as a 'throw over' inflatable, in the event that the davit system becomes inoperable.

18 When recovering a man overboard into a rescue boat, on which side of the boat would you attempt to bring the man from the water aboard?

Ans. The rescue boat should recover men from the water on the weather (windward) side, preferably recovering in a horizontal manner by means of a 'house recovery net' or similar apparatus.

19 Having launched an inflatable life raft, what are the main duties for the person in charge of the survival craft?

Ans. The person in charge of the life raft should:
1 cut the painter;
2 stream the sea anchor (drogue);
3 batten down the access points;
4 maintain the survival craft following the instructions.

20 When acting as coxswain of a lifeboat, what problems are likely if the boat is engaged in a transfer of personnel to a helicopter by the hi-line method?

Ans. Any operation with a boat and helicopter will involve considerable noise, which may make communication inside the craft and to the aircrew difficult. The rotors will also cause downdraft and rotor wash at the surface level, and this would make it difficult to steer the boat and keep a steady heading. Spray from the down draft must also be expected, which could well obscure windows of the cockpit of the surface craft and make visibility/steering and control difficult.

The coxswain would need to delegate the duties of dispatcher to another crew member if aircrew do not descend from the aircraft.

22

Appendix II: Officer of the Watch: Certificate of Competency

The headings to each section indicate to which examinations the following questions and answers are applicable. Candidates preparing for marine examinations are advised that nautical examiners are permitted to ask questions from lower grade qualifications if necessary to assess an understanding of any particular topic.

Example Seamanship, Oral Examination

1 When acting as a bridge Watch Officer, when would you expect to call the ship's Master to the navigation bridge?

Ans. The OOW should call the Master in any of the following circumstances:
- In the event of visibility becoming obscured to less than four miles.
- In the event of failure of any of the ship's navigational equipment.
- On any occasion that difficulty is experienced in maintaining the course.
- If traffic was causing concern and affecting the safe progress of the ship.
- In the event of sighting a landfall unexpectedly.
- In the event of failing to sight a landfall when expecting to.
- If soundings are seen to be shelving unexpectedly.
- On any occasion of encountering heavy weather.
- If sighting ice or receiving ice warning reports of ice on the vessel's track.
- In the event of management of watchkeepers becoming untenable.
- When sighting oil pollution on the surface.
- If a scheduled position fix is found to be suspect or unreliable.
- In any other emergency where the involvement of the Master is considered desirable or essential, e.g. fire on board.

2   While the vessel is at anchor, how would the OOW detect that the ship may be dragging her anchor?

Ans.   One of the main functions of the Watch Officer aboard the vessel at anchor is to ascertain the position of ship on a regular basis. This would be achieved by:

1   frequent checking of the visual anchor bearings (usually noted on the chart);
2   use of the radar by setting the range marker on a fixed land mass;
3   by operation and checking of the GPS;
4   in conjunction with a bridge relief, checking the vibration of the anchor cable caused by movement of the anchor on the sea bed;
5   a hand lead could also be set on the bottom from the bridge wing – if the line leads forward, towards the anchors, this also may indicate the vessel dragging her anchor.

NB. If dragging is detected or suspected the Master should be informed.

3   When on watch, how often would you check the compass?

Ans.   The compass is considered the most important instrument aboard the vessel. It would normally be checked by an azimuth or amplitude at least once every watch, and after every alteration of course, if practical.

A magnetic/gyro compass comparison would be made at the beginning and end of each watch period and at regular intervals during the watch period.

Additionally, a compass error would be ascertained as and when it may present itself, as with a transit or sector light use.

NB. All compass errors are recorded in the deviation record book.

4   What are your duties as a 3rd Officer during a lifeboat muster and drill?

Ans.   It is usual practice for the 3rd and 2nd Officers each to take charge of one or the other of the survival craft. The duty would entail making sure that all persons designated to the station are in attendance by use of a checklist.

The officer in charge would also check that each person has a lifejacket and immersion suit which is correctly donned and secure.

The boat may be turned out or even lowered and launched and the supervision of the movement of the boat would be carried out by the officer in charge.

During a drill it is an ideal time to train crew in the use of survival equipment and most company policies encourage this. Examples of suitable training activities could include: demonstrating the starting of the boat engine, the donning of an immersion suit correctly or a lecture on how to launch the inflatable life raft.

5   How is marine safety information and guidance promulgated through the maritime industries?

Ans.   The main supply of marine information is obtained through either MGNs, MSNs or MINs.

Additional information regarding the charts and nautical publications is obtained through the Admiralty, *Weekly Notices to Mariners*.

6   What does ECDIS stand for? Briefly, what do you know about it?

Ans.   ECDIS is the Electronic Chart Display and Information System. It employs an electronic navigation chart (ENC) and is a real-time display of a navigational chart without borders. It is displayed on a visual display unit (VDU) and is found extensively on ships with integrated bridge designs.

ECDIS works in conjunction with an ARPA overlay and two independent position-fixing methods (GPS). It is kept updated by a compact disc (CD) correction service.

7   What checks would you make during a drill with the self-contained breathing apparatus (B/A)?

Ans.   The B/A would be secured to a suitable crew member (clean shaven) and wearing the protective suit. The fittings, including the lifeline and back support plate, would be inspected for defects.

The contents of the bottle would be seen to be full by checking the air gauge meter. Once the air is turned on a whistle indication should be audible which is a check on the low air alarm warning. The mask would be donned and the tabs would be systematically tightened to provide an air seal around the wearer's face. This must be tested and is checked by briefly shutting off the air supply valve. This action would cause the mask to crush onto the face as a vacuum is established inside the mask. The air supply should be restarted, knowing that the mask has no side leaks and is providing a smoke/gas seal about the wearer's airways.

8   What is good holding ground for the anchor?

Ans.   The Master of a vessel about to go to anchor would carry out an anchor plan prior to letting the anchor go. One of the items of any anchor plan would include the nature and type of holding ground. The very best holding grounds are 'clay' and/or 'mud'.

9   How often would you check the sextant and what checks would you carry out?

Ans.   The sextant would be checked every time it is used. The checks would be for:
   • first adjustment – for the error of perpendicularity
   • second adjustment – for side error
   • third adjustment – for index error.

10   What would you use the dock water allowance for and how would you obtain it?

Ans.   The dock water allowance (DWA) is the amount that the vessel may legally submerge her summer loadline disc when loading in dock water of a lesser density than sea water.

The allowance is found by obtaining the density of the dock water by use of a hydrometer. The value of the DWA is then calibrated through the following formula:

$$DWA = \frac{1025 - \text{density number of dock water}}{25} \times FWA$$

11   When joining a new ship, where would you find information on any 'blind sectors' affecting the ship's radars?

Ans.   The vessel would be equipped with instrument maintenance manuals inclusive of radar performance and maintenance manuals. Should the ship's radars be affected by shadow sectors, details of these would be found inside the manuals.

It is also good practice to show a pictorial display of blind sectors and any shadow sectors on the bulkhead in the vicinity of the radar units.

12   What pyrotechnics are carried on the bridge of your vessel?

Ans.   The bridge will accommodate 12 bridge distress rockets, kept in a waterproof container.

In addition, two man-overboard bridge wing smoke floats are fitted for quick release and immediate operation. There are also four rocket line-throwing apparatus. Each line has a specification of 235 m length. Rocket lines are usually 4 mm and have a built-in permitted deflection of 10 per cent of the flight distance.

All ship's pyrotechnics inclusive of survival craft pyrotechnics are valid for a three-year period.

13   If the life rafts aboard your vessel are fitted with HAMAR disposable hydrostatic release units (HRU), how long can they remain *in situ* without being renewed?

Ans.   Two years.

14   What is the length of a shackle of anchor cable and if required to go to anchor how much cable would you need to pay out?

Ans.   A shackle length of cable is 27.5 m (or 90 ft or 15 fathoms).

When going to anchor, the scope (amount of cable) to use would depend on the depth of water. The minimum value of cable to use is usually based on four times the depth of water.

However, the actual amount of cable to deploy will depend on numerous factors, inclusive of the nature of the holding ground, tidal movement, weather conditions, length of stay and not least the Master's preference.

It is stressed that four times the depth is a minimum value and could well be influenced by the condition of the ship, loaded or light, as well as the above-stated criteria.

An anchor plan would be expected to take account of all such criteria.

15 What is stowage factor?

Ans. Stowage factor refers to a type of cargo and is defined as that volume occupied by unit weight of cargo. It is usually expressed as cubic metres per tonne (m³/tonne) or cubic feet per ton (cu. ft/ton). It does not take account of any space lost due to broken stowage.

16 Where would you find information about VTS, GMDSS and ship reporting systems?

Ans. Information on all these items is found in the Admiralty List of Radio Signals.

17 What are the regulatory requirements for rescue boats on board passenger vessels?

Ans. Passenger (Class 1) vessels must carry a rescue boat on either side of the vessel and the rescue boats must be fitted with a fixed communication system.

18 What liquid is found inside the compass bowl of the magnetic compass and why do we need it?

Ans. The older designed liquid compasses had a mixture of one part ethyl alcohol and two parts distilled water. The more modern magnetic compass will contain a clear oil such as 'Bayol'.

   The purpose of the liquid is to make the compass more 'dead beat' and thereby easier to steer by. Where a fine oil is used, friction is reduced between the compass card and the jewel pivot bearing by providing a lubricant. The card will also achieve maximum buoyancy and remain steady, even in bad weather.

19 While in port working general cargo, a small fire is discovered in number 3, lower hold. What action would you expect to take as the Cargo Watch Officer?

Ans. Any fire anywhere on board the vessel would generate the sounding of the ship's fire alarm. Once this is operational, the Officer of the Deck should stop all cargo operations and send all non-essential personnel ashore, e.g. stevedores, agents, etc.

   The following actions should also take place, but the operational order will depend on the availability of relevant personnel being on board:

1   The Fire Service ashore should be advised via VHF and the harbour/port control communication system.

2   The hatch on fire should be battened down as soon as practical and all ventilation to the affected hatch closed down.

3   The fire should be tackled immediately, or as soon after discovery as time permits, inclusive of effecting boundary cooling.

4   A head count of available crew should be made to assess who is ashore or who may be missing.

NB. Class 7 vessels do not have to have fixed communication systems in their rescue boats but must be provided with three walkie-talkie radios for use with survival craft under GMDSS requirements.

NB. All activities would be entered in the ship's log book when time permits.

5   The Chief Officer's messenger should be stationed at the head of the gangway to meet the Fire Service on arrival. The messenger should have ready for handover, the International Shore Connection and necessary cargo/personnel/stability and fire-fighting documentation, considered useful to the shoreside authorities.

20   For what type of activities would you require to draw a 'Permit to Work'. Who would issue this document and what does it contain?

Ans.   The Permit to Work is employed on board the vessel for any task which involves:
- entry into enclosed spaces;
- hot work or burning of any kind;
- activity above the deck level or overside;
- any electrical activity
- potentially hazardous cold work

The 'Permit' is issued by the Chief Officer in quadruplicate and effectively contains a checklist relevant to safety procedures of the respective task to be carried out.

23

Appendix III: Chief Officer (First Mate): Certificate of Competency

The headings to each section indicate to which examinations the following questions and answers are applicable. Candidates preparing for marine examinations are advised that nautical examiners are permitted to ask questions from lower grade qualifications if necessary to assess an understanding of any particular topic.

Example Seamanship, Oral Examination

1   When joining a vessel as Chief Officer, what would you take notice of when boarding the ship by the accommodation ladder?

Ans.   A Chief Officer would be expected to note that the access position of boarding is safe and conforms to the regulations regarding 'safe access', namely:

1   That the angle of the accommodation ladder is not more than 55° from the horizontal.

2   That the access point is monitored by gangway security.

3   That all safety equipment is correctly rigged and available, i.e. 'gangway net' rigged, lifebuoy close by, roller clear of obstructions, clean and free of grease or obstruction and well illuminated.

Additionally, the First Mate would observe the condition of the paintwork of the ship's sides, whether the moorings fore and aft are taut and whether the ship is loading or discharging cargo.

2   Your ship is about to enter 'drydock'. What documentation and plans are you most likely to require to be readily available?

Ans.

- The drydocking plan
- the 'shell expansion plan'
- the 'general arrangement plan'
- the Chief Officer's 'repair list'

- the stability information plans and general particulars of the vessel
- a tank arrangement plan
- the 'plug plan' if not incorporated into the 'drydock plan'
- the rigging plan
- the ship's fire-fighting arrangement
- cargo plan (if appropriate)
- relevant certificates for respective survey work.

3   What precautions would you take as the Chief Officer of a general cargo vessel if a heavy weather warning is received and bad weather is imminent?

Ans.   With heavy weather pending it would be prudent to secure the ship overall but specifically in the areas of stability, cargo, navigation and the deck.

Stability

Improve the 'GM' and remove any free surface effects if possible. Ballast the vessel down and make the ship as heavy as possible. Inspect and check the uppermost deck for watertight integrity, close all watertight doors throughout the vessel and pump out any swimming pool.

Cargo

Check all cargo lashings and harden up on any slack securings. Pay particular attention to any heavy lift cargoes and add additional lashings if appropriate. Thoroughly inspect the securings on deck cargoes and take up any slack on these.

Navigation

Consider re-routing the vessel in consultation with the ship's Master. Verify the ship's position, update the weather forecast and plot the position of the storm. Pass the position to a shoreside representative and revise the ETA.

Engage manual steering in ample time and reduce speed to prevent the vessel experiencing effects from pounding. Secure the navigation bridge against heavy rolling.

Deck

Rig lifelines fore and aft to ensure that access is not denied to any part of the vessel. Check securings on the gangways, anchors, survival craft and lifting appliances. Weather doors and ventilation systems should all be closed up if appropriate.

Clear the decks of all surplus gear and reduce deck working to a minimum by operating an internal work schedule. Warn all 'heads of department' of the impending weather and note all preparations in the log book.

4 When taking over as Chief Officer, what documents would you expect to receive from the outgoing Mate?

Ans. Handover procedures are usually carried out by means of a company checklist, and in the case of a Chief Officer the handover would include any or all of the following:

Stability documentation: all stability criteria, deadweight scale, cross curves, dynamical stability curve, tank disposition and capacities, statement of minimum bow height, damage stability information and the ship's general particulars.

Cargo details: a copy of the cargo plan, the cargo manifest, Mate's receipts, example loading plans, loadicator or computer software, bills of lading, cargo securing manual, any document of compliance for hazardous cargoes, register of lifting appliances and cargo-handling gear.

Certificates: safety equipment certificate, loadline certificate, sanitation certificate, safety ship construction, tonnage certificates, safety radio certificate, safe manning certificate, the safety management certificate and a copy of the document of compliance (ISM).

In order to carry out these duties additional information such as the planned maintenance schedule, the ship's key board and respective plans would be available in the Mate's office, e.g. fire-fighting arrangement, drydock plan, rigging plan, general arrangement, shell expansion plan, etc.

> NB. The task of the Chief Officer is one of being the ship's working manager and as such his or her responsibilities cover the stability of the vessel, the cargo details and the maintenance of the ship.

5 How often would you test the derrick/cranes aboard a general cargo vessel, and what constitutes the criteria of the test?

Ans. Derricks/cranes are checked every time they are used and undergo a detailed annual inspection. They are tested at five-yearly intervals, after installation when new and in circumstances following any repair.

The test will be a proof load test, where the rig will be sustained to the SWL + percentage tonnage to equal the proof load amount.

Proof load:

for derricks up to 20 tonnes SWL = 25 per cent excess of SWL

for derricks 20–50 tonnes SWL = 5 tonnes excess of SWL

for derricks 50+ tonnes SWL = 10 per cent excess of SWL

6 What type of operations would you enter on the work list prior to entering drydock with regard to keeping the anchors and associated equipment well maintained?

Ans. Drydock periods present an ideal opportunity to carry out maintenance tasks that are not possible when the ship is engaged in regular trading. Tasks on maintaining the anchors and anchor handling equipment would probably include the following items:

- Range the anchor cables in the bottom of the dock to permit inspection of all the joining shackles.

- Clean out the chain locker, inclusive of the mud box, while all the chain is out and clear of the locker.
- Inspect the locker and the paintwork condition and, if required, repaint the locker.
- Inspect the bitter end securing and test the release mechanism.
- Break the 'Kenter lugless joining shackles' on the floor of the dock and inspect for excessive wear and tear.
- Measure the cable size at respective lengths to ensure no excessive corrosion (use external callipers to assess cable size).
- If the situation is warranted, end for end shackle lengths of cable to ensure even wear and tear throughout the cable length (if end for ending, the cable will need to be re-marked).
- Inspect and renew brake liners on the windlass.
- Close inspection of the holding pins of the floating link of the band brake on the windlass.

7   In the event of having to abandon the vessel, what additional equipment would you order to be placed in the survival craft, assuming there is time?

Ans.   Any additional life-support items must be considered as useful in the event of having to evacuate from the vessel. In particular, the following specific items are suggested: SART, EPIRB, portable (walkie-talkie) VHF radio, extra food, water, blankets, medicines, additional immersion suits and TPAs. Torches and spare batteries.

8   What maintenance is carried out on your lifeboats?

Ans.   The lifeboats are maintained under the planned maintenance system of the ship, which is stipulated by the company. In addition, the lifeboats are visually inspected, the engines tried out and boats are moved from their chocks weekly.

A monthly inspection would include greasing of moving parts if required. Every three months the boats would be lowered into the water and manoeuvred.

A 12-monthly inspection of all LSA & FCA by the marine authority to comply with safety equipment certification regulation

The five-yearly test involves overloading the boat to 110 per cent to provide a dynamic test to the davit-launching appliances as per M 1655 and renewal of lifeboat falls.

9   What do you understand by the term 'risk assessment'?

Ans.   The term refers to a detailed and careful examination of a task or situation which could cause harm to person or persons directly involved or in close association with an operation. The assessment will classify the work and the nature of the operation to identify personnel at risk. It would expect to determine the type of risk, preferably a tolerable risk as opposed to an intolerable risk, taking account of such factors as: the experience of the personnel involved, the level of

NB. Normal practice in achieving this test is to use water bags in the boat to attain the required load of 110 per cent.

complexity of the work, external influences like the weather conditions, the time factor and period of daylight remaining, equipment required and condition of that equipment, etc.

Further reference would be sought from the CSWP.

10 When serving on a tanker, how are the tanks kept outside the flammable limit?

Ans. Tanks are kept within the safe limits by introducing inert gas. Once the gas is introduced into the tank that contains a hydrocarbon gas/air mixture the flammable range decreases to a point where the lower flammable limit (LFL) coincides with the upper flammable limit (UFL). This point corresponds to an oxygen content of approximately 11 per cent. No hydrocarbon gas/air mixture can burn at this oxygen level.

11 What precautions and procedures would you adopt when taking bunkers aboard your vessel?

Ans. Prior to bunkering it would be normal practice to establish the quantity of bunkers being taken and to identify the respective tank(s) where they are intended to be placed. Once established, the new stability condition and changes to the draughts would be calculated by use of the loadicator.

Plate 191 The passenger ship *Silver Wind* seen taking bunkers from a barge on her starboard side.

Checks and precautions would be carried out as per the company checklist and would include the following points:

1 Seal the uppermost deck and ensure all scuppers are blocked off.

2 Rig a pressurized hose overside and have extinguishers available at the manifold. Observe all standard fire precautions.

3 Establish a three-way communication link between the pumping station, the manifold and the tank sounding monitor.

4   Ensure a second means of access is available to and from the deck.

5   Display the 'bravo' flag by day or a 'red light' at night.

6   Display additional no-smoking signs, especially at the gangway and at the entrances to the manifold deck.

7   Place drip trays at any manifolds or pipe connections not provided for.

8   Advise the Master and the port authority of the operation.

9   Comply with all aspects of SOPEP (Ship's Oil Pollution Emergency Plan).

10   Start the pumping operation slowly and ensure no back pressure builds up.

11   Ensure an adequate number of personnel are available on deck, especially when topping off.

12   Make relevant entries into the log book.

13   Cause an entry to be made into the oil record book on completion.

NB. The circumstances of taking bunkers may vary depending on whether an offside bunker barge is being employed or bunkers are being delivered from a shoreside supply.

---

12   What security exists to prevent accidental release of the CO_2 total flood system into the engine room?

Ans.   All new gas flooding systems now have a two-lever operation.

The first lever action is to operate the distribution valve on the line to the engine room, while the second lever is the firing operation.

Additional to this, the remote station (usually in a glass case) in a position external to the engine room is kept locked. A small glass-protected key holder, close by, operating on the principle of 'in emergency – break glass, take key and unlock operational station', must be carried out to permit operation.

13   When acting as Chief Officer of a bulk carrier you are required to formulate a loading plan for the vessel. What factors would the loadicator system provide you with when the proposed plan is completed?

Ans.   The output from the loadicator would provide not only the disposition of bulk cargo quantities, but also the capacities of bunkers, ballast, fresh water and stores.

The output would also indicate the bending moment, shear force and torsional stresses that could be expected through the ship's hull. It would also indicate the loaded draughts, the trim and the GM, GZ factors affecting the positive stability of the vessel.

14 When formulating a muster list for emergency drills, what duties must be included in the display?

Ans. The muster list must show the allocated duties of each crew member and indicate the individual's relevant reporting station.

It should include the following:

- The clearing of personnel from passenger accommodation and the closing of watertight doors, fire doors and all ventilation ducts. All survival craft must have designated and certificated lifeboat men, and preparations to launch survival craft must be covered by respective listed duties.
- Specific duties must cover the operation of communication equipment, damage-control party and fire party activities, passenger control and the actual launching duties for ensuring survival craft become waterborne.
- The emergency signals for boat, fire and abandon-ship emergencies must be included on the list. The lists must be dated and signed by the Master and posted before the vessel puts to sea.

15 What is contained in the register of cargo-handling and lifting appliances?

Ans. The register is a file or files maintained by the Chief Officer which contains all the 'certificates' of: wires, shackles, blocks, cranes, derricks, chains, etc.

16 Following an engine-room fire your vessel is disabled in open waters and the Master has ordered the assistance of a tug. What type of towline would you expect to rig up for use with the towing vessel?

Ans. In open waters the towing options are to make use of the emergency towing arrangement (ETA). The alternative would be to rig a 'composite towline'.

The composite towline could be achieved by hanging the anchor off at the break of the forecastle head to bare the end of the cable. This would permit the anchor cable being led through the hawse pipe lead, then shackled onto the towing spring of the tug.

> NB. Considerable work would be involved in hanging the anchor off.

An alternative would be to leave the anchor on the cable and shackle the tug's towing wire directly to the anchor crown 'D' shackle. The anchor being left *in situ* may well dampen movement of the towline and reduce any 'yaw' effect on the vessel being towed.

17 In the event that your vessel had to enter drydock with cargo on board, what precautions would you expect to take?

Ans. Entering drydock with cargo on board is not a desired option and all effort would normally be made to discharge the cargo prior to entry. However, such handling costs would be expensive and the shipping company may insist that the cargo remains on board.

In such circumstances it would depend on the weight and distribution of the cargo, its value and whether it was pilferable.

In every case the main concern would be the additional stresses that would be incurred by the weight of the cargo and additional dock shores would need to be ordered in place to support the unexpected weight.

These shores would need to be placed before the dock is flooded in the majority of cases. To do this the drydock manager would require sight of not only the 'docking plan' but also the cargo plan.

Special cargo or valuable cargo would require lock-up stow and security services to prevent theft.

18   Prior to loading a heavy lift, what stability checks would be made by the ship's Chief Officer?

Ans.   Before loading a heavy lift it must be anticipated that the Chief Officer would carry out detailed checks regarding how the load will affect the vessel's stability. These checks would include a calculation to assess the maximum amount of heel that the vessel will incur when taking the load on the derrick/crane at its furthest outreach distance.

A check would also be made on the vessel's 'GM' to ensure that the GM value is adequate to cope with the rise of 'G' once the load has been accepted by the lifting gear. (Bearing in mind that the centre of gravity of the load becomes effective at the head of the derrick/crane once clear of the deck or ground.)

Tank systems would be noted and any free surface moments in slack tanks should be eliminated (press up or pump out when possible). Free surface effects will reduce the effective value of 'GM'.

NB. GM could be improved by filling double bottom tanks.

19   When the vessel is moored and lying to two anchors with a riding cable and a sleeping cable, how can you prevent the cables fouling if the vessel swings at the turn of the tide?

Ans.   The main concern with this type of either a 'running moor' or a 'standing moor' is a shift in the wind direction. The Master of the vessel would expect the ship to turn at approximately six-hourly intervals and at these times it would be prudent to give the vessel a sheer (with the rudder) to ensure the vessel turns in the desired direction. However, a change in the wind direction cannot be catered for and if the wind direction is changing, the sleeping cable must be recovered until the vessel re-aligns her heading. The Master would then be in a position to re-deploy anchors without having incurred a foul hawse.

20 Which NAVTEX messages can you not reject?

Ans. The coded messages that cannot be rejected on the receiver are: A, B, D and L.

Merchant Shipping Notice

No. M 1655: Five-Yearly Testing of Lifeboat and Rescue Boat Launching Appliances

Notice to shipowners, managers, Masters of merchant ships, ship-builders, shiprepairers, life-saving appliance testing houses, life-saving appliance manufacturers, certifying authorities

1 Regulation 13(8) of the Merchant Shipping (Life-Saving Appliances) Regulations 1986 and the Merchant Shipping Notice No. M 1546 require (among other things) that 'at least once every 5 years rescue boats and lifeboats shall be turned out and lowered when loaded with weights to simulate their full safe working load'.

2 The above test is not yet a SOLAS requirement; however, it should be noted that amendments to SOLAS Chapter III coming into effect in 1998 propose a similar test with 110 per cent loading, at least once every five years.

3 Such a test is intended to prove the adequacy of every part of the fully loaded launching system such as davits, winches and their foundations; falls, blocks, connecting loose gear such as shackles, rings; release gear hooks and their connections to the boat, etc. In addition it tests the tricing and bowsing arrangements and load distribution on the boat itself as would be imposed in a real evacuation situation.

4 Some shipowners have approached the MSA for guidance in conducting this test safely. The MSA has therefore consulted the relevant experts in the field and one method of carrying out the test safely is described in this paragraph. (However, see other main paragraphs of this Notice for possible variations.)

    4.1 The shipowner must *look ahead and plan* such tests. The launching device comprises many specialized components which require regular inspection, maintenance, replacement, overhaul etc., by specially trained personnel which should best be completed before the test. There may be a large number of boats to test (e.g. for large passenger ships), so a regular programme of tests spread over five years to cover all the boats may be essential.

4.2   Having decided to test a particular boat, *preliminary inspection* of key parts likely to be stressed during the test needs to be made. If specialized help from manufacturers is not available, the ship's Chief Engineer (or his delegate) should carry out such an inspection following the manufacturer's *manuals* and general engineering principles. In addition, the ship's *LSA maintenance logs* are to be scrutinized to confirm evidence of regular maintenance and history of any persistent problem or major fault. It is also necessary to carry out an *audit of certificates* on board for falls, blocks, loose gear, davits, winches, release gear hooks, lugs, etc. to confirm compliance with proof load requirements and identification of correct and adequate gear. Having been satisfied by these preliminary inspections and checks that a test can go ahead, a shore or floating *crane* with an SWL rating of at least 2.2 times the fully loaded boat weight (to absorb shock loads in emergency) is to be *arranged* for assisting in the test with safety.

24

Appendix IV: Ship's Master: Certificate of Competency

The headings to each section indicate to which examinations the following questions and answers are applicable. Candidates preparing for marine examinations are advised that nautical examiners are permitted to ask questions from lower grade qualifications if necessary to assess an understanding of any particular topic.

Example Seamanship and Duties, Respective Oral Examinations

1   Following the loss of a man overboard your vessel completes a Williamson turn and returns to the 'datum', but no sign of the man in the water is found. What search pattern would you adopt and what publication would you expect to consult?

Ans.   When searching for a man overboard, a 'sector search' is recommended. General reference on this search pattern would be found in Volume III of the IAMSAR manuals.

2   In the event of your vessel running aground, what action would you take as Master of the vessel?

Ans.   The general circumstances of the grounding scenario will differ in every case, but the general actions would probably include the following actions:

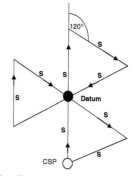

Where 'S' represents track space (distance or time variable) CSP represents commencement of search pattern

1   As soon as practical after the incident take the 'con' and raise the general alarm and ensure that all personnel are aware of the incident.

2   Order the engines to be stopped as soon as contact with the ground is made.

3   Order the Chief Officer to carry out a damage assessment to cover the essential items:

(i)   The watertight integrity of the hull.

(ii)   The engine room wet or dry condition.

(iii) A casualty report.

(iv) Any evidence of pollution.

4 Obtain a position of the incident on the chart prior to opening up communications.

5 Order the display of the aground navigational signals.

6 Order damage-control parties into respective duties to include further assessment of:

(i) a full set of internal soundings.

(ii) a comprehensive set of external soundings about the bow and the stern areas.

7 Open up communications to advise respective interested parties, e.g.:

• owners and charterers

• agents and local harbour authorities, if applicable

• MAIB and the MPCU, if appropriate

• underwriters and P & I Club (via owners)

• urgency or Mayday via Coastguard if appropriate

• tug assistance if required.

8 Deploy anchors to prevent accidental, unscheduled refloating.

9 Obtain a weather report.

10 Order tidal calculations to be made by the navigator to ascertain the state of the tide at grounding and when the next low/high water, heights and times are scheduled.

11 Carry out stability check and ballast movements to ease ship stresses.

3 What do you do with the ship's garbage?

Ans. Ships are now required to have a Garbage Management Plan and records must be kept of the disposal and treatment of all ship's garbage.

Entries are made into the 'garbage record book' which include the date and time of disposal of any garbage. The position of the vessel at the time of disposal, together with a description and quantity of the garbage disposed of.

Records must include any accidental discharge of garbage and receipts must be retained where garbage is sent ashore to be incinerated or deposited at a port reception facility.

All shipboard garbage records must be signed by the ship's Master.

4 What is the order of placing the correctors to the magnetic compass when engaged in the operation of correcting the compass?

Ans. Compass correction is usually carried out by a qualified compass adjuster, but these days such persons are often not available and it is left to the ship's Master to monitor the correction process.

It would be normal practice to place the correctors in the following order:

1   A correct length of 'Flinders bar' is placed.

2   The spheres (soft iron correctors) are then placed in the centre of the brackets provided (after placing the Flinders bar).

3   Heeling error corrector magnets are then placed (spheres and Flinders Bar must already be in place).

4   The fore and aft and the athwartships permanent correctors would be placed once heeling error has been corrected.

5   Adjust spheres (Kelvin's balls) for any residual coefficient 'D'.

5   Your vessel is experiencing sub-freezing air temperatures and taking ice accretion. What actions would you take?

Ans.   Ice accretion could destroy the positive stability of the ship and a prudent ship's Master would alter course to warmer latitudes and cause a change of wind direction affecting the vessel. A reduction of speed may be a viable alternative, or a combination of both course and speed.

In any event the Master would order the Chief Officer to engage the crew to shed the ice load by steam hose and/or shovels.

The sub-freezing air temperatures would also be reported to the Coastguard.

6   Following a fire in port the local Fire Service is called in to assist in the fire-fighting operation. Who retains the overall authority of fighting the fire?

Ans.   The ship's Master retains overall authority for all fire-fighting activity aboard the ship. However, it is pointed out that the expertise of the Fire Service Officers should be employed as appropriate and common sense must be seen to prevail.

7   Following a search for survivors you recover two persons from the water. What would you ask them as part of your debrief?

Ans.   Each would be asked his name and rank, together with their next of kin details. The rank of the individual will affect the nature of the questions within the debrief. They would invariably refer to:

- the name of the ship;
- the port of registry if known;
- the nature of the distress/situation;
- how many persons made up their ship's complement;
- whether any survival craft were launched;
- whether any other vessels were involved.

8   The vessel has an emergency source of power, either a generator or batteries. What does this emergency source of power supply?

Ans.   The emergency power source will be situated clear of the engine room and is usually established on one of the higher, more exposed decks. It would supply power to the emergency lighting circuits, essential navigation equipment inclusive of steering gear. Communication systems are also supplied, as are the lights covering the survival craft embarkation points. The emergency fire pump and watertight doors would also be connected to this power source.

9   Following a collision with another ship, what are the Master's expected legal duties?

Ans.   Following any collision incident, the Master of each vessel involved must legally:

1   Standby to render immediate assistance to the other ship.
2   Exchange voyage and ship details with the other Master or officer in charge.
3   Report the incident to the MAIB.
4   Cause an entry to be made in the official log book.

10   What are the Master's statutory duties inside known 'ice limits'?

Ans.   Under the SOLAS requirements, ship's Masters must report the sighting of any dangerous ice. The report must contain the ship's name and relevant details and include:

• position of the ice
• type of ice
• date and GMT of observation.

Additionally, Masters, when advised that ice is present on or near the intended ship's track, must alter course away from the reported danger and proceed at a moderate speed at night.

(Reference: *Mariner's Handbook* for content of ice reports and procedures.)

11   When formulating an anchor plan, what considerations would you expect to include?

Ans.   The anchor plan must take into consideration the following factors:

• The intended position of the anchor.
• The holding ground for the anchor to be good (mud or clay).
• No underwater obstructions on site.
• The amount of cable to use (scope).
• Which anchor to use (if going to a single anchor).
• What will be the state of the tide, ebbing or flooding, when anchoring?
• What will be the range of the tide?
• Is the anchorage area sheltered from prevailing weather?
• Is the anchorage clear of traffic focal points?
• What will be the swinging room of the vessel?
• Are position-fixing and monitoring methods available.
• What is the intended length of stay?
• What is the local weather forecast, and what is the long-term forecast?
• What are the maximum current effects in the area?

- The ship's course and approach speed when anchoring.
- Bridge procedures confirmed, e.g. manual steering engaged, anchors prepared, engine standby, etc.

12 Your vessel is a modern ferry, fitted with twin bow thrust units and twin controllable pitch propellers (CPPs). How would you bring the vessel to a 'Mediterranean moor'?

Ans. The vessel should approach parallel to the berth at a distance of about two ship's lengths off. The way should be taken off the vessel in a position where the bow is approximately one-third ship's length past the intended berthing position. Both anchors should be walked back and ready to let go from the brakes.

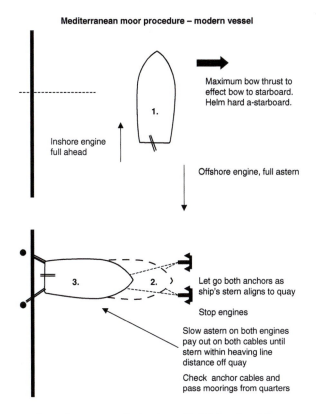

Mediterranean moor procedure – modern vessel

Maximum bow thrust to effect bow to starboard. Helm hard a-starboard.

Inshore engine full ahead

Offshore engine, full astern

Let go both anchors as ship's stern aligns to quay

Stop engines

Slow astern on both engines pay out on both cables until stern within heaving line distance off quay

Check anchor cables and pass moorings from quarters

13 What type of entries are made in the official log book?

Ans. The front cover of the official log book will contain the general particulars of the ship, namely port of registry and the respective tonnages. It will also contain the name and address of the vessel's owners, Master's name and certificate number, together with the date and place where the log book is opened and closed.

Inside the log book will be found a list copy of all the crew members and specific sections prior to the general narrative section. These specific sections include a record of:

- births and deaths that occur on board the vessel
- dates and times of emergency drills

- drills and inspections of ship's steering gear
- draughts and freeboards at departures from ports
- the ship's inspections.

Entries are made that are relevant in the life of the ship in the general section; if the Master is in any doubt as to whether an entry should be made or not he or she should make the entry on the basis that it can always be ignored.

14 Prior to departure, how would the Master know that the navigation charts have been corrected and are up to date?

Ans. The Master could inspect the 'chart correction log', which should provide details of the last effected corrections. Further inspection of the working charts, in the bottom left-hand corner, would reveal the last chart correction, consecutive number/date. ENLs have an inclusive log in account.

15 What are the influencing factors that would affect the Master's choice of route?

Ans. Many factors affect the route choice, not least the overall distance.

However, the final choice would be strongly influenced by which is the safest route and that which has adequate under-keel clearance with few or no navigation hazards.

Seasonal factors which may affect sensitive cargo could be a consideration and the charter party could also impose restrictions and limitations. Reference to ocean passages of the world and to shoreside routing services may also be an option.

Specific ships need to take specified routes, e.g. deep-draught ships require 'deep water routes', while ships without ice-strengthening would require 'ice-free routes'.

16 When passing a moored ship in close proximity, inside a canal, what precautions and factors would be taken into consideration?

Ans. Two ships in close proximity lend to the possibility of 'inter-action'. As such it would be prudent for my own vessel to reduce speed prior to drawing alongside the moored ship.

The prudent Master would take note of the condition of the mooring lines of the secured vessel and, if they are slack, then this fact would be noted in the deck log book. Interaction can cause vessels affected to 'range' on their moorings, leading to the possibility of the lines parting.

17 When would you have to swing the ship for checking the magnetic compass?

Ans. The compass would need to be checked by a ship 'swing' in the event that:
- it becomes unreliable;
- the ship undergoes repairs or alterations;
- if the vessel is struck by lightning;
- when magnetic cargoes are carried;
- if trading for long periods in high latitudes, e.g. Hudson Bay;
- every two years for retaining class.

Vessels with ECDIS have their charts corrected by a monthly CD. The system has a logging format to indicate that the chart corrections have been applied. This logging format can be checked at any time by the Master and Navigation Officers. The prudent Master would also check this issue with the ship's Navigation Officer.

18   If your vessel is requisitioned to act as a search unit in a search-and-rescue operation, how would you determine the track space of your search pattern?

Ans.   The choice of 'track space' for an operational search pattern would be determined by the Master of the search unit. However, the Master may be advised by external parties what type of search pattern and what track space to employ. The actual track space used can only be decided by the person who is actively engaged and recommendations by others may not be suited to the prevailing conditions in a different area.

Factors influencing the choice would include:
- the type of target and target definition presented;
- the state of visibility at the scene;
- day- or night-time search with or without search lights;
- the quality of the radar target likely to be presented;
- recommendations from MRCC or other interested parties;
- information from the target/meteorology table of the IAMSAR manual;
- the height of eye of own lookouts, above sea level;
- the speed of the search vessel;
- the number of search units deployed in the pattern;
- the remaining period of available daylight respective to the search area;
- the Master's experience.

19   While at sea in open waters the vessel suffers a steering gear failure. The OOW informs you, as the ship's Master, of the failure. What would you do and what are your options?

Ans.   In the event of failure of any of the ship's essential navigation equipment, I would expect to be called and it would be my duty to go immediately to the bridge and take the 'con' of the vessel. Once in control of the bridge and aware of the situation I would order the alarm cancelled and the second set of emergency steering motors to be engaged.

The following orders would then be issued:
1   Obtain a position on the chart.
2   Obtain a weather forecast if possible

Circumstances would dictate future activity, but in the event that the emergency steering motors have not returned the vessel to normal control, the following actions should take place immediately:
1   Display NUC shapes or lights.
2   Place engines on standby and stop the vessel, changing lights as appropriate.
3   Have a quartermaster stand by to engage manual steering.
4   Order ship's engineers to investigate steerage fault.
5   Cause an entry to be made in the log book.

The worst scenario is if the rudder has been lost. If this is the case then 'jury steering' by use of engines or drag weights (like anchors) might need to be considered. More likely communications would be brought into effect to arrange a tug under a towing contract. (Twin screw vessel could well steer by engines and continue to make way through the water under NUC signals.)

20   While lying at anchor your vessel is found to be dragging her anchor. What action would you take?

Ans. The Master would take the 'con' of the vessel and confirm the movement of the ship from the anchored position. He or she would order the anchor party forward to probably pay out more cable provided the increased scope did not compromise the swinging room.

If the vessel continued to drag her anchor a second anchor could be let go at short stay. If this failed, the prudent Master must consider picking up anchors and either seeking a more sheltered anchorage with improved holding ground, or steam up and down at slow speed until weather/current conditions improve.

25

Appendix V: Rule of the Road: Reference to the COLREGS and the IALA Buoyage System

The headings to each section indicate to which examinations the following questions and answers are applicable. Candidates preparing for marine examinations are advised that nautical examiners are permitted to ask questions from lower grade qualifications if necessary to assess an understanding of any particular topic.

Affecting All Mercantile Marine (Deck) Officers

1   When acting as the Officer of the Watch (OOW) your vessel is approaching an area of reduced visibility. What actions would you expect to take on the bridge?

Ans.   The changing visibility condition would generate the following actions:

1   Place the engines of the vessel on 'standby' and reduce the ship's speed.
2   Advise the Master of the reduced visibility condition.
3   Post extra lookouts on stations.
4   Commence sounding fog signals.
5   Carry out systematic plotting of any radar targets.
6   Switch on the navigation lights.
7   Engage the vessel in manual steering.
8   Close all watertight doors.
9   Place a position on the chart.
10   Enter a statement of actions in the log book.
11   Stop all loud working on deck.

NB. The company standing orders would normally have a checklist procedure for entering poor visibility. This would be followed to include all of the above.

2   When approaching a landfall the OOW sights a vessel aground at about four points of the compass, off the starboard bow. What action would be expected of the officer?

Ans.   The OOW on sighting the vessel aground would be expected to:

1   Stop the vessel immediately and take all way off the ship.
2   Advise the Master of the situation.
3   Obtain a chart assessment of the situation, to include own ship's position and the position of the vessel aground. Note the extent of the shoal and the current underkeel clearance.
4   Switch on the echo-sounder.
5   Establish communications, with station identification, with the vessel aground. (The ship would need to know the draught of the vessel aground and the time the ship ran aground.)

> NB. It must be anticipated that the Master would take the 'con' and navigate the vessel clear of the incident.

3   How would a tug and tow operation draw attention to the dangers of the towline to approaching vessels?

Ans.   Reference Regulation 36, by prudent use of a searchlight in the direction of the towline.

4   When switching on the ship's radars, how would you check that they are operating at peak performance?

Ans.   It would be expected practice to operate the 'performance monitor' of the radar set. Different manufacturers have differing styles of performance monitors: old-fashioned radar sets may see an interpretation of a plume or biscuit, whereas a more modern set would probably have an analogue bar indicator to provide visual performance criteria as compared to the radar performance manuals.

5   When acting as OOW and navigating in poor visibility, in open waters at reduced speed, you hear apparently four points off the starboard bow, a fog signal of one prolonged blast at intervals of not more than one minute. What are your immediate actions?

Ans.   The vessel should be stopped immediately and all headway taken off the ship.

Once the vessel is making no way over the ground own ship's fog signal should be changed from one prolonged blast to two prolonged blasts at intervals of not more than two minutes.

Lookouts should be increased on the affected side.

Radar inspection must be anticipated to try to establish a firm radar contact which will allow the target vessel to be systematically plotted. In the event that no radar contact is made the vessel should remain stopped and the frequency of own fog signal might be increased.

In the event that radar contact remains negative, a change in the radar range may be considered prudent, with possible application of the 'gain' and 'clutter' controls.

6   When acting as OOW, your vessel is proceeding eastward up the English Channel and you sight a south cardinal buoy right ahead. What action would you take?

Ans. Following a chart assessment and making sure that the buoy was not out of position, the ship's course should be altered to starboard and steered to pass south of the buoy.

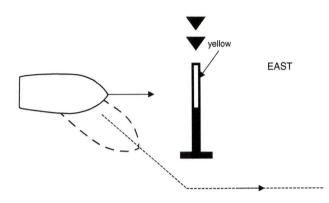

7   What type of radar signature would you expect to pick up from a SART?

Ans. A SART would give a radar signature of 12 small dashes from the position, received up to about a five-mile range from the SART. However, when closing to within metres of the SART this signature would expect to change to concentric circles around the position.

8   How would you describe the 'towing light'?

Ans. The towing light is of the same characteristics as the sternlight of the vessel, except that it is 'yellow' in colour. It would be carried over and above the sternlight in a towing operation and show over an arc of 135° (12 points of the compass) from two points abaft the beam through the stern to two points abaft the beam on the opposite side. The visible range would be three nautical miles.

9   When on passage through a traffic separation scheme the OOW reports a vessel displaying the code flag signal 'YG' (Yankee Golf). What is the significance of this signal, and what should the OOW do?

Ans. The flag signal means 'You appear not to be complying with the traffic separation scheme'. As such, the OOW would expect to immediately confirm the ship's position and check that the vessel is proceeding on the correct course and complying in every way with the traffic separation scheme (Regulation 10) procedures.

10   The regulations advise that a vessel aground may sound an appropriate whistle signal to provide additional warning of her predicament and the associated danger to other vessels. What signal would you expect this to be?

Ans. It is suggested that the signals:
- 'U' (Uniform): You are running into danger; or
- 'L' (Lima): You should stop your vessel instantly

would be considered appropriate for the general circumstances.

11 Sound signals may be supplemented by a light signal. This light signal is usually made by a manoeuvring light which is operational with the ship's whistle. Do ships have to carry the 'manoeuvring light', and if so what is its range?

Ans. Ships do not have to carry a manoeuvring light as the regulation only says MAY CARRY, not must carry. However, if the vessel does carry the manoeuvring light it will have a range of five nautical miles.

> NB. Ships must carry an Aldis lamp to give the required signal under the carriage of essential navigation equipment.

12 How will the OOW ascertain whether risk of collision exists between two vessels approaching one another?

Ans. Collision risk can be ascertained by the OOW by carefully taking a series of compass bearings and noting the range of the target on each bearing. If the bearings are staying constant and the range of the target is closing it can generally be assumed that risk of collision could exist.

> NB. On occasion the stated criteria may exist and risk of collision is not present, i.e. with large vessels.

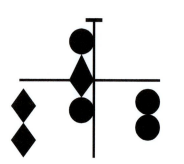

13 Your vessel is approaching a ship which displays the day signal shown on the left. What is this vessel and what would you do if it is seen right ahead?

Ans. The day signal would be displayed by a vessel engaged in dredging, surveying or underwater operations and restricted in ability to manoeuvre.

  If seen right ahead of my own vessel, my action would be to sound two short blasts on the ship's whistle and make a bold alteration to port, taking the unobstructed side, indicated by the 'black diamonds'.

> NB. Such a vessel must be expected to be encountered in shallow waters and as such the prudent navigator would expect to make a full chart assessment prior to making any alteration of course. The echo-sounder should also be operational in most circumstances.

14 When approaching a buoyed channel you see a sailing vessel moving outward to sea. The sailing vessel displays a day signal of a 'black cone' apex down, in the forepart where it can best be seen. What does this mean and what fog signal would be made by this vessel?

Ans. The black cone signal indicates the sailing vessel is underway with both sails and under power, with an engine operational.

For the purpose of the regulations, the vessel is considered a power-driven vessel and as such would exhibit the lights, shapes and sound signals as for a power-driven vessel of its length.

> NB. As soon as the vessel cuts the engine power, she would revert to being a sailing vessel and sound one prolonged blast followed by two short blasts at intervals of not more than two minutes, when in poor visibility making way under sail alone.

15 What classes of vessel are considered as being restricted in their ability to manoeuvre?

Ans. Under Rule 3 of the COLREGS there are six classes of vessel which fall into this category, namely, a vessel engaged in:
1. laying, servicing or picking up a navigation mark, submarine cable or pipeline;
2. dredging, surveying or underwater operations;
3. replenishment or transferring persons, provisions or cargo while underway;
4. launching or recovering aircraft;
5. mine clearance operations;
6. a towing operation such as severely restricts the towing vessel and her tow in their ability to deviate from their course.

16 Navigation sidelights are provided with screens to prevent the light from one side shining on the side of the other. Describe the screen?

Ans. The screen extends over a one-metre length, painted 'matt black' and is provided with a chock at the extremity.

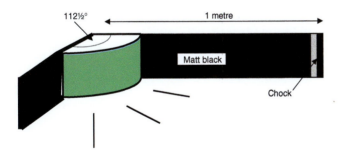

17 When two vessels are crossing, one vessel is considered the 'give way' vessel, while the other vessel is directed to 'stand on'. What do you understand by the terms:
(a) give way vessel,
(b) stand on vessel
within the context of the ColRegs?

Ans. The term (a) 'give way vessel' (Reference Regulation 16) should be taken to mean every vessel which is directed to keep out of the way of another vessel shall, as far as practical, take early and substantial action to keep well clear.

The term (b) 'stand on vessel' (Reference Regulation 17) should be understood as to mean that where one of two vessels is so directed to keep out of the way, the other shall keep her course and speed.

18 When navigating in coastal waters a warship is seen ahead of your own vessel to display the day signal as shown. What does this signify and what action would you take as OOW?

Ans. The signal represents a vessel engaged in 'mine clearance operations'. The OOW should stop and take all way off his or her own vessel. He or she should call the Master and advise of the mine clearance vessel.

With 'station identification' the warship should be contacted to ascertain if the operation is a training exercise or if it is an actual live mine clearing activity.

The vessel should also seek advice from the Commander of the warship as to where the clear water is indicated. On this advice a prudent alteration into clear waters is advised, passing no closer than 1,000 m to the warship.

19 When approaching an anchorage, another vessel displays the 'Y' flag of the International Code of Signals. What would you understand this to mean?

Ans. The single-letter meaning of (Y), Yankee flag, would indicate that the vessel displaying it is dragging its anchor.

20 What is the range of the 'navigation lights' shown by a power-driven vessel which is over 100 metres in length, underway and making way on the high seas?

Ans. Masthead steaming lights = six miles; port and starboard sidelights = three miles; sternlight = three miles.

21 Where would you find the direction of buoyage and what symbol is used to indicate it?

Ans. The direction of buoyage is found on the navigation chart and would be displayed by the following symbol:

22 Can a sailing vessel be constrained by her draught, within the meaning of the ColRegs?

Ans. No (Reference Regulation 3). The definitions within the regulations state that the term 'vessel constrained by her draught' means *a power-driven vessel* which, because of her draught in relation to the depth and width of navigable water, is severely restricted in her ability to deviate from the course she is following.

23 Which vessels must comply with the Regulations for the Prevention of Collision at Sea?

Ans. The regulations are applicable to all vessels upon the high seas and in all waters connected therewith navigable by sea-going vessels.

24 You sight a trawler engaged in fishing and a vessel constrained by its draught both on a collision heading with each other (there is no risk of collision with your own vessel). Which vessel of these two would keep out of the way of the other, and why?

Ans. Despite being constrained by its draught, the vessel constrained by its draught must still be proceeding at a safe speed in accord with Regulation 6. As such, it must be able to stop if necessary.

Therefore, the vessel so constrained by its draught would be deemed to be the give way vessel of the two.

NB. Although a power-driven vessel is directed to keep out of the way of a vessel engaged in fishing, this scenario is such that the fishing vessel (trawler) cannot 'stop' as she could become disabled with her trawl gear around the propeller.

---

25 While on watch you take a series of compass bearings of a sailing vessel which is four points abaft your own ship's beam. The bearings are steady and the range is closing, indicating that a risk of collision is clearly present. What action would you expect to take in these circumstances?

Ans. Under normal circumstances my own ship would expect to stand on and maintain course and speed. The reason for this is that the sailing vessel is overtaking from well abaft the beam and the regulations reflect that *any vessel overtaking*, must keep out of the way.

26 What type of craft would display a high-intensity, all-round, flashing red light with its regulatory navigation lights?

Ans. This flashing red light is exhibited by 'wing in ground' (WIG) craft when taking off, landing and in flight near the surface.

27 A vessel displays NUC signals. What will she show and what would be her fog signal?

Ans. A vessel not under command (NUC) must display two black ball shapes in a vertical line where they can best be seen. Each shape will be not less than 0.6 metre in diameter and separated by a tack line between the ball shapes of 1.5 metres.

The fog signal for a vessel NUC is one prolonged blast followed by two short blasts at intervals of not more than two minutes.

28 Can a trawler engage in fishing inside a traffic separation scheme?

Ans. Yes, a vessel can fish inside the scheme provided it does not impede the passage of any vessel following the lane.

The author would point out to readers that the rule of the road scenarios are depicted in general terms and the answers provided are offered as possible solutions. Every practical situation will, by the very nature and variance of general seamanship, be different and no one answer should be taken as being suitable for similar situations on other occasions.

Index